Human Genome
Epidemiology

HUMAN GENOME EPIDEMIOLOGY

A Scientific Foundation for Using Genetic Information to Improve Health and Prevent Disease

Edited by

MUIN J. KHOURY, M.D., Ph.D
Director, Office of Genomics and Disease Prevention
Centers for Disease Control and Prevention
Atlanta, Georgia

JULIAN LITTLE, Ph.D.
Chair of Epidemiology
Department of Medicine and Therapeutics
University of Aberdeen
Aberdeen, Scotland

WYLIE BURKE, M.D., Ph.D.
Chair, Department of Medical History and Ethics
University of Washington
Seattle, Washington

OXFORD
UNIVERSITY PRESS
2004

OXFORD

UNIVERSITY PRESS

Oxford   New York
Auckland   Bangkok   Buenos Aires   Cape Town   Chennai
Dar es Salaam   Delhi   Hong Kong   Istanbul   Karachi   Kolkata
Kuala Lumpur   Madrid   Melbourne   Mexico City   Mumbai
Nairobi   São Paulo   Shanghai   Singapore   Taipei   Tokyo   Toronto

Copyright © 2004 by Oxford University Press, Inc.

Published by Oxford University Press, Inc.
198 Madison Avenue, New York, New York, 10016
http://www.oup-usa.org

Library of Congress Cataloging-in-Publication Data
Human genome epidemiology:
a scientific foundation for using genetic information to
improve health and prevent disease/
edited by Muin J. Khoury, Julian Little, Wylie Burke.
p. cm.   Includes bibliographical references and index.
ISBN 0-19-514674-3
1. Genetic disorders—Epidemiology. 2. Medical genetics—Methodology.
3. Genomics.
I. Khoury, Muin J. II. Little, Julian. III. Burke, Wylie.
RB155.5.H86 2003   616'.042—dc21   2003048623

9 8 7 6 5 4 3 2

Printed in the United States of America
on acid-free paper

Preface

Advances in human genetics are expected to play a central role in medicine and public health in the 21st century by providing genetic information for disease prediction and prevention. Although human gene discoveries generate excitement and expectations, the translation of gene discoveries into meaningful actions to improve health and prevent disease depends on scientific information from multiple medical and public health disciplines. The field of epidemiology plays a central role in this effort. Epidemiology is often viewed as the scientific core of public health and involves the study of the distribution and determinants of health-related states or events in populations and the application of this study to control health problems. Epidemiologists determine risk factors for various diseases, identify segments of the population with highest risk to target prevention and intervention opportunities, and evaluate the effectiveness of health programs and services in improving the health of the population.

In this book, we show how the epidemiologic approach will play an important role in the continuum from gene discovery to the development and applications of genetic tests. We call this continuum human genome epidemiology (or HuGE) to denote an evolving field of inquiry that uses systematic applications of epidemiologic methods to assess the impact of human genetic variation on health and disease. Since most gene discoveries are based on studies of high-risk families or selected population groups, once disease genes are found, well-conducted epidemiologic studies are needed to quantify the population impact of gene variants on the risk for health outcomes and to identify and measure the impact of modifiable risk factors that interact with gene variants. Epidemiologic studies are also required to clinically validate new genetic tests, to monitor population use of genetic tests, and to determine the impact of genetic information on the health and well-being of different populations. The results of such studies will help medical and public health professionals integrate genetics into practice.

As more epidemiologic investigations of human genes are conducted and published, evidence needs to be integrated from different studies. Given the large numbers of genes that will be examined in relation to diseases, many spurious findings are likely to emerge. Variation in study design and execution will make it difficult to synthesize information across studies. The increasing number of human genome epidemiologic studies has uncovered the need for guidelines on synthesizing results, particularly for assessing the prevalence of gene variants, gene–disease associations, and gene–environment and gene–gene interactions, and for evaluating genetic tests.

Although none of the material presented here is novel, we have structured the book to allow readers to proceed systematically from the fundamentals of genome technology and discovery, to epidemiologic approaches to gene characterization, to evaluation of genetic tests and health services. We then illustrate these concepts in several disease-specific case studies. The book focuses on post-gene discovery (what do you do with a gene when you find one?) with an overview of emerging analytic methods for gene discovery. Part I describes genomic technologies and their applications, discusses evolving methods of gene discovery, and summarizes the current status of the ethical, legal, and social issues for conducting epidemiologic studies of the human genome (with emphasis on informed consent issues). Part II addresses epidemiologic approaches to the study of genotypes in populations and their relation to diseases, including the assessment of gene–gene and gene–environment interaction. This section also addresses issues of synthesis of these studies, including some methodologic standards. Part III deals with the application of epidemiologic methods to assessing genetic information for clinical and public health applications. The section lays an epidemiologic foundation for using population level information to assess individual risk for clinical use. We explore population-based concepts of the usefulness of genetic tests in populations and discuss the evaluation of genetic tests from a combined clinical-laboratory approach. We also explore an epidemiologic framework for the interface among genetics, pharmacology, and medicine. Lastly, we explore the integration of genetics into controlled clinical trials and the role of genetics in the development of clinical practice guidelines. Finally, Part IV uses case studies to illustrate concepts discussed in the first three sections in relation to specific disease examples, including gene–environment interactions (pesticides and oral contraceptive use), chronic diseases (colon cancer, Alzheimer disease, cardiovascular disease, breast cancer, and iron overload), occupational exposures (berylliosis), newborn screening issues (fragile X syndrome and hearing loss), and infectious disease (human immunodeficiency virus infection). These examples are by no means exhaustive, but they do illustrate the spectrum from single gene disorders to complex conditions, and the need for epidemiologic research to obtain population level information for developing health policy and practice. Indeed, for many of the case studies, the lack of solid epidemiologic data represents a primary barrier to developing appropriate health policies related to the use of genetic information. Because our knowledge is rapidly evolving for each one of these examples, most likely, the case studies will be outdated soon. Nevertheless, the basic methodologic foundation for translating gene discovery into usable clinical information will still apply.

Ultimately, a multidisciplinary approach is needed to fulfill the promise of the Human Genome Project in improving health. At the core of these disciplines, the epidemiologic approach will begin to fill immense gaps in our knowledge of population risk for various diseases in relation to genetic variation. This information is a necessary first step in the long road from gene discovery to medical and public health practice.

Atlanta, GA   M.J.K.
Aberdeen, UK   J.L.
Seattle, WA   W.B.

Acknowledgments

We are grateful to the following individuals for reviewing drafts of selected book chapters: Harry Campbell, David FitzPatrick, Seymour Garte, Scott Grosse, James E. Haddow, Robin Harbour, Neil Holtzman, Craig Hooper, Rolv T. Lie, Zosia Miedzybrodzka, Arno Motulsky, Thomas O'Brien, Scott Ramsey, Sonja Rasmussen, Duncan Shaw, Michael Steel, Karen Steinberg, Donna Stroup, James Tang, Emmanuela Taioli, Bruce Tempel, Paolo Vineis, Benjamin Wilfond, and Paula Yoon. In addition, we thank Karen Foster, Pamela Gillis Watson, Elizabeth Fortenberry, and Patti Seikus, with the Centers for Disease Control and Prevention editorial staff, for their literary support to this book; thanks also to Thelma Brown, without whom the book would have never been completed.

Contents

Contributors   xv

PART I   Fundamentals

1. Human Genome Epidemiology: Scope and Strategies   3
   Muin J. Khoury, Julian Little, and Wylie Burke

2. Emerging Genomic Technologies and Analytic Methods for Population-
   and Clinic-Based Research   17
   Darrell L. Ellsworth and Christopher J. O'Donnell

3. Approaches to Quantify the Genetic Component of and Identify Genes
   for Complex Traits   38
   Patricia A. Peyser and Trudy L. Burns

4. Ethical, Legal, and Social Issues in the Design and Conduct of Human
   Genome Epidemiology Studies   58
   Laura M. Beskow

PART II   Methods and Approaches I: Assessing Disease Associations and Interactions

5. Assessing Genotypes in Human Genome Epidemiology Studies   79
   Karen Steinberg and Margaret Gallagher

6. Statistical Issues in the Design and Analysis of Gene–Disease
   Association Studies   92
   Duncan C. Thomas

7. Facing the Challenge of Complex Genotypes and Gene–Environment
   Interaction: The Basic Epidemiologic Units in Case-Control and
   Case-Only Designs   111
   Lorenzo D. Botto and Muin J. Khoury

8. Inference Issues in Cohort and Case-Control Studies of Genetic Effects
   and Gene–Environment Interactions   127
   *Montserrat García-Closas, Sholom Wacholder, Neil Caporaso,
   and Nathaniel Rothman*

9. Applications of Human Genome Epidemiology to Environmental Health   145
   *Samir N. Kelada, David L. Eaton, Sophia S. Wang, Nathaniel R. Rothman,
   and Muin J. Khoury*

10. Reporting and Review of Human Genome Epidemiology Studies   168
    Julian Little

PART III   Methods and Approaches II:
Assessing Genetic Tests for Disease Prevention

11. Epidemiologic Approach to Genetic Tests: Population-Based Data for
    Preventive Medicine   195
    Marta Gwinn and Muin J. Khoury

12. Genetic Tests in Populations: An Evidence-Based Approach   207
    Paolo Vineis

13. ACCE: A Model Process for Evaluating Data on Emerging
    Genetic Tests   217
    James E. Haddow and Glenn E. Palomaki

14. The Interface between Epidemiology and Pharmacogenomics   234
    David L. Veenstra

15. Integrating Genetics into Randomized Controlled Trials   247
    John P.A. Ioannidis and Joseph Lau

16. Developing Guidelines for the Clinical Use of Genetic Tests:
    A U.S. Perspective   264
    Linda E. Pinsky, David Atkins, Scott Ramsey, and Wylie Burke

17. Using Human Genome Epidemiologic Evidence in Developing Genetics
    Services: The U.K. Experience   283
    Brenda J. Wilson, Jeremy M. Grimshaw, and Neva E. Haites

PART IV   Case Studies: Using Human Genome Epidemiology Information to Improve Health

18. Paraoxonase Polymorphisms and Susceptibility to Organophosphate Pesticides   305
   Kathryn Battuello, Clement Furlong, Richard Fenske, Melissa A. Austin, and Wylie Burke

19. Factor V Leiden, Oral Contraceptives, and Deep Vein Thrombosis   322
   Jan P. Vandenbroucke, Frits R. Rosendaal, and Rogier M. Bertina

20. Methylenetetrahydrofolate Reductase Gene (*MTHFR*), Folate, and Colorectal Neoplasia   333
   Linda Sharp and Julian Little

21. Apolipoprotein E and Alzheimer Disease   365
   Richard Mayeux

22. Immunogenetic Factors in Chronic Beryllium Disease   383
   Erin C. McCanlies, Michael E. Andrew, and Ainsley Weston

23. Fragile X Syndrome: From Gene Identification to Clinical Diagnosis and Population Screening   402
   Dana C. Crawford and Stephanie L. Sherman

24. The Connexin Connection: From Epidemiology to Clinical Practice   423
   Aileen Kenneson and Coleen Boyle

25. Genetic and Environmental Factors in Cardiovascular Disease   436
   Molly S. Bray

26. *BRCA1/2* and the Prevention of Breast Cancer   451
   Jenny Chang-Claude

27. The Role of Chemokine and Chemokine Receptor Genes in HIV-1 Infection   475
   Thomas R. O'Brien

28. Hereditary Hemochromatosis   495
   Giuseppina Imperatore, Rodolfo Valdez, and Wylie Burke

29. Genetic Testing of Railroad Track Workers with Carpal Tunnel Syndrome   511
   Paul A. Schulte and Geoffrey Lomax

Index   525

Contributors

MICHAEL E. ANDREW, PH.D.
Biostatistics and Epidemiology Branch
National Institute for Occupational
Safety and Health Centers for
Disease Control and Prevention
Morgantown, West Virginia

DAVID ATKINS, M.D., M.P.H.
Center for Practice and Technology
Assessment
Agency for Healthcare Research and
Quality
Rockville, Maryland

MELISSA A. AUSTIN, PH.D.
Institute for Public Health Genetics
Department of Epidemiology
University of Washington
Seattle, Washington

KATHRYN BATTUELLO, J.D., M.P.H.
Institute for Public Health Genetics
University of Washington
Seattle, Washington

ROGIER M. BERTINA, M.D., PH.D.
Thrombosis and Haemostatis Research
Center
Leiden University Medical Center
Leiden, Netherlands

LAURA M. BESKOW, M.P.H.
Department of Health Policy and
Administration
University of North Carolina at
Chapel Hill
School of Public Health
Chapel Hill, North Carolina

LORENZO D. BOTTO, M.D.
National Center on Birth Defects and
Developmental Disabilities
Centers for Disease Control and
Prevention
Atlanta, Georgia

COLEEN BOYLE, PH.D.
National Center on Birth Defects and
Developmental Disabilities
Centers for Disease Control and
Prevention
Atlanta, Georgia

MOLLY S. BRAY, PH.D.
Human Genetics Center
University of Texas Health Science
    Center
Houston, Texas

WYLIE BURKE, M.D., PH.D..
Department of Medical History and
    Ethics
University of Washington
Seattle, Washington

TRUDY L. BURNS, PH.D., M.P.H.
Division of Statistical Genetics
Department of Biostatistics
College of Public Health
University of Iowa
Iowa City, Iowa

NEIL CAPORASO, M.D.
Division of Cancer Epidemiology and
    Genetics
National Cancer Institute
National Institutes of Health
Bethesda, Maryland

JENNY CHANG-CLAUDE, M.D.
German Cancer Research Centre
Division of Clinical Epidemiology
Heidelberg, Germany

DANA C. CRAWFORD, PH.D.
Department of Genome Sciences
University of Washington
Seattle, Washington

DAVID L. EATON, PH.D.
Department of Environmental Health
Center for Ecogenetics and
    Environmental Health
University of Washington
Seattle, Washington

DARRELL L. ELLSWORTH, PH.D.
Gene and Drug Discovery Center
Windber Research Institute
Windber, Pennsylvania

RICHARD FENSKE, PH.D.
Department of Environmental Health
University of Washington
Seattle, Washington

CLEMENT FURLONG, PH.D.
Department of Genome Sciences and
    Medicine
University of Washington
Seattle, Washington

MARGARET GALLAGHER, PH.D.
Division of Laboratory Sciences
National Center for Environmental
    Health
Centers for Disease Control and
    Prevention
Atlanta, Georgia

MONTSERRAT GARCÍA-CLOSAS,
M.D., M.P.H., DR.P.H.
Division of Cancer Epidemiology and
    Genetics
National Cancer Institute
National Institutes of Health
Bethesda, Maryland

JEREMY M. GRIMSHAW, M.B.,
CH.B., PH.D., F.R.C.G.P.
Director, Clinical Epidemiology
    Program
Ottawa Health Research Institute
Director, Centre for Best Practice
Institute of Population Health
University of Ottawa
Ottawa, Canada

MARTA GWINN, M.D., M.P.H.
Office of Genomics and Disease
    Prevention
Centers for Disease Control and
    Prevention
Atlanta, Georgia

JAMES E. HADDOW, M.D.
Foundation for Blood Research
Scarborough, Maine

NEVA E. HAITES, PH.D., M.B.,
CH.B., F.R.C.PATH., F.R.C.P.
(EDIN), F.R.C.P. (LOND).
Department of Medical Genetics
University of Aberdeen
Aberdeen, Scotland

GIUSEPPINA IMPERATORE, M.D.,
PH.D.
Division of Diabetes Translation
National Center for Chronic Disease
 Prevention and Health Promotion
Centers for Disease Control and
 Prevention
Atlanta, Georgia

JOHN P.A. IOANNIDIS, M.D.
Department of Hygiene and
 Epidemiology
University of Ioannina School of
 Medicine, Greece
Department of Medicine
Tufts University School of Medicine
New England Medical Center
Boston, Massachusetts

SAMIR N. KELADA, M.P.H.
Department of Environmental Health
University of Washington
School of Public Health and
 Community Medicine
Seattle, Washington

AILEEN KENNESON, PH.D.
National Center on Birth Defects
 and Developmental Disabilities
Centers for Disease Control and
 Prevention
Atlanta, Georgia

MUIN J. KHOURY, M.D., PH.D.
Office of Genomics and Disease
 Prevention
Centers for Disease Control and
 Prevention
Atlanta, Georgia

JOSEPH LAU, M.D.
Division of Clinical Care Research
Department of Medicine
Tufts University School of Medicine
New England Medical Center
Boston, Massachusetts

JULIAN LITTLE, PH.D.
Epidemiology Group
Department of Medicine and
 Therapeutics
University of Aberdeen
Aberdeen, Scotland

GEOFFREY LOMAX, DR.P.H.,
M.P.H.
Occupational Health Branch
California Department of Health
 Services
Public Health Institute
University of California
Berkeley, California

RICHARD MAYEUX, M.D., M.SC.
The Taub Institute for Research on
 Alzheimer's Disease and the Aging
 Brain and
The Gertrude H. Sergievsky Center
College of Physicians and Surgeons
Columbia University
New York, New York

ERIN C. MCCANLIES, PH.D.
Biostatistics and Epidemiology Branch
National Institute of Occupational
 Safety and Health
Centers for Disease Control and
 Prevention
Morgantown, West Virginia

THOMAS R. O'BRIEN, M.D., M.P.H.
Viral Epidemiology Branch
Division of Cancer Epidemiology and
 Genetics
National Cancer Institute
National Institutes of Health
Bethesda, Maryland

CHRISTOPHER J. O'DONNELL, M.D.
M.P.H.
Cardiology Division, Department of
  Medicine
Massachusetts General Hospital
Harvard Medical School
Boston, Massachusetts

GLENN E. PALOMAKI, B.A.
Foundation for Blood Research
Scarborough, Maine

PATRICIA A. PEYSER, PH.D.
Department of Epidemiology
School of Public Health
University of Michigan
Ann Arbor, Michigan

LINDA E. PINSKY, M.D.
Department of Medicine
University of Washington
Seattle, Washington

SCOTT RAMSEY, M.D., PH.D.
Department of Medicine
University of Washington
Fred Hutchinson Cancer Research
  Center
Seattle, Washington

FRITS R. ROSENDAAL, M.D., PH.D.
Department of Clinical Epidemiology
Leiden University Medical Center
Thrombosis and Haemostatis Research
  Center
Leiden University Medical Center
Leiden, Netherlands

NATHANIEL ROTHMAN, M.D.,
M.P.H., M.H.A.
Division of Cancer Epidemiology and
  Genetics
National Cancer Institute
National Institutes of Health
Bethesda, Maryland

PAUL A. SCHULTE, PH.D.
Education and Information Division
National Institute for Occupational
  Safety and Health
Centers for Disease Control and
  Prevention
Cincinnati, Ohio

LINDA SHARP, M.SC.
Epidemiology Group
Department of Medicine and
  Therapeutics
University of Aberdeen
Aberdeen, Scotland

STEPHANIE L. SHERMAN, PH.D.
Department of Human Genetics
Emory University School of Medicine
Atlanta, Georgia

KAREN STEINBERG, PH.D.
Division of Laboratory Sciences
National Center for Environmental
  Health
Centers for Disease Control and
  Prevention
Atlanta, Georgia

DUNCAN C. THOMAS, PH.D.
Department of Preventive Medicine
Keck School of Medicine
Los Angeles, California

RODOLFO VALDEZ, PH.D.
Division of Diabetes Translation
National Center for Chronic Disease
  Prevention and Health Promotion
Centers for Disease Control and
  Prevention
Atlanta, Georgia

JAN P. VANDENBROUCKE, M.D.,
PH.D.
Department of Clinical Epidemiology
Leiden University Medical Center
Leiden, Netherlands

DAVID L. VEENSTRA, PHARMD., PH.D.
Pharmaceutical Outcomes Research
and
Policy Program
Institute for Public Health Genetics
University of Washington
Seattle, Washington

PAOLO VINEIS, M.D.
University of Torino and ISI
* Foundation*
Department of Biomedical Sciences
* and Human Oncology*
Torino, Italy

SHOLOM WACHOLDER, PH.D.
Division of Cancer Epidemiology and
* Genetics*
National Cancer Institute
National Institutes of Health
Bethesda, Maryland

SOPHIA S. WANG, PH.D.
Division of Cancer Epidemiology and
* Genetics*
National Cancer Institute
National Institutes of Health
Bethesda, Maryland

AINSLEY WESTON, PH.D.
Health Effects Laboratory Division
National Institute for Occupational
* Safety and Health*
Centers for Disease Control and
* Prevention*
Morgantown, West Virginia

BRENDA J. WILSON, M.B., CH.B., M.SC., M.R.C.P. (U.K.), F.F.P.H.M.
Department of Epidemiology and
* Community Medicine*
Univeristy of Ottawa
Ottawa, Canada

I
FUNDAMENTALS

1

Human genome epidemiology: scope and strategies

Muin J. Khoury, Julian Little, and Wylie Burke

The completion of the sequencing of the human genome is viewed as an important milestone in the history of biology and medicine (1,2). Many scientists believe that advances in human genetics and the sequencing of the human genome will play a central role in medicine and public health in the 21st century by providing genetic information for disease prediction and prevention (3,4). Indeed, we are confronted daily with one or more new gene discoveries claimed to be associated with increased risk for some disease and promising a sweeping change in the diagnosis, treatment, or prevention of that condition. Table 1.1 shows a sample of stories from Web-based headlines (5). These titles illustrate that gene discoveries involve a wide variety of diseases not normally considered "genetic," and often include information about interactions with nongenetic factors such as cigarette smoking and drugs (5). Although gene discoveries generate excitement and expectations, their contribution to disease prevention is not clear. The central theme of this book is that "translation" of gene discoveries into meaningful actions to improve health depends on scientific information from multiple medical and public health disciplines. The field of epidemiology plays a central role in this effort. In this book, we explore how the applications of the epidemiologic approach to the human genome in relation to health and disease (we call this human genome epidemiology, or HuGE for short) will form an important scientific foundation for using genetic information to improve health and prevent disease.

Vision of Genomic Medicine

Human genome discoveries have broad potential applications for improving health and preventing disease. For primary prevention, a better understanding of genetic effects and gene–environment interactions in disease processes will allow us to de-

3

Table 1.1 Selected Web-based News Stories Headlines Reporting on Association of Genes with Various Health Outcomes

Headline	Source	Date
Gene may trigger idiopathic epilepsy	Health News UK	March 2003
Flesh-eating disease linked to gene differences	New Scientist	November 2002
Genetic variant may impact smoking cessation	EurekAlert	November 2002
Genes influence heart disease risk Reuter's Health from fatty diet	October 2002	
Two genes linked to congestive heart failure	New York Times	October 2002
Genes may play a role in carpal tunnel syndrome	Reuter's Health	July 2002
Gene implicated in stress–alcohol connection	Reuter's Health	May 2002
Genetic variants put some patients at risk for particular drug reactions	ScienceDaily	April 2002
Osteoarthritis gene breakthrough	BBC News	April 2002

Source: Office of Genomics and Disease Prevention. Genomics Weekly Update, from http://www.cdc.gov/genomics/update/current.htm.

velop better interventions, such as avoidance of defined exposures and chemoprevention, and to identify subgroups who are candidates for the interventions. For secondary prevention, we may be able to develop new or tailor existing screening tests for early disease identification based on stratification by genotype and/or family history. For tertiary prevention and therapeutics, advances in human genetics could contribute to the development of better drugs and to tailoring drug use to maximize benefits and minimize harms.

Nevertheless, many authors have expressed skepticism regarding the value of human genome discoveries to health care and disease prevention (6–10). Concerns cited include the low magnitude of risk for common diseases associated with most genetic variants discovered thus far, the absence of interventions that are specific to different genotypes, and the potential for genetic labeling to cause personal and social harms. Furthermore, some public health professionals have expressed concern that too much emphasis on genomics can divert energy and resources from disease prevention strategies that have been proven to be effective on a population basis (e.g., diet, exercise, smoking cessation; 10,11). Nevertheless, leading scientists continue to predict the imminent integration of gene discoveries into medical practice. In 2001, Collins and McKusick stated: "By the year 2010, it is expected that predictive genetic tests will be available for as many as a dozen common conditions, allowing individuals who wish to know this information to learn their individual susceptibilities and to take steps to reduce those risks for which interventions are or will be available. Such interventions could take the form of medical surveillance, lifestyle modifications, diet, or drug therapy. Identification of persons at highest risk for colon cancer, for example, could lead to targeted efforts to provide colonoscopic screening to those individuals, with the likelihood of preventing many premature deaths" (12). Moreover, commercial marketing of a genomic approach to preventive medicine has already been implemented, albeit prematurely, both in Europe (13) and the United States (14).

This scenario of the practice of medicine is based on the assumption that the use of genetic information at multiple gene loci, perhaps in combination with a person's family history of disease, can stratify people according to their risks for various diseases for targeted medical and behavioral interventions. In one such hypothetical scenario (Table 1.2), John, a 23-year-old college graduate with a high cholesterol level and a paternal history of early onset of myocardial infarction, undergoes a battery of genetic tests. Because of his increased risk for coronary heart disease, a prophylactic drug chosen for its efficacy in people with John's genotype is prescribed to reduce his cholesterol level and the risk for coronary artery disease (15). His risk for colon cancer can be addressed by a program of annual colonoscopies starting at the age of 45, which provides an opportunity for early detection of colon cancer and secondary prevention of colon cancer mortality. Finally, his increased risk for lung cancer is addressed through behavior modification for smoking cessation. Although this hypothetical case is often used to illustrate future use of genetic information for prevention, it raises a number of issues. For example, it is not clear whether genetic information is really needed to target certain interventions, such as smoking cessation or treatment of hypercholesterolemia. In the case of smoking, testing for susceptibility may reduce the motivation for smoking cessation in those individuals who test "normal" for the genotype. Even when genetic information provides guidance—as is the case with early initiation of colorectal cancer screening—the need for a DNA-based test versus the more generic information provided by a family history is open to question.

Visions of genomic medicine also include gene therapy (12,16) and a whole generation of smart (or designer) drugs and vaccines that can be tailored to individual genetically mediated response (12). As with preventive strategies guided by genotype, potential uses of genomic information require rigorous evaluation.

How do we get from the sequencing of the human genome to using such knowledge to improve clinical care and disease prevention? Collins and McKusick outlined critical elements of the medical research agenda in the 21st century that include improved laboratory technology, clinical research, evaluation of biologic pathways, and improved drug investigation (pharmacogenomics; 12). However, there will be additional crucial missing gaps in our knowledge base that will be

Table 1.2 Vision of Genomic Medicine in the Year 2010: Results of Genetic Testing in a Hypothetical Patient

Condition	Genes	Relative Risk	Lifetime Risk (%)
Prostate cancer	HPC1, 2 and 3	0.4	7
Alzheimer disease	APOE, FAD3, XAD	0.3	10
Coronary artery disease	APOB, CETP	2.5	70
Colon cancer	FCC4, APC	4.0	23
Lung cancer	NAT2	6.0	40

Source: Adapted from Collins and McKusick (12).

filled by the classical public health sciences (17). This research effort is essential in determining the health risks associated with genetic variants and the potential for treatments and prevention. Data in both these areas are also crucial in determining the ethical, legal, and social implications of genetic information (18). Table 1.3 illustrates the convergence of several public health fields that will attempt to answer specific questions to allow the effective "translation" of gene discoveries into the realization of a vision of genomic medicine.

Obviously, the first question, What are the risks? is a crucial one that falls under the purview of epidemiology. In the following chapters, we explore how epidemiologic methods can begin to bridge the knowledge gaps to answer this question. The hypothetical relative risks and lifetime disease probabilities presented in Table 1.2 can be derived only from well-designed, population-based studies that assess disease risks and gene–gene and gene–environment interaction. When accurate risk estimates are obtained, other disciplines of medicine and public health will need to be applied to determine the appropriate use of such information, addressing questions related to public policy, economics, communication, and measurement of health outcomes (as shown in Table 1.3).

Epidemiology: What Are the Risks?

Epidemiology has been defined in many ways and is often viewed as the scientific core of public health (19). One widely used definition is "the study of the distribution and determinants of health-related states or events in populations and the application of this study to control health problems" (20). Epidemiologists determine risk factors for the occurrence of various diseases, identify segments of the population with highest risk to target prevention and intervention opportunities, and evaluate the effectiveness of health programs and services in improving the health of the population (21). Because of the observational nature of epidemiologic studies, the frequent inability to replicate associations across studies (22), and the inability to adjust for all potential confounding factors that are addressed in experimental designs, epidemiology has been occasionally viewed as having reached its limits (23).

Despite these perceived weaknesses, epidemiologic methods have grown steadily over the past 3 decades (24) and have become increasingly integrated with those of

Table 1.3 From Gene Discovery to Applications: Examples of the Role of the Public Health Sciences

Field	Question
Epidemiology	What are the risks?
Policy	What is the value-added of using genetic information to target interventions? (e.g., don't we want every one to stop smoking?)
Communication	What is the best way to communicate risk information?
Economics	Is it cost-effective to do genetic testing?
Outcomes research	How do we measure the impact of genetic information on health, personal, and social outcomes?

genetics in the discipline of genetic epidemiology, which seeks to find the role of genetic factors in disease occurrence in populations and families (25). Also, we have seen the emergence of molecular epidemiology that seeks to study disease occurrence using biologic markers of exposures, susceptibility and effects (26). Increasingly, scientific tools such as microarray technology and gene expression profiles (reviewed in Chapter 2) will be used to identify and measure the effects of various chemical and biologic exposures on health status in addition to the underlying genetic susceptibility to such exposures.

Most gene discoveries are based on studies of high-risk families or selected population groups. Once a gene–disease association is documented, well-conducted epidemiologic studies are needed to quantify the impact of gene variants on the risk for disease, death, and disability and to identify and measure the impact of modifiable risk factors that interact with gene variants. Epidemiologic studies are also required in the process of clinical validation of new genetic tests, and to monitor population use of genetic tests and determine the impact of genetic information on the health and well-being of different populations. The results of such studies will help medical and public health professionals design clinical trials, ultimately leading to benefits from medical, behavioral, and environmental interventions. To accomplish this goal, there must be collaboration among epidemiologists, geneticists, laboratory scientists, and medical and public health practitioners from government, professional, academic, industry, and consumer organizations. The rapid expansion in the number of reported gene–disease associations may lead to pressure to develop commercial tests before validation of research findings. Therefore, there is an urgent need for epidemiologic data generated from multidisciplinary collaboration as a basis for developing medical and public health policy. For example, the appropriate use of genetic testing is currently debated in relation to breast cancer and to *BRCA1/2* mutations (Chapter 26); Alzheimer disease and the apolipoprotein E-ϵ4 allele (Chapter 21); and iron overload and HFE gene mutations (Chapter 28). Rational and comprehensive health policies on the use of genetic tests will not be possible until robust population-based epidemiologic data are available regarding the frequency of newly discovered gene variants and associated disease risks.

What Is Human Genome Epidemiology?

As shown in Figure 1.1, epidemiology plays an important role in the continuum from gene discovery to the development and applications of genetic tests for diagnosing and treating various diseases, or even predicting and preventing future disease in asymptomatic persons. We call this continuum human genome epidemiology (or HuGE) to denote an evolving field of inquiry that uses systematic applications of epidemiologic methods and approaches to the human genome to assess the impact of human genetic variation on health and disease. The terms *genetics* and *genomics* are often used interchangeably but also could be confusing. Guttmacher and Collins refer to *genetics* as "the study of single genes and their effects" and *genomics* as "the study of not just of single genes but of the functions

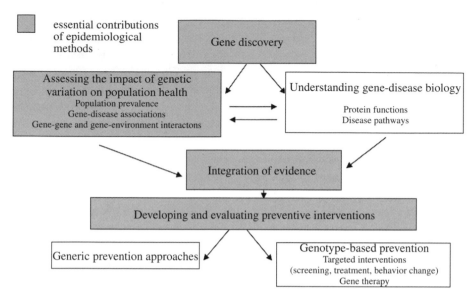

Figure 1.1 The role of epidemiology in the continuum from gene discovery to disease prevention.

and interactions of all genes in the genome" (27). Because of evolving technologies, we are now able to study many genes simultaneously. Our preference for *genomics* in the context of epidemiology reflects the idea that these methods can be applied to the whole human genome (28), rather than to one gene at a time. The intersection of epidemiology with genetics and genomics illustrates the need for common understanding of language and study designs (29).

As shown in Figure 1.1 and Table 1.4, epidemiologic study applications to the human genome include: (*1*) gene discovery (the traditional domain of genetic epidemiology; 25); (*2*) population risk characterization (the domain of molecular

Table 1.4 From Gene Discovery to Applications: The Continuum of Human Genome Epidemiology

Field	Application	Types of Studies
Genetic epidemiology	Gene discovery	Linkage analysis, family-based association studies
Molecular epidemiology	Gene characterization	Population studies to characterize gene prevalence, gene–disease associations, and gene–gene and gene–environment interaction
Applied epidemiology/ Health services research	Evaluating health impact	Studies to evaluate validity and utility of genetic information in clinical trials or observational clinical settings (clinical epidemiology) or population setting (public health epidemiology)

epidemiology) (26); and (*3*) evaluation of genetic information for diagnosis and prevention (including genetic tests and family history information), and for genome-based therapies (applied epidemiology and health services research) (30). The evaluation of genetic information can be done in predicting clinical outcomes (clinical epidemiology) or for population screening and public health assessment (public health epidemiology). The continuum of HuGE represents an extension of the concept of "genetic epidemiology with a capital E" proposed by Thomas (31), because it emphasizes that epidemiologic applications to the human genome go well beyond the quantitative and statistical methods of gene discovery and characterization. Throughout this book, we focus on how epidemiology, as a multidisciplinary field, has begun to address issues arising after gene discovery, with increasing emphasis on characterization of gene effects and genetic tests in populations (what do you do with a gene after you find one?). Table 1.5 shows examples of the types of studies encompassed in the spectrum of human genome epidemiology, including methods for gene discovery (such as linkage analysis), assessing the population prevalence of gene variants, evaluating genotype–disease associations, assessing the impact of gene–environment and gene–gene interaction on disease risk; evaluating the usefulness and impact of genetic information (i.e., tests) for individuals, families, and populations.

In the past few years, there has been a tremendous increase in the number and scope of human genome epidemiology in the peer-reviewed literature. However, most studies are still in the realm of gene discovery and/or genotype–disease associations. An analysis of abstracts of published HuGE papers for 2001 shows that of the 2402 published articles, 82% reported only on population prevalence of gene

Table 1.5 Types and Examples of Human Genome Epidemiology Studies

Type of Study	Study Designs	Examples
Gene discovery	Linkage analysis methods	Body mass index (53) Attention hyperactivity–deficit disorder (54) Type 2 diabetes (55) Celiac disease (56)
Gene characterization	Population prevalence	Prevalence of HFE mutations in the United States (57)
	Genotype–disease associations	Transforming growth factor beta-diabetic retinopathy (58)
	Gene–gene and gene–environment interaction	Multiple gene polymorphisms and dietary interactions in colorectal adenoma (59)
Application of genetic information	Evaluating validity and utility of genetic tests for clinical use	Assessment of interleukin-1 genotype testing for chronic periodontitis (60) Characteristics of individuals tested for *BRCA1/2* (61)
	Evaluating utility of genetic tests for population screening	Assessment of testing for HFE mutations in population screening for iron overload (62)

variants and gene-disease associations, 15% reported on gene–gene and gene–environment interactions, perhaps reflecting the difficulties in studying interactions, and 3% dealt with evaluation of genetic tests and population screening, reflecting a focus on different types of genes considered for clinical practice or population screening (32). Undoubtedly, in the next few years, the number of papers dealing with evaluation of genetic information in medicine and public health is likely to increase as many gene discoveries progress on the translation pathway. Epidemiologic studies of gene–environment interaction and genetic tests are bound to increase as more genes are discovered, characterized, and used to develop diagnostic and predictive tests.

Finally, we can view the role of epidemiology vis-à-vis the suggested paradigm shifts in biology and medicine that are occurring in conjunction with the human genome revolution (Table 1.6). Peltonen and McKusick (33) outlined the progressive shifts occurring in several areas. As shown in Table 1.6, post genome discovery, there will be an increasing shift from studying genetic diseases to studying all diseases, assessing gene products (proteomics), from mapping and sequencing to discovery of genetic variants, from studying single genes to multiple genes, from studying gene actions to studying gene regulations, and finally from diagnostic testing to testing for susceptibility. Epidemiologic methods should play a role in all these areas.

Need for Methodologic Standards in Epidemiologic Studies of the Human Genome

As more epidemiologic studies of human genes are conducted and published, it is important to integrate evidence from different studies. Given the large numbers of genes that will be examined in relation to numerous diseases, many spurious find-

Table 1.6 Role of Epidemiology in the Study of Human Diseases in Relation to Paradigm Shifts in Biology and Medicine

20th Century	21st Century	Role of Epidemiology
Genetic diseases	All diseases	Will use genetic variants routinely in the study of gene–environment interaction in disease
Genomics	Proteomics	Will use biological markers of exposure, susceptibility and outcomes
Gene mapping	Gene sequencing + variant discovery	Will evaluate allelic variants in relation to health outcomes
One gene	Multiple genes	Will assess complex genotypes
Gene action	Gene regulation	Will evaluate structural and regulatory genes as disease risk factors
Diagnostic testing	Susceptibility testing	Will provide a basis for assessing clinical validity of genetic tests for disease susceptibility

Source: Adapted from Peltonen and McKusick (33).

ings are likely to emerge. Moreover, variation in study designs and execution will make difficult the synthesis of information across studies. This is always compounded by the potential for publication bias, as positive gene–disease associations may be published preferentially (34). In a literature review of over 600 gene–disease associations, Hirschorn et al. documented that most reported associations are not robust. Of the 166 associations studied three or more times, only 6 were consistently replicated (34). Similarly, Ioannidis et al. conducted a meta-analysis of 370 studies addressing 36 genetic associations for various diseases. They showed that the results of the first study correlate only modestly with subsequent studies of the same association. The first study often suggests a stronger genetic effect than is found by subsequent studies (35).

In the past few years, several authors (36), using specified guidelines (37), have conducted systematic, peer-reviewed synopses of epidemiologic aspects of human genes, prevalence of allelic variants in different populations, population-based disease risk information, gene–environment interaction, and quantitative data on genetic tests and services. These reviews have uncovered the need for unified guidelines that can be used to synthesize results of the increasing number of such studies. In 2002, Cooper et al. proposed guidelines to promote the publication of scientifically meaningful disease association studies through the introduction of methodologic standards (with a focus on discovery of candidate genes) (38). Although several groups have addressed guidelines for the evaluation and synthesis of a number of areas (e.g., controlled clinical trials), no such recommendations exist that cover the full spectrum of HuGE studies. In an analysis of the epidemiologic quality of molecular genetic research, Bogardus et al. used seven methodologic standards to evaluate the quality of studies in four mainstream medical journals. They found that despite the major molecular genetic advances, 63% of the articles did not comply with two or more quality standards (39). This finding emphasizes the need for methodologic standards in reporting such studies. Based on an expert panel workshop held in 1997, Stroup et al. published a proposal for reporting results of meta-analyses of observational studies in epidemiology (40) but did not specifically address genetic studies. Bruns et al. provided a checklist for reporting studies of diagnostic accuracy of medical tests (41). A workshop sponsored by the National Cancer Institute led to a monograph on innovative study designs and analytic approaches to the genetic epidemiology of cancer (42). This series of articles was useful in outlining the spectrum of study designs in gene discovery and characterization in relation to disease, but the articles do not provide concrete guidance on the evaluation and synthesis of such studies. In 2001, an expert panel sponsored by the Centers for Disease Control and Prevention and the National Institutes of Health developed guidelines and recommendations for the evaluation and integration of data from human genome epidemiology with emphasis on studies of (*1*) prevalence of gene variants and gene–disease associations; (*2*) gene–environment and gene–gene interactions; and (*3*) evaluation of genetic tests. Conclusions and recommendations from this workshop have been published (43,44). Thus, progress is being slowly made in defining quality standards for genetic-epidemiologic research, but ongoing evaluation is needed to make sure that such guidelines are implemented.

Finally, there is increasing interest in the evaluation of genetic tests. The National Institutes of Health-Department of Energy Task Force on Genetic Testing (45) and the Secretary's Advisory Committee on Genetic Testing (46) have proposed detailed evaluation of new genetic tests. Many of the genetic tests that will emerge in the next decades will be used not only for diagnostic purposes but also to predict the risk of developing disease in otherwise healthy people and to make decisions about potentially preventive interventions or therapies. The utility of genetic tests in this context will depend heavily on the quality of epidemiologic information that summarizes the relation between genotypes and disease and how such a relation is modulated by the presence of other factors, such as drugs and environmental exposures, and how risk can be reduced by interventions. These concerns clearly apply to the hypothetical case scenario shown in Table 1.2. Such information will have to be based on properly designed epidemiologic information on genotype–disease associations and gene–gene and gene–environment interaction. In addition, genetic test reports will have to include specific information based on appropriate clinical and epidemiologic outcome studies on how the risk for these diseases can be reduced using medical, behavioral, and environmental interventions.

The HuGE Map Ahead

Because we recognized the immense gaps in our knowledge of the relation between genes and health outcomes, we organized this book to include chapters on epidemiologic methods and approaches that deal with the continuum from gene discovery to health care action. None of the material in the book is scientifically novel, but we have structured it in a way to allow readers to proceed systematically from the fundamentals of genome technology and gene discovery (Section I) to epidemiologic approaches to gene characterization (Section II) to evaluation of genetic tests and health services (Section III). These concepts are then illustrated in a series of disease-specific case studies (Section IV).

Section I (Fundamentals) describes genomic technologies and their applications (Chapter 2) and evolving methods of gene discovery (Chapter 3), and summarizes the current status of the ethical, legal, and social issues for conducting epidemiolog studies of the human genome (with emphasis on informed consent issues, Chapter 4).

Section II addresses epidemiologic approaches to the study of genotypes in populations (Chapter 5) and their relation to diseases, including the assessment of gene–gene and gene–environment interaction (Chapters 6–9). In addition, Chapter 10 addresses synthesis of these studies, including some methodologic standards.

Section III deals with the application of epidemiologic methods to assessing genetic information for clinical and public health applications. Chapter 11 lays an epi-

demiologic foundation for using population level information to make individualized risk predictions for clinical use. Chapter 12 explores population-based concepts
around the usefulness of genetic tests in populations. Chapter 13 discusses the evaluation of genetic tests from a combined clinical–laboratory approach. Because
gene–drug interactions are becoming more prominent in medical practice, Chapter
14 explores an epidemiologic framework for the interface between genetics, pharmacology, and medicine. Chapter 15 explores the integration of genetics into controlled clinical trials, and Chapters 16 and 17 focus on the role of genetics in clinical practice guideline development both in the United States and the United
Kingdom.

Finally, in Section IV case studies illustrate concepts discussed in the first three
sections in relation to specific disease examples, including gene–environment interactions (pesticides in Chapter 18 and oral contraceptive use in Chapter 19), chronic
diseases (colon cancer in Chapter 20, Alzheimer disease in Chapter 21, cardiovascular disease in Chapter 25, breast cancer in Chapter 26 and iron overload in Chapter 28), occupational exposures (berylliosis in Chapter 22, and testing of railroad
track workers, Chapter 29), newborn screening issues (fragile X syndrome in Chapter 23, and hearing loss in Chapter 24), and infectious disease (HIV in Chapter 27).
These examples are by no means exhaustive, but they do illustrate two points: (*1*)
the spectrum from single gene disorders to complex disorders that involve gene–
environment interactions and the application of genomic knowledge to disease prevention; and (*2*) the need for epidemiologic research to obtain population level information that is crucial for developing health policy and new approaches to practice. Indeed, for many of the case studies, the lack of such data represents a primary
barrier to developing appropriate health policies related to genetic risk information.
The readers will discover that our knowledge is rapidly evolving for each one of
these examples. Most likely, this information will be outdated soon. Nevertheless,
we hope that the basic methodologic foundations for how to approach the continuum from gene discovery to using genetic information will still apply. Throughout
the book, we assume that readers have introductory knowledge of the fields of epidemiology and genetics and we therefore mainly address the interface between the
two disciplines. For general readings on epidemiology and epidemiologic methods
many excellent books are available (e.g., refs. 24, 47–50). Also, we do not present
technical details of genome technologies or statistical methods of gene discoveries
with the use of family-based methods but merely an overview of these rapidly changing methods. Our central emphasis is on applying epidemiology to assess the role
of genetic variations in health outcomes and to evaluate how such information can
be used to improve health and prevent disease.

Finally, there is an increasing emphasis on using family history (with or without
the use of genetic testing) as a tool for disease prevention and public health (51).
Family history is a consistent risk factor for most common diseases and reflects
shared genetic and environmental factors. Family history is often used an initial tool
to genetic testing and targeted interventions. While this book does not cover fam-

ily history directly, we recognize that epidemiologic methods and approaches can be applied to the evaluation of family history with or without genetic testing. For further discussions about family history in the context of public health and disease prevention, the readers are referred to a series of workshop papers published in 2003 (52).

Ultimately, in order to fulfill the promise of the Human Genome Project in improving health, multidisciplinary medical and public health approaches are needed. At the core of these approaches is the simple question, What are the risks? followed by the question, What to do with numbers once you get them? To get there, we have a HuGE map to follow.

References

1. Venter JC, Adams MD, Myers EW, et al. The sequence of the human genome. Science 2001;291:1304–1351.
2. Lander ES, Linton LM, Birren B, et al. Initial sequencing and analysis of the human genome. Nature 2001;409:860–921.
3. Collins FS, Guttmacher AE. Genetics moves into the medical mainstream. JAMA 2001;286:2322–2324.
4. Guttmacher AE, Collins FS. Genomic medicine: a primer. New Engl J Med 2002;347:1512–1520.
5. Centers for Disease Control Office of Genomics and Disease Prevention. Genomics Weekly Update, Atlanta, 2002. http://www.cdc.gov/genomics/update/current.htm
6. Holtzman NA, Marteau TM. Will genetics revolutionize medicine? New Engl J Med 2000;343(2):141–144.
7. Zimmern R, Emery J, Richards T. Putting genetics in perspective. BMJ 2001;322: 1005–1006.
8. Willet WC. Balancing lifestyle and genomics research for disease prevention. Science 2002;296:695–698.
9. Evans JP, Skrzynia C, Burke W. The complexities of predictive genetic testing. BMJ 2001;322:1052–1056.
10. Vineis P, Schulte P, McMichael AJ. Misconceptions about the use of genetic tests in populations. Lancet 2001;357:709–712.
11. Rockhill B, Kawachi I, Colditz GA. Individual risk prediction and population-wide disease prevention. Epidemiol Rev 2000;22:176–180.
12. Collins FS, McKusick VA. Implications of the Human Genome Project for medical science. JAMA 2001;285:540–544.
13. Sciona. Discover the relationship between you and your genes, Accessed October, 2002, from http://www.sciona.com.
14. Genovations: the advent of truly personalized healthcare. Accessed October 2002, from http://www.genovations.com.
15. Collins FS. Shattuck lecture—medical and societal consequences of the Human Genome Project. N Engl J Med 1999;341:28–37.
16. Wadhwa PD, Zielske SP, Roth JC, et al. Cancer gene therapy: scientific basis. Annu Rev Med 2002;53:437–452.
17. Khoury MJ, Beskow L, Gwinn ML. Translation of genomic research into health care. JAMA 2001;285:2447–2448.
18. Burke W, Pinsky LE, Press NA. Categorizing genetic tests to identify their ethical, legal, and social implications. Am J Med Genet 2001;106:233–240.

19. Committee for the Study of the Future of Public Health, Institute of Medicine. Washington, DC: The Future of Public Health. National Academy Press, 1988.
20. Last JM. A Dictionary of Epidemiology, third edition. New York: Oxford University Press, 1995.
21. Koplan JP, Thacker SB, Lezin NA. Epidemiology in the 21st century: calculation, communication, and intervention. Am J Public Health 1999;89:1153–1155.
22. Ioannidis JP, Ntzani EE, Trikalinos TA, et al. Replication validity of genetic association studies. Nat Genet 2001;3:306–309.
23. Taubes G. Epidemiology faces its limits. Science 1995;269:164–165.
24. Rothman KJ, Greenland S. Modern Epidemiology, second edition. Philadelphia: Lippincott, Williams & Wilkins, 1998.
25. Khoury MJ, Beaty TH, Cohen BH. Fundamentals of Genetic Epidemiology. New York: Oxford University Press, 1993.
26. Schulte PA, Perera FP (eds.). Molecular Epidemiology: Principles and Practice. New York: Academic Press, 1993.
27. Guttmacher AE, Collins FS. Genomic medicine: a primer. New Engl J Med 2002;347:1512–1520.
28. Millikan R. The changing face of epidemiology in the genomics era. Epidemiology 2002;13:472–480.
29. Potter JD. At the interfaces of epidemiology, genetics and genomics. Nat Rev Genet 2001;2:142–147.
30. Brownson RC, Petitti DB. Applied Epidemiology: Theory to Practice. New York: Oxford University Press, 1998.
31. Thomas DC. Genetic epidemiology with a capital E. Genet Epidemiol 2000;19:289–300.
32. Centers for Disease Control. Database for the Human Genome Epidemiology Network. From http://www2.cdc.gov/nceh/genetics/hugenet/frmSearchMenu.asp.
33. Peltonen L, McKusick VA. Genomics and medicine. Dissecting human disease in the postgenomic era. Science 2001;291:1224–1229.
34. Hirschhorn JN, Lohmueller K, Byrne E, Hirschhorn K. Comprehensive review of genetic association studies. Genet Med 2002;4:45–61.
35. Ioannidis JP, Ntzani EE, Trikalinos TA, Contopoulos-Ioannidis DG. Replication validity of genetic association studies. Nat Genet 2001;29:306–309.
36. Khoury MJ, Little MJ. Human genome epidemiologic reviews:the beginning of something huge. Am J Epidemiol 2000;151:2–3.
37. American Journal of Epidemiology: Revised guidelines for publishing HuGE reviews. Am J Epidemiol 2000:131:4–6.
38. Cooper DN, Nussbaum RL, Krawczak M. Proposed guidelines for papers describing DNA polymorphism-disease associations. Hum Genet 2002;110:207–208.
39. Bogardus ST, Concato J, Feinstein AR. Clinical epidemiological quality in molecular genetic research: the need for methodologic standards. JAMA 1999;281:1919–1926.
40. Stroup DF, Berlin JA, Morton SC, et al. Meta-analysis of observational studies in epidemiology: a proposal for reporting. Meta-analysis of Observational Studies in Epidemiology (MOOSE) group. JAMA 2000;283:2008–2016.
41. Bruns DE, Huth EJ, Magid E, Young DS. Toward a checklist for reporting of studies of diagnostic accuracy of medical tests. Clin Chem. 2000;46:893–895.
42. National Cancer Institute. Innovative study designs and analytic approaches to the genetic epidemiology of cancer. Monographs of JNCI 1999;26:1–105.
43. Little J et al. HuGE paper 1. Am J Epidemiol 2002;156:300–310.
44. Burke W, Atkins D, Gwinn M, et al. Genetic test evaluation information needs of clinicians, policy makers and the public. Am J Epidemiol 2002;156:311–318.

45. Holtzman NA, Watson MS (eds). Promoting Safe and Effective Genetic Testing in the United States. Final Report of the Task Force on Genetic Testing (1997). National Human Genome Research Institute. From http://www.nhgri.nih.gov/ELSI/TFGT_final/.
46. Secretary's Advisory Committee on Genetic Testing. Enhancing the Oversight of Genetic Tests: Recommendations of the Secretary's Advisory Committee on Genetic Testing (SACGT) (2000). From http://www.od.nih.gov/oba/sacgt/reports/FINAL_SACGTreport713700correctedpage27.htm.
47. Lilienfeld DE, Stolley P. Foundations of Epidemiology, (third edition). New York: Oxford University Press, 1994.
48. Rothman KJ. Epidemiology: An Introduction. New York: Oxford University Press, 2002.
49. Kelsey JL, Whitemore AS, Evans AS, Thompson WD. Methods in Observational Epidemiology, second edition. New York: Oxford University Press, 1996.
50. Elwood JM. Critical Appraisal of Epidemiological Studies and Clinical Trials, second edition. Oxford, UK: Oxford University Press, 1998.
51. Yoon PW, Scheuner MT, Peterson-Oehlke P, et al. Can family history be used as a tool for public health and preventive medicine? Genet Med 2002;4:304–310.
52. Yoon PW, Scheuner M, Khoury MJ. Research priorities for using family history as tool for public health and disease prevention. Am J Prev Med (in press, 2003)
53. Wu X, Cooper RS, Borecki I, et al. A Combined Analysis of Genomewide Linkage Scans for Body Mass Index from the National Heart, Lung, and Blood Institute Family Blood Pressure Program. Am J Hum Genet 2002;70:1247–1256.
54. Fisher SE, Francks C, McCracken JT, et al. A genomewide scan for loci involved in attention-deficit/hyperactivity disorder. Am J Hum Genet 2002;70:1183–1196.
55. Lindgren CM, Mahtani MM, et al. Genomewide search for type 2 diabetes mellitus susceptibility loci in Finnish families: the Botnia study. Am J Hum Genet 2002;70:509–516.
56. Liu J, Juo SH, Holopainen P, et al. Genomewide linkage analysis of celiac disease in Finnish families. Am J Hum Genet 2002;70:51–59.
57. Steinberg KK, Cogswell ME, Chang JC, et al. Prevalence of C282Y and H63D mutations in the hemochromatosis (HFE) gene in the United States. JAMA 2001;285:2216–2222.
58. Beranek M, Kankova K, Bene P, et al. Polymorphism R25P in the gene encoding transforming growth factor-beta (TGF-beta1) is a newly identified risk factor for proliferative diabetic retinopathy. Am J Med Genet 2002;109:278–283.
59. Cortessis V, Siegmund K, Chen Q, et al. A case-control study of microsomal epoxide hydrolase, smoking, meat consumption, glutathione S-transferase M3, and risk of colorectal adenomas. Cancer Res 2001;61(6):2381–2385.
60. Greenstein G, Hart TC. A critical assessment of interleukin-1 (IL-1) genotyping when used in a genetic susceptibility test for severe chronic periodontitis. J Periodontol 2002;73:231–247.
61. Frank TS, Deffenbaugh AM, Reid JE, et al. Clinical characteristics if individuals with germline mutations in BRCA1 and BRCA2: analysis of 10,000 individuals. J Clin Oncol 2002;20:1480–1490.
62. Burke W, Thomson E, Khoury MJ, et al. Hereditary hemochromatosis: gene discovery and its implications for population-based screening. JAMA 1998; 280:172–178.

2

Emerging genomic technologies and analytic methods for population- and clinic-based research

Darrell L. Ellsworth and Christopher J. O'Donnell

With the completion of the human genome sequence and advances in technologies for genomic analysis, increasing attention is being focused on the discovery of new genes and genetic polymorphisms that may prove useful in the diagnosis, prevention, and treatment of complex human diseases. As new genes believed to influence disease are identified and genetic tests are developed to routinely assay polymorphisms within these genes, epidemiologists and health care professionals will need to become more knowledgeable about the capabilities and limitations of new methods for detecting genetic influences on disease, the ability of genetic tests to accurately predict disease, and the evolving implications of genetics for disease prevention and treatment. In this chapter, we discuss the (*1*) current methods for localizing genes that influence complex diseases; (*2*) emerging genomic technologies for complex disease research; (*3*) application of DNA sequence variation to predict disease risk in the general population; and (*4*) potential therapeutic utility of genetic information. Our purpose is to familiarize epidemiologists and health care providers with current techniques for applying genetic information to improve the diagnosis and treatment of complex human diseases.

The Human Genome Project

An historic joint announcement on June 26, 2000 by the International Human Genome Sequencing Consortium and Celera Genomics Corporation signified that the first "working draft" of the human genome sequence had been completed. The publicly funded Human Genome Project adopted a systematic approach to sequencing the human genome that involved breaking each chromosome into smaller fragments of DNA that could be manipulated in the laboratory, reconstructing the chromosome by organizing the fragments in the proper order, and then determining the DNA sequence of these ordered fragments (1). Conversely, Celera Genomics utilized a "whole genome shotgun" approach, in which all chromosomes were fragmented into many small pieces that were not ordered in any way, DNA sequence was obtained from the unordered fragments, and then the sequence of the genome was established by positioning all of the DNA sequences based on overlap (2).

Extracting and analyzing useful information from the primary sequence data, such as new genes, regulatory sequences, and genetic markers, are daunting and time-consuming tasks; therefore, a number of resources are available to facilitate easy access and widespread use of this valuable resource. Several Web sites have been established that display different views of the entire working draft sequence and provide useful tools for accessing important information: (1) the National Center for Biotechnology Information (NCBI) (http://www.ncbi. nlm.nih.gov/genome/guide/); (2) the European Bioinformatics Institute (EBI) (http://www.ensembl.org/); and (3) the University of California at Santa Cruz (http://genome.ucsc.edu/). Computer programs such as GENSCAN (3) have been developed to identify intron/exon structures and predict the occurrence of genes in genomic DNA sequence, and Web sites such as PipMaker (http:// bio.cse.psu.edu) (4) are available to compare genomic sequences from related species to identify conserved (functional) sequences (5). Over 30,000 genes have been predicted directly from the human genome sequence and through comparative genomics. Building on the successes of the Human Genome Project, the United States Department of Energy has initiated an ambitious "next phase" program known as "Genomes to Life," which will attempt to characterize gene regulatory networks, identify proteins that conduct critical life functions, and develop computational capabilities to model complex biologic systems (http:// doegenomestolife.org).

Methods for Gene Localization Emerging from the Human Genome Project

Microsatellite Markers and Linkage
Microsatellite markers, or short tandem repeats (STRs), usually contain two to four base pair sequences such as CA or GATA that are tandemly repeated a variable number of times. Microsatellites tend to exhibit high levels of polymorphism (variation in the number of repeated sequences), are widely distributed across the genome, and may be easily typed by the polymerase chain reaction (PCR) with high-throughput semi-automated systems (described in ref. 6). Microsatellite markers have been used in a variety of applications including forensic medicine and DNA fingerprinting (7), constructing genetic maps for genome mapping (http://gai.nci.nih.gov/CHLC/), and identifying chromosomal regions that may

contain genes contributing to complex diseases such as cardiovascular disease and diabetes.

The traditional approach for localizing genes contributing to complex disease susceptibility involves the simultaneous analysis of hundreds of microsatellite markers at regularly spaced intervals throughout the genome (a *genome-wide scan*) in families or affected sibling pairs (8,9). *Nonparametric linkage analyses* are then used to identify chromosomal *candidate regions* in which genes influencing the disease may reside (described in ref. 10). Although genome-wide linkage approaches using microsatellites have been popular over the past several years, such studies have limited power to detect genes with small to modest effects, and therefore, have met with limited success.

Single Nucleotide Polymorphisms and Association Methods

As abundant DNA sequences emerged from the Human Genome Project over the past several years, a gradual shift began to occur away from microsatellite markers and linkage approaches toward association-based methods for examining the genetics of complex diseases and identifying genetic influences on drug response. Polymorphisms characterized by variation at a single nucleotide, known as *single nucleotide polymorphisms* (SNPs), such as 5'-TCGAATTCAG-3' and 5'-TCGACTTCAG-3', are far more abundant than microsatellites—there are an estimated 7.1 million SNPs in the human genome with a minor allele frequency of at least 5% (11).

Single nucleotide polymorphisms are useful for examining human susceptibility to complex diseases and eventually may be important for locating disease-susceptibility genes using *genome-wide association* and *linkage disequilibrium* approaches. Most SNPs occur in noncoding sequences such as introns and regions between genes, but SNPs in the protein coding regions of genes (cSNPs), particularly nonsynonymous (amino acid altering) mutations, and in regulatory regions controlling gene expression are of particular interest in association studies attempting to define genetic influences on human disease. In their simplest form, association studies use a chi-square test to compare the frequencies of alleles for a single SNP in persons (cases) affected with disease with frequencies in matched controls who are disease free (see Chapter 6 for further discussion of analytic methods). If the frequency of one allele is significantly higher in the case group relative to the controls, then the polymorphism is considered to be associated with the disease. More complex statistical methods such as multiple logistic regression may be used in association studies to determine the influence of a SNP polymorphism on disease when traditional risk factors such as cholesterol and smoking are considered (12,13).

An important limitation of association studies attempting to relate variation at a single polymorphism to disease is that this approach does not account for other functional genetic variation within the gene. For example, numerous association studies have examined relationships between the angiotensin I-converting enzyme (ACE) insertion/deletion (I/D) polymorphism and different forms of cardiovascular disease. The D allele and DD genotype have been associated with various cardiovascular diseases, but the associations have not always been consistent. Most of these studies examined only the I/D polymorphism in intron 16 and did not account for genetic variation in other parts of the ACE gene. A recent survey of DNA sequence variation, however, identified many variants—74 SNPs and four insertion/deletion polymorphisms—throughout the gene (14) that also may influence risk of disease (Fig. 2.1).

The high levels of sequence variability throughout the ACE gene region illustrate the drawbacks of association studies attempting to relate a single genetic variant to disease. By examining only one variable site in a given candidate gene, genetic variation with a significant influence on disease risk may be overlooked. The "alleles" defined by a single polymorphism will likely include heterogeneous groups of variants (such as the ACE I and D alleles) and the distribution of these variants may

---

Figure 2.1 (A) Hypothetical gene region containing seven single nucleotide polymorphisms (SNPs) identified by the location of the nucleotide substitution relative to the transcription initiation site (+1) and the nature of the molecular change. Note that three polymorphisms occur in regions with known function, −438A/G occurs upstream from the transcription start site in the promoter region and +38Arg/Ser and +7178Pro/Ala cause amino acid substitutions in the corresponding protein, while four polymorphisms are located in regions believed to be nonfunctional, +2642C/T, +5795A/G, and +12491C/T occur in noncoding introns and +16435(+C) is an insertion polymorphism in the 3' flanking region. Ala, alanine; Arg, arginine; Pro, proline; Ser, serine. (B) Haplotypes designated by Roman numerals representing combinations of alleles across the seven polymorphic sites shown in (A) above that occur on the same chromosome. Haplotypes were constructed by observing transmission of the alleles from parents to offspring. The frequency of each haplotype in a hypothetical study population of 100 individuals is indicated in the column of values under N. (C) Network interrelating the haplotypes presented in (B) above. Circles containing Roman numerals represent haplotypes and the size of each circle indicates the relative frequency of each haplotype (large circles represent more frequent haplotypes, while small circles represent infrequent haplotypes). The lengths of the lines connecting the haplotypes reflect the number of genetic differences between them—cross bars show the actual mutational events. (D) Network of genetic variation throughout the angiotensin I-converting enzyme (ACE) gene. Alleles at polymorphic sites throughout the gene were organized into haplotypes, depicted as numbered circles where the size of each circle reflects the relative frequency of the haplotype in the sample population, and the lengths of the lines connecting haplotypes are proportional to the number of genetic differences between them (as in C above). The Alu element distinguishing the insertion (I) and deletion (D) alleles that is commonly examined in association studies occurs on the long branch separating the two primary groups of haplotypes. Note that genetic heterogeneity occurs within the I and D alleles, which may confound association studies examining relationships between the I and D alleles and complex human diseases. Adapted from Rieder et al. (14) with permission.

A

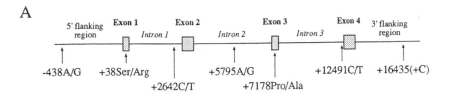

B

								N
I.	-438A	+38Ser	+2642C	+5795A	+7178Pro	+12491C	+16435C	10
II.	-438G	+38Ser	+2642C	+5795A	+7178Pro	+12491C	+16435C	92
III.	-438G	+38Arg	+2642C	+5795A	+7178Ala	+12491T	+16435C	76
IV.	-438G	+38Arg	+2642C	+5795A	+7178Ala	+12491T	+16435CC	4
V.	-438G	+38Arg	+2642T	+5795G	+7178Ala	+12491T	+16435C	18

C

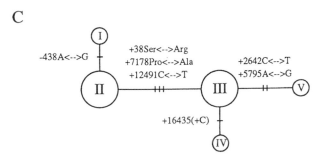

D

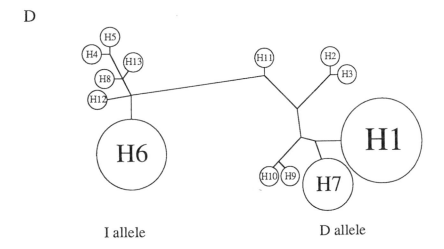

I allele   D allele

differ among racial/ethnic groups. Studies examining the role of genetic factors in susceptibility to complex human diseases should, therefore, examine associations between groups of polymorphisms within a gene (*haplotypes*) and differential risk of disease (for further discussions, see Chapters 6 and 10).

Although there is a large number of possible haplotypes that could be formed by the various combinations of SNPs, in reality only a limited number of haplotypes are commonly found for a given gene. Recently, there has been increasing focus on characterizing the haplotype structure of the entire genome as well as individual genes because haplotype variation better defines the full extent of genetic variability within individuals (15). Haplotype analyses are thus considered to be a more powerful tool for use in association studies of complex human diseases.

As mentioned above, SNPs and related haplotypes may be useful for identifying genes influencing susceptibility to disease, and this potential is receiving increasing attention from the scientific community as well as private industry (16). Large-scale association methods involving the simultaneous examination of hundreds of thousands of SNPs and corresponding haplotypes across the entire genome are being developed to localize novel genes contributing to complex diseases. The concept of a genome-wide association approach using SNP markers was initially viewed with great enthusiasm, because SNPs are far more numerous than microsatellites, SNPs may be assayed with high-throughput technologies, and association studies are believed to have greater power than conventional linkage approaches (17). The dynamic nature of the genome and phenotypic complexities of disease etiology, however, may make finding genes through genome-wide SNP analyses more difficult than originally anticipated (18,19). Common diseases are likely influenced by numerous genes (and hence many mutations), which may make it difficult to detect any one in a genome-wide SNP analysis. In addition, molecular recombination or crossing over may disrupt linkage disequilibrium among SNP polymorphisms and obscure associations with genes influencing disease risk.

Emerging Genomic Technologies for Complex Disease Research

Technologies for SNP Genotyping

Over 1.42 million candidate SNPs have already been identified and mapped in the human genome (20), but at present, our ability to identify SNPs greatly exceeds our ability to genotype SNPs. A number of non–microarray-based methods for high-throughput SNP genotyping are being developed, which soon may be applied to comprehensive association studies of SNP variation and whole-genome association approaches for localizing novel disease-susceptibility genes. For example, one type

of assay uses mass spectrometry to accurately measure the masses of short oligonucleotide primers that are extended by a single base (Fig. 2.2). Many polymorphisms in a variety of genes can be analyzed simultaneously by designing primers
with sufficiently different masses so that each primer–extended primer pair occupies a separate molecular weight region in the mass spectrum (21). A related but
more involved technique is also amenable to high-throughput SNP genotyping, because many reactions can be conducted in parallel (22,23). This methodology utilizes partially overlapping oligonucleotide probes and can discriminate SNP alleles
by a complex system of probe cleavage and detection with fluorescence energy
transfer (Fig. 2.3).

DNA Microarrays

Because of the molecular complexity of many human diseases, thousands of polymorphisms in many candidate genes may need to be assayed to have a clear understanding of genetic influences on disease risk. For example, there are an estimated 120,000 cSNPs in the human genome, 40% of which are expected to result
in an amino acid substitution. These estimates suggest that nearly 48,000 cSNPs
plus an unknown number of functional polymorphisms in regulatory regions controlling gene expression would need to be examined in a comprehensive association study (11). Fortunately, SNPs are also amenable to high-throughput analysis
on microarrays or DNA chips, which provide an alternative medium to the methods described above for genotyping SNP polymorphisms (24). The DNA chips contain arrays of DNA fragments (probes) that can be synthesized remotely then deposited onto the surface of the chip or can be synthesized directly on the chip by
using procedures similar to that used in manufacturing computer chips (Fig. 2.4).
On certain types of chips, the probes function by hybridizing to fluorescently labeled DNA that is applied to the surface of the chip. Hybridization material applied
to chips for genotyping may be generated by PCR amplification of protein coding
sequences or regions controlling gene expression and the probes may be designed
to assay relevant SNPs within these regions. With the use of DNA chip technology,
thousands of SNP polymorphisms in protein coding sequences or regions controlling gene expression can be analyzed simultaneously on a single chip, which can
be customized to assay genetic information on a large panel of candidate genes for
a specific disease (25).

   Customized DNA chips are already proving useful for quantifying genetic influences on complex diseases such as coronary heart disease. For example, a minia-
turized array format has been used to examine 12 common polymorphisms in eight
genes believed to influence myocardial infarction (MI) (26). In this study, the *4G*
allele of the plasminogen activator inhibitor gene and the *Pl^A2* allele of the glycoprotein IIIa gene were independently associated with increased risk of MI, but the
concurrent presence of both variants significantly increased risk of disease. Similarly, a low-density oligonucleotide probe array has been developed as a prelimi-

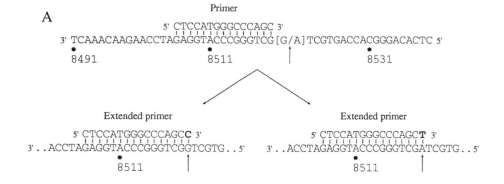

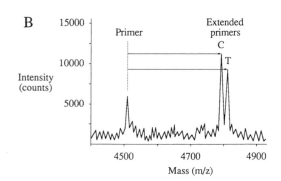

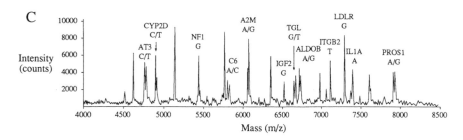

Figure 2.2 Schematic diagram of the PinPoint assay (21), which incorporates primer extension and matrix-assisted laser desorption ionization/time-of-flight mass spectrometry (MALDI-TOF MS) to perform high-throughput genotyping of single nucleotide polymorphisms (SNPs). The technique utilizes mass spectrometry to accurately measure the masses of short oligonucleotide primers extended by a single dideoxynucleotide. Careful attention to primer design permits multiple polymorphisms to be analyzed simultaneously. (A) Partial nucleotide sequence from intron 4 of the human antithrombin III (AT3) gene region (nucleotides 8491–8540; GenBank accession number X68793) showing a polymorphism denoted by the arrow at nucleotide position 8521. The polymerase chain reaction (PCR) is first used to generate millions of copies of the AT3 gene region containing the polymorphism, and then a complementary oligonucleotide primer is allowed to anneal to

nary screening tool to assay 12 polymorphisms in four genes contributing to familial hypertrophic cardiomyopathy (27).

Prior to the development of genotyping methods based on mass spectrometry or microarrays, it was difficult to simultaneously examine many genes and account for the additive effects of multiple polymorphisms on disease risk, but now these technologies are proving useful for simultaneously quantifying many genetic influences on complex diseases. The ability to examine polymorphisms in many genes simultaneously by using mass spectrometry and/or array-based methods has the potential to substantially improve the detection of genetic influences on disease risk resulting from interactions among multiple genes and many polymorphisms.

Expression Arrays

Polymorphisms in the protein coding regions of genes may alter the structure, and hence function, of a particular protein, and thereby influence onset and/or progression of disease. In contrast, DNA polymorphisms in important regulatory regions may influence disease by altering levels of gene expression in important organs such as the heart, and in tissues such as the vascular endothelium. The diverse and adaptable nature of microarray technologies makes it possible to assay gene expression in addition to genotyping polymorphisms. Expression arrays permit the simultane-

---

the amplified DNA of both alleles for a given locus immediately upstream (5') from the polymorphic site. The primer is then extended by a single dideoxynucleotide (ddNTP), which prevents any extension beyond the addition of a single dideoxynucleotide—the dideoxynucleotide added is determined by the sequence of the DNA. In this example, a dideoxycytosine (ddC) has been added to the primer bound to the DNA sequence on the left; while a dideoxythymine (ddT) has been added to the primer bound on the right. (B) Following primer extension, the reaction mixture is purified and analyzed by mass spectrometry. A popular technique known as MALDI-TOF MS uses laser optics to measure the masses of the extended primers and to genotype the polymorphic site by distinguishing primers that are extended by a different ddNTP. The mass spectrum shows unextended primer at a molecular weight of 4513.98 Da, primer extended with ddC at 4787.17 Da, and primer extended with ddT at 4802.18 Da. (C) Mass spectrum showing the simultaneous (multiplex) genotyping of 12 different SNPs from a variety of genes in a hypothetical individual. All primers have been carefully designed to have sufficiently different masses so that each primer—extended primer pair occupies a separate molecular weight region in the mass spectrum, and there is adequate resolution between the extended primers for a given locus. The genes and corresponding genotypes in this example are: AT3-antithrombin III, CT; CYP2D-cytochrome P450 2D6, CT; NF1-neurofibromatosis type I, GG; C6-complement component C6, AC; A2M-alpha-2-macroglobulin, AG; IGF2-insulin-like growth factor II, GG; TGL-triglyceride lipase, GT; ALDOB-aldolase B, AG; ITGB2-integrin B2 subunit, TT; LDLR-low-density lipoprotein receptor, GG; IL1A-interleukin 1-alpha, AA; PROS1-protein S alpha, AG.

Primary Invader Reaction

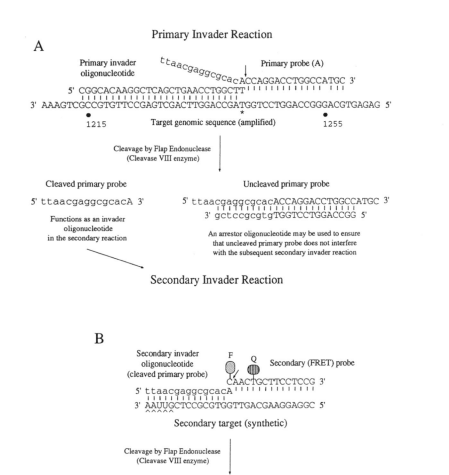

Primary invader oligonucleotide

Primary probe (A)

ttaacgaggcgcac

5' CGGCACAAGGCTCAGCTGAACCTGGCTT ACCAGGACCTGGCCATGC 3'
||||||||||||||||||||||||||||| ||||||||||||||| |||
3' AAAGTCGCCGTGTTCCGAGTCGACTTGGACCGATGGTCCTGGACCGGGACGTGAGAG 5'

• 1215

Target genomic sequence (amplified)

• 1255

Cleavage by Flap Endonuclease
(Cleavase VIII enzyme)

Cleaved primary probe

5' ttaacgaggcgcacA 3'

Functions as an invader
oligonucleotide
in the secondary reaction

Uncleaved primary probe

5' ttaacgaggcgcacACCAGGACCTGGCCATGC 3'
        |||||||||||||||||||||||
3' gctccgcgtgTGGTCCTGGACCGG 5'

An arrestor oligonucleotide may be used to ensure
that uncleaved primary probe does not interfere
with the subsequent secondary invader reaction

Secondary Invader Reaction

Secondary invader
oligonucleotide
(cleaved primary probe)

F   Q

Secondary (FRET) probe

CAACTGCTTCCTCCG 3'

5' ttaacgaggcgcacA ||||||||||||||||
    ||||||||||||||
3' AAUUGCTCCGCGTGGTTGACGAAGGAGGC 5'

Secondary target (synthetic)

Cleavage by Flap Endonuclease
(Cleavase VIII enzyme)

Fluorescent product resulting from
cleavage of secondary probe
containing the quenching dye (Q)

F

C

Signal detected with a
fluorescent plate reader

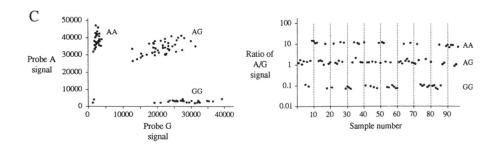

C

Probe A signal

Ratio of A/G signal

Sample number

Probe G signal

Figure 2.3 High-throughput genotyping of single nucleotide polymorphisms (SNPs) using Invader technology, which utilizes partially overlapping oligonucleotide probes and a cleavase enzyme for allelic discrimination coupled with a fluorescence resonance energy transfer detection system. (A) Partial nucleotide sequence from exon 1 of the human cytotoxic T-lymphocyte-associated protein 4 (CTLA4) gene region (nucleotides 1208-1264; GenBank accession number M74363) showing a polymorphism denoted by the asterisk at nucleotide position 1241. To conserve stored DNA samples, PCR may be used to generate millions of copies of the CTLA4 gene region containing the polymorphism. In the primary Invader reaction, a primary Invader oligonucleotide is designed to anneal to the amplified target sequence, so that the 3' base overlaps the polymorphic site to be genotyped, but is not complementary to (and thus will not base pair with) either allele of the polymorphism. A primary probe (specific for the "A" allele in this example) is designed so that approximately half of the probe is complementary to the target sequence and overlaps the primary Invader oligonucleotide at the polymorphic site, but the other half, the reporter arm, is not complementary to the target and is displaced like a flap by the Invader oligonucleotide. An enzyme (Cleavase VIII, or Flap Endonuclease) recognizes the three-dimensional structure formed by the primary Invader oligonucleotide-primary probe-target sequence complex and cleaves the primary probe on the 3' side of the polymorphic site. The reporter arm of the cleaved primary probe then functions as a secondary Invader oligonucleotide in a secondary reaction. Note: the enzyme will cleave the primary probe only if the primary probe is complementary to the target sequence at the base to be genotyped. Although the figure shows only the assay containing the target sequence and primary probe for the "A" allele, a separate reaction mixture containing a primary probe specific for the "G" allele (5'–ttaacgaggcgcac̲GCCAGGACCTGGCCATGC-3') would be needed to genotype both alleles of the polymorphism. (B) In the secondary Invader reaction, a synthetic target sequence and secondary fluorescence resonance energy transfer (FRET) probe are constructed so that the reporter arm of the cleaved primary probe will function as a secondary Invader oligonucleotide, displacing a portion of the overlapping FRET probe and forming a three-dimensional structure similar to that formed in the primary reaction. The FRET probe is labeled at the 5' end with a fluorescein compound (F) and internally with the quenching dye cy3 (Q). If the FRET probe remains uncleaved (because the primary probe was not cleaved in the primary reaction to provide the secondary Invader oligonucleotide), cy3 will quench the fluorescence of the fluorescein, but if the FRET probe is cleaved, the fluorescence of the fluorescein label will not be quenched and the signal can be detected with a standard fluorescence microtiter plate reader. Carats denote chemically modified bases. Note: although the figure shows only the assay containing the FRET probe designed to detect the "A" allele, a separate reaction mixture containing a different secondary target and a FRET probe (5'-T̲AACTGCTTCCTCCG-3'), labeled with a different fluorescein compound to detect the "G" allele, would be needed to genotype both alleles of the polymorphism. (C) In the left panel, gross fluorescent signal emitted by the cleaved secondary (FRET) probe for the "A" allele is plotted against fluorescent signal from the cleaved "G" allele probe. The plot shows four distinct clusters of points representing the three possible genotypes and assay failures near the origin. In the right panel, genotypes for 96 individuals are assigned by displaying the data as a ratio of the "A" signal to "G" signal. Adapted from Ryan et al. (22) and Mein et al. (23).

ous examination of tens of thousands of genes from various individuals reflecting different physiologic and/or pathologic states. Probes produced by PCR amplification of known gene sequences are arrayed on the surface of a chip or microscope slide and allowed to hybridize with fluorescently labeled *complementary DNA* (cDNA) representing the total messenger RNA (mRNA), and hence all of the ex-

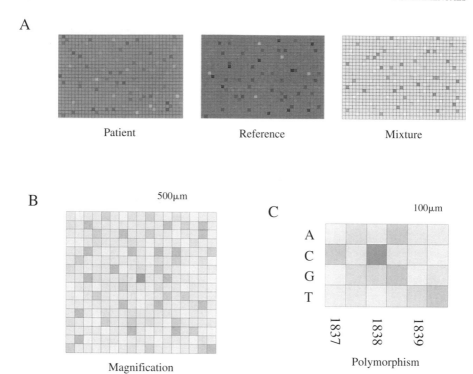

Figure 2.4 (A) Schematic representation showing black and white images of DNA chips from a two-color fluorescent assay system. The oligonucleotides on the chips function as probes by hybridizing to DNA (or RNA) that is applied to the surface of the chip if the probe and DNA sequences are complementary. The DNA (or RNA) being assayed is labeled with a compound that will fluoresce when exposed to ultraviolet (UV) light. DNA from a patient is labeled with a red fluorescent compound and hybridized to the probes on the surface of the chip marked "Patient;" while DNA from a healthy control is labeled with a green fluorescent compound and applied to the chip denoted "Reference." Both the patient and reference DNA samples are applied to the chip marked "Mixture." For each square in the array, the relative strength of the fluorescent signal for the patient and reference samples is then measured. Note that visualization procedures may use a variety of false coloring and contrast enhancing techniques. (B) Magnification of a section of the chip to which both patient and reference DNA have been applied. Note the positive signal near the center of the array. At this location, only the DNA from the patient hybridized to the probes, thus denoting a difference in sequence between the patient and the reference at this position. (C) Close-up of the area surrounding the positive signal from B demonstrating the detection of a T → C mutation in the patient at nucleotide position 1,838. Adapted from Hacia et al. (53) with permission.

pressed genes, from a test sample (Fig. 2.5). For each gene within the array, the intensity of the fluorescent signal will reflect the relative abundance of RNA transcripts for that gene in the total RNA pool, and thus will be indicative of expression levels (28).

Examinations of gene expression on a genomic scale have the potential to revolutionize the diagnosis and treatment of complex human diseases (29,30). For example, expression array technology has been used to examine differences in the expression of more than 10,000 genes in patients with dilated cardiomyopathy (DCM) or hypertrophic cardiomyopathy (HCM). Differences in expression patterns for genes such as αB-crystallin, β-dystrobrevin, calsequestrin, lipocortin, and lumican provide important insight into the physiologic and pathologic processes that lead to heart failure (31). Similar approaches profiling the expression of thousands of genes may permit: (*1*) identification of genes selectively expressed in important tissues such as cardiac muscle or atherosclerotic plaques; (2) examination of differences in gene expression between healthy and disease states, which may prove useful for characterizing new markers of subclinical disease; (*3*) assessments of exposure to atherogenic risk factors such as smoking; and (*4*) determination of characteristic patterns of gene expression induced by disease to identify and monitor genes that may be associated with disease onset and progression.

Despite the potential capabilities of gene expression assays, several factors are inhibiting chip technologies from becoming widely integrated into complex disease

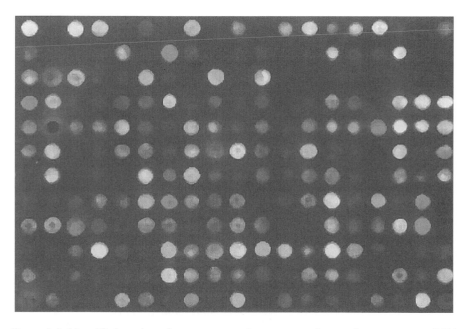

Figure 2.5 Magnified section of a gene expression assay used to evaluate messenger RNA (mRNA) abundance in a patient and healthy reference. Short oligonucleotides representative of thousands of genes are produced remotely and then deposited at high density on the surface of a glass slide. Complementary DNAs (cDNAs) representing the total mRNA, and hence all of the expressed genes, from a patient and a reference sample are labeled with different fluorescent compounds. The cDNA pools are then mixed and competitively hybridized to the oligonucleotides on the chip surface. For each gene within the array, the color and intensity of the fluorescent signal will reflect the relative abundance of RNA transcripts for that gene in the total RNA pool, and thus will be indicative of expression levels (28). Image courtesy of Mike Killeen, Packard BioChip Technologies.

research. These factors include high costs for specialized equipment and chip manufacture, insufficient standardization across technology platforms, and the need for invasive tissue sampling of a single tissue type. Before expression assays can become routine components of complex disease research, careful optimization of signal to noise and standardized image processing will be required to accurately detect differences in expression and to compare results among patients. Because a large proportion (approximately 85%) of mRNA transcripts are present in low abundance and subtle changes in gene expression may have important pathophysiologic consequences, considerable effort is being devoted to improve the reliability of assays for low abundance transcripts (32).

Proteomics

The field of proteomics is an emerging discipline that involves the study of proteins, rather than DNA or RNA, on a comprehensive scale. A variety of proteomic technologies is being developed to identify new proteins, measure levels of protein expression, and characterize post-translational modifications. Large-scale identification of proteins by biochemical methods has been conducted for several decades by *two-dimensional gel electrophoresis*, which has been used to create extensive catalogs of protein spots and to visualize patterns of differential protein expression, but the technique is limited by the number of proteins that can be resolved (usually no more than 1000 of the most abundant proteins) and difficulties in determining the identity of many proteins. A more recent technique known as *peptide-mass fingerprint (PMF) analysis* incorporates mass spectrometry to generate "protein fingerprints" (Fig. 2.6) and can be automated to permit high-throughput identification of low-abundance as well as high-abundance proteins (33,34). As the ultimate goal of proteomics is to transcend simple protein identification and cataloging to examine functional aspects of gene products (functional genomics) that cannot be addressed by DNA or RNA analysis, protein chips containing immobilized proteins in an array format are being developed to investigate post-translational modifications that affect protein function and protein–protein interactions.

Figure 2.6 Two-dimensional polyacrylamide gel electrophoresis (2D–PAGE) and peptide–mass fingerprint analysis for high-throughput identification of proteins in normal and diseased tissues. Proteins are extracted from the respective tissues and separated in one dimension on a pH gradient by their overall charge and then by molecular weight in a second (perpendicular) direction. A variety of techniques are available to visualize the separated proteins, which appear as a complex assortment of many spots. Sophisticated imaging and analysis software is often necessary to interpret these complex images because each spot represents a different protein (or modified protein). Proteins of interest, such as those differentially expressed in normal and diseased tissue, are then excised from the gels and enzymatically or chemically cleaved into smaller fragments (peptides). The resulting peptide mixture is analyzed by mass spectrometry, which produces a spectrum of peaks (a peptide-mass fingerprint), based on the masses of individual peptides, that is sufficient to unambiguously identify the differentially expressed proteins. Adapted from Pandey and Mann (34) with permission.

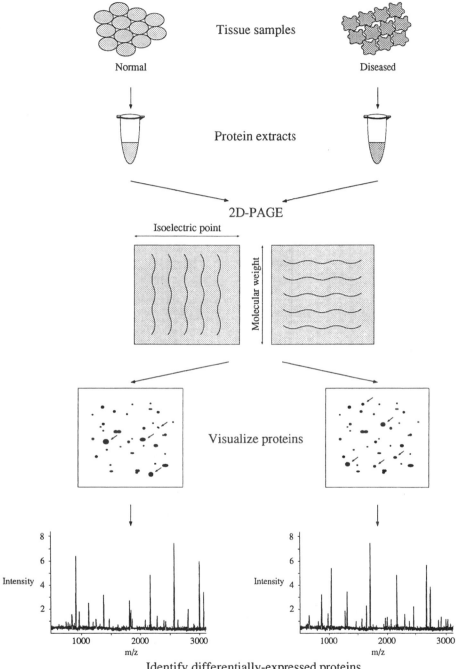

Tissue samples

Normal   Diseased

Protein extracts

2D-PAGE

Isoelectric point

Molecular weight

Visualize proteins

Intensity

Intensity

1000   2000   3000
m/z

1000   2000   3000
m/z

Identify differentially-expressed proteins
by mass spectrometry

Proteomics is currently receiving attention in the biotechnology and pharmaceutical industries because proteomic technology holds great potential for identifying novel small molecules that may have important therapeutic applications (35). From the perspective of complex disease research, proteomics may prove useful for defining molecular events that lead to dysfunction, identifying proteins characteristic of early phases of disease or diseased tissue, and accurate tracking of responses to drugs. For example, patients with dilated cardiomyopathy have been shown to exhibit abnormalities in the expression of contractile and cytoskeletal proteins, stress–response proteins, and mitochondrial polypeptides involved in energy production (reviewed in ref. 36). The ability to identify altered patterns of protein expression, which may not be detectable at the RNA level, may permit more accurate diagnosis of complex diseases such as MI or congestive heart failure and may eventually lead to the development of more effective pharmacologic or gene therapy interventions (37).

Applications in Genetic Epidemiology and Clinical Medicine

Basic Principles of Genetic Testing

The majority of genetic testing today is focused on screening newborns for disorders for which there are effective treatments (38) and on predicting disease in adult carriers of relatively rare Mendelian diseases. Genetic tests for 938 heritable disorders (as of March 2003) are available at http://www.genetests.org. Unlike Mendelian diseases, however, where patients who inherit disease-related alleles usually show manifestations of disease, common polymorphisms in many genes are believed to be associated with increased susceptibility to complex diseases, and this relationship is often expressed in terms of the probability or risk of developing disease. Therefore, genetic testing for complex diseases may be more informative for identifying patients at increased risk of disease, who would benefit most from prevention strategies such as reduction of behavioral risk factors, than for prediction or diagnosis of disease. For complex diseases such as coronary artery atherosclerosis and MI, a given genetic variant is rarely diagnostic of disease or highly predictive of disease onset. Variants that are associated with disease normally contribute only modestly to disease risk, and this contribution is often dependent on interactions with other genes and environmental or behavioral risk factors. As a result, the effects of a particular variant on disease may vary greatly by age and the presence of concomitant risk factors. Emerging information on the impact of genetic testing in human populations is available on the Centers for Disease Control and Prevention, Human Genome Epidemiology Network (HuGENet) website, http://www.cdc.gov/genomics/hugenet/default.htm.

The few genetic tests currently available for complex forms of coronary disease are research tests designed to investigate relationships between genetic variants and disease. Test results are generally not given to patients or their physicians because the tests are not highly predictive of disease and are usually performed in laboratories not subject to Clinical Laboratory Improvement Amendments (CLIA) regulation. Presymptomatic tests of single polymorphisms to predict future onset of diseases that are influenced by multiple genes and environmental factors will likely have low predictive value, and therefore, may not be particularly useful for population-based testing for complex diseases.

Predictive Medicine: Genetic Risk Factor Profiles and Risk Stratification

Genetic tests of *individual* polymorphisms may have low predictive value for complex human diseases, but population screening of many polymorphisms and gene–protein expression profiles may lead to informative assessments of disease risk by assimilating risk information from numerous disease-associated genes and biochemical pathways. As outlined above, technologies are being developed to simultaneously assay numerous SNP polymorphisms and/or gene expression profiles that may be associated with increased risk for disease. A multilocus genotyping assay has already been designed to interrogate 35 SNPs in 15 genes implicated in the development and progression of cardiovascular disease (39) and a 10,368-element microarray (the CardioChip) has been developed for expression profiling of genes that may influence myocardial development and cardiac function (40). The ability to simultaneously examine many polymorphisms and expression profiles will provide a comprehensive assessment of genomic influences on disease, which may not be highly predictive of disease, but is anticipated to provide useful information for assessing disease risk.

Therapeutic Applications of Genetic Information

Pharmacogenomic Implications

Genetic information resulting from the Human Genome Project coupled with recent advances in genome technology is causing a fundamental shift toward whole-genome approaches to drug discovery (*pharmacogenomics*). Current research is investigating the differential effects of drug exposure in individuals with different genotypic backgrounds (gene–drug interactions) and certain SNP genotypes are being examined as markers of increased or decreased response to specific drugs or as predictors of adverse effects. Drug discovery and development are being enhanced by high-throughput screening of genetic polymorphisms, which aids in the identification of new genes associated with disease and increases the number of potential drug targets. Examinations of molecular pathways and biochemical mechanisms of drug action by using gene expression arrays and proteomic techniques are proving critical to the development of new drugs (41–43). In addition, these technologies are being used to monitor changes in gene expression and modifications to proteins in response to drugs in order to predict therapeutic outcomes and potential adverse effects, guide drug selection, and help determine appropriate dosages (44). Pharmacogenomic approaches may thus provide the initial opportunities to exploit variation within the human genome for effective treatment of disease.

Prospects for Gene Therapy

Although traditional pharmacologic approaches have been highly effective in ameliorating the effects of known risk factors by controlling hypertension, preventing thrombosis, and lowering serum cholesterol levels, drug administration has been less successful in treating advanced disease. With recent advances in molecular genetic technologies, it is now feasible to transfer exogenous genetic material to somatic cells of the body for therapeutic purposes.

The majority of gene therapy protocols have focused on genetic disorders, such as familial hypercholesterolemia (45), that do not respond well to conventional treatments. The goal is to deliver a normal gene to somatic tissues or organs and then properly express that gene in order to reduce morbidity and mortality. Complex diseases such as atherosclerosis and congestive heart failure have much greater potential impact on public health, but are far more difficult to treat effectively via gene therapy (46). Despite many obstacles, gene transfer is being used to overexpress specific proteins that may inhibit atherosclerotic lesions or stabilize vulnerable plaque (47,48). Similarly, gene therapy may be beneficial in avoiding heart failure by modifying molecules that regulate myocardial function (49,50) and may ultimately prove useful for stimulating collateral artery development following an infarction or surgical intervention (51,52).

Gene therapy is currently limited because the vectors used to deliver genes to somatic cells may elicit an immune response, the ability to target specific tissues is often difficult, and the appropriate level of gene expression may not be sustainable over time. Moreover, additional data are needed to evaluate the short-term and long-term risk-to-benefit ratio of various gene therapy techniques. If the current safety concerns and other technical difficulties can be adequately resolved, gene therapy may find increasing application to complex diseases with a genetic basis. The ultimate success of gene therapy for complex diseases, however, may depend on the degree to which a given disease is genetically determined (*heritability*) and the influence of environmental factors on disease susceptibility. Genetic information from the Human Genome Project and advances in genomic/proteomic technologies are anticipated to aid in the identification of new therapeutic targets and promote the development of new strategies for molecular interventions.

Conclusions

With the accelerating pace of human genomic research, it will be advantageous for epidemiologists and health care providers to remain abreast of new developments in the discovery and characterization of genetic influences on complex human diseases. Technologies such as microarrays and proteomics have the potential to substantially improve the detection of genetic influences on disease risk; therefore, it is important to become familiar with the potential applications as well as limitations of these technologies. Genetic testing for many complex diseases is presently limited, but a dramatic increase in the number of genetic polymorphisms associated with disease risk, and hence the number of available genetic tests, is anticipated in the near future. Genetic tests are currently useful for predicting disease onset for a limited number of rare single-gene disorders; however, it remains to be seen if accurate and reliable tests can be developed for identifying patients at increased risk of developing complex diseases. Future population- and clinic-based research will be essential for defining how multi-gene tests will be used to accurately and reliably identify and treat patients at increased risk of developing complex diseases that commonly lead to increased morbidity and mortality in human populations.

References

1. International Human Genome Sequencing Consortium. Initial sequencing and analysis of the human genome. Nature 2001;409:860–921.
2. Venter JC, Adams MD, Myers EW, et al. The sequence of the human genome. Science 2001;291:1304–1351.
3. Burge C, Karlin S. Prediction of complete gene structures in human genomic DNA. J Mol Biol 1997;268:78–94.
4. Schwartz S, Zhang Z, Frazer KA, et al. PipMaker—A web server for aligning two genomic DNA sequences. Genome Res 2000;10:577–586.
5. Ellsworth RE, Jamison DC, Touchman JW, et al. Comparative genomic sequence analysis of the human and mouse cystic fibrosis transmembrane conductance regulator genes. Proc Natl Acad Sci U S A 2000;97:1172–1177.
6. Ellsworth DL, Manolio TA. The emerging importance of genetics in epidemiologic research. I. Basic concepts in human genetics and laboratory technology. Ann Epidemiol 1999;9:1–16.
7. Chakraborty R, Kidd KK. The utility of DNA typing in forensic work. Science 1991;254:1735–1739.
8. Hanis CL, Boerwinkle E, Chakraborty R, et al. A genome-wide search for human non-insulin-dependent (type 2) diabetes genes reveals a major susceptibility locus on chromosome 2. Nat Genet 1996;13:161–166.
9. Norman RA, Tataranni PA, Pratley R, et al. Autosomal genomic scan for loci linked to obesity and energy metabolism in Pima Indians. Am J Hum Genet 1998;62:659–668.
10. Ellsworth DL, Manolio TA. The emerging importance of genetics in epidemiologic research III. Bioinformatics and statistical genetic methods. Ann Epidemiol 1999;9:207–224.
11. Kruglyak L, Nickerson DA. Variation is the spice of life. Nat Genet 2001;27:234–236.
12. Ellsworth DL, Bielak LF, Turner ST, et al. Gender- and age-dependent relationships between the E-selectin S128R polymorphism and coronary artery calcification. J Mol Med 2001;79:390–398.
13. Feng D, Lindpaintner K, Larson MG, et al. Platelet glycoprotein IIIa *Pl*^{A} polymorphism, fibrinogen, and platelet aggregability: The Framingham Heart Study. Circulation 2001; 104:140–144.

14. Rieder MJ, Taylor SL, Clark AG, et al. Sequence variation in the human angiotensin converting enzyme. Nat Genet 1999;22:59–62.
15. Stephens JC, Schneider JA, Tanguay DA, et al. Haplotype variation and linkage disequilibrium in 313 human genes. Science 2001;293:489–493.
16. Risch N. Searching for genetic determinants in the new millennium. Nature 2000; 405:847–856.
17. Risch N, Merikangas K. The future of genetic studies of complex human diseases. Science 1996;273:1516–1517.
18. Pennisi E. A closer look at SNPs suggests difficulties. Science 1998;281:1787–1789. News.
19. Syvånen A-C, Landegren U, Isaksson A, et al. Enthusiasm mixed with scepticism about single-nucleotide polymorphism markers for dissecting complex disorders. Eur J Hum Genet 1999;7:98–101.
20. The International SNP Map Working Group. A map of human genome sequence variation containing 1.42 million single nucleotide polymorphisms. Nature 2001;409:928–933.
21. Ross P, Hall L, Smirnov I, et al. High level multiplex genotyping by MALDI-TOF mass spectrometry. Nat Biotechnol 1998;16:1347–1351.
22. Ryan D, Nuccie B, Arvan D. Non-PCR-dependent detection of the factor V Leiden mutation from genomic DNA using a homogeneous Invader microtiter plate assay. Mol Diagn 1999;4:135–144.
23. Mein CA, Barratt BJ, Dunn MG, et al. Evaluation of single nucleotide polymorphism typing with Invader on PCR amplicons and its automation. Genome Res 2000;10: 330–343.
24. Hacia JG. Resequencing and mutational analysis using oligonucleotide arrays. Nat Genet 1999;21(suppl.):42–47.
25. Sapolsky RJ, Hsie L, Berno A, et al. High-throughput polymorphism screening and genotyping with high-density oligonucleotide arrays. Genet Anal 1999;14:187–192.
26. Pastinen T, Perola M, Niini P, et al. Array-based multiplex analysis of candidate genes reveals two independent and additive genetic risk factors for myocardial infarction in the Finnish population. Hum Mol Genet 1998;7:1453–1462.
27. Waldmüller S, Freund P, Mauch S, et al. Low-density DNA microarrays are versatile tools to screen for known mutations in hypertrophic cardiomyopathy. Hum Mutat 2002; 19:560–569.
28. Duggan DJ, Bittner M, Chen Y, et al. Expression profiling using cDNA microarrays. Nat Genet 1999;21(suppl.):10–14.
29. Brown PO, Hartwell L. Genomics and human disease—variations on variation. Nat Genet 1998;18:91–93.
30. Gerhold D, Rushmore T, Caskey CT. DNA chips: promising toys have become powerful tools. Trends Biochem Sci 1999;24:168–173.
31. Hwang JJ, Allen PD, Tseng GC, et al. Microarray gene expression profiles in dilated and hypertrophic cardiomyopathic end-stage heart failure. Physiol Genomics 2002;10:31–44.
32. Second Workshop on Methods and Applications of DNA Microarray Technology. Getting hip to the chip. Nat Genet 1998;18:195–197.
33. Gevaert K, Vandekerckhove J. Protein identification methods in proteomics. Electrophoresis 2000;21:1145–1154.
34. Pandey A, Mann M. Proteomics to study genes and genomes. Nature 2000;405:837–846.
35. Dunn MJ. Studying heart disease using the proteomic approach. Drug Discov Today 2000;5:76–84.
36. Celis JE, Kruhøffer M, Gromova I, et al. Gene expression profiling: monitoring transcription and translation products using DNA microarrays and proteomics. FEBS Lett 2000;480:2–16.

37. Anderson NL, Matheson AD, Steiner S. Proteomics: applications in basic and applied biology. Curr Opin Biotechnol 2000;11:408–412.
38. Therrell BL Jr. Ed. Laboratory methods for neonatal screening. Washington, DC: American Public Health Association; 1993.
39. Cheng S, Grow MA, Pallaud C, et al. A multilocus genotyping assay for candidate markers of cardiovascular disease risk. Genome Res 1999;9:936–949.
40. Barrans JD, Stamatiou D, Liew C-C. Construction of a human cardiovascular cDNA microarray: portrait of the failing heart. Biochem Biophys Res Commun 2001;280:964–969.
41. Regalado A. Inventing the pharmacogenomics business. Am J Health Syst Pharm 1999;56:40–50.
42. Blower PE, Yang C, Fligner MA, et al. Pharmacogenomic analysis: correlating molecular substructure classes with microarray gene expression data. Pharmacogenomics J 2002;2:259–271.
43. Sehgal A. Drug discovery and development using chemical genomics. Curr Opin Drug Discov Devel 2002;5:526–531.
44. Debouck C, Goodfellow PN. DNA microarrays in drug discovery and development. Nat Genet 1999;21(suppl.):48–50.
45. Grossman M, Raper SE, Kozarsky K, et al. Successful *ex vivo* gene therapy directed to liver in a patient with familial hypercholesterolaemia. Nat Genet 1994;6:335–341.
46. French BA. Gene therapy and cardiovascular disease. Curr Opin Cardiol 1998;13: 205–213.
47. Feldman LJ, Isner JM. Gene therapy for the vulnerable plaque. J Am Coll Cardiol 1995;26:826–835.
48. Rader DJ. Gene therapy for atherosclerosis. Int J Clin Lab Res 1997;27:35–43.
49. Nuss HB, Johns DC, Kääb S, et al. Reversal of potassium channel deficiency in cells from failing hearts by adenoviral gene transfer: a prototype for gene therapy for disorders of cardiac excitability and contractility. Gene Ther 1996;3:900–912.
50. Maurice JP, Koch WJ. Potential future therapies for heart failure: gene transfer of b-adrenergic signaling components. Coron Artery Dis 1999;10:401–405.
51. Schneider MD. Myocardial infarction as a problem of growth control: cell cycle therapy for cardiac myocytes? J Card Fail 1996;2:259–263.
52. Sellke FW, Simons M. Angiogenesis in cardiovascular disease: current status and therapeutic potential. Drugs 1999;58:391–396.
53. Hacia JG, Brody LC, Chee MS, et al. Detection of heterozygous mutations in *BRCA1* using high density oligonucleotide arrays and two-colour fluorescence analysis. Nat Genet 1996;14:441–447.

3

Approaches to quantify the genetic component of and identify genes for complex traits

Patricia A. Peyser and Trudy L. Burns

Unlike simple traits, which tend to result from segregation of alleles at a single genetic locus (e.g., Huntington disease, cystic fibrosis), multifactorial or "complex" traits result from the combined effects of multiple genetic and environmental factors, potentially interacting with each other. These traits, which constitute the major causes of human morbidity and mortality, include many common congenital malformations (e.g., congenital heart defects, cleft lip with or without cleft palate), as well as most common diseases of adult life (e.g., type 2 diabetes mellitus, coronary heart disease).

Several criteria are consistent with possible genetic influences on traits with unknown etiology in humans. The criteria include the following:

- Association with a known genetic disease (often identified through Online Mendelian Inheritance in Man, described below)
- Association with a genetic marker, where genetic markers consist of specific nucleotide patterns
- Onset at a characteristic age without a known precipitating event
- Elevated risk in relatives compared with the general population (denoted by λ)
- Failure to spread to unrelated individuals (i.e., spouses), but with vertical transmission from parents to their offspring
- Greater concordance in monozygotic (MZ) twins than in dizygotic (DZ) twins, where MZ twin pairs share 100% of their genes in common and DZ twin pairs share, on average, 50% of their genes, like full siblings
- The incidence does not change within spans of generations

- Variation exists in accordance with genetic theory suggesting segregation (i.e., whether the observed pattern of disease in families is compatible with a Mendelian model of inheritance) or cosegregation with a linked marker, and
- Occurrence of a similar inherited condition in other mammals.

Many common, complex traits meet several of these criteria.

Before the 1990s, the genetic analysis of complex traits was largely limited to answering two basic questions: (*1*) is there familial aggregation of the trait? (*2*) if so, does any evidence indicate that genes are involved in determining susceptibility to the trait? These questions were usually answered without measured genotypes or genetic markers. The involvement of genes was inferred by comparing the fit of different etiologic models to the phenotypic patterns that were observed in families. The phenotype is an observable or measurable characteristic with an etiology resulting from the interaction between environments and genes.

Since the early 1990s, largely as a consequence of the Human Genome Project, it has been possible to begin to ask the next question: (*3*) can specific genes be identified that are involved in determining susceptibility to the trait? The goals of the Human Genome Project are to develop genetic maps, physical maps, and DNA sequence of the entire human genome. The official repository for the data resulting from this effort is the Genome Database (GDB) (http://www.gdb.org/), which contains many types of data including maps of genetic markers and the position of known genes on the genetic map. The availability of these data has motivated the expansion and development of analytic approaches (primarily linkage and association analyses, described below) that incorporate information on measured genotypes, and has greatly facilitated the mapping and identification of genes for a large number of disorders. These genes and disorders are cataloged in Online Mendelian Inheritance in Man (http://www.ncbi.nlm.nih.gov/Omim/) written and edited by Victor McKusick and his colleagues.

Most complex traits appear to have a heterogeneous etiology. Different genetic backgrounds may cause genetic susceptibility to the same disease. Genes that confer a high risk may be present in only a subset of affected persons. The trait may be caused by different genetic loci acting together, or a predisposing gene may be manifest only in the presence of a particular environmental exposure. Thus, identifying genetic factors for complex traits is proving to be quite a challenge. These complexities, to the extent they are recognized, should be taken into account in designing genetic epidemiologic investigations. For example, numerous risk factors exist for coronary heart disease (e.g., hyperlipidemia, hypertension), each of which has its own heterogeneous etiology (1). Therefore, it makes more sense to focus an investigation on hypertension, for example, rather than coronary heart disease and to focus only on families, for example, with at least two relatives having hypertension onset before age 60 years.

A number of alternative study designs have been used to conduct genetic epidemiologic investigations. These designs range from large multigenerational pedigrees to nuclear families to pairs of affected siblings to twin pairs to adoptees and their adoptive parents to unrelated cases and controls. Each of these designs has recognized strengths and weaknesses. For many traits, some designs may not be practical (e.g., if the trait has a late age of onset, large pedigrees might not be available). For many complex traits, however, any one of several designs would be feasible, and investigations into the optimal study design are a major focus of current investigations (2).

This chapter focuses on methodologic approaches that might be used to quantify the genetic component of, and to identify genes for, complex traits. We begin with approaches used when genes are not measured, then describe approaches when genes are measured. These topics are covered in considerably more detail in several books (3–9) as well as in review articles (10–12). Many specialized computer programs that have been written to investigate the relation between genotypes and complex traits are described elsewhere (6,9,11,12) and most are cataloged and available at http://linkage.rockefeller.edu.

Approaches When Genes Are Not Measured

Familial Aggregation

The initial focus of many genetic epidemiologic studies is to obtain a measure of familial aggregation or clustering. The degree of family clustering for a dichotomous trait can be expressed by λ_R, the risk to relative type R of an affected proband, compared with population risk. The proband is the first person in a pedigree to be identified clinically as having the trait of interest. R can represent parents, sibs, offspring or any other relative of a proband. When $\lambda_R = 1$, there is no familial aggregation. In general, a high value for λ_R suggests there might be a genetic component, although estimates of λ_R can be elevated because of common environmental factors or other nongenetic, e.g., cultural factors.

For example, a study was conducted in 179 African-American families ascertained through a proband with sarcoidosis. λ_R (95% confidence interval [CI]) was 2.2 (1.2, 3.9) for sibs, 2.8 (1.4, 5.0) for parents, and 2.5 (1.6, 3.7) for parents and sibs combined (all significantly higher than 1.0) (13). In addition, λ_R was higher for relatives of younger probands and for relatives of male probands (2.9 (95% CI, 1.5-5.1) and 4.0 (95% CI, 2.0-7.1), respectively). The investigators concluded that African-American sibs and parents of sarcoidosis probands are at a 2.5- to 3.0-fold higher risk for the disease compared to the general population of African Americans, and that there may be heterogeneity in disease risk among family members. Studies are now under way to identify sarcoidosis susceptibility genes using linkage and association analysis (13).

By comparison, λ_R for sibs of probands with Hirschsprung disease, the most common hereditary cause of intestinal obstruction, is estimated to be 187.5 (14) while λ_R for sibs of probands with breast cancer is estimated to be 2.0 based on many epidemiologic studies (15). Many other approaches are available to estimate familial aggregation for qualitative and quantitative traits and are reviewed in detail elsewhere (3).

Heritability

Investigations of the relative contributions of genes and environmental factors are often undertaken before DNA-based studies are initiated. The goal is to estimate the proportion of total phenotypic variance of a trait from all genetic effects. This proportion is often referred to as the "heritability" and denoted as "h^2." Often heritability is expressed as a percentage. An assumption is made that the genetic effects result from a large number of loci, each with small, equal, additive effects on the phenotype; these loci are often referred to as polygenes.

Several approaches can be used to estimate heritability, including regression analysis, twin concordance, comparisons of correlations between different types of relatives, variance components analysis, path analysis, and maximum likelihood methods (3,4,8,9). Heritability can be estimated from samples composed of extended pedigrees, nuclear families, siblings, adoptees and their biologic relatives, or twins (16).

Twins provide one of the simplest designs for estimating genetic contributions to a trait (17). Shared environment is assumed to be the same for both MZ and DZ twin pairs. Additional assumptions include no dominance, no epistasis (gene–gene interaction), and no gene–environment interaction or correlation. Dominance implies that the mean of the measurement for heterozygotes is not exactly the average of the two homozgote means. If the shared environmental effect is greater for MZ than for DZ pairs, which is very likely, then h^2 will be overestimated.

For quantitative traits, h^2 is usually estimated as $2(r_{MZ} - r_{DZ})$, where r_{MZ} and r_{DZ} are the intraclass correlations for the two different twin types. An example comes from a study by Austin et al. (18) of obesity measurements on participants in the Kaiser Permanente Women Twins Study. In a sample of 185 pairs of MZ twins and 130 pairs of DZ twins, high estimates of h^2 for the change in body mass index (BMI) [weight in kg/(height in m)^2] over a decade of follow-up were found. The estimates ranged from 57% to 86%, depending on the statistical method used to estimate h^2. All estimates of h^2 were significantly different from 0%. Austin et al. concluded that evidence existed for genetic influences on changes in BMI over a decade in adult women (18).

For qualitative traits, twin pairs are often ascertained only if one or both twins are affected. Pair-wise concordance is computed as the number of pairs of twins where both are affected divided by the total number of pairs of twins with at least one twin affected. The heritability can be estimated as $2(c_{MZ} - c_{DZ})$ where c_{MZ} and c_{DZ} are the twin concordance rates for MZ and DZ twins, respectively. A recent example is a study of Parkinson disease in white male twins from the National Academy of Science/National Research Council World War II Veteran Twins Registry (19). The sample consisted of 71 MZ pairs and 90 DZ pairs with at least one twin in each pair having been diagnosed with Parkinson disease. When the disease began after age 50, there was no evidence for the involvement of genetic factors. When the disease began by age 50 in at least one twin, the estimate of heritability was approximately 100%. Although the estimate was based on a small number of twins with disease onset by age 50 (4 MZ pairs and 12 DZ pairs), these findings suggest future research directions to understand the etiology of Parkinson disease. The many different methods of twin analysis are presented in detail elsewhere (20,21).

Variance components analysis provides an approach that can be applied to any study design to estimate genetic and environmental contributions to a quantitative trait. This approach was used to estimate the heritability of quantity of coronary artery calcification (CAC), a measure of coronary atherosclerosis, in 698 asymptomatic white adults from 302 families participating in the Rochester Family Heart Study (22). The level of CAC, y, for individual i was modeled as $y_i = \mu + g_i + e_i$ where μ is the trait mean. The remaining variables represent the random deviations from μ for individual i attributable to additive genetic and residual error effects, respectively. The effects of g_i and e_i have variances σ_g^2 and σ_e^2, respectively.

The covariances between relatives in quantity of CAC and the proportion of genes shared in common provide the basis for estimation of σ_g^2. The proportion of shared genes is $1/2$ for parents and offspring, $1/2$ for full sibs, $1/4$ for grandparents and grandchildren, $1/8$ for first cousins, 0 for unrelated spouses, and 0 for adoptees and their unrelated adoptive parents. Maximum likelihood methods were used to simultaneously estimate the mean and variances as well as the genetic effects (23). Genetic factors $(\sigma_g^2/(\sigma_g^2 + \sigma_e^2))$ explained 43.5% of the variation in CAC quantity before adjusting for any covariates ($p = 0.0007$). After adjusting for age, sex, fasting glucose, systolic blood pressure, pack years of smoking, and low-density lipoprotein cholesterol, 41.8% of the residual variation in CAC quantity was attributable to genetic factors ($p = 0.0003$). These findings provide evidence for the involvement of genetic factors in subclinical coronary atherosclerosis as measured by CAC quantity and suggest that unknown genes influencing CAC quantity are yet to be identified (22).

Complex Segregation Analysis

Phenotypes with evidence for a genetic component are often further investigated by using segregation analysis. Originally, segregation analysis was designed to test whether an observed mixture of phenotypes among offspring is compatible with Mendelian inheritance (e.g., autosomal recessive inheritance). Over time, segregation analysis has broadened to encompass the fitting of more general models of inheritance to pedigrees of arbitrary structure. The ultimate goal is to test for compatibility with Mendelian expectations by estimating parameters of a given model of inheritance. The strategy involves fitting various alternative models to family data, where the models include both genetic and nongenetic factors to explain the distribution of phenotypes, and selecting the model that best "explains" the observed data.

Complex segregation analysis was used to study the role of genetic and environmental factors in determining variability in BMI in 1302 relatives identified through 284 school children (i.e., probands) from Muscatine, Iowa (24,25). In each of the

284 families, the probands, their parents, their siblings, a related aunt or uncle, and a first cousin were examined. The BMI levels were first adjusted for variability in age, by sex, and by relative type. There was significant familial aggregation of adjusted BMI in pedigrees, as indicated by inter- and intraclass correlation coefficients, excluding the school children used to identify families. The estimated correlations were 0.17 between spouse pairs, 0.22 between parents and offspring, 0.35 among siblings, 0.09 between uncles and aunts and their nephews and nieces, and 0.08 among first cousins. Only the correlation among first cousins was not significantly different from 0.0.

The observed distribution of adjusted BMI was assumed to result from the following: a single genetic or nontransmitted environmental factor with a major effect on adjusted BMI levels, the additive allelic effects of a large number of polygenes, the effects of shared environments (defined by living in the same household at the time of study), and individual-specific environmental influences. Complex segregation analysis was used to test a series of 12 different models that represent combinations of these genetic and environmental factors in the determination of the adjusted BMI levels. The model that gave the best fit to the data assumed a recessive allele was associated with higher adjusted BMI levels. The frequency of the allele in the population was estimated to be 0.345. Approximately 6% of the population were homozygous for the recessive allele and 37% were heterozygous. The estimated mean BMI for the homozygotes was 34.8 kg/m^2; the estimated mean was 23.8 kg/m^2 for the other two genotypes. Based on the parameter estimates for the recessive model, 35% of the variability in BMI resulted from segregation at a single genetic locus, 42% from polygenes, and 23% from nongenetic factors. For spouses living in the same household, their shared environment accounted for 12% of the variation in their BMI levels. For siblings living in the same household, their shared environment accounted for 10% of their variation in BMI. Although shared environments contributed to variation in BMI, more than 75% of the variation was explained by genetic factors in these families (25).

Studies of other traits associated with obesity have confirmed a role for genetic factors (26). Complex segregation analysis has been applied to a number of other complex traits, for example, traits associated with diabetes (27,28). More details and other examples of complex segregation analysis are presented elsewhere (4,5,9,10).

Approaches When Genes Are Measured

Genetic Markers

A genetic marker is a polymorphic DNA or protein sequence derived from a specific chromosomal location. A polymorphic locus is one where the most common allele has a frequency no higher than 99% (9). Before 1960, the only available markers were those of various blood groups. Then electrophoretic mobility variants of serum proteins and the human leukocyte antigen (HLA) tissue types became available for study. DNA restriction fragment length polymorphisms (RFLPs), available since 1975, were the first generation of DNA markers. Variable number of tandem repeat (VNTR) polymorphisms, with dinucleotide, trinucleotide, and tetranucleotide repeats, represented the next generation of DNA markers. The most recently developed DNA markers, greatly facilitated by the Human Genome Project, are single nucleotide polymorphisms (SNPs) (29). Genetic markers have been extremely useful in identifying genes related to disease. At least one disease-related mutation has been identified for more than 1000 genes (30).

Model-Based Linkage Analysis

Positional cloning, which is the isolation and cloning of a disease gene after determining its approximate physical location in the human genome, was introduced in the 1980s. At that time, the first step in positional cloning was to use model-based linkage analysis to localize the disease-causing gene to a region of one million base pairs (1 Mb) or larger by using DNA markers. These DNA markers had no known function (anonymous markers) and were dispersed throughout the genome. The goal was to identify marker loci that had a low probability of recombination with the unknown disease locus to narrow the section of the genome that needed to be investigated with molecular techniques. Model-based linkage analysis is most successful when a disease has a known mode of inheritance.

Genetic linkage occurs when two loci are physically connected on the same chromosome at a distance from one another at less than 50% recombination. Recombination is the occurrence among offspring of new combinations of alleles, resulting from crossovers that occur during meiosis. For example, assume that a father has genotype D_1D_2 at the dominant trait locus and A_1A_2 at a marker locus (see Fig. 3.1). On the basis of information from the father's parents, we can determine that

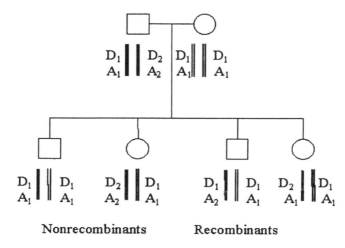

Figure 3.1 Family where some siblings are nonrecombinants and some siblings are recombinants.

one of his chromosomes has the D_1 and A_1 alleles, while the other chromosome has the D_2 and A_2 alleles. We would expect his children who are nonrecombinants to have received a chromosome from their father that had either D_1 and A_1 or D_2 and A_2 at the two loci as shown in Figure 3.1. If crossing over occurred between the D locus and the A locus during meiosis, we would see chromosomes that were either D_1 and A_2, or D_2 and A_1 in recombinant offspring as shown in Figure 3.1. The parameter of interest is the recombination frequency theta (Θ), which is the proportion of meioses in which recombination between two loci are observed. The recombination frequency is used to estimate genetic distances between loci. When loci are unlinked by virtue of being far apart on the same chromosome or on different chromosomes, then $\Theta = 0.5$ and all four allele combinations D_1 / A_1, D_1 / A_2, D_2 / A_1, and D_2 / A_2 are equally likely.

In model-based linkage analysis, a model of inheritance for the disease of interest must be specified. In addition, estimates of the allele frequencies for the disease gene and the marker, and the penetrance associated with each genotype at the disease locus must be specified. Penetrance is the proportion of persons with a disease-causing genotype who manifest the disease. For example, if 8 of every 10 people who inherit the disease genotype develop the disease, the penetrance is 80%. This disease genotype has reduced or incomplete penetrance. For quantitative traits, the penetrance is defined in terms of the means and standard deviations of the quantitative trait distributions for each genotype at the trait locus. Information about mutation rates and phenocopies also can be incorporated into model-based linkage analysis. A phenocopy resembles a trait produced by a specific gene, but it is due to nongenetic factors. The null hypothesis is that a disease-susceptibility gene is not linked to a specific marker ($\Theta = 0.5$). The alternative hypothesis is that they are linked ($\Theta < 0.5$). A ratio is computed of the likelihood of observing the pedigree under the alternative hypothesis ($\Theta < 0.5$) divided by the likelihood of observing the pedigree under the null hypothesis ($\Theta = 0.5$).

If the pedigree data argue against linkage of the two loci, then the ratio will be less than 1 (because the denominator will be greater than the numerator). This ratio is really an odds and it is customary to take the logarithm of the odds to obtain a "logarithm of the odds score" (LOD score). A LOD score is computed for each family in a sample and then summed across all families. By convention, a LOD score of 3.0 is taken as evidence for linkage (1000 to 1 odds), and a LOD score of -2 (100 to 1 odds) is considered evidence against linkage. This convention is now being debated (8,31).

LOD score linkage analysis has been in use in humans for almost 50 years, starting with Morton (32,33) and continuing through the present. In a recent study in Chinese families with cleft lip with or without cleft palate (34), the highest LOD score from model-based linkage analysis was 1.91 (with $\Theta = 0.20$). The LOD score analysis suggested heterogeneity, implying that different regions of the genome are involved in the etiology of cleft lip with or without cleft palate in different families. In the period of time between the studies conducted by Morton and Marazita and colleagues (34), many LOD score analyses have been conducted of such diseases as cystic fibrosis, breast cancer, Huntington disease, and Alzheimer disease (6,35).

Linkage analysis is more efficient if more than two marker loci are analyzed simultaneously. Most current studies use multipoint mapping in which the recombination frequencies among three or more loci are estimated simultaneously. This approach is useful to establish the order of linked loci. It is also helpful when individual genetic markers contain limited information about meiosis.

One major disadvantage of the LOD score linkage approach is that the values for all the parameters that describe the genetic model of the trait (mode of inheritance, frequencies, and penetrances) must be specified. One interesting application of LOD score linkage analysis avoids having to specify the penetrance for each genotype. This approach focuses only on the affected persons in a pedigree. The penetrance parameters are set to a very low value (e.g., 0.001), and thus the trait information for unaffected persons is not used. The disease genotype of the affected persons is inferred to include the susceptibility allele, but the disease genotypes of the unaffected persons are irrelevant. Marker information is used for all persons if it is available. Although the general approach of using only affected pedigree members has the advantage of not requiring information on penetrance, all the other parameters of the genetic model for the trait of interest must be specified.

Model-Free Linkage Analysis

Traditional LOD score methods of linkage analysis have proven to be powerful for identifying genes involved in simple Mendelian traits where the mode of inheritance is relatively clear and a genetic model of inheritance can be specified. The identification of genes for complex traits has more often used analytical approaches that do not require the specification of a genetic model. These so-called model-free (nonparametric, robust) methods can be used to screen a number of loci or even the entire genome. Model-free linkage methods are based on the identification of chromosomal segments (alleles or haplotypes) shared by relatives, most often relatives who have the qualitative trait of interest. This contrasts with the traditional LOD score method, which is based on the identification of loci with alleles that tend to cosegregate with the trait in families. A haplotype is the allelic constitution at multiple linked loci on a single chromosome.

The concept of two persons sharing alleles at a genetic locus that are identical by descent is central to the application of model-free methods of linkage analysis. Assume a marker locus with four different alleles denoted A_1, A_2, A_3, and A_4. Two copies of the same allele at this locus, say allele A_3, in two different persons, regardless of whether they are related, are said to be identical by state (IBS), that is, they appear to have the same DNA sequence, to be the same form of the gene. In Figure 3.2 in family A, the A_3 alleles in the two siblings are IBS.

In two genetically related persons, the A_3 alleles are said to be identical by descent (IBD) if they were inherited from a common ancestor. In Figure 3.2, in fam-

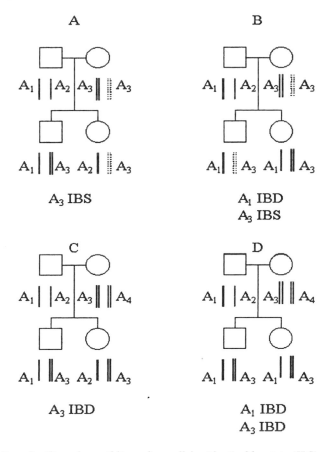

Figure 3.2 Four families where siblings share alleles identical by state (IBS) or identical by descent (IBD).

ily C and in family D the A_3 alleles in the two siblings are IBD. If two siblings with genotypes A_1A_3 and A_2A_3 at the locus have parents with genotypes A_1A_2 and A_3A_3, then the A_3 alleles in their children are IBS but they may not be IBD, because they may not be the same ancestral copy of the A_3 allele (see Fig. 3.2, family A and family B). If the two children have genotypes A_1A_3, then the children are IBD for the A_1 allele as shown in Figure 3.2 family B. However, if the parents' genotypes are A_1A_2 and A_3A_4, then the A_3 alleles in their children are IBD, because each offspring inherited the same A_3 allele from the same common parent (see Figure 3.2, family D). In family D in Figure 3.2, the A_1 allele in the children is also IBD. Determination of the IBD status of alleles for a pair of relatives usually requires that the genotypes of common ancestors are also available.

Model-free linkage analysis methods can be applied to identify candidate chromosomal regions for either qualitative or quantitative traits, and for study designs that range from sibling pairs to large extended pedigrees. However, these methods have most often been applied to the analysis of data obtained from sibling pairs, in which case the basic concept can be summarized as follows: the greater the proportion π of alleles IBD that a sib pair shares at a marker locus,

- the more likely the sibs are to be concordant for a qualitative trait if the trait is affected by a locus linked to the marker locus, or
- the smaller should be the difference between the sibs' quantitative trait values if these are affected by a locus linked to the marker locus.

The unconditional probability that a relative pair shares no, one or two alleles IBD at a genetic locus is displayed in Table 3.1. At any randomly chosen genetic locus, two siblings are expected to share no, one or two alleles IBD with probabilities $^1/_4$, $^1/_2$, and $^1/_4$, respectively. On average, they are expected to share one allele IBD ($^1/_4 \times 0 + ^1/_2 \times 1 + ^1/_4 \times 2 = 1$ allele IBD). If, for example, two siblings both affected with a disease are typed for a genetic marker that is tightly linked to the disease gene, the siblings are expected to share more than one allele IBD.

The affected sib-pair design is the simplest and most common design used to obtain data for a model-free linkage analysis of a qualitative trait (36–42). In this design, the most basic data for n affected sib-pairs consist of the number of pairs that share no, one or two alleles IBD. Early methods for analyzing these data focused on:

1. the proportion of pairs that share two alleles IBD (39), which is expected to be $^1/_4$ under the null hypothesis of no linkage,
2. the mean number of alleles shared IBD in the entire sample (43), which is expected to be one under the null hypothesis, or
3. the goodness of fit of the observed proportion of pairs that share no, one or two alleles (44), which is expected to be $^1/_4$, $^1/_2$, and $^1/_4$, respectively, under the null hypothesis.

Blackwelder and Elston (45) examined the properties of these alternative tests in a simulation study and found that the most powerful test of linkage in a variety of situations was the one based on the mean number of marker alleles shared IBD. In

Table 3.1 The Unconditional Probability of a Relative Pair Sharing No, One and Two Alleles Identical by Descent at a Genetic Locus

	EXPECTED PROPORTION OF IBD SHARING		
Relative Pair	0 Alleles	1 Allele	2 Alleles
MZ twins			1
Full siblings	$^1/_4$	$^1/_2$	$^1/_4$
Parent-child		1	
Grandparent-grandchild	$^1/_2$	$^1/_2$	
First cousins	$^3/_4$	$^1/_4$	

some instances, more than one affected sib pair from a family may be available for inclusion in the data set. Blackwelder and Elston also demonstrated that incorporating information from all possible sib pairs as though they came from separate families is valid, although it has been suggested that additional power may be gained if different weights are given to the different sibship sizes (45,46).

If the marker locus is not very informative, an unambiguous determination of the sharing of no, one or two alleles IBD for a pair of siblings (or other relative pairs) may not be possible even when parental genotypes are available. Various algorithms have been proposed (47,48) that use the genotypic information from all available relatives to estimate IBD status. Several strategies have been developed for conducting multipoint analysis (31,49,50). Algorithms for obtaining estimates of multipoint IBD sharing for genome searches are also contained in the computer programs, e.g., MAPMAKER/SIBS and GENEHUNTER (http://linkage.rockefeller. edu) that are currently used to conduct a model-free linkage analysis.

The sib-pair design also can be used to investigate linkage for quantitative traits (51,52). Inference is based on the correlation between similarity of the sibs with respect to the trait and their similarity with respect to one or more marker loci. Genotypic information from parents is again used to estimate the proportion of alleles each sib pair shares IBD at a locus. In the original formulation of this analytical approach (47), the squared trait difference for each sib pair was regressed on the estimated proportion of alleles shared IBD by each pair at a typed locus. Based on this model, a significant negative regression coefficient reflected the fact that sib pairs who shared a higher proportion of alleles IBD, had smaller trait differences, which suggested linkage. In a more recent formulation (52) the dependent variable is $[(\text{trait}_{\text{sib1}} - \text{mean})(\text{trait}_{\text{sib2}} - \text{mean})]^2$, and a positive regression coefficient suggests linkage. Both of these formulations and several others that involve various weighting schemes for multiple sib pairs from the same family are implemented in the SIBPAL2 program of the S.A.G.E. package (53).

Numerous examples of genome scans aimed at identifying candidate regions for complex traits are appearing in the literature. For example, Ghosh et al. (54) conducted a genome scan for type 2 diabetes susceptibility genes in 719 affected sib pairs from 478 Finnish families. The Finnish population is believed to be relatively genetically isolated with a small number of founders. Such isolated populations may offer an advantage for gene mapping because of a more homogeneous genome and shared environment. Anonymous markers were typed on average every 8 centi-Morgans (cM) throughout the genome. One cM is approximately equal to a recombination frequency of 1%. The strongest evidence for linkage was seen on chromosome 20 (three loci), followed by chromosomes 11, 2, 10, and 6. However, the highest LOD score on chromosome 20 was 2.15, which does not reach a genome-wide significance level. On the other hand, these candidate regions are consistent with results from similar analyses conducted in other populations, and thus they are encouraging. In the same population sample, an analysis of diabetes-associated quantitative traits (e.g., fasting insulin, fasting glucose, fasting C-peptide, BMI) was conducted by using quantitative trait locus (QTL) linkage analysis (55). This analysis identified additional regions on chromosomes 1, 3, 7, 17, and 19 that may contain diabetes susceptibility loci.

Genes responsible for blood pressure determination and hypertension in humans are largely unidentified, although up to 50% of the variation in blood pressure has been attributed to genetic factors. Genetic studies of similar traits in animal models, e.g., hypertension in the rat, may provide important leads for the identification of susceptibility loci in humans. In the rat, a blood pressure gene, the angiotensin I-converting enzyme gene (ACE), has been identified on chromosome 10. A homologous region (containing a similar DNA sequence) exists on human chromosome 17. Julier et al. (56) analyzed 21 microsatellite markers that spanned the homologous chromosome 17 region in 518 hypertensive sib pairs, and found significant evidence for a human hypertension susceptibility locus. The strongest evidence of linkage, however, was found with two markers D17S183 and D17S934, 18 cM proximal to the human ACE locus. This suggests that in the region of homology, a gene other than ACE may be involved in the etiology of human hypertension. It may also help to explain why candidate gene studies focusing specifically on the ACE locus have been largely negative.

It has been suggested that the power to detect linkage for a quantitative trait can be increased by using selected samples (57). Xu et al. (58) employed a design that identified sib pairs that were extremely discordant (57) for blood pressure, top versus bottom age-specific decile, as well as highly concordant and minimally concordant sib pairs to conduct a genome scan (58). The sib pairs were identified after a community-based survey of more than 200,000 Chinese adults. Although no chromosomal region satisfied the 5% genome-wide significance level, candidate regions were identified on chromosomes 3, 11, 16, and 17 for systolic blood pressure, and on chromosome 15 for diastolic blood pressure.

These examples demonstrate the challenge and the complexity of studies of complex traits. Follow-up of the analytical results from the three studies used as examples in this section might involve typing additional marker loci in an attempt to narrow candidate regions, the identification of additional sib pairs to increase the sample size, the identification of a new sample to try to replicate the suggestive linkage results, more careful definition of the phenotype or identification of intermediate phenotypes, and association studies focused on candidate genes that are located in candidate regions.

Association Analysis

Association analysis provides an alternative to linkage analysis approaches for identifying specific genes that may be involved in the etiology of a trait. Association studies attempt to identify susceptibility genes by comparing allele or genotype frequencies between persons with and without a qualitative trait, or by estimating the portion of the variability in a quantitative trait that is associated with genotypic variability at the typed locus. Most association studies are conducted using either a case-control design or a family-based design where the "control" group is constructed from the genotypes of family members (59). The genetic polymorphisms that are typed for an association analysis are usually present in or very close to candidate genes, i.e., genes thought to be directly involved in the pathogenesis of the trait. Association studies can be performed for any random DNA polymorphism; however, they are most meaningful when applied to functionally significant variations in genes having a clear biologic relation to the trait.

A *true* association at the population level between a typed locus and a trait can arise under either of two circumstances: (*1*) when the functional genetic variant has actually been typed (measured genotype); in this situation, it is possible to directly test the effect of the measured genotype on the trait; or (*2*) when the typed locus is in linkage disequilibrium with the susceptibility locus (60,61). Linkage disequilibrium is the nonrandom association between the alleles at two genetic loci. In the context of an association study, this means that the trait-associated locus that was typed is linked to a susceptibility locus, and there is a significant association of the alleles at the two loci such that the trait-associated allele is actually a surrogate for the susceptibility allele at the linked susceptibility locus. The degree of linkage disequilibrium between two linked loci decays over time in relation to the recombination frequency, i.e., the farther apart the two linked loci, the greater the chance of recombination during meiosis and the more quickly the allelic association will disappear. If two loci are very closely linked, disequilibrium may persist for long periods of time, and it may be possible to take advantage of such allelic associations to infer the presence of trait susceptibility loci from association studies (62). Linkage disequilibrium is detectable only when a significant proportion of the apparently independent chromosomes that we examine from a population are actually copies of the same ancestral chromosome (63). It is not an inevitable consequence of close linkage. Therefore, the same allele will be present in apparently unrelated cases, i.e., these apparently unrelated cases often carry alleles that are IBD.

A major challenge in conducting any case-control association study is the choice of appropriate study groups. If possible, cases should be identified such that the group is enriched for disease susceptibility alleles, e.g., incident cases who also have a family history of the disease, if the aim is detection of association rather than estimating the effect in the population. In most studies, the control group consists of unaffected persons, who are unrelated to the case group, i.e., the cases and controls are independent, unrelated persons. Ideally, the control group should be sampled from the same genetic population as the cases to avoid confounding because of population stratification. Population stratification can occur when cases and controls differ in their ethnic backgrounds, and both the disease risk and the genotype frequencies at a typed locus differ among ethnic groups. To avoid potential confounding from population stratification, controls can be matched to cases for ethnic background, as well as for factors such as age and sex (for further discussions, see Chapters 6 and 8).

The analytical focus of case-control data is on the comparison of allele or genotype frequencies between the case and control groups, which can be accomplished by using such tests as chi-square statistics or logistic regression analysis. For example, a 3-year nationwide case-control study of orofacial clefts (cleft lip with or without cleft palate, and isolated cleft palate) in Denmark focused on associations with two putative susceptibility loci, *MSX1* and *TGFB3* (64). A significant association was identified between carriers of the "2" allele at the *TGFB3* locus and isolated cleft palate, with an associated odds ratio versus noncarriers of 2.13 (95% CI, 1.12-4.19).

The problem of population stratification also can be addressed by an alternative study design with family-based controls, usually the parents or siblings of the case. With parental genotypic information, a fictitious control group can be formed from the parental alleles at the typed locus that were not transmitted to the case (65,66). Stratification is reduced or eliminated because one person, a parent, contributes the case and control alleles.

The basic premise is that a parent with a heterozygous genotype consisting of one allele that is associated with the trait and one allele that is not associated with the trait will transmit the associated allele more often than expected to their case-child. The transmission-disequilibrium test (TDT) (67) compares the frequencies of alleles that were transmitted from heterozygous parents to their case-children to the frequencies of alleles that were not transmitted. In the simplest instance of a locus with two alleles, where allele A_1 represents the associated allele, the data from 2n parents of n cases can be displayed as in Table 3.2. Data from homozygous parents (in cells a and d) do not contribute to the analysis of transmission disequilibrium. McNemar's test statistic $(b - c)^2 / (b + c)$ can be used to evaluate the null hypothesis of equal transmission of the matched pairs of alleles from heterozygous parents as reflected in cells b and c. If allele A_1 is actually associated with the trait, the expectation is that b will be much higher than c, and the null hypothesis will be rejected. The TDT is actually a test of the linkage of the typed locus to a susceptibility locus. However, the TDT will only detect linkage if the two loci are also associated, that is, in linkage disequilibrium. With linkage equilibrium, b and c in Table 3.2 would not differ across families, because the allele in coupling with the disease gene in each family would be random, preventing the detection of linkage using the TDT (see also Chapter 6).

In many instances, DNA may be impossible to obtain from both parents of a case, e.g., when studying a trait with a late age of onset. In such instances, unaffected

Table 3.2 Data for the Transmission-Disequilibrium Test

Transmitted Allele	NONTRANSMITTED ALLELE		
	A_1	A_2	Total
A_1	a	b	w
A_2	c	d	x
Total	y	z	2n

siblings, rather than parents, can be used as controls (68). If only a few families have a missing parent, guidelines exist with regard to the inclusion of data from those families in a TDT analysis (69). The TDT is adequate when the typed locus has only two alleles or when an *a priori* hypothesis exists that a particular allele is strongly associated with the trait. In the later instance, the other alleles can be combined in the analysis as represented in Table 3.2; however, an *a priori* hypothesis may not exist for a multi-allelic locus, and the TDT has been extended to accommodate the analysis of such loci (70,71).

The TDT has been extended to the analysis of quantitative traits (72–75). Analysis of variance and covariance approaches also can be used to estimate a component of the total quantitative trait variance that represents the effect of genetic variability at a typed locus. These approaches can be applied to quantitative traits measured in samples of unrelated persons (76), as well as families (77).

Although most association studies to date have focused on candidate genes, a complete genome-wide scan to identify markers in linkage disequilibrium is now a theoretical possibility. Such a scan would require several thousand SNPs. The analysis would involve thousands of hypothesis tests, each carrying an independent risk for a false-positive result. So only extremely strong associations would be statistically significant after correction for the multiple tests.

Association studies also allow for the investigation of possible gene–environment or gene–gene interaction effects. For example, Caspi et al. (78) identified an interaction effect between the gene that codes for monoamine oxidase A (*MAOA*) and childhood abuse that was related to a history of antisocial behavior. A positive history was most likely in abused males with the genotype for low *MAOA* activity. In the absence of abuse, males with the low-activity genotype were no more likely to display antisocial behavior than males with the high-activity genotype (78).

Conclusions

The approaches to quantify the genetic component of and identify genes for complex traits are constantly evolving. Some of the emerging issues include meta-analysis to increase power, the exploration of appropriate significance levels to use, and use of phenotype information to stratify samples (5,7,8). More studies are focusing on gene–gene interactions (79) as well as gene-environment interactions (80) instead of solely on gene trait associations. In addition, newer technologies such as gene expression profiles are being used to identify the role of many genes simultaneously (81). There is a need to begin to identify the functional effects on gene expression or protein function for any genes that are implicated for complex traits (82).

Ultimately, we will gain more understanding of genetic risk for complex traits. Understanding such risks will impact individuals, their families, and the population. It is not too early to consider the population impact of genetic tests, one of the goals of the Human Genome Epidemiology (HuGE) Network (83,84), and to make sure that appropriate protections are in place to prevent discrimination as well as to allow access to appropriate health care. Hopefully, the information gained about genes for complex traits can be used to prevent disease and improve health.

Acknowledgments

The authors thank Lillian Brown for her assistance with the figures. This work was supported, in part, under a cooperative agreement from the Centers for Disease Control and Prevention (CDC) through the Association of Schools of Public Health (ASPH). Grant Number U36/CCU300430-22. The contents of this chapter are solely the responsibility of the authors and do not necessarily represent the official views of CDC or ASPH.

References

1. Peyser, PA. Genetic epidemiology of coronary artery disease. Epidemiol Rev 1997;19: 80–90.
2. Gu C, Rao DC. Optimum study designs, in Genetic Dissection of Complex Traits. Rao DC, Province MA (eds). San Diego: Academic Press, 2001, pp. 439–457.
3. Khoury MJ, Beaty TH, Cohen BH. Fundamentals of Genetic Epidemiology. New York: Oxford University Press, 1993.
4. Weiss KM. Genetic Variation and Human Disease. Principles and Evolutionary Approaches. Cambridge: Cambridge University Press, 1993.
5. Haines JL, Pericak-Vance MA (eds). Approaches to Gene Mapping in Complex Human Diseases. New York: Wiley-Liss, 1998.
6. Ott J. Analysis of Human Genetic Linkage. Third edition. Baltimore: The Johns Hopkins University Press, 1999.
7. Bishop T, Sham P (eds). Analysis of Multifactorial Disease. San Diego: Academic Press, 2000.
8. Rao DC, Province MA (eds). Genetic Dissection of Complex Traits. San Diego: Academic Press, 2001.
9. Elston RC, Olson J, Palmer L. Biostatistical Genetics and Genetic Epidemiology. New York: John Wiley & Sons, Inc., 2002.
10. Khoury MJ, Risch N, Kelsey JL (eds). Genetic Epidemiology. Epidemiol Rev 1997; 19:1–185.
11. Ellsworth DL, Manolio TA. The emerging importance of genetics in epidemiologic research II. Issues in study design and gene mapping. Ann Epidemiol 1999;9:75–90.
12. Ellsworth DL, Manolio TA. The emerging importance of genetics in epidemiologic research III. Bioinformatics and statistical genetic methods. Ann Epidemiol 1999;9: 207–224.
13. Rybicki BA, Kirkey KL, Major M, et al. Familial risk ratio of sarcoidosis in African-American sibs and parents. Am J Epidemiol 2001;153:188–193.
14. Chakravarti A, Lyonner S. Hirschsprung disease, in The Metabolic and Molecular Bases of Inherited Disease, eighth edition. Scriver CR, Beaudet AL, Sly WS, Valle D (eds). New York: McGraw-Hill, 2001, pp. 6231–6255.
15. Pharoh PD, Day NE, Duffy S, et al. Family history and the risk of breast cancer: a systematic review and meta-analysis. Int J Cancer 1997;71:800–809.
16. Rice TK, Borecki IB. Familial resemblance and heritability, in Genetic Dissection of Complex Traits. Rao DC, Province MA (eds). San Diego: Academic Press, 2001, pp. 35–44.
17. Martin NG, Eaves LJ, Kearsey MJ, Davies P. The power of the classical twin study. Heredity 1978;40:97–116.

18. Austin MA, Friedlander Y, Newman B, et al. Genetic influences on changes in body mass index: a longitudinal analysis of women twins. Obes Res 1997;5:326–331.
19. Tanner CM, Ottman R, Goldman SM, et al. Parkinson disease in twins: an etiologic study. JAMA 1999;281:341–346.
20. Neale MC. Twin analysis. In: Biostatistical Genetics and Genetic Epidemiology. Elston RC, Olson J, Palmer L. New York: John Wiley & Sons, Inc., 2002, pp. 743–756.
21. Hopper JL. Twin concordance. In: Biostatistical Genetics and Genetic Epidemiology. Elston RC, Olson J, Palmer L. New York: John Wiley & Sons, Inc., 2002, pp. 756–759.
22. Peyser PA, Bielak LF, Chu JS, et al. Heritability of coronary artery calcium quantity measured by electron beam computed tomography in asymptomatic adults. Circulation 2002;106:304–308.
23. Almasy L, Blangero J. Multipoint quantitative-trait linkage analysis in general pedigrees. Am J Hum Genet 1998;62:1198–1211.
24. Burns TL, Moll PP, Lauer RM. The relation between ponderosity and coronary risk factors in children and their relatives. The Muscatine ponderosity family study. Am J Epidemiol 1989;129:973–987.
25. Moll PP, Burns TL, Lauer RM. The genetic and environmental sources of body mass index variability: the Muscatine ponderosity family study. Am J Hum Genet 1991; 49:1243–1255.
26. Collier DA, Treasure F. Genes and environment in eating disorders and obesity, in: Analysis of Multifactorial Disease. Bishop T, Sham P (eds). San Diego: Academic Press, 2002, pp. 251–290.
27. Fogarty DG, Hanna LS, Wantman M, et al. Segregation analysis of urinary albumin excretion in families with type 2 diabetes. Diabetes 2000;49:1057–1063.
28. Pihlajamaki J, Austin MA, Edwards K, Laaskso M. A major gene effect on fasting insulin and insulin sensitivity in familial combined hyperlipidemia. Diabetes 2001;50: 2396–2401.
29. Sachidanandam R, Weissman D, Schmidt SC, et al. A map of human genome sequence variation containing 1.42 million single nucleotide polymorphisms. Nature 2001;409: 928–933.
30. Peltonen L, McKusick VA. Genomics and medicine. Dissecting human disease in the postgenomic era. Science 2001;291:1224–1229.
31. Lander E, Kruglyak L. Genetic dissection of complex traits: guidelines for interpreting and reporting linkage results. Nat Genet 1995;11:241–247.
32. Morton NE. Sequential tests for detection of linkage. Am J Hum Genet 1955;7:277–318.
33. Morton NE. The detection and estimation of linkage between the genes for elliptocytosis and the RH blood type. Am J Hum Genet 1956;8:80–96.
34. Marazita ML, Field LL, Cooper ME, et al. Genome scan for loci involved in cleft lip with or without cleft palate, in Chinese multiplex families. Am J Hum Genet 2002; 71:349–364
35. Risch N. Evolving methods in genetic epidemiology. II. Genetic linkage from an epidemiologic perspective. Epidemiol Rev 1997;19:24–32.
36. Penrose LS. The detection of autosomal linkage in data which consist of pairs of brothers and sisters of unspecified parentage. Ann Eugen 1935;6:133–148.
37. Penrose LS. A further note on the sib-pair linkage method. Ann Eugen 1946;13:25–29.
38. Penrose LS. The general sib-pair linkage test. Ann Eugen 1953;18:120–144.
39. Day NE, Simons MJ. Disease susceptibility genes—their identification by multiple case family studies. Tissue Antigens 1976;8:109–119.
40. de Vries RR, Fat RF, Nijenhuis LE, van Rood JJ. HLA-linked genetic control of host response to Mycobacterium leprae. Lancet 1976;ii:1328–1330.

41. Suarez BK, Rice JP, Reich T. The generalized sib pair IBD distribution: its use in the detection of linkage. Ann Hum Genet 1978;42:87–94.

42. Holmans P. Affected sib-pair methods for detecting linkage to dichotomous traits: a review of the methodology. Hum Biol 1998;70:1025–1040.

43. Green JR, Woodrow JC. Sibling method for detecting HLA-linked genes in disease. Tissue Antigens 1977;9:31–35.

44. Weitkamp LR, Stancer HC, Persad E, et al. Depressive disorders and HLA: a gene on chromosome 6 that can affect behavior. N Engl J Med 1981;305:1301–1306.

45. Blackwelder WC, Elston RC. A comparison of sib-pair linkage tests for disease susceptibility loci. Genet Epidemiol 1985;2:85–97.

46. Suarez BK, Hodge SE. A simple method to detect linkage for rare recessive diseases: an application to juvenile diabetes. Clin Genet 1979;15:126–136.

47. Haseman JK, Elston RC. The investigation of linkage between a quantitative trait and a marker locus. Behav Genet 1972;2:3–19.

48. Amos CI, Dawson DV, Elston RC. The probabalistic determination of identity-by-state sharing for pairs of relatives from pedigrees. Am J Hum Genet 1990;47:842–853.

49. Risch N. Linkage strategies for genetically complex traits. I. Multilocus models. Am J Hum Genet 1990;46:222–228.

50. Risch N. Linkage strategies for genetically complex traits. II. The power of affected relative pairs. Am J Hum Genet 1990;46:229–241.

51. Penrose LS. Genetic linkage in graded human characters. Ann Eugen 1937;8:233–237.

52. Elston RC, S Buxbaum S, Jacobs KB, Olsen JM. Haseman and Elston revisited. Genet Epidemiol 2000;19:1–17.

53. S.A.G.E. Statistical Analysis for Genetic Epidemiology. Computer program package available from the Department of Epidemiology and Biostatistics, Rammelkamp Center for Education and Research, MetroHealth Campus, Case Western Reserve University, Cleveland. 2001.

54. Ghosh S, Watanabe RM, Valle TT, et al. The Finland-United States investigation of non-insulin-dependent diabetes mellitus genetics (FUSION) Study. I. An autosomal genome scan for genes that predispose to Type 2 diabetes. Am J Hum Genet 2000;67:1174–1185.

55. Watanabe RM, Ghosh S, Langefeld CD, et al. The Finland-United States investigation of non-insulin-dependent diabetes mellitus genetics (FUSION) Study. II. An autosomal genome scan for diabetes-related quantitative trait loci. Am J Hum Genet 2000;67: 1186–1200.

56. Julier C, Delepine M, Keavney B, et al. Genetic susceptibility for human familial essential hypertension in a region of homology with blood pressure linkage on rat chromosome 10. Hum Mol Genet 1997;6:2077–2085.

57. Risch N, Zhang H. Extreme discordant sib pairs for mapping quantitative trait loci in humans. Science 1995;268:1584–1589.

58. Xu X, Rogus JJ, Terwedow HA, et al. An extreme sib-pair genome scan for genes regulating blood pressure. Am J Hum Genet 1999;64:1694–1701.

59. Schaid DJ, Rowland C. Use of parents, sibs, and unrelated controls for detection of associations between genetic markers and disease. Am J Hum Genet 1998;63:1492–1506.

60. Lander ES, Schork NJ. Genetic dissection of complex traits. Science 1994;265:2037–2048.

61. Almasy L, MacCluer JW. Association studies of vascular phenotypes: how and why? Arterioscler Thromb Vasc Biol 2002;22:1055–1057.

62. Jorde LB. Linkage disequilibrium and the search for complex disease genes. Genome Res 2000;10:1435–1444.

63. Jorde LB, Carey JC, Bamshad MJ, White RL. Medical genetics, 2nd edition. St. Louis: Mosby, Inc., 1999.

64. Mitchell LE, Murray JC, O'Brien S, Christensen K. Evaluation of two putative susceptibility loci for oral clefts in the Danish population Am J Epidemiol 2001;153:1007–1015.
65. Falk CT, P Rubinstein P. Haplotype relative risk: an easy reliable way to construct a
proper control sample for risk calculations. Ann Hum Genet 1987;51:227–233.
66. Schaid DJ, Sommer SS. Comparison of statistics for candidate-gene association studies
using cases and parents. Am J Hum Genet 1994;55:402–409.
67. Spielman RS, McGinnis RE, Ewens WJ. Transmission test for linkage disequilibrium:
the insulin gene region and insulin-dependent diabetes mellitus (IDDM). Am J Hum Genet
1993;52:506–516.
68. Spielman RS, Ewens WJ. A sibship test for linkage in the presence of association: the
sib transmission/disequilibrium test. Am J Hum Genet 1998;62:450–458.
69. Curtis D, Sham PC. A note on the application of the transmission disequilibrium test
when a parent is missing. Am J Hum Genet 1995;56:811–812.
70. Sham PC, Curtis D. An extended transmission/disequilibrium test (TDT) for multi-allele
marker loci. Ann Hum Genet 1995;59:323–336.
71. Spielman RS, Ewens WJ. The TDT and other family-based tests for linkage disequilibrium and association. Am J Hum Genet 1996;59:983–989.
72. Allison DB. Transmission-disequilibrium tests for quantitative traits. Am J Hum Genet
1997;60:676–690.
73. Rabinowitz D. A transmission disequilibrium test for quantitative trait loci. Hum Hered
1997;47:342–350.
74. Xiong MM, Krushkal J, Boerwinkle E. TDT statistics for mapping quantitative trait loci.
Ann Hum Genet 1998;62:431–452.
75. George V, Tiwari HK, Zhu X, Elston RC. A test of transmission/disequilibrium for quantitative traits in pedigree data, by multiple regression. Am J Hum Genet 1999;65:236–245.
76. Boerwinkle E, Chakraborty R, Sing CF. The use of measured genotype information in
the analysis of quantitative phenotypes in man. I. Models and methods. Ann Hum Genet
1986;50:181–194.
77. George VT, Elston RC. Testing the association between polymorphic markers and quantitative traits in pedigrees. Genet Epidemiol 1987;4:193–20.
78. Caspi A, McClay J, Moffitt TE, et al. Role of genotype in the cycle of violence in mal-
treated children. Science 2002;297:851–854.
79. Small KM, Wagoner LE, Levin AM, et al. Synergistic polymorphisms of beta1- and al-
pha2C-adrenergic receptors and the risk of congestive heart failure. N Engl J Med
2002;347:1135–1142.
80. Humphries SE, Talmud PJ, Hawe E, et al. Apolipoprotein E4 and coronary heart disease
in middle-aged men who smoke: a prospective study. Lancet 2001;358:115–119.
81. Beer DG, Kardia SL, Huang CC, et al. Gene-expression profiles predict survival of patients with lung adenocarcinoma. Nat Med 2002;8:816–824.
82. Boerwinkle E, Hixson JE, Hanis CL. Peeking under the peeks: following up genome-
wide linkage analyses. Circulation 2000;102:1877–1878.
83. Khoury MJ, Dorman JS. The Human Genome Epidemiology Network. Am J Epidemiol
1998;148:1–3.
84. Little J, Bradley L, Bray MS, et al. Reporting, appraising, and integrating data on genotype prevalence and gene-disease associations. Am J Epidemiol 2002;156:300–310.

4

Ethical, legal, and social issues in the design and conduct of human genome epidemiology studies

Laura M. Beskow

Human genetic variation provides an important tool for refining epidemiologic research. Although investigators have long recognized that disease generally results from a constellation of host- and environment-specific factors, they have tended to focus on the environment because of scientific and technologic limitations. Environmental exposures alone, however, whether chemical, physical, infectious, nutritional, social, or behavioral, often poorly predict who will develop disease—particularly the common, complex diseases of public health interest. Understanding the impact of environmental factors among people who have specific genetic susceptibilities offers the possibility of more effective and targeted medical and public health interventions. A critical question is: How can we advance this public health research agenda while protecting and respecting the participants in such research?

Ethical principles generally require that epidemiologists working with human participants submit their research protocols for independent review, obtain informed consent, protect privacy and maintain confidentiality, and safeguard the rights and welfare of the individuals and groups they study (1). How these obligations are best addressed depends on the nature of the risks and benefits associated with the research. Whether and how the addition of a genetic component to an epidemiologic study changes the balance of risks and benefits merit further analysis. Many argue that genetic information is fundamentally similar to other kinds of health information (2). Even so, society invests enormous power in the concept of genetics, and notions of genetic determinism and genetic reductionism have significant negative implications for public health and prevention messages (3). Combined with the history of eugenics and other research abuses in the United States and around the world, clarifying the duties of investigators to participants in population-based research involving genetics is important. Although much has been written about ethical, legal, and social issues associated with genetic testing (4,5), and about ethical issues in epidemiology (6,7), less attention has been given to issues that arise specifically in the context of genetic epidemiology.

This chapter explores ethical, legal, and social issues in human genome epidemiologic research, highlighting differences in the risks and benefits of genetic research among high-risk families compared to population-based settings. This exploration draws on US guidelines for research involving human tissue, such material being necessary for conducting genetic analyses in any context (See also ref. 8 for an international perspective).

Protection of Human Participants in Research

Regulations and institutional rules for the protection of human research participants derive from a number of sources. Among the best known of these is the *Belmont Report* (9), which identifies three ethical principles particularly relevant to research involving human participants: respect for persons, beneficence, and justice. Federal policy for the protection of human research participants incorporates these principles in two basic protections: oversight by an Institutional Review Board (IRB) and requirements for informed consent. These policies, codified by the Department of Health and Human Services in the Code of Federal Regulations, Title 45, Part 46, Subpart A (45 CFR 46), are often called the "Common Rule" because an identical set of regulations has been adopted by a number of federal agencies. A diverse array of research is subject to the Common Rule, involving varying combinations of risks and benefits. The regulations therefore provide for exemption, waiver, or alteration of the requirements for review and informed consent when appropriate.

Family-Based Studies versus Population-Based Studies Involving Genetics

Genetic research is typically considered a sensitive enterprise because much of it has been conducted among families with a heavy burden of disease, e.g., the investigation of *BRCA1* mutations among families with multiple members affected with breast or ovarian cancer. Family studies provide a unique framework for investigating highly penetrant gene variants, which are those that lead to disease expression a majority of the time and thus produce marked familial aggregation. However, this kind of research poses the risk of psychological and social harms, such as insurance and employment discrimination, social stigmatization, familial disruption, and psychological distress, which could result from uncovering genetic information that has significant implications for the health of the individual and her family. These concerns are magnified when only limited or unproven measures are available for prevention or treatment. Thus, recommendations for the protection of research participants developed in this context usually call for close IRB oversight, detailed informed consent disclosures, and counseling by a genetic counselor or medical geneticist (10).

Highly penetrant gene variants ideally should also be studied in a population-based setting. Although *BRCA1* mutations are thought to account for less than 5% of all breast cancer (11), quantifying their impact in the general population could help scientists understand the risks, mechanisms, and natural history of breast cancer in general. The most important contribution of population-based research, however, will be to elucidate the interactions between common, lower-penetrance gene variants and modifiable environmental factors that influence the probability of disease. Identifying these kinds of genetic factors may help illuminate the biologic processes responsible for disease, leading to new avenues for prevention or treatment. Increasing knowledge about gene–environment interactions may one day enable us to provide individuals with more accurate predictions of their disease risk as well as more effective and tailored interventions—such as more frequent or earlier medical surveillance, lifestyle or dietary modifications, or targeted drug therapy (12).

To move this research agenda forward, it is important to have guidelines to protect participants from risks specifically associated with integrating genetic variation into population-based research. Such research would be expected to have meaningful public health implications but involve a low probability of physical, psychological, or social harm to individual research participants. Guidelines developed in the context of family-based research generally fail to distinguish between research on gene variants known to carry high risk for disease and research on lower-risk genotypes that would allow a better understanding of disease processes and the role of environmental exposures (13). As noted by Clayton et al. (14), the risks involved in identifying high-penetrance mutations must be distinguished from the risks of identifying "common alleles that are neither necessary nor sufficient for the development of disease." The uniform application of guidelines for research on high-risk genotypes to all research with a genetic component, without adjustment for the type or degree of risk involved, could render many studies of gene–environment interaction impossible to conduct and deprive society of important and beneficial knowledge.

Application of Guidelines for Research Involving Human Tissue

One approach to clarifying the obligations of investigators to participants in population-based research involving genetics is to draw on guidelines for research involving human tissue. Genetic analysis in any context requires human biologic material. Numerous collections of such material already exist; public health examples include residual dried bloodspots from newborn screening (15) and specimens collected as part of population-based studies, such as the National Health and Nutrition Examination Survey (NHANES) (16) and the Framingham (17) and Jackson (18) heart studies. New public health collections will be created at an increasing rate as their value for genetic research is recognized; for example, the storage of specimens collected during the investigation of infectious diseases (19).

Many groups have published statements regarding the appropriate research use of human tissue (14,20,21,22). The National Bioethics Advisory Commission

(NBAC) examined this topic in its August 1999 report, *Research Involving Human Biological Materials: Ethical Issues and Policy Guidance* (23). This report interpreted several concepts and terms in the Common Rule and recommended ways to both strengthen and clarify existing regulations with regard to research using human tissue.

The following discussion does not provide a comprehensive review of the extant literature on human tissue research, but uses NBAC recommendations to highlight some of the ethical, legal, and social issues that arise specifically in the context of research on the interactions between lower-penetrance gene variants and environmental factors.

Issues in Human Genome Epidemiologic Research

Is the Research Subject to Human Participants Regulations?

For investigators conducting research using human biologic materials, knowing whether and how policies intended to protect human participants apply may sometimes be difficult. With certain exceptions described below, the Common Rule applies to virtually all research involving human participants conducted or supported by the federal government (45 CFR 46.101(a)). Non–federally funded research may be controlled by federal policy if the research is subject to regulation by the Food and Drug Administration, or if it is conducted at an institution that has voluntarily agreed to apply Common Rule requirements to all its research.

"Human participant" is defined as a living individual about whom an investigator conducting research obtains (*1*) data through intervention or interaction with the individual, or (*2*) identifiable private information (45 CFR 46.102(f)). Population-based research in which investigators collect *new* biologic samples for genetic analysis clearly entails an interaction with individual participants. Population-based research using *existing* biologic samples may not require an intervention or interaction but may involve identifiable information to correlate genetic findings with health outcomes. For instance, investigators used samples and health data from the Atherosclerosis Risk in Communities cohort to measure the associations between apolipoprotein E polymorphisms, physical and lifestyle factors, and the prevalence of significant carotid artery atherosclerosis (24). Still other kinds of population-based research can be conducted using existing biologic materials with no identifiable information and no interaction with participants. One example is a study in New York for which researchers used anonymized residual newborn blood spots to determine in different populations the prevalence of two interleukin-4 receptor polymorphisms thought to be associated with asthma (25). In all of these studies, the risks to participants are primarily related to the disclosure of private information, and the level of protection needed is directly related to the ease and likelihood of identifying the sources of the samples. NBAC's report thus recommends that the Common Rule be interpreted such that the fol-

lowing categories of biological samples may be used without informed consent or
IRB review:

Samples That Are Unidentified. According to NBAC, unidentified (or
"anonymous") samples are those created from specimens for which identifiable personal information was never collected or was not maintained and cannot be retrieved
by the repository. They advise that research using unidentified samples does not
meet the definition of human participants research and is not regulated by the Common Rule (NBAC Rec. 1A).

Samples That Have Been Unlinked through a Sound Process. Research
conducted with unlinked (or "anonymized") samples is human participants research
and is regulated by the Common Rule but is eligible for exemption from IRB review pursuant to 45 CFR 46.101(b)(4) (NBAC Rec. 1B). According to this section
of the federal regulations, research activities that are eligible for exemption include:

> Research involving the collection or study of *existing* data, documents, records,
> pathologic specimens, or diagnostic specimens, if these sources are *publicly
> available* or if the information is recorded by the investigator in such a manner
> that *subjects cannot be identified, directly or through identifiers* linked to the
> subjects [emphasis added].

See "Unlinking Samples" for further discussion of risks and potential benefits associated with anonymizing samples.

Samples That Are Publicly Available. Research using coded or directly
identified samples is human participants research, and because participants can be
identified, it is not eligible for the above-referenced exemption unless the specimens
are publicly available (NBAC Rec. 1C). Generally, the justification for exempting
research using publicly available data or specimens is that investigators are not obtaining private information when examining resources that are already in the public domain, and any harm from its disclosure has already occurred. However, exactly what constitutes "publicly available" under the Common Rule has been a matter
of some confusion. For example, some may consider NHANES specimens to be a
public resource because they are collected and banked by an agency of the federal
government. However, samples from NHANES are in fact available only to quali-
fied researchers and federal laws strictly protect the confidentiality of all NHANES
health data and samples (16). NBAC concludes that in the case of human biologic
materials, the term *publicly available* should be defined literally as only those materials available to the general public (p. 59). They also note that any data that emerge
from genetic analysis of stored tissue were not previously "existing" in any genuine
sense, much less publicly available. Thus, the exemption for research using existing specimens from a publicly available source will rarely, if ever, apply to population-based research involving genetics.

Samples from Participants Who Are Deceased. Because the Common Rule defines a human participant as a living person, there is no human participant if the source of the biologic material is deceased. No IRB review or informed consent is required for research using samples from participants who are deceased, but NBAC suggests that because research on such samples may have implications for living relatives, investigators should to the extent possible plan their studies to avoid harms to these individuals (NBAC Rec. 18).

Although NBAC notes *legal* possibilities of exemption under the Common Rule, many institutions require some form of IRB review even for technically exempt studies. The goal of IRB oversight is to ensure that risks to participants are minimized, that risks are reasonable in relation to anticipated benefits, that selection of participants is equitable, and that informed consent procedures are adequate (45 CFR 46.111(a)). It can be problematic to charge researchers with deciding that their own projects do not need IRB review; in some cases, the potential for conflict of interest or lack of detailed knowledge about federal regulations or local policies may exist (26). The Common Rule provides for an expedited review procedure, under which the IRB chairperson or designee can exercise all of the authorities of the IRB except disapproving the research (45 CFR 46.110).

Unlinking Samples

Because the risks of genetic research are related to the disclosure of private information, unlinking biologic samples from identifying information essentially removes the potential for harm. Truly "anonymizing" genetic information may not be possible because one could theoretically always match a known person to his or her research sample. However, at present, it would be extremely difficult to match a research sample back to a particular person without identifiers or another DNA sample from that person.

However, unlinking samples also raises a number of concerns. These include the administrative cost of making the samples completely unidentifiable (27), the possibility of compromising the value of the research (28), and the potential ethical objection that investigators have the opportunity to seek consent but instead unlinked the samples (14). These concerns may be especially relevant for population-based research involving genetics. For example, the value of some studies may be reduced if the only way to render the samples unidentifiable is to limit linkage to demographic and exposure data. An investigation of an outbreak of leptospirosis among triathletes (29,30) offered a unique situation in which a number of exceedingly healthy people were nearly uniformly exposed to contaminated water, thereby facilitating research on genetic susceptibility factors. However, consent for such research had not been obtained and all information regarding previously identified environmental risk factors (swim time and approximate amount of water ingested) would have had to be dropped to completely unlink the samples.

According to NBAC, when a researcher proposes to create unlinked samples from identifiable materials already under his or her control, the research may be exempt from IRB review if (*1*) the process used to unlink the samples will be effective, and (*2*) unlinking the samples will not unnecessarily reduce the value of the research (NBAC Rec. 3). NBAC did not put forward a preferred method for unlinking samples but instead chose the word "effective" to allow for IRB judgment in ensuring the appropriateness of any method to reduce the risk of harm. There is a continuum between directly identified and completely anonymous tissues or data, many points on which can deter the ability to identify participants whose tissue or data remain in some sense linked (21) (see Privacy and Confidentiality below). One should not assume that risk can or must be completely eliminated. As Buchanan (27) notes, "It is a mistake to proceed on the assumption that the goal is to develop policies that reduce the risk . . . to zero, as if risk reduction were costless," i.e., as if the costs of reducing risk did not include setbacks to important and morally legitimate interests, including improving health and preventing harm through the application of scientific knowledge in health care.

Assessing Risk

The *Belmont Report* formulated two complementary expressions of the ethical principle of beneficence: (*1*) do not harm, and (*2*) maximize possible benefits and minimize possible harms (9). A major responsibility of IRBs is to assess the risks and anticipated benefits of proposed research, both to participants and to society (32), and a part of this assessment is to determine whether the research presents greater than minimal risk. Minimal risk is a pivotal concept with regard to eligibility for expedited review and to the possibility of waiver or alteration of consent requirements.

The Common Rule defines "minimal risk" to mean that the probability and magnitude of harm anticipated in the research are not greater than those ordinarily encountered in daily life or during the performance of routine examinations (45 CFR 46.102(i)). It can be difficult for IRBs and investigators to operationalize this definition and in particular to quantify the risks associated with health information—including genetic information—relative to the risks of daily life. NBAC discusses this difficulty at some length, but nonetheless believes "most research using human biological materials is likely to be considered of minimal risk because much of it focuses on research that is not clinically relevant to the sample source (p. 67)." Likewise, most population-based research involving genetics will not be expected to reveal clinically relevant information. We are in the infancy of the "genetic revolution," and much is unknown. Establishing associations between genes and disease in the general population begins with quantifying statistical relationships, and even those that appear to be significant cannot be applied with any precision to particular individuals. The interpretation of epidemiologic data requires a chain of evidence to evaluate the presence of a valid association and to support a considered judgment as to cause and effect (33), and any single study is but one component.

In addition, many population-based studies will focus on lower-penetrance gene variants. Lower-penetrance gene variants by definition lead to smaller increases in relative risk for disease and, absent misplaced notions of genetic determinism, a corresponding decrease in the probability of harms stemming from the misuse of the information. Steinberg et al. (34) note that testing identifiable specimens for relatively benign polymorphisms that have a small impact on disease risk should entail less risk for loss of insurance, psychological distress, or social stigmatization than testing for the genetic mutations that almost ensure development of a serious disease. Others similarly argue that risk factors such as blood pressure or cholesterol level show comparable patterns of incomplete penetrance, and there is little reason that risk factors based on DNA should not be treated the same way (35). Thus, in risk assessment, it is important to assess not only the likelihood that individual results could be inappropriately disclosed, but also the probability and magnitude of the harm that could realistically result if such disclosure occurred. NBAC suggests that IRBs consider the following questions when determining the extent to which the source of a sample could be harmed (p. 67):

- How easily identifiable is the source?
- What is the likelihood that the source will be traced?
- If the source is traced, what is the likelihood that persons other than the investigators will obtain information about the source?
- If noninvestigators obtain information regarding the source, what is the likelihood that harm will result?

NBAC recommends that IRBs operate on the presumption that research on coded samples is of minimal risk when adequate safeguards are in place to protect confidentiality (NBAC Rec. 10).

Privacy and Confidentiality
According to the Office for Human Research Protections' IRB Guidebook (32), privacy can be defined as having control over the extent, timing, and circumstances of sharing oneself (physically, behaviorally, or intellectually) with others. Confidentiality pertains to the treatment of information that an individual has disclosed in a relationship of trust with the expectation that it will not be divulged to others in ways that are inconsistent with the understanding of the original disclosure without permission.

Risks that arise from the potential for invasion of privacy or breach of confidentiality may pose the possibility of serious harm, as could happen if sensitive personal information is disclosed to third parties who misuse the information to discriminate against or stigmatize research participants. These kinds of concerns are often expressed with regard to genetic information, but they are also of concern with regard to many other kinds of health information, e.g., a diagnosis of angina or diabetes, HIV status, or history of treatment for drug addiction. Informational risks may also take the form of "wrongs," rather than harms, such as treating people solely as a means to an end (36). Wrongs are often perceived to be less grave than harms, although they may infringe on the ethical principles of respect for persons and beneficence. When addressing harms and wrongs, it is important to try to craft a balance that not only respects individual interests in privacy and avoiding discrimination but also considers societal interests in clinical benefit, research, and public health (2).

Assessing the adequacy of safeguards to protect the privacy and confidentiality of all health information, including genetic information, requires attention to a number of factors. These include physical, technologic, organizational, and administrative practices and procedures, as well as legal protections (e.g., Certificates of Confidentiality). Detailed manuals are available from agencies such as the National Center for Health Statistics of the Centers for Disease Control and Prevention that have extensive experience in collecting and maintaining confidential health information (37). Absolute privacy of health information can never be ensured, even with maximal security protections, because no system can safeguard against access by those who are authorized to use the data system (2). Consent documents should provide details about measures that will be taken to protect privacy, as well as disclose that confidentiality cannot be guaranteed.

Group Harms

The ethical principle of justice and corresponding federal policy require that research participants be selected in an equitable manner (45 CFR 46.111(a)(3)), so that the burdens and benefits of research are fairly distributed. Investigators conducting population-based research involving genetics may recruit participants from particular groups of people because of differences in disease prevalence. Genetic research has been conducted among Native Americans, for example, because of the high prevalence of type II diabetes in this population (38). Study findings may one day benefit the groups that are the focus of such research, as well as allow a more comprehensive evaluation of the ethical, political, and social factors that influence health in human populations (39). At the same time, when the groups studied are socially defined (e.g., by race or ethnicity), research on genetic susceptibilities could be used to rationalize prejudices and perpetuate discrimination against or stigmatization of the group as a whole. This raises the possibility of harm to participants and nonparticipants alike, even when the increase in disease risk for the individual is small. As the *Belmont Report* points out, "Injustice may appear in the selection of subjects, even if individual subjects are selected fairly by investigators and treated fairly in the course of research. This injustice arises from social, racial, sexual and cultural biases institutionalized in society (9)."

Current regulations for protecting research participants address risks and benefits to identifiable individuals. IRBs may consider group harms and should consider consulting group members about cultural and other issues the research may raise. Foster et al. (40,41) described processes for communal discourse as a supplement to informed consent, although others have criticized the premise of "groups as gatekeepers" and pointed out logistic problems, particularly when research takes place outside the context of small groups that have a well-defined leadership structure (42,43). Although communities may be consulted, the burden of considering group implications falls primarily on the participants themselves (44), and NBAC recommends that reasonably foreseeable risks to groups should be disclosed in informed consent documents (NBAC Rec. 19).

Informed Consent

The process of developing and implementing informed consent procedures is intended to ensure that each prospective participant receives sufficient information, in a comprehensible manner and under conditions free of coercion and undue influence, to make an autonomous choice to participate or not participate in a research project. It also reminds researchers of their ethical obligations to participants (45). Once research is deemed to involve human participants, informed consent is required unless all four of the following criteria are met (45 CFR 46.116(d)):

The Research Involves No More Than Minimal Risk to the Participants. See "Assessing Risk" above.

The Waiver or Alteration Will Not Adversely Affect the Rights and Welfare of the Participants. Given the difficulties of interpreting minimal risk, the application of this criterion is vital in determining the appropriateness of a waiver (46). NBAC elaborates a number of ways in which the rights of research participants could be compromised (p. 68). For instance, participants' interests in controlling sensitive information and not being discriminated against could be jeopardized in some studies if existing samples are used without consent. Both participants' and nonparticipants' interests may be harmed by research that explicitly compares, for example, ethnic or racial groups; see Group Harms above. These concerns do not mean such studies should not be undertaken, but that they should be considered carefully before a waiver of consent is granted (NBAC Rec. 11).

The Research Could Not Practicably Be Carried out without the Waiver or Alteration. As with "publicly available," defining "practicably" can be difficult. Depending on the prevalence of the gene variant, the frequency of exposure to environmental risk factors, and the magnitude of the effect to be detected, gene–environment interaction studies may require hundreds or thousands of biologic samples (47). Existing stores of specimens represent a potentially valuable and efficient resource. Requiring without exception that the sources of these samples be recontacted to gain informed consent for each use could delay or halt some important research. At the same time, obtaining informed consent is a reflection of the ethical principle of respect for persons and the conviction that individuals who are capable of self-determination should be treated as autonomous agents (9). NBAC believes that the criteria of minimal risk and rights and welfare should be the compelling considerations in determining the appropriateness of waiving consent (p. 69). They suggest (NBAC Rec. 12) that IRBs may presume that meeting consent requirements would be impracticable when using *existing* coded or identified materials and when the research poses minimal risk. For *new* materials collected after the adoption of NBAC recommendations, IRBs should apply the recommended informed consent process and their usual standards for the practicability requirement.

Some disagree with the portion of this recommendation dealing with existing materials because of the challenges in evaluating minimal risk, and because it forfeits the opportunity for feedback from research participants in cases in which it *is* practicable to obtain consent (46). Grizzle et al. (22), for example, agree that IRBs should be permitted broader latitude to waive the requirement for informed consent—provided that written nondisclosure, confidentiality, and security policies have been approved—but that such a waiver should be granted on a case-by-case basis.

Whenever Appropriate, the Participants Will Be Provided with Additional Pertinent Information after Participation. NBAC finds that this debriefing requirement, which has a historic basis in "deception studies" used in behavioral sciences, is usually not relevant to research involving human biologic materials and may even cause harm (NBAC Rec. 13).

Consent Disclosures

The Common Rule specifies eight required elements of information (and six optional elements) that must be conveyed to prospective participants (45 CFR 46.116). Because genetic research evolved in the context of family studies of highly penetrant genetic mutations, recommendations concerning specific disclosures have understandably focused on the consequences of discoveries that have implications for the health of individual participants and their families. For example, the American Society of Human Genetics in 1996 recommended that disclosures to individuals considering participation in genetic studies should include, among other things, the potential implications of research results, the possibility of unexpected findings such as medical risks, carrier status, risks to offspring, and misidentified parentage, and the possibility of adverse psychologic sequelae, disruption of family dynamics, social stigmatization, and discrimination (10).

In studies of diseases involving low-penetrance gene variants and environmental exposures, where the possibility of confirmed and clinically relevant findings are remote, requiring some of these disclosures may be inappropriate. The investigator's charge is to neither understate nor overstate the risks involved (48), and tailoring the content of informed consent disclosures to convey risks and benefits that are material to the research at hand will promote prospective participants' ability to make informed choices about entering the study.

Consent for Storage and Future Research

When human biologic materials are collected, whether in a research or clinical setting, asking participants for their consent is appropriate if secondary uses are foreseen (23,32). Consent to future research is meaningful, however, only if participants appreciate the types of studies that may be conducted (23). NBAC recommends that consent forms be developed that provide a sufficient number of options to help prospective participants understand clearly the decision they are asked to make and presents six possible options (NBAC Rec. 9):

- Refusing use of their biological materials in research;
- Permitting unidentified or unlinked use only;
- Permitting coded/identified use for one study only;
- Permitting coded/identified use for one study and contact regarding further studies;
- Permitting coded/identified use for any studies related to the condition for which the sample was originally collected; or
- Permitting coded/identified use for any future study.

In some situations, offering this number of relatively imprecise options may be prohibitively complex. One alternative is for consent documents to state that investigators would like to store remaining samples for future research, describe plans for such research to the extent they are known, and provide participants the opportunity to consent or refuse (44). The description of future research plans should include whether the samples will be unlinked, coded, or directly identified. It could also include whether studies using the samples will be confined to a particular disease or class of diseases, who (or at least what types of investigators) will have access to the samples, and whether these investigators would be given unlinked or coded samples. Storing samples for future research is essentially a separate project, and thus consent forms should expressly state the right to refuse to have one's sample stored irrespective of the decision to participate in the immediate project (48).

The matter of consent for any future study will no doubt continue to engender controversy. One NBAC commissioner voiced concerns regarding the last two options set forth in Recommendation 9 because he believes neither adequate IRB review nor informed consent is possible in the context of unspecified future studies entailing unknown risks and benefits (p. 65). Public input about the individual and societal risks and benefits of population-based research involving genetics would be a valuable contribution to this debate. Little information is available about the opinions of the public at large or of minority groups about research uses of human tissue (49), and specifically about the storage and use of biologic samples for gene–environment interaction research and the application of genetics in disease prevention. In addition, as Feinleib (50) notes with regard to how explicit the description of proposed research must be to qualify for fully informed consent: "This is an eminently researchable area in its own right and would give valuable insight into how far it is desirable to go on providing full disclosures, what aspects are most persuasive to those who consent to participate, and what are the concerns of those who decline."

Reporting Research Results

Whether and when to disclose individual research results to participants are additional areas of debate. Although it is essentially universally accepted that individuals should have access to their own medical information, several key characteristics differentiate medical information from research data (51). However, some IRBs have held that investigators are obligated to offer participants in genetic research their individual results. This stance is perhaps based in part on justifiable concerns arising in the context of family-based research, but it can create serious problems when applied to some population-based studies.

First, the clinical validity of the results will not be known until a chain of evidence regarding risk associations has been established. Grizzle et al. (22) state, "A single research project does not establish irrefutable scientific fact, and the results of a single investigation have no applicability to an individual patient. Disclosure of a single research project's results to a patient is at best not beneficial, and at worst could be misleading or even harmful." Similarly, Armstrong (52) notes, "By definition, the study of low-penetrance genes seeks to find small effects that may be observed only in conjunction with certain exposures or in certain population subgroups. Dismissing results that are inconsistent or effect sizes that are small may miss the very truth that genetic epidemiology seeks to discover. At the same time, spurious associations are possible and even likely when hundreds of genotypes can be determined for a subject with only a single sample of blood or saliva."

Second, without independent confirmation, the analytic validity of the results may be in question. Current regulations require that results given to patients be performed in a laboratory certified under the Clinical Laboratory Improvement Amendments (CLIA). Because CLIA contains an exemption for "research laboratories that test human specimens but do not report patient specific results for the diagnosis, prevention or treatment of any disease" (42 CFR 493.3(b)(2)), many research laboratories are not so certified. This creates a quandary for investigators if they are expected to offer results. If a researcher discloses an individual's result specifically in response to a request under the Privacy Act or other applicable law (see below), that is generally not considered a "report for diagnosis" because it is being disclosed to comply with the law and not for the medical purposes set out under the exemption. However, the quandary remains with regard to any routine expectation that research results be offered when the research lab is not CLIA certified.

Third, when research involves existing biologic materials and consent has been waived, offering results is especially problematic. Genetic test results should never be given to a participant who does not want them (44). Results should not be returned unless a process of informed consent is in place that includes the opportunity for an informed decision not to receive results.

Finally, creating an ethical or legal obligation to provide research results to participants could confuse the role of the researcher, especially if the researcher is not a physician (31). Physicians are obligated to act in the best interests of their patients. To the extent that generalizable knowledge is generated and consequently the

standard of care is or may be altered, the researcher's "obligation" to participants—
to conduct good science and to disseminate his or her findings widely—is satisfied
(44).

According to NBAC, disclosure to individuals the results of research on their biologic materials should be an exceptional circumstance and should occur only when
all of the following apply (NBAC Rec. 14):

- the findings are scientifically valid and confirmed,
- the findings have significant implications for the participant's health concerns,
  and
- a course of action to ameliorate or treat these concerns is readily available.

NBAC further recommends that the investigator in his or her research protocol
should describe anticipated research findings and situations that might lead to a decision to disclose the findings to participants, as well as plans for how to manage
such disclosure (NBAC Rec. 15). They also recommend that when research results
are disclosed to a participant, appropriate medical advice or referral should be provided (NBAC Rec. 16).

Attempting to define the exceptional circumstances under which an after-the-fact
determination to offer results might be made could prove problematic. Informing
participants about these exceptions would be difficult, and researchers, IRBs, and
participants are apt to disagree about what constitutes a finding sufficiently certain
or significant to merit disclosure (44). One alternative is for investigators to communicate at the outset their plans for informing or not informing participants of their
results and then not deviate from that position (53). These plans can be developed
by assessing at the beginning of the research project the likelihood that clinically
relevant information will result, based on existing evidence and the aims of the study
(44). Merz et al. (31) state that "when information of potential clinical relevance is
a possible result of research activities, then a decision must be made prior to performing the research about whether to de-link the tissues for study or to secure detailed informed consent from subjects, specifically addressing the disclosure and use
of such information." Similarly, Fuller et al. (51) recommend:

> If a policy mandating return of clinically meaningless data were implemented,
> associated costs and personnel might provide an obstacle to doing the study.
> Thus, in the absence of clinical validity there should not be an absolute re-
> quirement for data to be returned to subjects. For research in which data are not
> provided to subjects, the researcher should demonstrate absence of clinical va-
> lidity, an IRB should be required to review and approve the exception, and the
> informed consent document should state explicitly that the data will not be re-
> turned to the research subject.

When individual research results are not disclosed, participants could be given
the option to receive an aggregate report of overall study results (14). In the rare

event that results unexpectedly have clinical significance, participants could still receive through the aggregate report any recommendation to consider testing for a particular trait in a clinical laboratory, without revealing individual results. In addition, participants who consent to storage and use of their biologic sample for future research could be given the option to receive aggregate reports about studies conducted with tissue from the "bank" where their sample is stored (44). The challenge will be to find ways of presenting research findings in lay language and to be clear about any clinical implications and their meanings in different populations.

This approach may need to be modified in certain instances to meet applicable laws. For example, when a study is subject to the Privacy Act, the Act provides individuals the right to review and get copies of their information (5 USC 552(a)(d)(1)). The Privacy Act applies only to situations in which records are maintained by a federal agency in a system of records. The term "system of records" refers to a group of records under the control of a federal agency from which information is retrieved by the name of the individual, identifying number, or some other identifying particular. When research is subject to the Privacy Act, it would be appropriate to inform individuals in advance of their right to see their information. However, although the Privacy Act permits an individual access to his or her records upon request, it does not command affirmative steps to disclose results absent such a request.

Commercialization

Some genetic studies may have the potential to result in commercial gain, for example from gene patents and genetic tests, and thus may create situations in which researchers have interests in conflict with those of research participants. In *Moore v. Regents of the University of California* (54), a patient suffering from hairy cell leukemia consented to have his spleen removed to slow progress of the disease. The patient's physician was later awarded a patent that included claims on a cell line derived from the patient's spleen. The patient filed suit, alleging, among other things, conversion (deprivation of another's property without his authorization) and lack of informed consent. The California Supreme Court held that the use of excised human cells in medical research does not amount to conversion and that the plaintiff had no rights in the patent. However, the Court noted that when soliciting consent a physician has a fiduciary duty to disclose all information material to the patient's decision and concluded that this disclosure must include personal interests unrelated to the patient's health, whether research or economic, that may affect the physician's medical judgment.

Although the potential for commercial gain may be remote in many population-based studies involving genetics, some participants may object to supporting any such gain or may be opposed in principle to the use of patents to secure intellectual property rights related to the human genome (45). Thus, any possibility of commercial gain should be disclosed in consent documents, along with a statement about whether participants would share in any profits (14).

Conclusions

The ability to use genetic information to benefit human health requires population-based data concerning the prevalence of gene variants, their associations with disease, and their interactions with modifiable risk factors (55). As gene discovery continues apace, effective interventions based on knowledge of gene–environment interaction must follow close behind. Without interventions, the risks of genetic information could outweigh the potential benefits for individuals and society, undermining the ultimate value of the Human Genome Project and public trust in the research enterprise.

Recommendations for protecting participants in genetic research have evolved in the context of studies on highly penetrant mutations in families with a heavy burden of disease. These recommendations are based on justifiable concerns, which all epidemiologists would share, about the misuse of highly predictive genotypes (13). However, the risks associated with population-based research on lower-penetrance gene variants and their interactions with environmental exposures may oftentimes be lower in probability and magnitude than those associated with family studies.

This chapter highlighted some of the ethical, legal, and social issues relevant to the design and conduct of human genome epidemiologic research. It explored these issues using ethical principles, federal policy for the protection of human research participants, and guidelines for research involving human tissue, particularly NBAC recommendations. In its report on human biologic materials, NBAC confined its recommendations to interpretations of existing federal regulations, but subsequently undertook a comprehensive assessment of the basic purpose, structure, and implementation of research oversight and recommended broad, strategic changes to the oversight system (56). Regardless of any regulatory or procedural changes that may occur, however, the same ethical principles will apply. Epidemiologists, genetics professionals, policy makers, and the general public can work together to develop recommendations for the protection of participants in population-based research that involves genetics. By protecting participants from the risks specific to such research, we promote our ability to fulfill the promise of genetic information to improve health and prevent disease.

Acknowledgments

Thanks to Dr. Muin Khoury and Dr. Marta Gwinn for their contributions to earlier versions of this paper. This project was supported under a cooperative agreement from the Centers for Disease Control and Prevention through the Association of Teachers of Preventive Medicine.

References

1. Beauchamp TL. Moral foundations. In: Ethics and Epidemiology. Coughlin SS, Beauchamp TL, eds. New York: Oxford University Press, 1996;24–52.

2. Gostin LO, Hodge JG. Genetic privacy and the law: an end to genetics exceptionalism. Jurimetrics Journal 1999;40:21–58.

3. Rothenberg KH. Breast cancer, the genetic "quick fix," and the Jewish community. Health Matrix Journal of Law-Medicine 1997;7:97–124.

4. Andrews LB, Fullarton JE, Holtzman NA, Motulsky AG (eds). Assessing Genetic Risks: Implications for Health and Social Policy. Washington DC: National Academy Press, 1994.

5. Holtzman NA, Watson MS (eds). Promoting Safe and Effective Genetic Testing in the United States: Final Report of the Task Force on Genetic Testing. Baltimore MD: Johns Hopkins University Press, 1998.

6. Beauchamp TL, Cook RR, Fayerweather WE, et al. Ethical guidelines for epidemiologists. J Clin Epidemiol 1991;44(Suppl I):151S–69S.

7. Council for International Organizations of Medical Sciences. International guidelines for ethical review of epidemiological studies. Law Med Health Care 1991;19:247–258.

8. World Health Organization 2002. Ethical issues in genetic research, screening and testing with particular reference to developing countries, in Genomics and World Health, Geneva: WHO, pp. 147–173.

9. National Commission for the Protection of Human Subjects of Biomedical and Behavioral Research. The Belmont Report: Ethical Principles and Guidelines for the Protection of Human Subjects of Research. Washington DC: US Government Printing Office, 1979.

10. American Society of Human Genetics. Statement on informed consent for genetic research. Am J Hum Genet 1996;59:471–474.

11. Coughlin SS, Khoury MJ, Steinberg KK. BRCA1 and BRCA2 gene mutations and risk of breast cancer: public health perspectives. Am J Prev Med 1999;16:91–98.

12. Collins FS, McKusick VA. Implications of the Human Genome Project for medical science. JAMA 2001;285:540–544.

13. Hunter D, Caporaso N. Informed consent in epidemiologic studies involving genetic markers. Epidemiology 1997;8:596–599.

14. Clayton EW, Steinberg KK, Khoury MJ, Thomson E, Andrews L, Kahn MJ, Kopelman LM, Weiss JO. Informed consent for genetic research on stored tissue samples. JAMA 1995;274:1786–1792.

15. McEwen JE, Reilly PR. Stored Guthrie cards as DNA "banks." Am J Hum Genet 1994;55:196–200.

16. National Center for Health Statistics. NHANES III DNA Specimens Guidelines for Proposals. Jun 26, 2001. Accessed Aug 24, 2002, from http://www.cdc.gov/nchs/about/major/nhanes/dnafnlgm2.htm.

17. Voelker R. Two generations of data aid Framingham's focus on genes. JAMA 1998;279:1245–1246.

18. Mitka M. New heart study a legacy for the future. JAMA 2000;283:38, 41, 44.

19. McNicholl JM, Cuenco KT. Host genes and infectious diseases: HIV, other pathogens, and a public health perspective. Am J Prev Med 1999;16:141–154.

20. American College of Medical Genetics Storage of Genetics Materials Committee. Statement on storage and use of genetic materials. Am J Hum Genet 1995;57:1499–1500.

21. Office for Protection from Research Risks. Issues to Consider in the Research Use of Stored Data or Tissues. Nov 1997. Accessed Aug 24, 2002 from http://ohrp.osophs.dhhs.gov/humansubjects/guidance/reposit.htm.

22. Grizzle W, Grody WW, Noll WW, Sobel ME, Stass SA, Trainer T, Travers H, Weedn V, Woodruff K. Recommended policies for uses of human tissue in research, education, and quality control. Ad Hoc Committee on Stored Tissue, College of American Pathologists. Arch Pathol Lab Med 1999;123:296–300.

23. NBAC. Research Involving Human Biological Materials: Ethical Issues and Policy Guidance—Vol. I. Rockville, MD: National Bioethics Advisory Commission, 1999.

24. de Andrade M, Thandi I, Brown S, Gotto AJr, Patsch W, Boerwinkle E. Relationship of the apolipoprotein E polymorphism with carotid artery atherosclerosis. Am J Hum Genet 1995;56:1379–1390.

25. Caggana M, Walker K, Reilly AA, Conroy JM, Duva S, Walsh AC. Population-based studies reveal differences in the allelic frequencies of two functionally significant human interleukin-4 receptor polymorphisms in several ethnic groups. Genet Med 1999; 1:267–271.

26. Merz JF. IRB review: necessary, nice, or needless? Ann Epidemiol 1998;8:479–481.

27. Buchanan A: An ethical framework for biological samples policy. In: Research Involving Human Biological Materials: Ethical Issues and Policy Guidance—Vol. II. Rockville, MD: National Bioethics Advisory Commission, 2000.

28. Human Genome Organisation Ethics Committee. Statement on DNA sampling: control and access. 1998. Accessed August 24, 2002, from http://www.hugo-international.org/hugo/sampling.html.

29. Outbreak of acute febrile illness among athletes participating in triathlons—Wisconsin and Illinois, 1998. MMWR Wkly Rep 1998;47:585–588.

30. Update: leptospirosis and unexplained acute febrile illness among athletes participating in triathlons—Illinois and Wisconsin, 1998. MMWR 1998;47:673–676.

31. Merz JF, Sankar P, Taube SE, Livolsi V. Use of human tissues in research: clarifying clinician and researcher roles and information flows. Journal of Investigative Medicine 1997;45:252–257.

32. Office for Human Research Protections. Protecting Human Research Subjects: Institutional Review Board Guidebook. 1993. Accessed August 24, 2002, from http://ohrp.osophs.dhhs.gov/irb/irb_guidebook.htm.

33. Hennekens CH, Buring JE, in Epidemiology in Medicine. Mayrent SL, ed. Boston MA: Little, Brown & Co. 1987.

34. Steinberg KK, Sanderlin KC, Ou CY, Hannon WH, McQuillan GM, Sampson EJ. DNA banking in epidemiologic studies. Epidemiol Rev 1997;19:156–162.

35. Bell J. The new genetics in clinical practice. BMJ 1998;316:618–620.

36. Capron AM. Protection of research subjects: do special rules apply in epidemiology? J Clin Epidemiol 1991;44:81S–89S.

37. NCHS. Staff Manual on Confidentiality. Hyattsville, MD: National Center for Health Statistics; DHHS Publ No. (PHS) 84–1244, 1984.

38. Hanson RL, Ehm MG, Pettitt DJ, Prochazka M, Thompson DB, Timberlake D, Foroud T, Kobes S, Baier L, Burns DK, Almasy L, Blangero J, Garvey WT, Bennett PH, Knowler WC. An autosomal genomic scan for loci linked to type II diabetes mellitus and body-mass index in Pima Indians. Am J Hum Genet 1998;63:1130–1138.

39. Horgan J. The End of Science: Facing the Limits of Knowledge in the Twilight of the Scientific Age. New York: Broadway Books, 1997.

40. Foster MW, Eisenbraun AJ, Carter TH. Communal discourse as a supplement to informed consent for genetic research. Nat Genet 1997;17:277–279.

41. Foster MW, Bernsten D, Carter TH. A model agreement for genetic research in socially identifiable populations. Am J Hum Genet 1998;63:696–702.

42. Juengst ET. Groups as gatekeepers to genomic research: conceptually confusing, morally hazardous, and practically useless. Kennedy Inst Ethics J 1998;8:183.

43. Reilly PR. Rethinking risks to human subjects in genetic research. Am J Hum Genet. 1998;63:682–685.

44. Beskow LM , Burke W, Merz JF, Barr PA, Terry, S, Penchaszadeh VB, Gostin LO, Gwinn M, Khoury MJ. Informed consent for population-based research involving genetics. JAMA. 2001;286:2315–2321.
45. Reilly PR, Boshar MF, Holtzman SH. Ethical issues in genetic research: disclosure and informed consent. Nat Genet 1997;15:16–20.
46. Woodward B. Challenges to human subject protections in US medical research. JAMA 1999;282:1947–1952.
47. Yang Q, Khoury MJ: Evolving methods in genetic epidemiology. III. Gene-environment interaction in epidemiologic research. Epidemiol Rev 1997;19:33–43.
48. Centers for Disease Control and Prevention. Consent for CDC research: a reference for developing consent forms and oral scripts. Nov 1998. Accessed August 24, 2002, from http://www.cdc.gov/od/ads/hsrconsent.pdf.
49. Wertz DC: Archived specimens: a platform for discussion. Community Genet 1999; 2:51–60.
50. Feinleib M. The epidemiologist's responsibilities to study participants. J Clin Epidemiol 1991;44:73S–79S.
51. Fuller BP, Ellis Kahn MJ, Barr PA, Biesecker L, Crowley E, Garber J, Mansoura MK, Murphy P, Murray J, Phillips J, Rothenberg K, Rothstein M, Stopfer J, Swergold G, Weber B, Collins FS, Hudson KL. Privacy in genetics research. Science 1999;285: 1360–1361.
52. Armstrong K. Genetic susceptibility to breast cancer: from the roll of the dice to the hand women were dealt. JAMA 2001;285:2907–2909.
53. Holtzman NA, Andrews LB. Ethical and legal issues in genetic epidemiology. Epidemiol Rev 1997;19:163–174.
54. Moore v. Regents of the University of California. 1990;793 P.2d 479.
55. Khoury MJ, Dorman JS. The Human Genome Epidemiology Network (HuGE Net). Am J Epidemiol 1998;148:1–3.
56. National Bioethics Advisory Commission. Ethical and Policy Issues in Research Involving Human Participants—Vol. I. Bethesda, MD: National Bioethics Advisory Commission; 2001.

II

METHODS AND APPROACHES I: ASSESSING DISEASE ASSOCIATIONS AND INTERACTIONS

5

Assessing genotypes in human genome epidemiology studies

Karen Steinberg and Margaret Gallagher

As the Human Genome Project provides the foundation for understanding the genetic basis of common disease (1), population-based genetic studies will provide the information needed for the practical application of genetic risk factors to clinical and public health practice. To this end, researchers have begun collecting specimens for molecular analyses in epidemiologic studies and surveys in order to identify genetic risk factors for disease (2). Genomic markers including restriction fragment length polymorphisms (RFLPs), short tandem repeats (STRs, also called microsatellites), insertion–deletion polymorphisms, single nucleotide polymorphisms (SNPs) and groups of markers inherited together on one chromosome as haplotypes are being used to locate disease-associated genetic loci, and studies of the association between these loci and disease are elucidating the genetic basis for disease. Once risk-associated genotypes are identified, the validity of genetic testing for screening and clinical practice must be assessed. This includes analytic and clinical validity of genotyping methods. Here we address factors to be considered in choosing appropriate specimens for epidemiologic studies, some quality assurance issues, and analytic validity. Clinical utility, which is the ultimate standard by which to evaluate case management on the basis of the result of a laboratory assay, is addressed elsewhere in this volume.

Specimen Selection

Factors to be considered in choosing appropriate specimens for epidemiologic studies, include cost, convenience of collection and storage, quantity and quality of DNA, and the ability to accommodate future needs for genotyping. Four types of specimens are commonly collected in epidemiologic studies with a genetics component:

(*1*) dried blood spots, (*2*) whole blood from which genomic DNA is stored or extracted using either anticoagulated or clotted blood, or buffy coats, (*3*) whole blood from which lymphocytes are isolated and immediately immortalized or cryopreserved for later immortalization, and (*4*) buccal epithelial cells.

Blood spots are a stable, inexpensive source of DNA, useful for genotyping polymorphisms for association studies (2). The stability of stored blood spots makes them a potential specimen source for population-based studies. Although specimens collected in newborn screening programs can serve as samples from which to determine population gene frequencies, use of these specimens in any way other than anonymously is problematic because the specimens may not have been collected with adequate informed consent (3,4). Blood spots can be collected without a phlebotomist and safely transported by regular mail. In general, genotyping one locus requires from about 10 ng to as little as 2.5 ng per SNP given current technology, so that scores to hundreds of genotypes could be obtained from one blood spot (Table 5.1). With the advent of multiplex testing (genotyping several loci in one assay), and whole genome amplifications, these numbers can be increased (15).

Whole blood provides high quality genomic DNA in microgram quantities sufficient for current applications including genome scans, polymorphism discovery, and genotyping loci. Numerous genotyping methodologies have been developed and are based on hybridization (e.g., arrays), ligation (e.g., oligonucleotide ligation assay), polymerization (e.g., minisequencing) or cleavage (e.g., RFLP). Quantities of DNA ranging from 100 μg to 400 μg can be obtained from 10 mL of whole blood, and approximately 200 μg from 1 mL of buffy coat. Blood is most often collected using ethylenediaminetetraacetic acid (EDTA), although anticoagulants including heparin and acid citrate dextrose (ACD) have also been used. Cells can be stored in anticoagulated whole blood, in clots, or in buffy coats. Guidelines for obtaining these specimens are available (16). Polypropylene rather than glass containers should be used to store frozen blood, and blood should be divided into aliquots to prevent freeze-thaw cycles. Although evidence exists to suggest that lymphocytes can be transformed with EBV after cryopreservation (17), and optimization of transformation of small numbers of cryopreserved lymphocytes is an active area of research, at the time of writing, there is insufficient evidence to be confident that lymphocytes stored in whole blood for years can be consistently transformed.

For genotyping large numbers of polymorphisms requiring microgram quantities of DNA, EBV-transformed lymphocytes provide an unlimited source of high-quality genomic DNA. Although transformed lymphocytes may provide specimens for functional studies, properties of EBV-transformed lymphocytes may be different from those of untransformed lymphocytes, and lymphocyte gene expression may not be representative of the expression patterns in the target tissue of interest. Because of the expense associated with establishing and maintaining immortalized cell lines, many investigators are attempting to cryopreserve lymphocytes for later immortalization of selected specimens in nested-case control studies. However, the expense of establishing and maintaining cell cultures has resulted in a trend toward

Table 5.1 Comparison of Specimens for DNA Banking for Epidemiologic Studies

Specimen Type	DNA Yield	Advantages	Disadvantages
Blood spots	12–42 ng/μL (adults)[a] 43–78 ng/μL (neonates)[a] 1/4 inch punch from 75 μL volume yields about 12 μL of blood[b]	Small sample size Ease of sample collection Ease of shipping (regular mail) Stability and low cost storage Offers a source for study of exogenous or endogenous compounds other than DNA Genotyping generally requires 10 ng/genotype, and with current technology as little as 2.5 ng per SNP, so that scores to hundreds of genotypes could be obtained from one blood spot	Low DNA yield: may not be suitable for whole genome amplification. Nonrenewable Smaller amplicons
Whole blood (anticoagulated or blood clots)	100–400 μg/10 mL[c] ~200 μg/mL[c]	Relatively low-cost storage (−80°C) Yields large quantities of high-quality, genomic DNA	Invasive sample collection Shipping (special requirements)
Buffy coat		Offers a source for study of exogenous or endogenous compounds other than DNA	Nonrenewable
Transformed lymphocytes	10[6] cells = 6 μg[c] 1–2 × 10[6] cells = 5–10 μg	Renewable source of DNA Yields large quantities of high-quality, genomic DNA	Labor intensive preparation High-cost storage (liquid nitrogen-and periodic reculture) Does not offer a source for study of exogenous or endogenous compounds other than DNA or RNA

(continued)

Table 5.1 Comparison of Specimens for DNA Banking for Epidemiologic Studies (continued)

Specimen Type	DNA Yield	Advantages	Disadvantages
Buccal cells	49.7 μg mean; 0.2–134 μg range (mouthwash-total DNA)[d] 12–60 μg range (mouthwash–total DNA)[e] ~16–30 μg median; 1–290 μg range (mouthwash-hDNA)[f] 32 μg median; 4–196 μg range (mouthwash-hDNA)[g] 1–1.6 μg/2 cytobrushes⁻ median; 6 ng–13 μg range(hDNA)[e] 1–2 μg/cytobrush (total DNA)[h] 1–2 μg/ swab (total DNA)[h]	Non-invasive collection Ease of sample collection and shipping (allows participant to collect and mail specimen). Genotyping generally requires 10 ng/genotype, and with current technology as little as 2.5 ng per SNP, so that many thousands of genotypes could theoretically be obtained from a buccal cell specimen	Low DNA yield: not in general use for whole genome amplification Highly variable yield Does not offer a source for study of exogenous or endogenous compounds other than DNA or RNA Bacterial contamination must be addressed

[a]Ref. 5.
[b]Ref. 6.
[c]Refs. 7–10.
[d]Ref. 11.
[e]Ref. 12.
[f]Ref. 13.
[g]Ref. 14.
[h]Ref. 10.

storing whole blood for obtaining large quantities of genomic DNA. When lymphocytes are held in culture, they should be monitored for contamination with mycoplasma, bacteria, and fungi, and original specimens (e.g. blood spots, whole blood, or extracted DNA) should be maintained for identity checks. From 5 μg to 10 μg of DNA can be obtained from $1-2 \times 10^6$ cells.

Buccal cells can be obtained for DNA isolation using cytobrushes, swabs, or oral lavage. Although there are few systematic studies that compare yield of human DNA (hDNA) from buccal cells (excluding bacterial contamination) using different collection methods, there is a growing consensus that the use of mouthwash used to obtain cells yields more and higher quality DNA, in the range of 5 μg to 100 μg, than swabs or cytobrushes, which yield DNA in the range of 1 μg to 2 μg per cytobrush or swab. However, swabs or cytobrushes are necessary for collecting specimens from infants and small children (Table 5.1).

Quality Control for Molecular Methods

We define quality control as the inclusion of characterized specimens in analytic runs to ensure the correct performance of a method and the quality of the resulting data. This discussion does not include the broader issue of quality assurance which subsumes quality control and includes standards of professional qualifications for personnel performing and interpreting genetic tests as well as standards for interpretation in the clinical context. Quality control generally entails the inclusion of positive and negative controls, reagent blanks, and duplicates in analytical runs to assess the precision of a method within a laboratory. Proficiency testing (PT), or external quality assessment (EQA) as it is also known, includes the external component of quality control in which unknown specimens from either a commercial source or outside laboratory are analyzed to assess consistency and accuracy among laboratories. We first discuss several ongoing programs to standardize and assure the quality of genetic testing through published recommendations or regulations. We then give specific recommendations for some of the more commonly used molecular methods.

Guidelines and Regulations

In practice, research studies that will report clinically relevant results should use laboratories that are held to the highest standard of practice. In the United Kingdom, the common practice is not to report results of diagnostic relevance generated as a part of research, but to have the test repeated by a diagnostic laboratory on a fresh specimen. The UK studies that are clinically based are expected to assure that laboratories performing tests are diagnostic laboratories or have equivalent standards of practice.

In the United States, a distinction is made between the quality control requirements for laboratories that perform tests for which results are reported for clinical use and those that perform tests as a part of research and for which results are not reported. Laboratories performing the former tests are regulated under the Clinical Laboratory Improvement Amendments (CLIA) of 1988 (http://www.hcfa.gov/medicaid/clia/cliahome.htm) and the latter are not (18). However, because results of genetic testing done as a part of clinical or epidemiologic research are sometimes reported to participants, this distinction cannot always be easily made. Further, CLIA may usefully provide guidelines for genetic testing done purely for research and not reported for clinical use. In addition to CLIA guidelines, guidelines are made available by states, such as New York (http://www.wadsworth.org/labcert/clep/Survey/Standards.pdf), and private organizations, such as the College of American Pathologists (CAP) (http://www.cap.org/html/ftpdirectory/checklistftp.html) and the American College of Medical Genetics (ACMG) (http://www.kumc.edu/gec/prof/acmg.html). Manuals which provide detailed discussions of quality control for molecular methods include NCCLS[1] (19) and that of Saunders and Parkes. (20)

Distinctions between requirements for quality assurance for laboratories that report results and those that do not notwithstanding, the quality of research data that will be the foundation for clinical practice depends on implementation of quality control in research as well as clinical laboratories. Further, quality standards are usually promulgated for laboratories that report results rather than those that do not. Therefore, any discussion of quality control should include regulations and recommendations intended for clinical laboratories. In this regard, most developed countries have systems for accrediting laboratories on the basis of government regulations or professional guidelines. However, most are still in the process of developing specific standards for molecular genetic laboratories.

The Centers for Disease Control and Prevention (CDC), as a part of its mandate to implement CLIA, has funded contract studies to produce recommendations for performance evaluation and quality assurance. An example is available of these recommendations for quality control strategies for laboratory genetic tests, including nucleic acid amplification by polymerase chain reaction (PCR), DNA sequencing, Southern blot analysis, and fluorescence in situ hybridization (FISH) (Table 5.2),

Table 5.2 Critical Steps in Molecular Methods Requiring Quality Control

Molecular Method[a]	Critical Step[a]	QC Method
Nucleic acid amplification PCR	Specimen acquistion	Identity check (e.g., microsatellites) and barcoding
(Other methods such as ligase chain reaction and cloning are not included)	DNA isolation	Assess yield by quantitation Assess quality by electrophoresis, endonuclease digestion, PCR
	PCR reagents Detection	Verify performance of reagents and primers Monitor sensitivity and specifity using positive controls, standards, negative controls

[a]Modified from CLIAC.

(http://www.phppo.cdc.gov/dls/genetics/qapt.asp). The ACMG Laboratory Practice Committee has also published practices standards for clinical genetics laboratories that prescribe general guidelines for laboratories and specific guidelines for molecular genetics, as well as cytogenetics, including FISH, and biochemical genetics, which is in most respects the same as for clinical chemistry laboratories except in the more extensive interpretation that is required for results of biochemical genetic testing (http://www.faseb.org/genetics/acmg/stds/copyrite.htm). ACMG guidelines for molecular genetic methods include details on quality control for DNA preparation, probe/primer/locus documentation, assay validation, Southern blot analysis, and PCR methods including containment and amplification conditions, product detection and analysis, and use of controls and standards.

Proficiency testing should be a component of all laboratory quality control programs. The College of American Pathologists provides a voluntary Laboratory Accreditation Program, and CAP and the ACMG jointly operate PT programs in genetics that provide materials for approximately 17 different mutations that cause single-gene disorders including cystic fibrosis, factor V Leiden deficiency, Duchenne muscular dystrophy (DMD)/Becker, rhesus monkey antigen D (RhD), Prader-Willi/Angelman syndrome, Huntington disease, fragile X syndrome, hereditary hemochromatosis, hemoglobin S/C (sickle cell disease), myotonic dystrophy, type 1 (DM1), Friedrich ataxia, prothrombin, spinocerebellar atrophy, spinal muscular atrophy, methylene tetrahydrofolate reductase, *BRCA1* and *BRCA2*, and multiple endocrine neoplasia (MEN)2 (21).

CLIA requires laboratories performing tests that are not included in available PT programs to have a system for verifying the accuracy of the test results at least twice a year. Although laboratory participation in the CAP Molecular Genetics survey is currently voluntary, many laboratories performing DNA-based genetic testing elect to participate in CAP surveys to meet the CLIA quality assurance requirement. The National Institute of Standards and Technology provides human DNA standard reference materials for forensic as well as clinical applications which include standards for RFLPs, STRs, and amplification and sequencing of mitochondrial DNA (22).

In the absence of PT materials for gene variants of interest, which is the rule rather than the exception in the research setting, exchange of specimens among laboratories is an acceptable means to test consistency among laboratories (23–25). Accuracy can also be assessed in this way when the PT materials have been well-characterized by a reference method.

In conjunction with the quality assurance efforts of individual nations including those of Europe, Australia, Japan, Korea, Mexico, New Zealand, the United States and others, the Organisation for Economic Co-operation and Development (OECD) held a workshop in Vienna in 2001 "to consider whether the approaches of OECD member countries for dealing with new genetic tests are appropriate and mutually compatible" (http://www1.oecd.org/dsti/sti/s_t/biotech/act/gentest.pdf). One of the main considerations of the workshop was the development of international best practice policies for analytic and clinical validation of genetic tests. The EQA/PT component of quality assurance provides a means to measure laboratory results against an external gold standard. Because external quality assurance (EQA)/proficiency testing includes the laboratory's ability to interpret results in a clinical context as well as accurate test performance, EQA has been developed in a disease-specific fashion. In the United Kingdom EQA includes workshops held by representatives of participating laboratories to develop and publish best-practice guidelines which are made available by the Clinical Molecular Genetics Society (CMGS) (http://www.cmgs.org). Guidelines for 10 disorders were available in 2001 including breast cancer, Huntington disease, fragile X syndrome, Prader-Willi/Angelman syndrome, Charcot-Marie-Tooth disease, retinoblastoma, Duchenne muscular dystrophy, cystic fibrosis, Friedrich ataxia, and Y-chromosome microdeletions. These guidelines serve as the nucleus for guidelines funded by the European Commission and published by the European Molecular Genetics Quality Network (EMQN) (http://www.emqn.org/emqn.htm). In Europe, compliance with guidelines is still voluntary but could ultimately be required for accreditation for service as is true in the United States through CLIA and in the United Kingdom through Clinical Pathology Accreditation (http://www.cpa-uk.co.uk/) (OCED document pp. 41–42).

Because most of the quality assurance schemes in the above references are designed for clinical genetic tests, they are often disease specific. They do, however, provide many of the necessary generic components of quality control for molecular laboratories (e.g., guidelines for PCR) making them useful for the validation phase of test development before clinical testing is available and for genetic tests for rare disorders that are done in only a few laboratories.

Specific Recommendations

Genetic material is analyzed as part of epidemiologic studies for generally two purposes: (*1*) to test the significance of an association between a gene variant and a disease and (*2*) to use gene variants as markers for mapping other gene variants that are causal in disease. Methods most commonly used to localize gene variants associated with disease take advantage of the sequence variation (or polymorphisms) in populations. The most commonly used polymorphic markers include microsatellites and SNPs. Because of the large number of individuals who must be genotyped for a large number of polymorphisms in these studies (26), new methods are being developed to accommodate high-throughput analyses, to facilitate assay design, and to reduce costs. The newer methods often include array technology sometimes coupled to a mass spectrophotometric detection system.

Because of the rapid and continuing proliferation of molecular methods used in the research setting, and because others have furnished more detailed guidelines for quality control in genetic testing (19,20,27), we focus the remainder of our discussion on DNA extraction and characterization and analytic validity, because both are fundamental to all DNA-based methods. In most cases, DNA must be extracted and amplified before automated sequencing or polymorphism identification is done.

With regard to DNA amplification and genotyping, ideally, each step in the analysis should be performed in duplicate, from extraction and PCR to genotyping in order to determine the precision of the method. Reagent blanks should be included in all runs to identify the presence of contamination and obviate false positive results. If well-characterized control specimens are unavailable, DNA with the sequence of interest may be substituted to assure efficiency of the methods; however, this approach would be problematic when using array technology. In all cases, other than the so-called closed systems in which amplification and genotyping occur in one vessel, pre- and postamplification of DNA must be carried out in separate work areas to prevent contamination of specimens which can cause false positive results. Movement of specimens should be in one direction, from specimen preparation to PCR to genotyping with careful physical separation of sample preparation from extracted DNA and PCR reactions (28). Reagents should be made from molecular biology grade chemicals and reagent-quality water. Before they are judged acceptable, new reagents should be tested in the same assays with reagents currently in use that have been validated (29).

DNA Extraction and Characterization

Visvikis et al. (9) have divided issues related to DNA extraction into three steps comprising whole blood preservation, extraction procedures, and storage of DNA. DNA is stable in whole blood at room temperature for about 24 hours with only slight decreases in stability within 72 hours. Specimens held from 4 to 8 days before DNA extraction should be held at +4°C. Optimal yield is obtained from whole blood specimens that are processed before freezing. Extraction methods include use of (*1*) enzymes (including proteinase K and RNAse) (30), (*2*) organic solvents or organic solvents with enzymes (31), (*3*) salt precipitation (32,33), and (*4*) resins or affinity gels, which are the basis for many commercial kits. After extraction, DNA is resuspended in a buffer such as Tris buffer. The quality of extracted DNA is assessed by its yield, molecular weight, purity, and the ability to serve as a substrate for PCR and restriction enzymes (Table 5.2).

Yield and purity are most commonly estimated by using the ratio of optical absorbance at 260 nm to absorbance at 280 nm. Although convenient and usually sufficiently accurate for most applications, this method does not distinguish double-stranded DNA (dsDNA) from single-stranded DNA (ssDNA), does not distinguish between DNA and RNA, and is relatively insensitive. Also, contaminants may cause interference. If interfering substances are present, DNA can be more precisely quantified using one of the dsDNA-binding dye methods such as PicoGreen (Molecular Probes Inc, Eugene OR, http://www.probes.com), which is minimally affected by ssDNA, RNA, or protein.

Molecular weight is most commonly determined with the use of electrophoresis. Electrophoresis employs an electric current and a sieving matrix to separate molecules on the basis of their charge and size. Agarose is used for separation and sizing of large DNA fragments, and polyacrylamide gel can be used for smaller DNA fragments. In either case, the gel matrix acts as a molecular sieve that causes the separation of DNA fragments on the basis of size. The DNA fragments of standard molecular weight, obtained from commercial sources or developed and character-ized in the laboratory performing the tests, are included for comparison.

The DNA should be tested to confirm that it can serve as a substrate for restric-tion enzymes such as EcoR1 or Hind III. In the case of cell lines, the source of the DNA should be confirmed by comparing patterns of microsatellites between the processed specimen and an aliquot that was saved, for example as a blood spot, for identification purposes. Long-term storage of extracted DNA should be done at tem-peratures of $-20°C$ or $-70°C$, although DNA may be stable in suitable buffers at $4°C$ for years.

Analytic Validity

As is the case for all laboratory methods, analytical validity for DNA-based tests is the probability that a test will be positive when a target sequence is present (sensi-tivity) and that the test will be negative when that target sequence is absent (speci-ficity) and that the results using the same target sequence will be consistently re-produced (*precision, reproducibility, or reliability.*). These characteristics should be determined for each method used for genotyping or sequencing.

Analytic sensitivity can also be measured as the lowest concentration of the tar-get sequence that can be distinguished from background signal or noise and defines the limit of detection in the assay. In the case of genotyping, detection limits ap-proaching a single molecule are possible. The more sensitive an assay, the less likely false negative results will be obtained.

Analytic specificity is the probability of a positive result will occur only in the presence of the target sequence being measured. The more specific an assay, the less likely false-positive results will be obtained.

Summary

The type of specimen collected in epidemiologic studies will depend on the costs, study needs, and the laboratory experience and technology available to the investi-gators. Given that current technology can analyze a SNP in as little as 2.5 ng of DNA, all of the specimens described above should allow hundreds to thousands of analyses (2).

Genomic DNA extracted from whole blood for immediate use or storage assures that sufficient material will be available for most current and future molecular ap-plications at a cost that is sustainable. Blood spots are appropriate when protocols call for easier collection and room temperature, low-cost storage. Buccal cells al-low noninvasive collection that can be self-administered, and specimens can be mailed. Even though these specimens provide limited amounts of DNA with wide interindividual variation when buccal cells are collected, they can provide material sufficient for genotyping scores to thousands of loci. When a large amount of DNA is needed, such as for repeated or collaborative studies, lymphocytes might be transformed, provided that funding is sufficient. Lymphocytes should also be transformed when studies of gene expression using RNA or protein are needed. In this situation, the investigators should be aware of the potential for alteration of normal gene expression in cells that are transformed or otherwise manipulated in vitro. Although cryopreservation and later transformation of selected specimens could reduce the number of specimens to be transformed, the high costs of maintaining the cell lines that are created later is still a factor, and there are too little data to confirm that this strategy would ensure viable cell cultures upon transformation.

Although the emphasis on innovation in molecular genetic research makes quality control a moving target, basic rules of quality control can help assure quality results. Quality control measures generally include use of control materials, duplicate specimens, blanks, and proficiency testing. Most developed countries have guidelines or regulations for laboratory accreditation which include recommendations for technical proficiency, but most are still working to develop specific guidelines for genetic testing. These guidelines are generally the same for clinical and research laboratories. Proficiency testing is also an essential component of laboratory quality control. Currently available proficiency testing materials are designed for the more commonly performed clinical DNA-based tests. Comparable materials are usually not available for DNA-based tests being performed in epidemiologic studies of association between gene polymorphisms and disease. In this case, laboratories performing the tests can exchange material for external quality assurance. Nonetheless, quality control of molecular genetic methods is essential whether tests are performed for clinical decision making or to serve as the basis for hypothesis testing in research.

Notes

[1]On the NCCLS website (http://www.nccls.org/, accessed March 24, 2003), it is stated that the acronym NCCLS used to stand for National Committee for Clinical Laboratory Standards, but NCCLS is not a global organization and develops consensus documents for additional audiences beyond the clinical laboratory community. Therefore, the organization should be referred to by the acronym NCCLS.

References

1. Collins FS, Patrinos A, Jordan E, Chakravarti A, Gesteland R, Walters L. New goals for the U.S. Human Genome Project:1998–2003. Science 1998;282:6829.
2. Steinberg K, Beck J, Nickerson D, et al. DNA Banking for epidemiologic studies: a review of current practices. Epidemiology 2002;13:246–254.
3. Shafer FE, Lorey F, Cunningham GC, et al. Newborn screening for sickle cell disease: 4 years of experience from California's newborn screening program. J Ped Hematol Oncol 1996;18:36–41.
4. Therrell BL Jr., Hannon WH, Pass KA, et al. Guidelines for the retention, storage, and use of residual dried blood spot samples after newborn screening analysis: Statement of

the council of regional networks for genetic services. Biochem Mol Med 1996;57: 116–124.

5. Jinks DC, Minter M, Tarver DA, et al. Molecular genetic diagnosis of sickle cell disease using dried blood specimens on blotters used for newborn screening. Hum Genet 1989;81:363–366.

6. Mei JV, Alexander JR, Adam BW, et al. Use of filter paper for the collection and analysis of human whole blood specimens. J Nutr 2001;131:1631S–1636S.

7. Madisen L, Hoar DI, Holroyd CD, et al. DNA banking: the effects of storage of blood and isolated DNA on the integrity of DNA. Am J Med Genet 1987; Jun;27(2):379–390.

8. Cushwa WT, Medrano JF. Effects of blood storage time and temperature on DNA yield and quality. Biotechniques 1993 Feb;14(2):204–207.

9. Visvikis S, Schlenck A, Maurice M. DNA extraction and stability for epidemiological studies. Clin Chem Lab Med. 1998;36:551–555.

10. Whole blood, buffy coat, and cell values are a range derived from expected yields given by manufacturer. See www.gentra.com/product/puregene, www.qiagen.com/catalog/chapter-06, www.shpromega.com, and www.epicentre.com/catalogue.

11. Lum A, Le Marchand L. A simple mouthwash method for obtaining genomic DNA in molecular epidemiological studies. Cancer Epidemiol Biomarkers Prev 1998 7:719–724.

12. Heath EM, Morken NW, Campbell KA, Tkach D, Boyd EA, Strom DA. Use of buccal cells collected in mouthwash as a source of DNA for clinical testing. Arch Pathol Lab Med 2001;125(1):127–133.

13. Garcia-Closas M, Egan KM, Abruzzo J, Newcomb PA, Titus-Ernstoff L, Franklin T, Bender PK, Beck JC, Le Marchand L, Lum A, Alavanja M, Hayes RB, Rutter J, Buetow K, Brinton LA, Rothman N. Collection of genomic DNA from adults in epidemiological studies by buccal cytobrush and mouthwash. Cancer Epidemiol Biomarkers Prev 2001;10:687–696.

14. Feigelson HS, Rodriguez C, Robertson AS, et al. Determinants of DNA yield and quality from buccal cell samples collected with mouthwash. Cancer Epidemiol Biomarkers Prev 2001;10:1005–1008.

15. Hawkins TL, Detter JC, Richardson PM. Whole genome amplification—applications and advances. Curr Opin Biotechnol 2002;13(1):65–67.

16. Austin MA, Ordovas JM, Eckfeldt JH, Tracy R, Boerwinkle E, Lalouel J-M, Printz M. Guidelines of the National Heart, Lung, and Blood Institute Working Group on Blood Drawing, Processing, and Storage for Genetic Studies. Am J Epidemiol.1996; 144:437–441.

17. Beck JC, Beiswanger CM, John EM, et al. Successful transformation of cryopreserved lymphcytes: a resource for epidemiological studies. Cancer Epidemiol Biomarkers Prev 2001;10:551–554.

18. Andrews LB, Fullarton JE, Holtzman NA, eds. Assessing Genetic Risk: Implications for Health and Social Policy. Washington, DC: Washington National Acad Press. 1996; 116–145.

19. Altmiller DH, Gordon J, Grody WW, Matteson KJ, Murphy PD, Noll WW, Richards CS, Wesolowski A, Winn-Deen ES, Yoder FE. *Molecular Diagnostic Methods for Genetic Diseases; Approved Guideline.* NCCLS document MM1-A (ISBN 1-56238-395-7). NCCLS, Wayne, (PA) 2000;20:14–62.

20. Saunders G, Parkes H. Eds. Analytical molecular biology: quality and validation. Cambridge: RSC, 1999:1–72.

21. College of American Pathologists 2002; Molecular Genetics Survey, ACMG/CAP Molecular Genetics MGL1, MGL2, MGL3. Surveys & Educational Anatomic Pathology Programs. p. 149.

22. Levin BC, Cheng H, Kline MC, et al. A review of the DNA standard reference materials developed by the National Institute of Standards and Technology. Fresenius J Anal Chem 2001;370:213–219.
23. Pillai SD, Ricke SC. Strategies to accelerate the applicability of gene amplification protocols for pathogen detection in meat and meat products. Crit Rev Microbiol 1995;21(4):239–261.
24. Hurst CJ, Schaub SA, Sobsey MD, et al. Multilaboratory evaluation of methods for detecting enteric viruses in soils. Appl Environ Microbiol 1991 Feb;57(2):395–401.
25. Coon JS, Deitch AD, de Vere White RW, Koss LG, Melamed MR, Reeder JE, Weinstein RS, Wersto RP, Wheeless LL. Check samples for laboratory self-assessment in DNA flow cytometry. The National Cancer Institute's Flow Cytometry Network experience. Cancer 1989;63:1592–1599.
26. Kruglyak L, Nickerson DA. Variation is the spice of life. Nat Genet 2001;27:234–236.
27. Nollau P, Wagener C. International Federation of Clinical Chemistry Scientific Division Committee on Molecular and Biology Techniques. Clin Chem 1997;43:1114–1128.
28. Reichelderfer PS, Jackson JB. Quality assurance and use of PCR in clinical trials. PCR Methods Applications. 1994;4:S141–S149.
29. NCCLS document C3-A3. Preparation and Testing of Reagent Water in the Clinical Laboratory; Approved Guideline, third edition (Vol. 17, No. 18) 1997.
30. Lachaud L, Chabbert E, Dubessay P, Reynes J, Lamothe J, Bastien P. Comparison of various sample preparation methods for PCR diagnosis of visceral leishmaniasis using peripheral blood. J Clin Microbiol 2001;92:613–617.
31. Gross-Bellard M, Oudet P, Chambon P. Isolation of high-molecular-weight DNA from mammalian cells. Eur J Biochem 1973 Jul 2;36(1):32–38.
32. Miller SA, Dykes DD, Polesky HF. A simple salting out procedure for extracting DNA from human nucleated cells. Nucleic Acids Res 1988;16:1215.
33. Lahiri DK, Bye S, Nurnberger JI Jr, et al. A non-organic and non-enzymatic extraction method gives higher yields of genomic DNA from whole-blood samples than do nine other methods tested. J Biochem Biophys Methods 1992;25:193–205.

6

Statistical issues in the design and analysis of gene–disease association studies

Duncan C. Thomas

Genetic epidemiology comprises two broad types of activity that entail the use of biologic specimens: gene discovery and gene characterization. Studies of familial aggregation and segregation analysis are aimed at establishing a genetic component to a disease and inferring the mode of inheritance, but do not use any DNA analysis. Once evidence for the existence of one or more major genes has been found, geneticists use linkage analysis to localize these genes by identifying genetic markers at known chromosomal locations that appear to be transmitted within families in a manner that parallels the transmission of the disease. Once genes are localized in this manner, association studies can be used either to test hypotheses about possible candidate genes within the region or to further localize the region using linkage disequilibrium. Association studies are also used once a causal gene has been cloned to characterize its age-specific penetrance function and interactions with other factors. Following a brief review of methods for gene discovery to set the stage for a unified approach to gene discovery and gene characterization, the remainder of this chapter focuses on design and analysis issues in these various types of association studies. Throughout, we limit attention to binary disease traits, although many of the design and analysis issues also apply to continuous traits.

Although there is obviously a continuous spectrum of gene effects, we are accustomed to thinking in terms of two general types of genes that are potentially detectable by genetic epidemiologists: "major susceptibility genes" having a high penetrance, such mutations usually being rare in the general population; and "common low penetrance genes," such as those involved in metabolic activation and detoxification of carcinogens, DNA repair, and other complex pathways involving multiple genes and interactions with environmental agents. It is increasingly being recognized that the classical positional cloning approaches (i.e., linkage analysis) are more effective for discovery of major susceptibility genes than common low penetrance genes, and genome-wide association studies are now being suggested as an approach to detection of the latter type (1).

A variety of study designs are available for these various purposes. Linkage analysis of necessity requires family studies, typically either sib-pair designs (affected sib pairs for a binary disease trait) or extended pedigrees. Association studies, on the other hand, can use either family designs or population-based designs involving unrelated individuals. These various options will be discussed in greater detail below, with the suggestion that it is possible to design efficient population-based family studies that can be used for both linkage and association. We will also discuss the use of collections of "high risk" families in gene characterization studies and discuss a general population-based framework for discovery and characterization. Finally, we touch briefly on a number of modeling issues and future challenges that are likely to arise in studying complex diseases

Methods for Gene Discovery

Linkage analysis entails a search across the genome for markers that are associated with the disease *within families*, i.e., that there is a tendency for pairs of affected individuals to have the same alleles at that marker locus. Between families, however, different alleles may be associated with the disease, because the marker allele *per se* has no causal role in the disease; the marker simply travels with the disease allele from parents to offspring because the two are close together on a chromosome. Thus, it is possible for a marker to be linked to a disease gene but not associated with it. Once one or more markers in a region have been found to be linked, additional nearby markers are then tested in an effort to localize the disease gene more precisely. Efficient multistage testing strategies for such "genome scans" have been discussed by Brown et al. (2) and Elston et al. (3). Multipoint linkage analyses might use several markers jointly for greater precision in estimating that location than can be obtained from a series of "two-point" linkage analyses one marker at a time (4).

Linkage analyses can be model-based (parametric or "lod score") or model-free ("nonparametric"). The lod score method is based on the likelihood of the observed marker and disease data in a pedigree under a model for the distribution of the unobserved disease gene. Typically, the parameters of such a model (e.g., the age- and sex-specific penetrance function and disease allele frequency) are assumed to be known from earlier segregation analyses and the likelihood is maximized with respect to the recombination fraction θ (or the location of the disease gene in a multipoint analysis). This approach can be applied to nuclear families or extended pedigrees. By modeling the conditional distribution of markers given disease phenotypes, no assumption about the families having been ascertained in any systematic statistical sampling manner is needed; indeed, heavily loaded families identified through genetic counseling clinics are often used for this purpose and are typically the most informative for linkage analysis, even though they would be in no sense population based. The lod score method is the most powerful approach if the genetic model is correctly specified, but can lose power or even produce false evidence of linkage under some kinds of misspecification. In contrast, nonparametric approaches do not require any assumptions about the genetic model and thus are robust to model misspecification, but they are generally less powerful than lod score methods; furthermore, the choice of the optimal nonparametric test will still depend upon the presumed mode of inheritance. Nonparametric methods are based on a comparison of the proportion of alleles shared "identical by descent" (IBD) by pairs of affected relatives against the proportion expected based solely on their relationship (e.g., one-quarter of sibling pairs would be expected to share zero, one-half to share one, and one-quarter to share two alleles IBD). This approach is most commonly applied to affected sib pairs (where possible their parents are also included to aid in the determination of IBD status), although it is possible to include other relative types in other forms of analysis.

The limit of resolution of linkage analysis is generally felt to be not much smaller than about 1 cM (1% recombination, corresponding to roughly one million base pairs [bp]), even with the largest pedigree studies. Thus, other techniques are needed to further localize a disease gene before undertaking massive sequencing in search of mutations. These techniques typically entail use of a very dense panel of markers, perhaps on the order of 10 kb (0.01 cM), which can be used in various ways. The simplest of these is to search for markers one-by-one that appear to associated with the trait across the population, a phenomenon known as linkage disequilibrium (LD). An LD can arise in a number of ways—new mutations, genetic drift, admixture, etc.—but will tend to decay across generations G at a rate $(1 - \theta)^G$; thus, after many generations from the event that created the LD initially, only very nearby pairs of loci will remain associated. The extent of detectable LD in various human populations and its usefulness as a mapping tool constitute an active area of research, but in most outbred populations it is generally believed to be very short range, ~100 kb or less. (In recently admixed populations, LD will generally extend over a much wider interval, making them potentially useful for genome scans [5]; LD tends to be larger in magnitude and more consistent in population isolates, making them more useful for fine mapping [6].) Even within a region of significant LD, its magnitude will be extremely variable across pairs of nearby loci, owing to chance mutation, recombination, and coalescent events in the ancestral history of the population. Therefore, there is now great interest in using haplotypes—sequences of adjacent marker alleles on a single chromosome—as a tool for fine mapping. A variety of approaches have been proposed for doing this, including searching for segments that are frequently shared by pairs of cases (7–10), association between specific haplotypes and disease (11,12), or using coalescent methods to model the ancestry and evolution of mutation-carrying haplotypes (13–15).

Designs for Association Studies

Whether the aim is fine mapping by LD, testing an association with a candidate gene, or characterizing a cloned gene, a number of different study designs could be used, their relative merits depending upon the context. The most important distinction between these designs concerns whether families or unrelated individuals are studied. We therefore begin by discussing the standard population-based epidemiologic case-control and cohort designs using unrelated individuals and then survey a range of family-based designs—case-control, case-parent trios, kin-cohort, and use of heavily loaded pedigrees.

Population-Based Case-Control and Cohort Designs

The majority of disease traits studied in genetic epidemiology are relatively rare, so that case-control designs are natural to consider. The design of such studies is essentially no different for a genetic risk factor than for environmental risk factors and follows well recognized principles discussed in standard epidemiologic textbooks (16–18). Thus, for example, cases should be representative of all cases in the population and controls should be representative of the source population of cases. This is most easily accomplished in situations where a population-based disease registry exists, such as the Surveillance, Epidemiology, and End Results (SEER) registries in the United States, and where there is some means of sampling from the total population. The latter is more difficult in the United States, although some other countries maintain voter registration or other databases that are available for epidemiologic research. Absent such a register, some imagination may be needed to construct a suitable control selection procedure: techniques such as neighborhood censuses, random digit dialing, sampling from birth registries (for childhood diseases) or Medicare files (for diseases of old age), prepaid health maintenance organization rosters, or hospital controls have been used in various epidemiologic studies and their advantages and disadvantages have been widely discussed. Cases and controls are frequently individually or stratum-matched on potential confounding variables, such as age, gender, race, and possibly other established risk factors. See Chapter 8 in this volume for further discussion of some of the issues of validity and efficiency that can arise in such studies. In some respects, one of the major drawbacks of case-control studies in conventional risk factor epidemiology—recall bias due to retrospective collection of exposure information—is not as much a concern in genetic epidemiology, since a subject's constitutional genotype does vary over time and is not subject to the vagaries of an individual's memory. (Of course, phenotypic assays of genotype could be distorted by the disease process, and other confounding or modifying factors could be misclassified.) Indeed, Clayton and Mc-Keigue (19) have argued that because the transmission of genes from parents to offspring is random, a gene association study carries the same interpretability in terms of causality as a randomized control trial, at least in terms of freedom from bias and residual confounding. Thus such associations would reflect a causal effect of either the variant under study or a nearby one in linkage disequilibrium with it. This "Mendelian randomization" argument is directly applicable to the case-parent trio design discussed below, but they also apply it to ordinary case-control and cohort designs involving unrelated individuals. However, this extension of the principle requires that one address the problem of population stratification using one of the approaches discussed below.

Cohort studies have well recognized advantages and disadvantages (19). For the purpose of genetic epidemiology, few investigators would contemplate initiating a new prospective cohort study for any but the most common diseases, but there are now in excess of a million persons enrolled in various existing cohorts for whom biologic specimens have already been obtained and stored. Some of these cohorts have already accrued decades of follow-up time and represent a rich resource for genetic association studies (20). The cost of genotyping everyone in a large cohort would likely be prohibitive, even with recently developed high throughput technologies, but this can be avoided by using nested case-control (21) or case-cohort (22) designs. These entail comparison of cases arising in the cohort with a sample of suitably selected controls drawn from the cohort, thereby capitalizing on the inferential advantages of a cohort design at greatly reduced cost. The design of such nested studies is in principle no different when studying a genetic association than any other risk factor and nested studies are discussed in standard textbooks (23) and review articles (24). However, a number of options for efficient sampling are available, such as multistage sampling (25–28) and countermatching (29,30).

Multistage sampling might entail an initial random sampling of cases and controls on whom a surrogate for some risk factor (e.g., family history as a surrogate for genotype) is obtained. Subjects are then subsampled using this information for the more expensive determination of genotype and perhaps other risk factors. Counter-matching aims to improve the efficiency of a matched case-control design by increasing the proportion of pairs that are discordant for the risk factor(s) of interest through systematically mismatching them on a surrogate for the factor; for example, in a genetic study, a case with a positive family history might be matched with a family-history-negative control and vice versa. The inherent bias in both these designs is then accounted for by including suitable weights in the analysis.

Whether case-control, cohort, or any of these nested designs are used, any association study based on unrelated individuals potentially suffers from a form of confounding known in the genetics literature as "population stratification" (31). If the population comprises two or more subgroups with different allele frequencies and different baseline rates of disease, then confounding can occur (see Fig. 6.1), leading to increased risk of false positive associations, and biasing relative risk estimates upwards or downwards, depending upon the direction of these two associations. If these subgroups were identifiable, standard techniques such as matching or statistical adjustment could be used to control this problem—indeed, epidemiologic studies are routinely matched or adjusted for "race/ethnicity." The difficulty is that even within the broad categories of race/ethnicity that are conventionally used, there can be strong gradients in allele frequencies and baseline risks. Some authors (Chapter 8, ref. 32–34) have questioned the practical importance of this concern, at least for studies of common polymorphisms in non-Hispanic whites of European descent, ar-

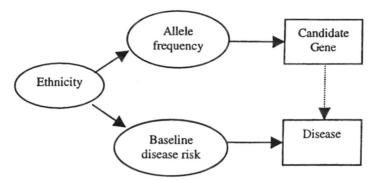

Figure 6.1 Schematic representation of population stratification as a form of confounding by ethnicity.

guing that any correlation between baseline rates and allele frequencies that would give rise to confounding tends to disappear as larger numbers of subgroups are combined. Hence, they argue that adherence to standard principles of sound epidemiologic study design should be adequate to address it. However, in multiethnic populations, particularly heavily admixed populations such as African-Americans or Hispanics or those with a high prevalence of multiracial individuals as in Southern California, individuals can be difficult to classify or match appropriately (35). A different approach to this problem which is gaining some theoretical attention (but remains to be applied widely), known as "genomic control," is based on using a panel of markers unlinked to the gene under study to infer the hidden population structure and adjust for it. One such approach uses the distribution of test statistics for the markers to estimate an inflation factor by which to adjust the naïve chi square test for the over-dispersion caused by population stratification (36,37). Another approach uses Bayesian clustering or latent class analysis methods to assign cases and controls probabilistically to strata defined by their markers and then perform a stratified analysis within these strata (38,39).

Although not the focus of this chapter, all the designs discussed in this section and the family-based designs section which follows can also be used for testing gene–environment and gene–gene interactions (see Chapter 8 for more discussion of this topic). Another approach to testing such interaction effects is known as the case-only or case-case design (40–42) (see also Chapter 7). In this approach, a series of unrelated cases is used and the association between genotype and environment (or two genes) is tested. If the two factors were independently distributed in the source population, then any such association in cases would be evidence of departure from a multiplicative model for their joint effects on disease risk. Of course, with this design it is not possible to test the main effects of either factor, and careful consideration is needed to judge whether the assumption of independence in the source population is tenable (43), but if it is, the design is more powerful for testing that interaction than a conventional case-control design.

Family-Based Designs

The use of family-member case-control designs is appealing because family members have a common gene pool and hence the problem of population stratification is overcome by matching. The two main variants of this idea involve the use of sibling controls or case-parent-trios (also known as "parental controls" or "pseudo-siblings"). In a case-sibling study, affected individuals are the cases and their unaffected siblings the controls, the data being analyzed as a matched case-control study using standard conditional logistic regression methods. (Unaffected cousins could also be used instead of siblings, although the protection against population stratification would no longer be absolute, since they would have only one pair of grandparents in common and the other grandparents might come from different ethnic groups.) In the case-parent-trio design, cases and their parents are genotyped, but the parents themselves are not used as controls; instead, one forms the set of hypothetical "pseudo-siblings" comprising the three other genotypes that could have been transmitted from the parents; the case-pseudosib sets are then analyzed as a 1:3 matched case-control design using conditional logistic regression (44). A variant of this design, known as the "Transmission-Disequilibrium Test (TDT)" (45), compares each of the two alleles for the case separately against the other allele not transmitted by that parent as two independent contributions to a 1:1 matched case-control comparison of alleles rather than a single genotype comparison. The two procedures are mathematically equivalent under a multiplicative model for allele contributions to risk, but would differ under a dominant or recessive model. Both the case-sib and the case-parent-trio designs test an alternative hypothesis of linkage and association, i.e., they will detect associations only with causal genes or with genes that are in linkage disequilibrium with a causal gene.

There are a number of drawbacks to the use of these designs. Because cases and their siblings are more likely to share genotypes (and environmental factors), their comparison tends to be less efficient than using unrelated controls (depending upon the genetic model, about 50% as efficient, meaning that double the sample size would be required to obtain the same statistical precision or power). For gene–environment interactions, however, the use of sib controls can be more efficient (46). In general, siblings should have attained the age of the case's diagnosis to rule out the possibility that he or she might still have been affected prior to the case (47), effectively limiting the pool of potential controls to older siblings for many cases; this lack of comparability on birth order, however, risks introduction of other biases, particularly if time-dependent environmental variables are to be included in the model or if a substantial proportion of cases need to be excluded for lack of a suitable sib control (46). There are other subtleties if multiple cases or multiple controls are selected from the same family, since the possible permutations of disease status against genotypes are not equally likely under the null hypothesis that the gene is not itself causal but is linked to a causal gene (48). Furthermore, the familial relationships among cases and controls must not give away who must have been the case: for example, if cousins were used as controls and one was drawn from each side of the family, it would be obvious who was the case because he or she would be the only one who was a blood relative of both of the other two. While some authors have advocated limiting such comparisons by selecting a single case and a single control from each family, for example, taking the pair with the maximally different genotypes (48), there has been a rapidly developing literature on valid family-based-association-tests (FBATs) that would exploit all the possible comparisons within a family (49–51).

The case-parent-trio design generally does not suffer from the efficiency loss that the case-sib design does, and indeed can be more powerful than using unrelated controls for a recessive gene (46). However, it does require that both parents be available for genotyping, which makes it difficult to apply to diseases of middle or old age. Although some information is contained in the transmission from a single parent, if that is all that is available, care must be taken to avoid bias, since the subset of transmissions for which parental sources can be inferred unambiguously is not random (52,53). As with the case-sib design, families with multiple cases do not contribute independent information under the null hypothesis of linkage but no association, so more sophisticated techniques are required (54). Although the trio design cannot be used to test for the main effects of environmental factors, it can in principle test for gene–environment interactions by comparing the genetic relative risks in exposed and unexposed cases (55). This comparison, however, involves an assumption that genes and environments are independently distributed within families (i.e., conditional on parental genotypes) (42,56); an assumption that is similar to that for the case-only design, but somewhat weaker because it applies within families, not between families, and thus would not be influenced by such factors as family history that could potentially induce such an association. See references 57 and 58 for a log-linear models approach to case-parent-trio data, with particular application to testing maternal genotype effects and imprinting. For example, birth defects could involve a direct or interactive effect of maternal alleles, in which case the deleterious allele would tend to have a higher frequency in mothers than fathers.

A more complex design is the case-control-family study. Here, population-based series of unrelated cases and controls are identified, possibly matched on various factors as in a traditional case-control study, and their family members are also recruited as study participants. While more commonly used for testing familial aggregation (59) or segregation analysis (60) without use of any molecular data, the design can also be used for testing candidate gene associations or characterizing cloned genes (61,62). Used in this way, it can be seen as an extension of the kin-cohort design discussed below, but its real advantage lies in its population-based nature and its ability to serve as the basis for an integrated approach to gene discovery and gene characterization.

The "kin cohort" design (63) entails ascertainment of a series of probands unselected with respect to family history and obtaining their genotypes and their family history in first-degree relatives (but relatives' genotypes are not needed in this approach). The probands themselves could be affected or unaffected and need not necessarily be representative of the population (provided they are not biased with respect to family history). For example, in a study of the penetrance of the *BRCA1* and *BRCA2* ancestral mutations in Ashkenazi Jews, Struewing et al. (64) enrolled volunteers from Jewish community organizations in the Washington, D.C. area. The cumulative incidence curves in first-degree relatives of carrier and noncarrier probands are then estimated using standard Kaplan-Meier survival analysis methods. Because first-degree relatives of carriers have a roughly 50% probability of carrying the same mutation while first-degree relatives of noncarriers have only half the population probability of being a carrier, it is then possible to decompose the observed cumulative incidence curves into their constituent penetrance curves (cumulative incidence in carrier and noncarrier individuals) by a simple algebraic manipulation.

This design has been extended in a number of ways. The relatively simple analysis described above does not exploit all the information in the sample, so Gail et al. proposed a maximum likelihood analysis, similar to segregation analysis but conditioning on the observed genotypes (65,66). With the use of this likelihood, it then becomes straightforward to extend the design to include more distant relatives, as well as measured genotype information on other relatives; they call this general approach the "genotyped proband design." Siegmund et al. (67) have considered the question of which members of a pedigree would be the most informative to genotype and concluded that for a common low penetrance dominant gene, genotyping additional relatives per family was more efficient than genotyping a single proband in a larger number of families (for the same total genotyping costs), while for a rare major dominant gene, the reverse was true.

Since in the process of gene discovery, the heavily loaded families typically used are not representative of all cases in the population, it is natural to inquire whether any useful information about penetrance or modifying factors can be obtained at the gene characterization phase. Their great advantage, beyond simply the cost efficiency from having already collected the pedigree information and biologic specimens, is that such families will tend to have a higher prevalence of mutations, particularly for rare high penetrance genes, so that smaller sample sizes should be required. On the other hand, since such families were collected specifically because they had many cases, a naïve analysis that ignored the ascertainment process would greatly overestimate both penetrance and allele frequency, compared to their true values in the general population. While in principle, one might be able to construct a maximum likelihood analysis which would be conditioned on the ascertainment scheme, in practice such collections seldom can be described in terms of any well-defined statistical sampling plan, and even if they were, the analysis of complex sampling schemes would likely be computationally intractable. Fortunately, an alternative approach is available that theoretically should allow valid estimates of population parameters even from samples that are not population based. Known as the "mod score" (68–70) or "retrospective likelihood" (72) approach, the analysis is based on the conditional likelihood of the measured genotypes, given the observed distribution of phenotypes in the family. By conditioning on the phenotypes in this manner, their ascertainment is automatically controlled for, assuming that families were ascertained solely on the basis of their phenotypes, not their genotypes. This is the approach that was used in the initial estimation of *BRCA1/2* penetrance from the Breast Cancer Linkage Consortium families (72,73), which led to an estimate of risk of breast or ovarian cancer by age 70 of 83%. Subsequent estimates based on the kin-cohort and population-based case-control-family designs have been substantially lower (62,64,74). This difference cannot be explained simply as an artifact of ascertainment bias because of the use of the mod score approach to analyze the high-risk families. However, by limiting that analysis to the linked families (done to address the problem of genetic heterogeneity, i.e., some families' disease being due to genes other than the one under study), the assumption that families were ascertained solely on the basis of their phenotype was violated; this has been shown to lead to upwardly biased estimates of penetrance (75). Other explanations that have been offered to explain the discrepancy between the clinic-based and population-based estimates are that the former may also be segregating other modifying factors (other genes or environmental factors), leading to truly higher penetrance in such families, or that the penetrance varies by specific mutations, with the more commonly occurring mutations in the general population (e.g., the Ashkenazi founder mutations) having lower penetrance than those occurring in the heavily loaded families.

Integrated Designs for Discovery and Characterization

With this brief tour of approaches to discovering and characterizing genes, we now turn to the question of whether it is possible and efficient to try to design a resource that can be used for both purposes. The experience from the use of heavily loaded clinic-based collections of pedigrees to estimate *BRCA1/2* penetrance should be somewhat cautionary about the limitations of relying exclusively on series that are not population-based. On the other hand, they are arguably the most efficient way to assemble pedigrees that are highly informative for linkage analysis. In an attempt to bridge this gap, Zhao et al (76,77) have proposed a general framework based on the case-control-family design described above. Since the initial ascertainment of families is population based, there would be no difficulty in estimating population parameters from such a design. Of course, the yield of rare major genes would be relatively low, but multistage sampling of probands based on family history as discussed earlier (27) could be used to hone in on the families most likely to be segregating mutations, and this could be extended following the principles of "sequential sampling of pedigrees (78)." Here, the basic idea is that at each stage of pedigree extension, one is entitled to use all the phenotype information already collected systematically as well as knowledge of the pedigree structure (but not anecdotal information about phenotypes) in branches not yet explored, to decide whether and in what direction to extend the pedigree; once extended, *all* the phenotype information obtained must then be included in the analysis, whether additional cases were identified or not. Following these simple rules, Cannings and Thompson (78) show that the likelihood for the pedigree need be conditioned solely on the initial ascertainment of probands, not on all the decisions made subsequently.

Still, for mapping a very rare gene, it is unclear whether this process can yield a sufficient number of highly informative pedigrees, even using the most efficient approaches to multistage sampling and sequential extension of pedigrees, without requiring enrollment of a prohibitive number of probands. For genes with mutations that are not extremely rare, however, there is great merit in this approach, as it will not only provide a basis for mapping genes and then characterizing them *in the same sample,* but it will also provide a resource for continuing the search for additional genes after some have been discovered. For example, Antoniou et al. (79) and Cui et al. (80), using such approaches, have provided evidence for an additional major gene for breast cancer, possibly a more common recessive gene, after removing the families attributable to *BRCA1* and *BRCA2*. Their approaches differ somewhat, Antoniou et al. fitting a multilocus model which includes the measured genes and all families in the analysis, and Cui et al. excluding the families known to be segregating one of the two measured genes. On the basis of such segregation analysis results, one might then feel confident to launch a further genome scan to localize such a gene, now using the more powerful LOD score approaches which require a population-based estimate of the genetic model. Absent such knowledge, one would be forced to use the affected sib-pair approach, first screening all pairs to exclude those that were carrying a known mutation.

It is this general philosophy that underlies the establishment of the Cooperative Family Registries for Breast and Colorectal Cancer Research of the National Cancer Institute (NCI) (81–84). In order to address the aims of both discovery and characterization, this multicenter resource comprises population-based and clinic-based series of families. The population-based series are ascertained through affected probands from population-based cancer registries, stratified in various ways. Some are unselected consecutive series, some restricted or sampled by age, race, or family history in first-degree relatives; a few registries have used multistage sampling (28,85). The clinic-based registries are intended to provide a large series of multiple-case families for gene discovery purposes, but would not be included in analyses aimed at characterization, except perhaps with the use of the mod-score approach. Whatever the mode of ascertainment, all probands provide a standardized risk factor questionnaire, including extended family history, and blood samples that are being stored for genotyping and creation of cell lines. Participating centers differ in the specifics of their protocols for developing extended pedigrees, but in general as many surviving family members (affected and unaffected) as possible are enrolled as participants, providing the same risk factor information and blood samples, which are also being stored. To date, over 6000 breast and 6000 colorectal cancer families have been enrolled, comprising over 100,000 individuals in each registry. Depending upon the specific scientific aims, these families might be sampled in various ways for genotyping. A variety of studies aimed at using this resource for gene discovery and characterization are currently underway.

Models for Complex Diseases

Whether parametric linkage analysis or association analysis is planned, some form of statistical model of penetrance is needed. Among the complexities that must be considered are variable age at onset; the role of polygenes, other major genes, and environmental factors, including their possible interactions; residual familial aggregation due to unmeasured factors; and heterogeneity of effect for genes with multiple mutations or polymorphisms. One might also wish to take account of somatic events, such as loss of heterozygosity, genomic instability, DNA methylation, and gene expression data. Genome-wide association studies are also being proposed as a means of gene discovery, perhaps requiring something of the order of a million statistical tests (1), introducing yet another level of statistical complexity. In this brief section, we can only outline a general approach to model building, leaving the details to other papers.

For binary disease traits with variable age at onset, the techniques of survival analysis provide a natural framework for modeling penetrance. Letting $\lambda(t)$ denote the incidence rate of disease at age t ("hazard function") and $S(t) = \exp(-\int_0^t \lambda(u)\,du)$ the probability of surviving to age t free of disease ("survival function"), then the likelihood contribution for a case diagnosed at age t is $\lambda(t)\,S(t)$ and the contribution for a subject last seen at age t disease free at that time is simply $S(t)$. If we assumed that, conditional on all the measured risk factors, the outcomes of all subjects $i = 1, \ldots, n$ were independent, then the overall likelihood of the data would be simply

$$L = \prod_{i=1}^{n} \lambda_i(t_i)^{d_i}\, S_i(t_i)$$

where d_i is an indicator for affected, a value of 1, or not, a value of 0. The conditional independence assumption would not pose any difficulty for unrelated individuals (e.g., a population-based case-control study), but is more problematic for family data. If not all family members have been genotyped for a major gene, then a likelihood contribution for the entire family must be constructed by summing over the possible genotypes of all the untyped individuals that are compatible with the available genotype information on other family members. This is essentially a segregation analysis, but conditional on partially measured genotype information (62). Additional familial dependencies might be caused by other as yet unidentified genes, by unmeasured environmental factors, or by correlated measurement errors in measured risk factors. Such dependencies might be taken into account by using regressive models (86), latent variable approaches like frailty models (87,88), or marginal models using Generalized Estimating Equations methods (90).

By whatever means the likelihood is constructed, a model is needed for the hazard function in relation to the various measured risk factors, genetic and environmental. One possibility is the proportional hazards model (90), which might be written as

$$\lambda(t,G,Z) = \lambda_0(t) \exp(\beta_G + Z'\gamma + \ldots)$$

where G represents the major gene(s), β_G the log relative risk associated with genotype G, Z the measured environmental covariates, $\lambda_0(t)$ an unspecified function representing the baseline risk as a function only of age, and " . . ." indicates the possibility of adding additional interaction terms (e.g., gene–environment, gene–gene, gene–age, etc.). However, a number of major genes such as *BRCA1* seem to have much stronger effects at younger ages on a relative risk scale. While this could be addressed by adding age $\times$ genotype interaction terms, it might be preferable to re-formulate the model as

$$\lambda(t,G,Z) = \lambda_G(t) \exp(Z'\gamma + \ldots)$$

i.e., with separate age-specific baseline rates for each genotype, but still assuming that environmental factors acted multiplicatively on these baseline rates, unless specific interaction terms were added to the model. In either of these approaches, the form of the baseline rates might be left completely unspecified, as in the Cox partial likelihood approach (91), or some parametric form could be adopted; for example, the S.A.G.E. package assumes a logistic distribution for the ages at onset amongst the affected, coupled with a logistic model for the lifetime risk of disease, either of which could depend upon genotype and/or covariates, as in an application to smoking-gene interactions for lung cancer (92). Other mathematical models might also be considered for the joint effects of age and genotype, such as an additive model of the form $\lambda(t,G) = \lambda_0(t) + \beta_G$ or an accelerated failure time model of the form $S(t,G) = S_0(t\ e^{\beta}{}_G)$. For example, Peto and Mack (93) have suggested that the rate of breast cancer in co-twins of affected twins or of second cancer in the contralateral breast is virtually constant as a function of age or time since diagnosis of the first, suggesting an additive model for genetic effects might be appropriate.

The coding of β_G would depend upon what is assumed about dominance. For a dominant gene, with wild-type allele a and mutant allele A, one would set $\beta_{aa} = 0$ and constrain $\beta_{aA} = \beta_{AA}$; likewise, for a recessive gene, one would set $\beta_{aA} = 0$; for a codominant gene, one would estimate both β_{aA} and β_{AA}. For a multiallelic gene, one might have many more parameters to estimate. Most analysis of *BRCA1* penetrance have treated all mutations as equivalent, but there is some evidence that different mutations confer different risks of breast versus ovarian cancer (94), and it remains an open question whether certain common polymorphisms in the gene also have an effect on penetrance (95). For genes like *BRCA1* with hundreds of rare mutations, the prospects of ever having direct estimates of penetrance for any one of them are virtually nonexistent, so some kind of modeling approach is needed to test for systematic influences of broad classes of mutations (truncating or not, by location, etc.) as well as random between-mutation heterogeneity in effect. Hierarchical models (96) provide a natural framework for addressing such questions. Bayesian approaches to smoothing the effects of many haplotypes within a gene (sequences of alleles on a single chromosome) have also been suggested (12). This entails the use of a multilevel model, in which the first level would be a conventional logistic model for disease as a function of a set of relative risks for all possible haplotypes, and the second level would be a model for the prior means and covariances of haplotype relative risks in terms of their structural similarities to each other.

Increasingly, gene characterization efforts have been directed towards trying to understand complex pathways involving multiple genes and multiple exposures jointly, particularly for common polymorphisms in low-penetrance "metabolic" genes. For example, hypothesized causes of colorectal polyps and cancer include polycyclic aromatic hydrocarbons (PAHs) and heterocyclic amines (HCAs), which derive from tobacco smoke and well-done red meat (97). The metabolic activation and detoxification of these compounds are regulated by a number of genes, including several cytochrome P450 enzymes (such as *Cyp1A1* and *Cyp1A2*), various glutothione-S-transferases (such as *GSTm3*), N-acetyl-transferases (*NAT1* and *NAT2*), and microsomal epoxide hydrolase (*mEH*, aka *EPHX1*) (98). The complexity of these pathways makes it difficult to examine the effects of these exposures or these genes one at a time, or even in pairwise interactions, without allowing for the influence of the other factors, but the problems of sparse data and multiple comparisons preclude standard approaches based on multiway stratification. Cortessis and Thomas (99) have proposed a Bayesian approach to such problems with the use of physiologically based pharmacokinetic (PBPK) models. In essence, the approach entails estimating the concentrations of the various intermediate metabolites for each subject, as a function of the measured exposures, and a set of unmeasured metabolic rates, which are in turn determined by the subject's genotypes at the relevant loci, and relating the estimated concentrations of the relevant metabolites to the disease risk. The distributions of the various individual parameters are determined by a set of population parameters that are the primary object of inference, e.g., regression coefficients for the contributions of exposures to pathways or of pathways to disease, means and variances of metabolic rates as a function of genotype, etc.

Summary

Both population-based and family-based designs have their uses in testing candidate gene associations and characterizing genes once their causal connection to a disease has been established. Appropriately designed, such studies can also be a useful resource for discovering other genes that may also be involved. Nonmendelian disorders may involve a complex interplay between multiple genes and multiple environmental factors, as well as age and other time-dependent factors, requiring sophisticated methods of analysis. While survival analysis techniques, such as Cox regression can provide a flexible framework for empirical modeling of penetrance functions, mechanistic models such as PBPK models for complex metabolic networks can also be useful. Stochastic models of carcinogenesis, which have long been used to describe exposure–time–response relationships for environmental exposures, might usefully be extended to incorporate the influence of germline mutations or such epigenetic phenomena as microsatellite instability and DNA methylation.

References

1. Risch N, Merikangas K. The future of genetic studies of complex human diseases. Science 1996;273:1616–1617.
2. Brown D, Gorin M, Weeks D. Efficient strategies for genomic searching using the affected-pedigree-member method of linkage analysis. Am J Hum Genet 1994;54:544–552.
3. Elston R, Guo X, Williams L. Two-stage global search designs for linkage analysis using pairs of affected relatives. Genet Epidemiol 1996;13:535–558.
4. Kruglyak L, Lander E. Complete multipoint sib-pair analysis of qualitative and quantitative traits. Am J Hum Genet 1995;57:439–454.
5. Stephens J, Briscoe D, O'Brien S. Mapping by admixture linkage disequilibrium in human populations: limits and guidelines. Am J Hum Genet 1994;55:809–824.
6. Jorde L. Linkage disequilibrium as a gene-mapping tool. Am J Hum Genet 1995;56:11–14.
7. Houwen R, Baharloo S, Blankenship K, et al. Genome screening by searching for shared segments: mapping a gene for benign recurrent intrahepatic cholestasis. Nature Genet 1994;8:380–386.
8. Qian D, Thomas D. Genome scan of complex traits by haplotype sharing correlation. Genet Epidemiol 2001;21:S582–S587.
9. Te Meerman G, Van Der Meulen M. Genomic sharing surrounding alleles identical by descent effects of genetic drift and population growth. Genet Epidemiol 1997;14:1125–1130.
10. Bourgain C, Genin E, Holopainen P, et al. Use of closely related affected individuals for the genetic study of complex diseases in founder populations. Am J Hum Genet 2001;68:154–159.
11. Chiano M, Clayton D. Fine genetic mapping using haplotype analysis and the missing data problem. Ann Hum Genet 1998;62:55–60.
12. Thomas D, Morrison J, Clayton D. Bayes estimates of haplotype effects. Genet Epidemiol 2001;21(suppl 1):S712–S717.
13. McPeek M, Strahs A. Assessment of linkage disequilibrium by the decay of haplotype sharing, with application to fine-scale genetic mapping. Am J Hum Genet 1999;65:858–875.
14. Morris A, Whittaker J, Balding D. Bayesian fine-scale mapping of disease loci, by hidden Markov models. Am J Hum Genet 2000;67:155–169.
15. Niu T, Qin ZS, Xu X, Liu JS. Bayesian haplotype inference for multiple linked single-nucleotide polymorphisms. Am J Hum Genet 2002;70:157–169.
16. Breslow NE, Day NE. Statistical methods in cancer research: I. The analysis of case-control studies. Lyon: IARC Scientific publications, 1980.
17. Rothman KJ, Greenland S. Modern Epidemiology. Philadelphia: Lippencott-Raven, 1998.
18. Klienbaum DG, Kupper LL, Morgentern H. Epidemiologic Research: Principles and Quantitative Methods. Belmont, CA: Lifetime Learning Publications, 1982.
19. Clayton DG, McKeigue PM. Epidemiological methods for studying genes and environmental factors in complex diseases. Lancet 2001;358:1357–1360.
20. Langholz B, Rothman N, Wacholder S, Thomas D. Cohort studies for characterizing measured genes. Monogr Natl Cancer Inst 1999;26:39–42.

21. Mantel N. Synthetic retrospective studies and related topics. Biometrics 1973;29:479–86.
22. Prentice R. A case-cohort design for epidemiologic studies and disease prevention trials. Biometrika 1986;73:1–11.
23. Breslow NE, Day NE. Statistical methods in cancer research. II. The design and analysis of cohort studies. Lyon: IARC Scientific Publications, 1987.
24. Thomas DC. New approaches to the analysis of cohort studies. Epidemiol Rev 1998;14:122–134.
25. White J. A two stage design for the study of the relationship between a rare exposure and a rare disease. Am J Epidemiol 1982;1982:119–128.
26. Breslow N, Cain K. Logistic regression for two-stage case-control data. Biometrika 1988; 75:11–20.
27. Whittemore A, Halpern J. Multi-stage sampling in genetic epidemiology. Statistics in Medicine 1997;16:153–167.
28. Siegmund K, Whittemore A, Thomas D. Multistage sampling for disease family registries. Monogr Natl Cancer Inst 1999;26:43–48.
29. Langholz B, Borgan O. Counter-matching: a stratified nested case-control sampling method. Biometrika 1995;82:69–79.
30. Andrieu N, Goldstein A, Langholz B, Thomas D. Counter-matching in gene-environment interaction studies: efficiency and feasibility. Am J Epidemiol 2001;153:265–274.
31. Lander ES, Schork NJ. Genetic dissection of complex traits. Science 1994;265: 2037–2048.
32. Caparaso N, Rothman N, Wacholder W. Case-control studies of common alleles and environmental factors. Monogr Natl Cancer Inst 1999;26:25–30.
33. Wacholder S, Rothman N, Caporaso N. Population stratification in epidemiologic studies of common genetic variants and cancer: quantification of bias. J Natl Cancer Inst 2000;92:1151–1158.
34. Wacholder S, Rothman N, Caporaso N. Counterpoint: Bias from population stratification is not a major threat to the validity of conclusions from epidemiologic studies of common polymorphisms and cancer. Cancer Epidemiol Prev Biomarkers 2002;11:513–520.
35. Thomas D, Witte J. Population stratification: A problem for case-control studies of candidate gene associations? Cancer Epidemiol Prev Biomark 2001;11:505–512.
36. Devlin B, Roeder K. Genomic control for association studies. Biometrics 1999;55: 997–1004.
37. Reich DE, Goldstein DB. Detecting association in a case-control study while correcting for population stratification. Genet Epidemiol 2001;20:4–16.
38. Pritchard JK, Stephens M, Rosenberg NA, Donnelly P. Association mapping in structured populations. Am J Hum Genet 2000;67:170–181.
39. Satten GA, Flanders WD, Yang Q. Accounting for unmeasured population substructure in case-control studies of genetic association using a novel latent-class model. Am J Hum Genet 2001;68:466–477.
40. Umbach D, Weinberg C. Designing and analysing case-control studies to exploit independence of genotype and exposure. Statistics in Med 1997;16:1731–1743.
41. Khoury M, Flanders W. Nontraditional epidemiologic approaches in the analysis of gene-environment interaction: case-control studies with no controls! Am J Epidemiol 1996; 144:207–213.
42. Weinberg C, Umbach D. Choosing a retrospective design to assess joint genetic and environmental contributions to risk. Am J Epidemiol 2000;152:197–203.
43. Albert P, Ratnasinghe D, Tangrea J, Wacholder S. Limitations of the case-only design for identifying gene-environmental interactions. Am J Epidemiol 2001;154:687–693.
44. Self SG, Longton G, Kopecky KJ, Liang KY. On estimating HLA/ disease association with application to a study of aplastic anemia. Biometrics 1991;47:53–61.

45. Spielman RS, McGinnis RE, Ewens WJ. Transmission test for linkage disequilibrium: The insulin gene region and insulin-dependent diabetes mellitus (IDDM). Am J Hum Genet 1993;52:506–516.
46. Witte JS, Gauderman WJ, Thomas DC. Asymptotic bias and efficiency in case-control studies of candidate genes and gene-environment interactions: basic family designs. Am J Epidemiol 1999;148:693–705.
47. Lubin JH, Gail MH. Biased selection of controls for case-control analysis of cohort studies. Biometrics 1984;40:63–75.
48. Curtis D. Use of siblings as controls in case-control association studies. Ann Hum Genet 1997;61:319–333.
49. Horvath S, Laird N. A discordant-sibship test for disequilibrium and linkage: No need for parental data. Am J Hum Genet 1998;63:1886–1897.
50. Laird N, Horvath S, Xu X. Implementing a unified approach to family-based tests of association. Genet Epidemiol 1998;19 (suppl):S36–S42.
51. Kraft P. A robust score test for linkage disequilibrium in general pedigrees. Genet Epidemiol 2001;21(suppl 1):S447–S452.
52. Curtis D, Sham PC. A note on the application of the transmission disequilibrium test when a parent is missing. Am J Hum Genet 1995;56.
53. Schaid DJ, Li H. Genotype relative-risks and association tests for nuclear families with missing parents. Genet Epidemiol 1997;14:1113–1118.
54. Martin E, Kaplan N, Weir B. Tests for linkage and association in nuclear families. Am J Hum Genet 1997;61:439–448.
55. Schaid D. Case-parents design for gene-environment interaction. Genet Epidemiol 1999;16:261–273.
56. Thomas D. Re: "Case-parents design for gene-environment interaction" by Schaid. Genet Epidemiol 2000;19:461–463.
57. Weinberg CR, Wilcox AJ, Lie RT. A log-linear approach to case-parent—triad data: assessing effects of disease genes that act either directly or through maternal effects and that may be subject to parental imprinting. Am J Hum Genet 1998;62:969–978.
58. Wilcox AJ, Weinberg CR, Lie RT. Distinguishing the effects of maternal and offspring genes through studies of case-parent triads. Am J Epidemiol 1998;148:893–901.
59. Claus E, Risch N, Thompson W. Age at onset as an indicator of familial risk of breast cancer. Am J Epidemiol 1990;131:961–972.
60. Claus E, Risch N, Thompson W. Genetic analysis of breast cancer in the cancer and steroid hormone study. Am J Hum Genet 1991;48:232–242.
61. Hopper J, Chenevix-Trench G, Jolley D, et al. Design and analysis issues in a population-based, case-control-family study and the Co-operative Family Registry for Breast Cancer Studies (CFRBCS). Monogr Natl Cancer Inst 1999;26:95–100.
62. Hopper J, Southey M, Dite G, et al. Population-based estimate of the average age-specific cumulative risk of breast cancer for a defined set of protein-truncating mucations in BRCA1 and BRCA2. Cancer Epidemiol Biomark Prevent 1999;8:741–747.
63. Wacholder S, Hartge P, Struewing J, et al. The kin cohort study for estimating penetrance. Am J Epidemiol 1998;148:623–630.
64. Struewing J, Hartge P, Wacholder S, et al. The risk of cancer associated with specific mutations of BRCA1 and BRCA2 among Ashkenazi Jews. NEJM 1997;336:1401–1408.
65. Gail M, Pee D, Benichou J, Carroll R. Designing studies to estimate the penetrance of an identified autosomal dominant mutation: cohort, case-control, and genotype-proband designs. Genet Epidemiol 1999;16:15–39.
66. Gail M, Pee D, Carroll R. Kin-cohort designs for gene characterization. Monogr Natl Cancer Inst 1999;26:55–60.

67. Siegmund K, Morrison J, Gauderman W. Who should be genotyped for estimating gene main effects in family-based disease registries? Abstract. Genet Epidemiol 2001;21:176.
68. Risch N. Segregation analysis incorporating linkage markers. I. Single-locus models with an application to type I diabetes. Am J Hum Genet 1984;36:363–386.
69. Clerget-Darpoux F, Bonaiti-Pellie C, Hochez J. Effects of misspecifying genetic parameters in lod score analysis. Biometrics 1986;42:393–399.
70. Hodge S, Elston R. Lods, Wrods, and Mods: The interpretation of lod scores calculated under different models. Genet Epidemiol 1994;11:329–342.
71. Kraft P, Thomas DC. Bias and efficiency in family-matched gene-characterization studies: Conditional, prospective, retrospective, and joint likelihoods. Am J Hum Genet 2000;66:1119–1131.
72. Easton D, Bishop D, Ford D, Crockford G. Genetic linkage analysis in familial breast and ovarian cancer results from 214 families. Am J Hum Genet 1993;52:678–701.
73. Ford D, Easton D, Bishop D, Narod S, Goldgar D. Risks of cancer in BRCA1-mutation carriers. Lancet 1994;343:692–695.
74. Fodor FH, Weston A, Bleiweiss IJ, et al. Frequency and carrier risk associated with common BRCA1 and BRCA2 mutations in Ashkenazi Jewish breast cancer patients. Am J Hum Genet 1998;63:45–51.
75. Siegmund K, Gauderman W, Thomas D. Gene characterization using high risk families: a sensitivity of the MOD score approach. Am J Hum Genet 1999;65:A398. Abstract 2251.
76. Zhao LP, Hsu L, Davidov O, Potter J, Elston R, Prentice RL. Population-based family study designs: an interdisciplinary research framework for genetic epidemiology. Genet Epidemiol 1997;14:365–388.
77. Zhao L, Aragaki C, Hsu L, et al. Integrated designs for gene discovery and characterization. Monogr Natl Cancer Inst 1999;26:71–80.
78. Cannings C, Thompson E. Ascertainment in the sequential sampling of pedigrees. Clin Genet 1977;12:208–212.
79. Antoniou A, Pharoah P, McMullan G, Day N, Ponder B, Easton D. Evidence for further breast cancer susceptibility genes in addition to BRCA1 and BRCA2 in a population-based study. Genet Epidemiol 2001;21:1–18.
80. Cui J, Antoniou, AC, et al. After BRCA1 and BRCA2—what next? Multifactorial segregation analyses of three-generation, population-based Australian families affected by female breast cancer. 2001;68:420–431.
81. Seminara D, Obrams G. Genetic epidemiology of cancer: A multidisciplinary approach. Genet Epidemiol 1994;11:235–254.
82. Ziogas A, Gildea M, Cohen P, et al. Cancer risk estimates for family members of a population-based family registry for breast and ovarian cancer. Cancer Epidemiol Prev Biomark 2000;9:103–111.
83. Peel DJ, Ziogas A, Fox EA, et al. Characterization of hereditary nonpolyposis colorectal cancer families from a population-based series of cases. J Natl Cancer Inst 2000;92:1517–1522.
84. Daly MB, Offit K, Li F, et al. Participation in the cooperative family registry for breast cancer studies: issues of informed consent. J Natl Cancer Inst 2000;92:452–456.
85. Haile R, Siegmund K, Gauderman W, Thomas D. Study design issues in the development of the University of Southern California consortium's colorectal cancer registry. Monogr Natl Cancer Inst 1999;26:89–93.
86. Bonney G. Regressive models for familial and other binary traits. Biometrics 1986;42:611–625.
87. Hougaard P, Thomas D. Frailty. In: Elston RC, Palmer L, Olsen JH, eds. Encyclopedia of Genetics. Oxford: Oxford University Press, 2002:277–283.

88. Siegmund K, Todorov A, Province M. A frailty approach for modeling diseases with variable age of onset in families: the NHLBI Family Heart Study. Stat Med 1999;18: 1517–1528.

89. Zhao LP, Hsu L, Holte S, et al. Combined association and segregation analysis of case-control family data. Biometrika 1998;85:299–315.

90. Cox D. Regression models and life tables (with discussion). J R Statist Soc 1972; 34:187–220.

91. Li H, Thompson E. Semiparametric estimation of major gene and family-specific random effects for age of onset. Biometrics 1997;53:282–293.

92. Sellers T, Bailey-Wilson J, Elston R, et al. Evidence for mendelian inheritance in the pathogenesis of lung cancer. JNCI 1990;82:1272–1279.

93. Peto J, Mack T. High constant incidence in twins and other relatives of women with breast cancer. Nat Genet 2000;26:411–414.

94. Easton D, Ford D, Bishop D. Breast and ovarian cancer incidence in BRCA1-mutation carriers. Am J Hum Genet 1995;56:265–271.

95. Durocher F, Shattuck-Eidens D, McClure M, et al. Comparison of BRCA1 polymorphisms, Rare sequence variants and/or missence mutations in unaffected and breast/ovarian cancer populations. Hum Mol Genet 1996;5:835–842.

96. Witte JS. Genetic analysis with hierarchical models. Genet Epidemiol 1997;14: 1137–1142.

97. Potter J. Colorectal cancer: molecules and populations. JNCI 1999;91:916–932.

98. Cortessis V, Siegmund K, Chen Q, et al. A case-control study of microsomal epoxide hydrolase, smoking, meat consumption, Glutathione S-Transferase M3, and risk of colorectal adenomas. Cancer Res 2001;61:2381–2385.

99. Cortessis V, Thomas DC. Toxicokinetic genetics: An approach to gene-environment and gene-gene interactions in complex metabolic pathways, in Mechanistic Considerations in the Molecular Epidemiology of Cancer. Bird P, Boffetta P, Buffler P, Rice J, eds. Lyon, France: IARC Scientific Publications, 2003, in press.

7

Facing the challenge of complex genotypes and gene–environment interaction: the basic epidemiologic units in case-control and case-only designs

Lorenzo D. Botto and Muin J. Khoury

In this chapter, we focus on fundamental units of epidemiologic analysis of studies that relate health outcomes with complex genotypes and gene–environment interaction. The goal is to offer a practical perspective that researchers might find useful as they design, analyze, and present their studies, with emphasis on case-control and case-only designs. In the first part of the chapter, we focus on case-control studies and their core information (1). In particular, we illustrate ways in which such core information can be clearly presented to provide the fundamental measures of effect and impact, including the relative risks for the multiple factors under study (alone and jointly); the interaction effects; the exposure frequencies; and the attributable fractions.

In the second part of the chapter we discuss the potential role of well-designed disease registries as adjuncts or antecedents of case-control studies, and suggest that they might be particularly useful in the study of complex genotypes and interaction. In particular, we discuss the notion that a disease registry, approached through a case-only perspective, might be scanned for complex genotypes ranked by potential attributable fraction for the disease. Finally, we discuss the advantages and challenges of these approaches and their possible integration in studying the causation of common multifactorial conditions.

We view the perspective presented in this chapter as complementary to the discussion in other sections of the book, in which methodologic aspects of the detection of joint effects and interaction are systematically presented. The approaches discussed here, particularly those related to the case-only analysis of disease registries, could enhance but not replace other strategies for the study of complex genotypes and gene–environment interaction

Investigating Interaction in Epidemiology

Investigating genetic and gene-environment interaction in epidemiology raises definitional, methodologic, and practical questions. The meaning, measurement, and modeling of the effect of multiple factors, the biologic significance of epidemiologic assessment of interaction, and the appropriateness of specific study designs are but a few topics that continue to engender considerable debate (2–4). We will note briefly only two such issues for their relevance in this discussion of genetic factors and interaction.

First, bias and confounding in case-control studies (the type of study discussed here in some detail), though always a concern, can likely be decreased more easily when assessing genetic factors compared, for example, to environmental factors such as diet or lifestyle (4). For example, genotype can in principle be measured more precisely and objectively, than can, for example, smoking or folic acid intake, which are commonly assessed based on a subject's recall and may be imprecise or biased by disease status. Thus, exposure misclassification, both differential and nondifferential, should decrease with a corresponding improvement in the precision and validity of risk estimates. Also, the stability of genotype over time is particularly valuable in case-control studies in which the factors under study are measured months or years after disease onset. Finally, genotypes for a given set of alleles are likely to distribute randomly in the population (Mendelian randomization), reducing the likelihood of spurious gene–environment or gene–gene associations (at unlinked loci) (4). Genetic substructure in the population remains a concern, but researchers have suggested strategies that take such substructure into account, such as the use of a panel of unrelated markers (5,6). These considerations, combined with the known statistical efficiency of case-control studies, have revived the interest in case-control studies as powerful tools for the study of the effect of genotype on disease risk (4) and in part prompted our emphasis on such studies.

The second aspect of interaction that has discussed extensively relates to which measures of effect are most informative or useful. For example, in the case of two dichotomous factors, one could estimate the effect of each factor alone as well as the joint effect. One could also estimate the departure of the joint effect from specific models of interaction (e.g., additive or multiplicative). It can be useful to note that the relation between individual and joint effects can take different forms (7), which can depend on the biologic mechanism underlying the interaction. However, it has been noted that predicting the biologic mechanism from such epidemiologic data is difficult and perhaps not productive (2).

With more than two factors under study, summary measures of interaction and statistical models become more complicated, and the ability to present the data and the primary measures of effect acquires renewed value. The explosive growth of genetic technology and the ever expanding catalogue of human genes (8,9) are already leading to studies of increasing complexity. For example, the risk for venous thrombosis is already being studied in relation to variants of the factor V, prothrombin, and 5,10 methylenetetrahydrofolate reductase (MTHFR) genes, as well as to blood homocysteine levels and oral contraceptive use (10–13). Similarly, the risk for spina bifida is being studied in relation to variants of folate-related genes (e.g., MTHFR, cystathione-beta-synthase, methionine synthase, and methionine synthase reductase) and blood levels of selected vitamins (folate, B_{12}) (14–17). Even, and perhaps particularly, in such complex settings, an appreciation of the basic analytic unit of epidemiologic analysis should help researchers develop a consistent starting point for data presentation and assessment.

Population-Based Case-Control Studies and the 2 × 4 Table

The simplest case of interaction is perhaps that of two dichotomous factors (e.g., presence or absence of a genotype, use or nonuse of a pill). For illustration, we present data from case-control settings in which we assume the ideal conditions of an unbiased, unconfounded, population-based, incident-case study. We will further assume that the study's odds ratios are valid estimations of relative risks.

Data from such case-control study can be presented in a 2 × 4 table (Table 7.1). The same reference group is used to compute three odds ratios (each factor alone and jointly). Such odds ratios are the basic, direct measures of association.

Such presentation has several advantages (Table 7.2). The role of each factor is independently assessed both in terms of association and of potential attributable fraction. In addition, the odds ratios can be examined to assess their general relation (7) and formally evaluated in terms of departure from specified models of interaction (most commonly multiplicative or additive). The table also provides the distribution of the exposures among controls and helps evaluate the dependence of factors in the underlying population (provided the controls are representative of such population). Finally, a case-only odds ratio can be easily derived and used as a comparison with findings from case-only studies in the literature.

The 2 × 4 table approach to presenting genetic and gene–environment interactions is appealing for several reasons.

- It is efficient: it summarizes, without loss of detail, seven 2 × 2 tables and generates a comprehensive set of effect estimates that none of the latter, individually, can match.
- It highlights potential sample size issues: cell sizes are directly presented, and confidence intervals show their effect on statistical power.
- It emphasizes effect estimation over model testing: the relative risk estimates associated with the joint and individual exposures are the primary elements of an interaction, whereas departures from specific models of interactions are derived parameters and explicitly labeled as such.

Table 7.1 Layout for a Case-Control Study Assessing the Effect of a Genotype and an Environmental Factor

G	E	Cases	Controls	Odds Ratio	Contrast		Main Information
+	+	a	b	ah/bg	A vs. D	A	Joint genotype and environmental factor vs. none
+	−	c	d	ch/dg	B vs. D	B	Genotype alone vs. none
−	+	e	f	eh/fg	C vs. D	C	Environmental factor alone vs. none
−	−	g	h	1		D	Common reference

Other Measures	Odds Ratio	Main Information
Case-only odds ratio	ag/ce	Departure from multiplicative model of interaction
Control-only odds ratio	bh/df	Independence of factors in population
Multiplicative interaction	A/(B*C)	Deviation from multiplicative model of interaction
Additive interaction	A − (B + C − 1)	Deviation from additive model of interaction
Stratified 1-a	ad/bc	Association with environmental factor among people *with* genotype
Stratified 1-b	eh/fg	Association with environmental factor among people *without* genotype
Stratified 2-a	af/be	Association with genotype among people *exposed* to environmental factor
Stratified 2-b	ch/dg	Association with genotype among people *not exposed* to environmental factor

G, genotype; E, environmental factor.

Table 7.2 Advantages of the 2 × 4 Table in the Study of Gene–Environment Interactions

1. The primary data are displayed clearly and completely.

2. The primary measures of association-relative risk estimates for each factor alone and for the joint exposure are readily generated. Because they use the same reference group, these estimates can be compared.

3. Attributable fractions can be computed separately for each exposure alone and for the joint exposure

4. Relative risk estimates can be used to assess the relation between the joint exposure and the individual exposures. For example, the departure from additive or multiplicative models of interactions can be readily derived from the table.

5. Risk estimates stratified by either exposure can also be calculated if needed.

6. For case-control studies, the case-only and the control-only odds ratios can be easily computed. For adequately chosen control groups, the control-only odds ratio estimates exposure dependencies in the underlying population.

In summary, the table provides the simplest epidemiologic equivalent of the general statement that all effects on human health are attributable to the joint effect of genes and the environment. Indeed, it can be argued that the 2 × 4 table (and not the 2 × 2 table) is the fundamental unit of epidemiologic analysis.

A Simple Application of the 2 × 4 Table

We illustrate the 2 × 4 table approach by using data from a case-control study of venous thromboembolism in relation to factor V Leiden and oral contraceptive use (18). When the original data are so rearranged (Table 7.3), one can clearly appreciate certain key aspects of the interaction:

- The marginal and joint effects. For example, the odds ratio associated with factor V Leiden and oral contraceptive use alone (6.9 and 3.7, respectively) can be contrasted with that associated with the combined exposure (34.7).
- The potential attributable fractions. Provided the associations are causal, one can note the potential public health relevance of the findings: the computation of attributable fractions for two or more factors was developed by several authors and has been summarized (19). The relatively high frequency in the population of the gene variant (2.4% among controls) and of the joint exposure (1.2%) translates into considerable population attributable fractions for thromboembolic disease (5.5% and 15.7%, respectively).

One can contrast such a presentation with a stratified analysis, in which the association between the oral contraceptive use and venous thrombosis is assessed separately among those with and without the factor V Leiden polymorphism (Table 7.4). The latter approach does not provide information on individual and joint effects immediately and tends to emphasize departure from a specific (multiplicative) model of interaction. The 2 × 4 table does not have such limitation and provides the data to test for other nonmultiplicative models as well.

Table 7.3 Analysis of Oral Contraceptive Use, Presence of Factor V Leiden Mutation, and Risk for Venous Thromboembolism

Factor V[a]	OC[a]	Cases	Controls	Odds Ratio	95% CI	AF-Exp (%)[a]	AF-Pop (%)[a]	Exposure Frequency in Controls (%)
+	+	25	2	ORge 34.7	7.83–310.0	97.1	15.7	1.2
+	–	10	4	ORg 6.9	1.83–31.80	85.6	5.5	2.4
–	+	84	63	ORe 3.7	2.18–6.32	73.0	39.6	37.3
–	–	36	100	Ref				59.2
Total		155	169					

[a]Factor V: +, presence of factor V Leiden mutation (heterozygotes and homozygotes)

–, absence of factor V Leiden mutation

OC: +, current use of oral contraceptives

–, no current use of oral contraceptives

AF-Exp (%): Attributable Fraction (percent) among exposed cases

AF-Pop (%): Attributable Fraction (percent) among all cases in the population

Note: the departure of the observed from the expected effect of the joint exposure depends on the definition of no interaction, as shown below for simple additive and multiplicative definitions.

	Expected OR-GE	Departure from expected
Additive	$(3.7 + 6.9) - 1 = 9.6$	$34.7 - 9.6 = 25.07$
Multiplicative	$(3.7 * 6.9) = 25.7$	$34.7/25.7 = 1.4$

Source: Modified from Vandenbroucke et al., 1994. Lancet 344:1453–1457.

Table 7.4 Comparing the Stratified and Case-only Approaches with the 2 × 4 Approach[a]

COMPARISON WITH STRATIFIED ANALYSIS

		FACTOR V PRESENT		FACTOR V ABSENT		Ratio of Odds Ratios
		Cases	Controls	Cases	Controls	
Oral contraceptive use	+	25	2	84	63	
	−	10	4	36	100	
Odds ratio (95%CI)		5.0 (0.8–31.8)		3.7 (2.2–6.1)		1.4

CASE-ONLY AND CONTROL-ONLY ODDS RATIOS

Case-only odds ratio:	(25*36)/(10*84) = 1.1		
Control-only odds ratio	(2*100)/(4*63) = 0.8		1.4

[a]The data are from Table 7.3.
Source: Note that ratios of odds ratios are identical to departure from multiplicative model (Table 7.3).

A further assessment of the data from the 2 × 4 table involves the relation of the factors separately among cases and controls (Table 7.4). Conceptually, one can split vertically the case-control study into a case-only study and a control-only study and examine the respective odds ratios. The case-only design in itself is an efficient and valid approach to screening for interaction, provided that the fundamental assumption of independence of exposure and genotype in the population is justified (20,21). The potential role of such studies in the epidemiologic approach to complex diseases has been reviewed (22,23) and will be examined later in connection with the discussion of disease registries. Also the association of risk factors among controls (control-only odds ratio) can provide useful information, namely the dependencies of the risk factors (genetic or environmental) in the underlying population. Detecting such dependencies is important both as a clue for a biologic relation between alleles at the loci under study and as a test of the key assumption in the interpretation of case-only data.

Three Factors: The 2 × 8 Table

The points underscored by the 2 × 4 table are even clearer for three factors—three genes, three environmental factors, or a combination of genetic and environmental factors. With three dichotomous factors, the exposure combinations become 8 (2^3). Although more complex, such a table still shows the primary epidemiologic parameters (odds ratios and attributable fractions) associated with each factor and combination of factors. Because all refer to the same reference group, the relations between these measures are immediately evident; if needed, one can also assess which model of interaction best fits the data. Methodologic issues, such as sample size and exposure dependencies among the controls, can also be assessed with relative ease. The contrast with classic stratified analysis is even greater than in the case of two factors. To present such stratified analysis, a minimum of four tables is needed; because they have different reference groups, the four odds ratios would not be directly comparable; and the overall interpretation of the study is less immediately clear.

Increasing Complexity

The 2×4 or the 2×8 table, though simple, may adequately summarize some, but not all epidemiologic relations. Issues that come into play in more complex situations include the following.

- The number of factors can increase. Even for dichotomous factors, the number of exposure combinations grows quickly (2^n for n factors), and the corresponding table rapidly becomes unwieldy.
- The relation between exposure and outcome can be other than dichotomous. For example, the relation can be graded or continuous (dose–response) as occurs with smoking and lung cancer or with obesity and hypertension. In the general case of n exposures each with its dose–response curve, the response surface is best described as a general n-dimensional manifold which may not be meaningfully summarized by few discrete odds ratios.
- As more factors are involved, their interaction may not adequately described by simple multiplicative or additive models.

These limitations highlight two issues that will increasingly confront epidemiologists as they try to unravel the web of interaction in disease causation. First, new or improved epidemiologic methods may be needed to deal with such complex situations. For example, researchers have suggested using a variety of regression models, including hierarchical models, and neural networks, traditionally used in modeling the probability of clinical outcomes (24,25), to the study multiple factors and interaction (26–29). So far, these approaches have limitations: the output of regression models, for example, is model-dependent; neural networks, though in general less dependent on prior model specification (26–28), may be limited in their ability explicitly to estimate dependencies among risk factors (26,27).

The second issue relates to sample size. As the number of factors under study increases, so do the strata that have to be defined within the study. With a fixed total number of subjects, increasing the number of factors quickly reduces per-stratum size and the associated statistical power. Thus, negative findings should be carefully interpreted. Strategies to deal with this issue include conducting well-designed collaborative studies that increase sample size but also deal effectively with extraneous genetic heterogeneity.

In conclusion, researchers are challenged to apply epidemiologic methods to increasingly complex data on multiple factors and interaction. Carefully conducted collaborative studies may provide adequate sample size. A clear presentation and analysis of the core elements of these interactions (the data distribution and the primary measures of association) may increase the information that can be extracted from the data. In this sense, the 2×4 table and its immediate extensions are fundamental, simple, and useful tools to documenting and studying gene–environment interaction.

Disease Registries and Case-Only Designs

Population-based case-control studies are fundamental tools in etiologic studies, particularly for their ability to provide key parameters of the human genome epidemiology of many conditions (4,30). The challenges of case-control studies, particularly the recruitment of an adequate set of control subjects, and the refinement of case-only approaches suggest novel approaches in studying the role of complex genotypes in disease etiology. The availability of well-designed disease registries provides a practical setting for case-only studies of common conditions such as certain cancers and birth defects. Such case-only approaches cannot replace but rather enhance traditional case-control (or cohort) studies, particularly in three key areas:

- Scanning for genotypes that potentially contribute the most to disease in a population.
- Evaluating etiologic heterogeneity and genotype–phenotype correlations among subsets of cases.
- Detecting supramultiplicative effects of interacting alleles.

Scanning Genotypes by Potential Contribution to Disease in the Population

Studying the role of complex genotypes, i.e., the interaction of multiple alleles at multiple loci, presents numerous challenges, including the large number of possible allele combinations. In theory, m alleles at n loci can generate m^n combinations (haplotypes): with 10 loci, two alleles can generate in excess of 1000 combinations, and three alleles nearly 60,000 combinations.

Given their potentially large number, which allele combinations should one look at first? One approach is to focus first on allele combinations that potentially contribute to the largest proportion of disease in a population or, in epidemiologic terms, on those with the highest potential population-attributable fraction. It is easy to show that even though one cannot determine relative risks in case-only studies, one can estimate the upper limit of a genotype's attributable fraction. If causality is assumed, such potential maximum attributable fraction is simply the frequency of the genotype among cases. This relation is intuitively obvious, since if x percent of a random series of cases has a particular exposure, then at most x percent can be caused by that exposure. The formal relation

$$F_c = AF*(OR/OR - 1)$$

derives directly from Miettinen's formula for attributable fraction (31).

Thus, attributable fraction (AF) is, at most, as high as the fraction of cases with the exposure—in this case the genotype—of interest (F_c) but never higher, regardless of how high the odds ratio or relative risk. The equation also illustrates the non-linear relation between odds ratio and AF, implying that variations in the upper range of odds ratios translate into progressively smaller changes in AF (for variations in the odds ratio between 10 and 1000, the fraction of exposed cases differs from the AF by less than one part in 10).

One might argue that when the genotype frequency in the population is unknown, little should be inferred from genotype frequencies among cases. However, in the case of complex genotypes such a relation becomes interesting because, under the hypothesis of no effect, few subjects are expected to have any given (complex) genotype, defined as a certain combination of alleles at a number of loci. More precisely, that number decreases multiplicatively with the number of loci considered concurrently (Fig. 7.1). For example, with five loci and one common variant allele per locus with a frequency in the population of 10%, one would expect that by chance alone, the genotype with the five variant alleles would be found in 0.10^5 or one in 100,000 people. The practical usefulness of such consideration is that researchers can expect that complex genotypes observed with some frequency among cases, even in a small percent of cases, might be likely candidates for further study. Thus complex genotypes are one specific scenario where examining case-only frequencies might help focus the search for allele combinations with a potentially significant role in disease causation.

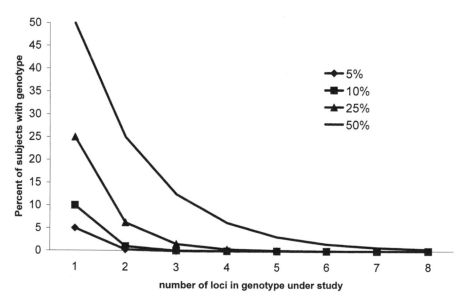

Figure 7.1 Expected proportion of people with a complex genotype, by increasing number of loci examined concurrently (assuming independence and equal genotype frequency at each locus). The expected proportion falls quickly.

Examining Homogeneous Subset and Determinants of Severity or Phenotype

The case-only approach to the analysis of complex genotypes could also help define smaller, more homogenous subsets distinguished by phenotype, disease progression, or severity (20). These more homogeneous subsets can be compared with respect to genotypes to study the possible relation between genotype and outcome. For example, one might separate cases of first occurrence of venous thrombosis from those of recurrence, or cases of myocardial infarction by age of onset. Such analyses can provide clues to the genetic heterogeneity underlying common disorders and help relate genotypic variation to clinically relevant differences in outcome.

Searching for Supramultiplicative Interactions

The analysis of disease registries in a case-only fashion can provide some indication of interaction among alleles by using, for example, log-linear models. Log-linear models have been used to test for higher order associations in a multiplicity of settings, including associations between structural anomalies in the same baby (32), between maternal and fetal genotypes and disease (33–35), and between genotype markers and disease (36,37). In conjunction with prior information about linkage between alleles, the results of log-linear modeling, for example, can provide some indication of whether the joint effect of certain allele combinations differs from that expected under a multiplicative null hypothesis, i.e., whether the joint effect equals the product of each allele's effect alone. In this respect, such an approach is a natural extension of the well-known case-only odds ratio (21), which measures the deviation from simple multiplicative effects of two factors, and is subject to similar interpretations and limitations (22,38). Among the limitations of log-linear modeling are its sensitivity to sparse data, which is a real concern in the analysis of complex genotypes, and its assumption of a log-linear relation between factors. Moreover, marginal effects of each allele cannot be measured. Nevertheless, the context of complex genotypes with relatively common susceptibility alleles is precisely where one might expect to find significant, supramultiplicative interactions if such genotypes contribute to disease.

Limitation of the Scanning Disease Registries Using a Case-Only Approach

The main thrust of this discussion of case-only designs (Table 7.5) is that, in the context of the study of complex genotypes in disease etiology, well-designed disease registries can be informative and relatively inexpensive resources that could complement and enhance the value of traditional case-control or cohort studies. Provided the key assumptions of case-only studies hold, such assessment of disease registries could provide researchers with clues on the health effects of certain complex genotypes, including their potential contribution to disease in the population, their involvement in significant supramultiplicative interaction, and their relation to

Table 7.5 Advantages and Disadvantages of Using Case-only Studies to Screen for
Complex Genotypes in Disease Etiology

Advantages/Disadvantages

CAN BE USED TO	CONSIDERATIONS
—Screen for genotypes with highest potential impact on disease	Provides upper limit of attributable fraction associated with genotype
—Screen for supramultiplicative interactions between multiple loci	Tools include, for example, log-linear analyses and other methods of clustering
—Provide clues to etiologic heterogeneity and determinants of phenotypes and severity	Shows variations in genotype frequencies and interactions by different disease phenotypes, severity, or age at onset
CAN IMPROVE	
—Speed of study	Particularly useful in conditions for which well-designed disease registries are available
—Precision of estimates	Eliminates controls and associated variance
—Validity of findings	Assumes no population stratification
—Efficiency of subsequent studies	Contributes to efficient case-control design by highlighting notable case- and population subsets, factors with highest potential impact, and potential sample size issues
BUT HAVE DISADVANTAGES	
—Validity sensitive to assumptions	Results are exquisitely sensitive to independence of factors in population
—Limited information	Provides few data on marginal effects, relative and absolute risks, and nonmultiplicative interactions

case subgroups with distinctive etiology and severity of outcome. It should be noted
that such approach does not pursue gene discovery in the manner of a genome scan.
Rather, it uses known allelic variation at candidate loci as a starting point to examine the potential contribution to disease etiology.

However, the assessment of disease registries using case-only methods is not an
alternative to traditional studies that use population controls. Its limitations, which
stem from the limits of the case-only design, should be recognized clearly:

- Case-only studies offer no information about marginal risks for specific genotypes.
- They assess only deviations from purely multiplicative interactions, which is
  only one of the possible scenarios in which different alleles at different loci interact to modulate disease risk (7). Important genetic effects, such as strong effects from single gene variants, might not generate a signal. Other complex scenarios that defy facile conclusions include interactions of gene variants that
  increase disease risk with others that reduce risk.
- The validity of interaction assessment in case-only studies is exquisitely sensitive to independence assumptions for the factors in the population (39). One
  might imagine combinations that could be expected to violate that independence, whether among genotype and environmental factors (e.g., cigarette

smoking and genes involved in detoxification, due to selective attrition in the population), or else among different genes. Also, independence among gene combinations might be violated if population stratification induces correlation between genes (even if the genotypes occur independently within each sub-population), though this problem might be solved by appropriate stratification. Alternatively, dependencies between loci might occur if the two loci are on the same chromosome, even in populations with random mating, if mutations are relatively recent. So far, these appear mostly to be theoretical concerns, for lack of empirical evidence for such dependencies. Recent data for example suggest that this is not a problem for the more commonly studied metabolic genes (40).

- Case-only studies also do not provide full information on the attributable fraction for gene combinations, and they only estimate their upper limit.
- Case-only studies require that cases represent a random or unselected series of cases, as could be assembled by a population-based registry. Series assembled from tertiary centers might be subject to selection forces that might preclude valid inferences.

At the same time, one should note the potential advantages in speed, efficiency, and precision of the two-tiered approach that begins with case-only studies and uses their findings to design further studies. Such advantages include the following:

- Researchers could complete their studies more rapidly, by examining existing or easily developed case groups, as might be derived from population-based disease registries. Several such registries, for example, already exist for many conditions including cancers and birth defects.
- The resources otherwise used to enroll and study convenient controls could be used instead to expand the spectrum of candidate genes and alleles among increasing numbers of cases.
- The effect estimates could gain in precision (no variance associated with controls) and validity (no population stratification).
- Subsequent studies might be more efficient. For example, case-control studies might use evidence from case-only studies to decide on reasonable sample sizes (for cases and controls) that might vary by ethnicity or disease subgroup.

In a broader perspective, one should realistically approach such screening of case-only studies within the well-known expectations of a screening process. Along with valid results, both false positives and false negatives will occur, the former, for example, if linkage disequilibrium unrelated to disease were present.

Final Considerations

It is tempting to speculate to what extent the conceptual framework presented here can be transferred from genomics to proteomics. Whereas the ability to detect multiple genetic variants with a single functional test would appear to increase researchers' ability to examine an ever-widening web of metabolic networks, the independence assumptions might be commonly violated with proteomics, because of feedback regulation systems governing the transcription of genes into proteins.

Recruiting sufficient numbers of study participants remains a basic issue. Although analytic techniques such as multifactor dimensionality reduction (41) are being suggested as possible enhancements in the study of complex genotypes, sample size requirements remain an inescapable challenge for researchers.

Finally, case-only approaches in no way diminish but in fact underscore the need to tackle and solve the complex legal, ethical, social, and practical issues of selecting, recruiting, and testing representative samples of the population for genetic studies. As researchers realize the synergy between traditional and nontraditional studies, we should encourage a concerted effort at developing, on the one hand, well-designed disease registries, and on the other, representative samples from well-defined populations that are large and accessible.

Summary

- In the study of interaction, it is useful to evaluate and present information on both the marginal and joint exposures (gene–environment combinations). Departure from specified models of interaction can be informative but should not be the sole focus of the analysis. Key information for each term of the interaction includes the frequency of the gene–environment exposure (or the complex genotype) in the reference population, and the disease-associated relative risks and attributable fractions.
- The appreciation of certain epidemiologic units of analysis can facilitate the systematic assessment and clear presentation of data on multiple factors and interaction. In population-based case-control studies, particularly in their simplest forms (with two dichotomous factors), one such unit of analysis is the 2×4 table. Though more complex situations can require more complex approaches, the 2×4 table in many situations can provide a useful starting point for data assessment, presentation, and analysis.
- Population-based disease registries can be important research resources. Using analytic approaches derived from case-only methods, researchers could scan such registries for complex genotypes and other exposure combinations associated with the highest attributable fraction for disease. Such analysis could also provide clues on the presence of supramultiplicative interaction, as well as of determinants of disease severity and phenotype among population subgroups.
- Case-control and case-only studies are best viewed as complementary rather than alternative approaches to the assessment of interaction. Appreciating the basic units of epidemiologic analysis within each study design and using both designs synergically can contribute to the efficient and systematic assessment of the role in disease etiology of multiple factors, complex genotypes, and interactions.

References

1. Botto LD, Khoury MJ. Commentary: facing the challenge of gene-environment interaction: the 2 × 4 table and beyond. Am J Epidemiol 2001;153:1016–1020.
2. Thompson WD. Effect modification and the limits of biological inference from epidemiologic data. J Clin Epidemiol 1991;44:221–232.
3. Greenland S, Rothman KJ. Concepts of interaction, in Modern Epidemiology. Greenland S, Rothman KJ, eds. Philadelphia: Lippincott, 1998:329–342.
4. Clayton D, McKeigue PM. Epidemiological methods for studying genes and environmental factors in complex diseases. Lancet 2001;358:1356–1360.
5. Pritchard JK, Stephens M, Rosenberg NA, Donnelly P. Association mapping in structured populations. Am J Hum Genet 2000;67:170–181.
6. Satten GA, Flanders WD, Yang Q. Accounting for unmeasured population substructure in case-control studies of genetic association using a novel latent-class model. Am J Hum Genet 2001;68:466–477.
7. Khoury MJ, Adams MJ, Jr., Flanders WD. An epidemiologic approach to ecogenetics. Am J Hum Genet 1988;42:89–95.
8. Hamosh A, Scott AF, Amberger J, Valle D, McKusick VA. Online Mendelian Inheritance in Man (OMIM). Hum Mutat 2000;15:57–61.
9. Collins FS, Patrinos A, Jordan E, Chakravarti A, Gesteland R, Walters L. New goals for the U.S. Human Genome Project: 1998–2003. Science 1998;282:682–689.
10. Gerhardt A, Scharf RE, Beckmann MW, et al. Prothrombin and factor V mutations in women with a history of thrombosis during pregnancy and the puerperium [see comments]. N Engl J Med 2000;342:374–380.
11. Akar N, Akar E, Akcay R, et al. Effect of methylenetetrahydrofolate reductase 677 C-T, 1298 A-C, and 1317 T-C on factor V 1691 mutation in Turkish deep vein thrombosis patients. Thromb Res 2000;97:163–167.
12. Martinelli I, Taioli E, Bucciarelli P, et al. Interaction between the G20210A mutation of the prothrombin gene and oral contraceptive use in deep vein thrombosis. Arterioscler Thromb Vasc Biol 1999;19:700–703.
13. Cattaneo M, Chantarangkul V, Taioli E, Santos JH, Tagliabue L. The G20210A mutation of the prothrombin gene in patients with previous first episodes of deep-vein thrombosis: prevalence and association with factor V G1691A, methylenetetrahydrofolate reductase C677T and plasma prothrombin levels. Thromb Res 1999;93:1–8.
14. Botto LD, Yang Q. 5,10-Methylenetetrahydrofolate reductase gene variants and congenital anomalies: a HuGE review. Am J Epidemiol 2000;151:862–877.
15. Christensen B, Arbour L, Tran P, et al. Genetic polymorphisms in methylenetetrahydrofolate reductase and methionine synthase, folate levels in red blood cells, and risk of neural tube defects. Am J Med Genet 1999;84:151–157.
16. Shaw GM, Rozen R, Finnell RH, Wasserman CR, Lammer EJ. Maternal vitamin use, genetic variation of infant methylenetetrahydrofolate reductase, and risk for spina bifida. Am J Epidemiol 1998;148:30–37.
17. Wilson A, Platt R, Wu Q, et al. A common variant in methionine synthase reductase combined with low cobalamin (vitamin B12) increases risk for spina bifida. Mol Genet Metab 1999;67:317–323.
18. Vandenbroucke JP, Koster T, Briet E, Reitsma PH, Bertina RM, Rosendaal FR. Increased risk of venous thrombosis in oral-contraceptive users who are carriers of factor V Leiden mutation. Lancet 1994;344:1453–1457.
19. Rockhill B, Newman B, Weinberg C. Use and misuse of population attributable fractions. Am J Public Health 1998;88:15–19.
20. Begg CB, Zhang ZF. Statistical analysis of molecular epidemiology studies employing case-series. Cancer Epidemiol Biom Prev 1994;3:173–175.

21. Piegorsch WW, Weinberg CR, Taylor JA. Non-hierarchical logistic models and case-only designs for assessing susceptibility in population-based case-control studies. Stat Med 1994;13:153–162.
22. Khoury MJ, Flanders WD. Nontraditional epidemiologic approaches in the analysis of gene-environment interaction: case-control studies with no controls. Am J Epidemiol 1996;144:207–213.
23. Yang Q, Khoury MJ. Evolving methods in genetic epidemiology. III. Gene-environment interaction in epidemiologic research. Epidemiol Rev 1997;19:33–43.
24. Ioannidis JP, McQueen PG, Goedert JJ, Kaslow RA. Use of neural networks to model complex immunogenetic associations of disease: human leukocyte antigen impact on the progression of human immunodeficiency virus infection. Am J Epidemiol 1998;147:464–471.
25. Marchevsky AM, Patel S, Wiley KJ, et al. Artificial neural networks and logistic regression as tools for prediction of survival in patients with Stages I and II non-small cell lung cancer. Mod Pathol 1998;11:618–625.
26. Duh MS, Walker AM, Ayanian JZ. Epidemiologic interpretation of artificial neural networks. Am J Epidemiol 1998;147:1112–1122.
27. Tu JV. Advantages and disadvantages of using artificial neural networks versus logistic regression for predicting medical outcomes. J Clin Epidemiol 1996;49:1225–1231.
28. Warner B, Misra M. Understanding neural networks as statistical tools. The American Statistician 1996;50:284–293.
29. Aragaki CC, Greenland S, Probst-Hensch N, et al. Hierarchical modeling of gene-environment interactions: estimating NAT2 genotype-specific dietary effects on adenomatous polyps. Cancer Epidemiol Biom Prev 1997;6:307–314.
30. Khoury MJ, Little J. Human genome epidemiologic reviews: the beginning of something HuGE. Am J Epidemiol 2000;151:2–3.
31. Miettinen OS. Proportion of disease caused or prevented by a given exposure, trait or intervention. Am J Epidemiol 1974;99:325–332.
32. Beaty TH, Yang P, Khoury MJ, Harris EL, Liang KY. Using log-linear models to test for associations among congenital malformations. Am J Med Genet 1991;39:299–306.
33. Wilcox AJ, Weinberg CR, Lie RT. Distinguishing the effects of maternal and offspring genes through studies of "case-parent triads". Am J Epidemiol 1998;148:893–901.
34. Shields DC, Kirke PN, Mills JL, et al. The "thermolabile" variant of methylenetetrahydrofolate reductase and neural tube defects: An evaluation of genetic risk and the relative importance of the genotypes of the embryo and the mother. Am J Hum Genet 1999;64:1045–1055.
35. Shields DC, Ramsbottom D, Donoghue C, et al. Association between historically high frequencies of neural tube defects and the human T homologue of mouse T (Brachyury). Am J Med Genet 2000;92:206–211.
36. Huttley GA, Wilson SR. Testing for concordant equilibria between population samples. Genetics 2000;156:2127–2135.
37. Khamis HJ, Hinkelmann K. Log-linear-model analysis of the association between disease and genotype. Biometrics 1984;40:177–188.
38. Yang Q, Khoury MJ, Sun F, et al. Case-only design to measure gene-gene interaction. Epidemiology 1999;10:167–170.
39. Albert PS. Limitations of the case-only design for identifying gene-environment interactions. Am J Epidemiol 2001;154:687–693.
40. Garte S, Gaspari L, Alexandrie AK, et al. Metabolic gene polymorphism frequencies in control populations. Cancer Epidemiol Bioma Prev 2001;10:1239–1248.
41. Ritchie MD, Hahn LW, Roodi N, et al. Multifactor-dimensionality reduction reveals high-order interactions among estrogen-metabolism genes in sporadic breast cancer. Am J Hum Genet 2001;69:138–147.

8

Inference issues in cohort and case-control studies of genetic effects and gene–environment interactions

Montserrat García-Closas, Sholom Wacholder,
Neil Caporaso, and Nathaniel Rothman

Classic epidemiologic studies, i.e., prospective cohort, case-control, and cross-sectional studies, are the primary source of knowledge about causal relationships between environmental exposures, broadly defined, and disease in human populations. Consideration of common genetic polymorphisms in epidemiologic studies can enhance our understanding of the relationship between environmental exposures and disease by (1): (*1*) providing mechanistic insights into disease etiology when the effects of established risk factors are evaluated among people with different genetic variants, (*2*) uncovering effects of environmental exposures on disease risk when the effect is mainly or only present in small subgroups of the population that are genetically susceptible to that exposure, and (*3*) discovering new etiologic pathways to disease that are mediated by alleles found to be associated with disease. Linkage analysis and positional cloning have made very important contributions including the discovery of rare genetic variants associated with large effects on disease (e.g., *BRCA1* and *BRCA2* and breast and ovarian cancer risk and β-amyloid precursor protein [*APP*] and Alzheimer disease). However, classic epidemiologic studies are better suited to evaluate relatively common genetic variants associated with smaller effects on complex diseases, and to evaluate how genetic variants modify the effects of environmental factors (2–4).

This chapter focuses on prospective cohort studies and traditional case-control studies with unrelated controls. The advantages and disadvantages of these two types of study designs have been discussed elsewhere (3,5–7). Genetic polymorphisms in cohort studies are usually evaluated with the use of the nested case-control design, including case-cohort. Therefore, our discussion on cohort studies will refer to aspects relevant to nested case-control studies rather than studies of the full cohort. Family-based study designs to evaluate gene–disease associations such as case-control studies with related controls (e.g., parents, siblings, or cousins) and kin-cohort design are discussed in Chapter 7.

Cohort and case-control studies of genetic polymorphisms and gene–environment interactions are susceptible to the well-established biases from classic epidemiology (8). In addition, they are susceptible to other sources of biases specific to the collection and analysis of biologic specimens (Table 8.1). Totally avoiding bias in observational research such as epidemiology is unattainable. Instead, careful consideration of epidemiologic principles aimed at minimizing the magnitude of potential biases is critical. Unfortunately, important design considerations have been sometimes obscured by the challenges of collecting biological specimens. In addition, these studies have frequently suffered from low statistical power and data over-interpretation, especially when subgroup analyses and multiple comparisons are performed. These methodologic shortcomings have contributed to frequent lack of replication of study results (4), leading to skepticism by some investigators of the ability of traditional epidemiologic study designs to identify and characterize genetic determinants of disease risk.

The objective of this chapter is to review aspects of the most common type of biases in cohort and case-control studies, i.e., selection bias, information bias, and confounding, and highlight the aspects that are especially relevant in studies of com-

Table 8.1 Sources of Biases in Epidemiologic Studies of Genetic Polymorphisms

	SOURCE OF BIAS	
Type of Bias	Study Recruitment and Questionnaire	Collection and Analysis of DNA
Selection	Loss to follow-up in cohort studies Poor control selection in case-control studies Incomplete case ascertainment	DNA collection: —Refusal or inability to provide biologic specimens —Insufficient amount of DNA limits the number of assays being performed in subsets of subjects
Information	Errors in questionnaire: —Nondifferential error due to problems in questionnaire design —Differential error or recall bias in case-control studies due to cases changing habits or recalling differently than controls	Errors handling specimens such as mislabeling of vials Errors in genotype assays
Confounding	Unaccounted factors associated with the disease and exposure of interest (that are not an intermediate step in the exposure's pathway to disease)	—Unaccounted alleles associated with the disease in linkage disequilibrium with the allele under study —Unaccounted variation in ethnic backgrounds of cases and controls, when ethnic groups have different rates of disease and different frequency of allelic variants (population stratification)

mon genetic polymorphisms. In addition, we address the problem of high rates of false positive results due to subgroup analyses, multiple comparisons, and low statistical power. Although this chapter emphasizes the theoretical principles that should be considered in the design, conduct, and interpretation of epidemiologic studies, it should be recognized that researchers often make compromises between theoretical principles and practical considerations.

Selection Bias

Selection bias occurs when control subjects are not representative of the *study base* from which the cases arise, or when not all the cases from the study base are identified. As a consequence, the distribution of genetic or environmental factors of interest among case patients and control subjects is not comparable (9). The study base can be thought as the experience of the population that is the source of the case patients during the time period when they are eligible to become cases for the study (10). Efforts to achieve complete case ascertainment are the main means to reduce the impact of selection bias introduced during the selection of case patients. The principles of control selection in traditional case-control studies have been extensively discussed in a series of papers by Wacholder et al. (10–12). In this section, we summarize these principles and highlight some key aspects in control selection that have often been neglected in studies of genetic polymorphisms.

The study base in a prospective cohort study is simply the experience of the cohort members during the period of study. Nested case-control studies sample control subjects from an existing cohort, thus providing a roster for random selection of control subjects and minimizing selection bias due to unwillingness to participate in the study. The major challenge in these studies is to achieve complete case ascertainment and complete follow-up of cohort members. Complete case ascertainment usually requires high quality disease registries that cover the geographic areas where cohort members live or active follow-up of cohort members.

Stand-alone case-control studies do not always have a readily available roster of the study base that helps identify eligible cases and sample controls randomly. In addition, obtaining high participation rates is more difficult than in nested studies because eligible subjects are not already part of an existing cohort. The study base in "population-based" case-control studies is usually defined geographically and temporally by the investigator, i.e., subjects living in a certain geographic area during a particular time period. Sampling of controls from the population is greatly facilitated when there is a complete listing of the study base, such as a complete census of the population or electoral lists. When such a roster does not exist, investigators need to rely on alternative sampling methods, such as random digit dialing and neighborhood controls, to try to obtain a random sample of the study base. Complete case ascertainment cannot be achieved through active follow-up as in cohort studies, and thus requires good disease registries and/or the collaboration of all or at least most hospitals that diagnose patients from the study base.

The "hospital-based" case-control design, where both case patients and control subjects are selected from hospitals, has been popular in studies of genetic polymorphisms, because it facilitates subject enrollment and collection of biologic specimens. Unfortunately, this study design can be more prone to bias than population-based studies especially since it is not always carefully conducted. Case ascertainment is facilitated in hospital-based case-control studies, where eligible cases are defined as patients attending one or a few hospitals that participate in the study. The study base then becomes the experience of all subjects who would be diagnosed at the study hospital/s had they developed the study disease during the study time period (10). Hospital-based studies select controls among subjects attending the study hospitals with other conditions or diseases avoiding the need for population rosters. However, this assumes that subjects attending the study hospitals with other conditions have similar referral patterns to individuals with the disease under study. Assessment of the validity of this assumption requires knowledge of both the catchment area and the referral patterns for both the case and control diseases that can be difficult to obtain. Hospital-based studies have often used groups of "healthy subjects" who are relatively easy to approach, such as blood donors or laboratory or hospital workers, as control subjects for case patients originating from populations that often have very different characteristics (e.g., different age and gender distribution, ethnic background, socioeconomic status, and behavioral characteristics such as smoking). This is a clear example of a serious violation of the study base principle. The degree of bias on the estimates of effect derived from such studies will depend on the magnitude of the difference between the distribution of genetic polymorphisms and other factors of interest among the "healthy subjects" and the actual study base for the cases.

Another key assumption for the validity of hospital-based studies is that the diagnosis used to determine the inclusion of control subjects is unrelated to the exposures of interest, since this would distort the exposure distribution (11). This assumption can be hard to meet in studies of complex diseases such as cancer with many factors of potential interest. Thus, it is recommended that in studies of complex diseases, controls be selected from several diseases or conditions to reduce the impact of a single control disease being related to the factors of interest (12–14). Wacholder et al. discuss the appropriate exclusion criteria in control selection for estimating main and subgroup effects as well as additive and multiplicative gene–environment interactions (12–14).

An important advantage of population-based studies (including studies nested in prospective cohorts and population-based case-control studies) is that the rate of disease in the source population is estimable, and thus, these studies can provide estimates of absolute risk and population attributable risk in addition to estimates of relative risk. In contrast, the size of the study base, which provides the denominator of the rate of disease in the source population, is often unknown or inestimable in hospital-based studies. As a consequence, hospital-based studies often are limited to providing estimates of relative risk and cannot provide estimates of absolute risk and population-attributable risk.

Collection of genomic DNA by using mailed, self-collected protocols such as buccal cell collection (15), should facilitate the collection of DNA in population-based studies. One limitation of the collection of buccal cells is that the DNA yield is substantially lower than in blood collections, and it varies widely among individuals (15). As a consequence, only a limited number of genotype assays can be performed in subjects with low DNA yields, raising the possibility of bias if low DNA yields are related to factors under study. Methods that amplify very small amounts of DNA and increase the number of genetic assays in subjects with low yields, such as whole genome amplification (16), may ameliorate this potential limitation.

Lack of participation biases study results only if the reason for nonparticipation is directly or indirectly related to the factors under study. Because this condition is often hard to prove, especially when many factors are under study, high participation rates become the only means to ensure comparability of participants and nonparticipants. When participation rates are lower than desirable, it is important to try to obtain at least basic risk factor information from nonparticipants and use this information to identify differences between participants and nonparticipants. Genetic polymorphisms are probably less likely to be related to reasons to be in the study than demographic or behavioral risk factors for disease, however this has been largely unexplored.

Exposure and Genotype Measurement Error

Another important potential source of bias in epidemiologic studies is *information bias* as a consequence of errors in measuring factors of interest. When the factors being measured are categorical, such as genotype determinations, measurement error is usually referred to misclassification. There are many sources of measurement error or misclassification in epidemiologic studies of environmental and genetic risk factors. Errors can be introduced throughout each phase of a study, from questionnaire design and administration, biologic sample collection, processing and storage, sample labeling, DNA extraction and storage, laboratory assays, through data coding, data entry, and analysis. Including duplicate samples for quality control when performing genotype assays and checking if the control population is in Hardy-Weinberg equilibrium can help identify problems such as artifactual errors in genotyping, and therefore it is recommended that they be done routinely.

Measurement error or misclassification can distort the distribution of the factor of interest, bias estimates of effect on disease risk, and result in decreased study power (or increased sample size needs) to detect such an effect (17–19). Generally but not universally, when the distribution of measurement error does not depend on disease status, i.e., *nondifferential* error, the effect for the mismeasured environmental or genetic factor is underestimated (17,20,21). In contrast, the direction of the bias from errors that depend on disease status, i.e., *differential* errors, is more difficult to predict (17). In the case of interaction effects between two or more factors, both nondifferential and differential errors can result in over- or underestimation of the interaction parameter (18). Under some reasonable conditions, i.e., independence of the environmental and genetic factor among the control population and independence of exposure misclassification and genetic factors, both differential and nondifferential misclassification of a binary factor tend to underestimate a multiplicative interaction effect (22). However, no corresponding theorem applies for additive interaction effects, and thus the impact of misclassification on additive interaction effects is more difficult to predict (23).

Even small to modest errors in measuring environmental and/or genetic factors may result in substantial increases in sample size requirements to attain adequate statistical power to detect the individual effects or interactions between these factors (23). For example, the sample size required to detect an additive interaction between the *NAT2* genotype and an environmental exposure, such as smoking, when both the genotype and the exposure are perfectly measured is 1296 cases and similar number of controls (assuming a two-sided test with α-level = 5%, power = 80%, 1:1 case:control ratio, prevalence of the genotype and exposure of 50% and 25%, respectively, odds ratio of 1.0 for the effect of the slow acetylators among unexposed subjects, odds ratio of 2.0 for the effect of the exposure among intermediate/fast acetylators, and odds ratio of 3.0 for the exposure–genotype joint effect). If the exposure is measured with 80% sensitivity and 100% specificity and the genotype is perfectly measured, the sample size increases by about 30% to 1815 cases and similar number of controls. The sensitivity and specificity for the NAT2 acetylation phenotype as predicted from a genotype assay that determines the three most common *NAT2* alleles compared to an assay that determines 26 *NAT2* alleles has been found to be 93% and 100%, respectively, in a Caucasian population (24). Despite the small magnitude of the error, using the 3-allele rather than the 26-allele assay would further increase the required sample size by an additional 15% to 2137 cases and a similar number of controls. In addition to the increase in sample size, the estimates of effect for the *NAT2* genotype, the exposure, and the interaction parameter would be biased due to the presence of misclassification in the genotype and exposure determination. The required sample size to detect interactions between common genetic polymorphisms and environmental exposures tends to be very large (25–27). Therefore, efforts to collect almost perfect data on genotypes and highly accurate data on environmental factors are critical for the valid and efficient assessment of gene–environment interactions.

The magnitude of the bias introduced to measures of effect by misclassification depends on the type and magnitude of the interaction being evaluated, the prevalence of the environmental and genetic factors, and the misclassification probabilities (sensitivity and specificity for binary factors) (23). These parameters vary across study populations, e.g., differences in genotype prevalence between racial groups. Therefore, differences in study population characteristics and measurement instruments may contribute to explain disparate findings in the literature (28).

Confounding

The term *confounding* is used in epidemiology to denote some distortion of the estimate of effect of an exposure on disease risk, due to an extraneous factor that: (*1*) is a risk factor for the disease, (2) is associated (may or not be a causal association) with the exposure of interest in the source population, and (3) is not an intermediate step in the causal pathway between the exposure of interest and disease (8). Important confounding bias on estimates of effect can be taken into account and be reduced or eliminated by matching or by stratification and adjustment in data analysis, when strong potential confounders are identified and accurately measured. Unknown or hard to measure confounders sometimes can be taken into account by adjusting for a surrogate of the confounder. For instance, investigators often adjust for socio-economic status, not because it is a direct cause of disease, but because it is a surrogate for risk factors that could be related to the exposure of interest and thus act as confounders. Similarly, investigators often adjust for ethnic background of study participants because ethnicity is a surrogate for genetic and/or environmental risk factors for disease (responsible for observed differences in disease rates across ethnic groups) that can also be related to the exposure of interest.

When the exposure under study is a genetic factor, ethnicity is not only a surrogate for unknown or unaccounted risk factors, but also can be associated with the genetic factor, since allele frequencies often vary across ethnic groups. Therefore, failure to adequately account for ethnicity in the study design or analysis can bias estimates of allele effects on disease risk, even if the unknown/unaccounted risk factors for disease are not directly related to the allele of interest. This special type of confounding bias by ethnicity has been referred in the genetics field as *population stratification* (see Fig. 8.1 for a schematic representation of classical confounding and population stratification) (29). This potential for bias has been the tar-

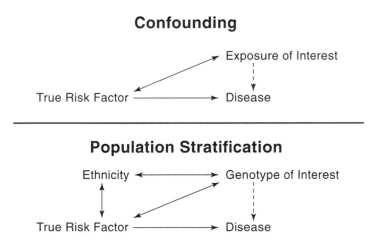

Figure 8.1 Schematic drawing of classical confounding and population stratification. In population stratification, ethnicity acts as a surrogate for an environmental or genetic factor that varies with ethnicity and with risk of disease. Broken unidirectional arrow indicates confounded or biased association. Solid unidirectional arrow indicates direction of causal relationship. Solid bi-directional arrow indicates correlation that may or may not be causal. Reprinted from reference (32).

get of strong criticism of classical epidemiologic study designs to evaluate allelic effects on disease risk (30).

Wacholder et al. (31,32) note that several conditions must be met in order to have substantial bias in the effects of common genetic polymorphisms on the risk of a particular disease due to population stratification:

1. There must be substantial variation across ethnicities in the frequency of the variant allele being considered.
2. There must be substantial variation across ethnicities in disease rates, *after* adjustment for risk factors that were collected in the study; typically, but certainly not always, these will reduce the inter-ethnic differences in disease rates.
3. The allele frequencies must track with the adjusted disease rates across ethnicities, for reasons other than the effects of the allele of interest.
4. Adjustment for ethnicity using ethnic information collected from study participants is not able to reduce bias to an acceptable level.

Investigators concerned about this potential bias have proposed greater use of alternative study designs to assess allele–disease associations that use unaffected relatives of case patients as control subjects, rather than unrelated control subjects as is common in classical epidemiology (30,33). (Chapter 7) Advocates of choosing control subjects because they are related to case patients, such a parents, siblings, or cousins, claim that this ensures that both groups of subjects have the same ethnic background, and thus eliminates the potential for population stratification (29,34,35). Others have argued that use of family controls is usually not necessary since the impact of this potential bias on estimates of allelic effects is likely to be small in well-conducted epidemiological studies of non-Hispanic Caucasians of European descent (3,31,32). The potential for bias in other ethnic groups has been less explored. Particularly, Wacholder et al. (32) have shown that the magnitude of this potential bias is greatest when the ethnic admixture in the population under study is composed of only two or three ethnic groups and tends to be less severe in more diverse populations, where ethnicity is more difficult to determine. Wacholder et al. (31) further argue that the bias is likely to be smaller for estimates of interaction effects between genetic and environmental factors than for allelic main effects, since the direction and magnitude of the bias present within each stratum are likely to be similar and thus tend to cancel out. Thus, the potential for what is likely to be a small bias does not justify the use of related controls in studies of genetic factors (31).

In fact, the use of related controls might have important biases, inefficiencies, and logistical challenges of its own that need to be considered (31,36,37). For instance, in studies of adulthood diseases such as many cancers, parents and to a lesser extend siblings, are likely to be unavailable for study. This can be an important source of bias, since the reasons for availability of a sibling or parent can be related to the exposures of interest, e.g., fertility, social relationships, or socioeconomic and occupational status related to residential mobility or migration (31). In addition, studies of unrelated controls can be more powerful to evaluate genetic effects than studies using unaffected siblings as controls (38,39). Furthermore, overmatching on environmental risk factors can reduce the power to evaluate effects of environmental exposures in studies that use related controls (31). Study designs using unaffected family controls may have a role in certain situations, such as in studies of an interaction between an environmental factor and a rare allele (40) when evaluating the effect of an environmental exposure in a subgroup defined by a rare allele (31), or in studies of perinatal or childhood diseases (35). However, the reasons to use family controls in these instances are not to account for population stratification.

New methods to control for population stratification that use characteristics of the genome from study participants, in addition to self-reported information on ethnic background of study participants, have been proposed (14,41–43). These methods might be proven to be superior to adjustment by self-reported ethnicity if the risk factors responsible for unaccounted variations in disease rates across ethnicities are genetic rather than environmental, since genome-based methods should be better surrogates for genetic risk factors than self-reported ethnicity (31). However, self-reported ethnicity is likely to be a better surrogate for environmental risk factors than genome-based methods, since people's behavior is likely to be closer to the ethnicity they identified themselves with, than to the ethnic composition of their genome (31). Therefore, for diseases for which variations in rates across ethnicities are thought to be mainly environmental, such as cancer, self-reported ethnicity might be a better method to control for population stratification than newer genome-based methods (31).

Subgroup Analysis and Multiple Comparisons

There are many population subgroups that one might want to consider in epidemiologic studies, such as subgroups defined by the presence or absence of an allelic variant, environmental exposure such as smoking and diet, gender, and age. Although evaluating population subgroups might uncover small effects present only in subgroups, this practice can increase the chance of false-positive findings; a simple dichotomization gives three chances of a positive result—overall and in each subgroup. An additional problem with subgroup analysis is that a significant finding in one subgroup and not in the other is often misinterpreted as the presence of effect modification or interaction by the stratifying factor. Only proper assessment of statistical interaction or effect modification by performing adequate tests of interaction or estimating the interaction parameters and their confidence interval should be used to make conclusions about the presence or absence of an interaction. The number of false-positive findings from subgroup analysis can be reduced when considering subgroups with biologic plausibility, designing studies with enough subjects in subgroups of interest, and by careful data analysis and conservative interpretation.

The use of methods to adjust for making multiple comparisons and reduce the rate of false-positive findings has been the topic of intense debate in the epidemiologic literature (44–48). The main argument against the use of adjustments for multiple comparisons in epidemiology has been the increased probability of missing important findings not expected by the investigator (44–46). It has also been argued that the aim of statistical analysis in observational epidemiology is to detect and describe patterns in the data by estimating measures of effect and their confidence intervals, rather than by testing statistical hypotheses, let alone by making adjustments when many tests are performed (45,46). Many epidemiologists would agree with these arguments when well-defined scientific hypothesis are evaluated. Study of genetic polymorphisms with strong *a priori* biologic plausibility for being associated with disease, such *NAT2* genotype and risk of bladder cancer (49) would be an example of this type of hypotheses. However, the use of high throughput, economical techniques to determine genetic polymorphisms, gives investigators the exciting opportunity to evaluate associations between thousands of alleles and disease risk. Although the vast majority of these alleles will not be causally associated with disease, "screening" large numbers of alleles can lead to the identification of important alleles that would be missed by a narrower exploration that focused only on alleles that investigators believe could be related to disease. The main problem with this approach is that because the prior probability of a positive finding or probability of the alternative hypothesis being true is very low, it can lead to a very high proportion of false-positive findings, unless a more conservative standard than an α-level of 0.05 is used (50,51).

Two mutually exclusive outcomes are possible when a null hypothesis (H_0), such as H_0: genotype X is not associated with disease, is rejected: (*1*) H_0 is rightly rejected (i.e., genotype X is truly associated with disease), or (*2*) H_0 is wrongly rejected (i.e., genotype X is not truly associated with disease). The probability of rightly rejecting H_0 is the study power (i.e., probability of rejecting H_0 when Ha is true) and the probability of wrongly rejecting H_0 depends on the α-level or size of the test (i.e., probability of rejecting H_0 when H_0 is true). Thus, for any given power, lowering the α-level will decrease the chances of wrongly rejecting H_0 and, consequently, will reduce the proportion of false-positive findings (i.e., the percentage of the time when H_0 is wrongly rejected). This relationship can be easily derived by Bayes Theorem (see Appendix). Figure 8.2 shows the inverse relationship between the proportion of false-positive findings and the prior probability of the alternative hypothesis in a two-sided test, when the study power is fixed at 80% and the α-level takes the values of 0.05, 0.01, and 0.001. By comparing the three different curves, one can clearly see how decreasing the α-level results in substantial reduction of the proportion of false-positive findings, especially when the prior probability is low.

Similarly, for any given α-level, increasing the study power will increase the chances of rightly rejecting H_0 and, consequently, will reduce the proportion of false-positive findings (Appendix; Figure 8.3). Thus, there are two related conse-

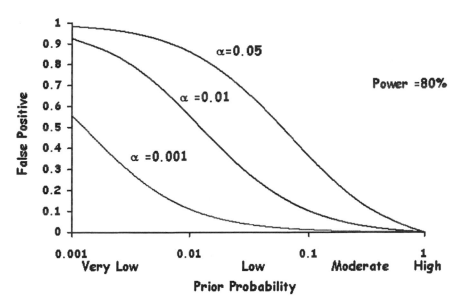

Figure 8.2 Relationship between percentage of false positive findings and prior probability of the alternative hypothesis, for a two-sided test with 80% power and α-level values of 0.05, 0.01, and 0.001.

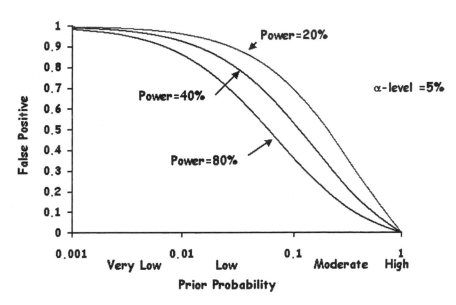

Figure 8.3 Relationship between percentage of false positive findings and prior probability of the alternative hypothesis, for a two-sided test with α-level of 0.05, and power values of 20%, 40%, and 80%.

137

quences of small, underpowered studies: reduced probability of detecting a true association between a factor of interest and disease, and consequently, a higher proportion of false-positive findings than larger studies with adequate study power. The later characteristic of underpowered studies is often not appreciated, and it is likely to explain in part why small studies tend to generate results that are not replicated in subsequent larger studies.

A strategy to reduce the number of comparisons when evaluating multiple candidate genes could be to group polymorphisms that are thought to have similar effects. Examples include grouping allelic variants of a single gene with similar phenotype, grouping allelic variants from different genes acting in a common pathway, or constructing individual scores for a particular pathway based on the number of allelic variants that an individual has in that pathway. This approach makes the strong assumption of identical effects for the grouped allelic variants. However, it might prove useful in some situations where there is substantial knowledge on the function of the allelic variants under study, so that groups can be formed with the use of strong biologic priors.

Several methods can be used to adjust for multiple comparisons and thus reduce the proportion of false-positive findings. One of the simplest methods to correct p-values for multiple comparisons is the Bonferroni method. This method consists on setting the desired significance level for all tests being performed (or probability of making at least one type I error after performing all tests), and then divide this total error probability equally among all tests performed. This is accomplished by setting the significance level in each individual test to the overall significance level divided by the number of tests performed. This method treats all hypotheses being evaluated equally without taking into account the prior probability of each hypothesis. As a consequence, the most promising hypotheses, such as allelic variants with known and relevant functional significance, are being penalized with low power to be detected because of the evaluation of many other less promising hypotheses. In spite of being one of the most conservative methods, the increase in sample size required to maintain a given statistical power after Bonferroni correction is not as large as one might believe (52). Figure 8.4 depicts the increase in sample size requirements to detect a genotype odds ratio of 1.5 with a two-sided test with 5% alpha-level and 80% power, after adjusting for multiple comparisons by using the Bonferroni method. In this example, testing 10,000 genotypes would only require a threefold increase in sample size over uncorrected tests. An alternative method is to vary the p-value needed to call each test significant according to the prior probability of a true effect (51). This approach is similar to what investigators do informally when they give more weight to findings for hypotheses for which they have higher priors.

Other approaches used to take into account multiple comparisons are the semi-Bayes and empirical-Bayes methods that, unlike the test-oriented methods, also provide "corrected" point estimates and confidence intervals (53,53–55). A requisite for these methods is that the parameters being estimated can be grouped into sets

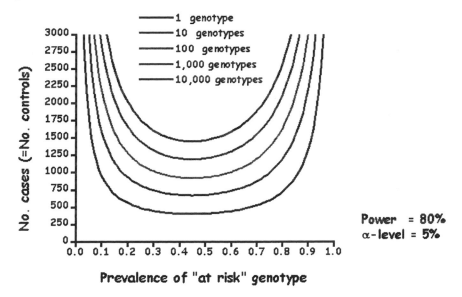

Figure 8.4 Number of cases (= no. of controls) required to detect a genotype–disease odds ratio of 1.5 for varying prevalence values of the "at risk" genotype, after adjusting for multiple comparisons using the Boferroni correction (assuming a two-sided test with 80% power and overall α-level of 0.05).

within which all parameters can be considered similar or "excheangable" (55). The multiplicity of comparisons being made within each exchangeable group of parameters is used to try to improve individual estimates of effect within that group. Multiple genetic polymorphisms with low prior knowledge about their functional relevance could be considered to be in the same exchangeable group, since most of them will not be truly associated with disease. Similarly, polymorphisms in related genes that are thought to have similar effects on disease could also be grouped together. Semi-Bayes adjustments use *prior* information about the magnitude of effects being estimated in one exchangeable group (i.e., a likely range of effects) to shrink outlying estimates within that group. On the other hand, empirical Bayes adjustments use information about the observed effects rather than *prior* information to shrink outlying estimates (55). Specifically, the outlier estimates are shrunk towards the geometric mean of the observed effects within an exchangeable group, to a larger or lesser degree based on their observed variance of the estimates. As Steenland et al. pointed out, empirical Bayes adjustments attempt to anticipate the "regression to the mean" of outlier observations that tends to occur when new data are obtained (55). When investigators have strong prior information about the effects being estimated, such as a reasonable range in which they are likely to fall, semi-Bayes adjustment might be preferable to empirical-Bayes adjustments (55,56). Empirical and semi-Bayes estimators can be obtained with the use of hierarchical, multilevel regression models (56,57,58–61).

Any method used to control for study-wise error rate will inevitably reduce study power for a given sample size. Using different approaches with different sets of assumptions in a single study can be useful in assessing the robustness of the results. In addition, coordinating analysis through consortiums of comparable studies will increase the total number of subjects studied, perhaps compensating for the loss in power due to the use of methods to control for multiple comparisons. Replication of results in studies conducted in different populations should also reduce the probability that study results are due to bias specific to a particular study. Lack of replication of allele effects across study populations could be due to the presence of linkage disequilibrium between the allele under study and the true disease allele. Thus, inconsistency of allele findings could indicate that the allele being evaluated is not a true disease allele, but instead is in linkage disequilibrium with the disease allele in some but not all populations.

Concluding Remarks

Adherence to the principles of epidemiologic studies and rigorous statistical analysis and interpretation of results is critical to allow correct inferences from epidemiology studies. In addition to the general principles covered in this chapter, inferential issues specific to the use of new techniques to study genetic susceptibility to disease, such as use of whole genome linkage disequilibrium studies (62), haplotype tagging SNPs (63,64), and pooling DNA from different individuals to screen for relevant variants (65–67) will also need to be considered. Epidemiologic studies that measure susceptibility genes should provide opportunities for detecting lower levels of risk due to certain environmental exposures, to illuminate pathways of action that may help identify previously unsuspected carcinogens, and to detect gene–environment interactions that may give rise to new public health and clinical strategies aimed at preventing and controlling disease.

References

1. Rothman N, Wacholder S, Caporaso NE, et al. The use of common genetic polymorphisms to enhance the epidemiologic study of environmental carcinogens. Biochim Biophys Acta, 2001;1471:C1–C10.
2. Risch NJ. Searching for genetic determinants in the new millennium. Nature 2000; 405:847–856.
3. Caporaso N, Rothman N, Wacholder S. Case-control studies of common alleles and environmental factors. J Natl Cancer Inst Monogr 1999;26:25–30.
4. Cardon LR, Bell JI. Association study designs for complex diseases. Nat Rev Genet 2001;2:91–99.
5. Langholz B, Rothman N, Wacholder S, et al. Cohort studies for characterizing measured genes. J Natl Cancer Inst Monogr 1999;26:39–42.
6. Clayton D, McKeigue PM. Epidemiological methods for studying genes and environmental factors in complex diseases. Lancet 2001;358:1356–1360.

7. Wacholder S, Garcia-Closas M, Rothman N. Study of genes and environmental factors in complex diseases. Lancet 2002;359:1155.

8. Rothman KJ, Greenland S. Modern Epidemiology. Philadelphia: Lippincott-Raven, 1998.

9. Miettinen OS. The "case-control" study: valid selection of subjects. J Chronic Dis 1985; 38:543–548.

10. Wacholder S, McLaughlin JK, Silverman DT, et al. Selection of controls in case-control studies. I. Principles. Am J Epidemiol 1992;135:1019–1028.

11. Wacholder S, Silverman DT, McLaughlin JK, et al. Selection of controls in case-control studies. II. Types of controls. Am J Epidemiol 1992;135:1029–1041.

12. Wacholder S, Silverman DT, McLaughlin JK, et al. Selection of controls in case-control studies. III. Design options. Am J Epidemiol 1992;135:1042–1050.

13. Wacholder S, Chatterjee N, Hartge P. Joint effects if genes and environment distorted by selection bias: implications for hospital-based case-control studies. Cancer Epidemiol Biomarkers Prev 2002;11:885–889.

14. Pritchard JK, Donnelly P. Case-control studies of association in structured or admixed populations. Theor Popul Biol 2001;60:227–237.

15. Garcia-Closas M, Egan KM, Abruzzo J, et al. Collection of genomic DNA from adults in epidemiological studies by buccal cytobrush and mouthwash. Cancer Epidemiol Biomarkers Prev 2001;10:687–696.

16. Zheng S, Ma X, Buffler PA, et al. Whole genome amplification increases the efficiency and validity of buccal cell genotyping in pediatric populations. Cancer Epidemiol Biomarkers Prev 2001;10:697–700.

17. Armstrong BK, White E, Saracci R. Exposure measurement error and its effects, in Principles of Exposure Measurement in Epidemiology, Armstrong BK, White E, Saracci R (eds.). New York: Oxford University Press, 1992, pp. 49–75.

18. Greenland, S. The effect of misclassification in the presence of covariates. Am J Epidemiol 1980;112:564–569.

19. Flegal KM, Brownie C, Haas JD. The effects of exposure misclassification on estimates of relative risk. Am J Epidemiol 1986;123:736–751.

20. Dosemeci M, Wacholder S, Lubin JH. Does nondifferential misclassification of exposure always bias a true effect toward the null value? Am J Epidemiol 1990;132:746–748.

21. Wacholder S, Hartge P, Lubin JH, et al. Non-differential misclassification and bias towards the null: a clarification. Occup Environ Med 1995;52:557–558.

22. Garcia-Closas M, Thompson WD, Robins JM. Differential misclassification and the assessment of gene-environment interactions in case-control studies. Am J Epidemiol 1998;147:426–433.

23. Garcia-Closas M, Rothman N, Lubin J. Misclassification in case-control studies of gene-environment interactions: assessment of bias and sample size. Cancer Epidemiol Biomarkers Prev 1999;8:1043–1050.

24. Deitz AC, Garcia-Closas M, Rothman N, et al. Impact of misclassification in genotype-disease association studies: example of N-acetyl 2 (NAT2) smoking and bladder cancer. Proc Am Assoc Cancer Res 2000;41:559.

25. Garcia-Closas M, Lubin JH. Power and sample size calculations in case-control studies of gene-environment interactions: comments on different approaches. Am J Epidemiol 1999;149:689–692.

26. Hwang SJ, Beaty TH, Liang KY, et al. Minimum sample size estimation to detect gene-environment interaction in case-control designs. Am J Epidemiol 1994;140:1029–1037.

27. Khoury MJ, Beaty TH, Hwang SJ. Detection of genotype-environment interaction in case-control studies of birth-defects—how big a sample-size. Teratology 1995; 51:336–343.

28. Rothman N, Stewart WF, Caporaso NE, Hayes RB. Misclassification of genetic suscep-tibility biomarkers: implications for case-control studies and cross-population compar-isons. Cancer Epidemiol Biomarkers Prev, 1993;2:299–303.
29. Lander ES, Schork NJ. Genetic dissection of complex traits. Science 1994;265:2037–2048.
30. Thomas DC, Witte JS. Population statification: a problem for case-control studies of can-didate-gene associations? Cancer Epidemiol Biomarkers Prev 2002;11(6):505–512.
31. Wacholder S, Rothman N, Caporaso NE. Counterpoint: bias from population stratifica-tion is not a major threat for the validity of conclusions from epidemiologic studies of common polymorphisms and cancer. Cancer Epidemiol Biomarkers Prev, In Press, 2002.
32. Wacholder S, Rothman N, Caporaso N. Population stratification in epidemiologic stud-ies of common genetic variants and cancer: quantification of bias. J Natl Cancer Inst 2000;92:1151–1158.
33. Khoury MJ, Yang QH. The future of genetic studies of complex human diseases: An epi-demiologic perspective. Epidemiology 1998;9:350–354.
34. Spielman RS, Ewens WJ. The TDT and other family-based tests for linkage disequilib-rium and association. Am J Hum Genet 1996;59:983–989.
35. Wilcox AJ, Weinberg CR, Lie RT. Distinguishing the effects of maternal and offspring genes through studies of "case-parent triads." Am J Epidemiol 1998;148:893–901.
36. Caporaso N. Chapter 4. Selection of candidate genes for population studies. IARC Sci Publ 1999;148:23–36.
37. Gauderman WJ, Witte JS, Thomas DC. Family-based association studies. J Natl Cancer Inst Monogr 1999;26:31–37.
38. Teng J, Risch N. The relative power of family-based and case-control designs for link-age disequilibrium studies of complex human diseases. II. Individual genotyping. Ge-nome Res 1999;9:234–241.
39. Li Z, Gail MH, Pee D, Gastwirth JL. Statistical properties of Teng and Risch's sibship type tests for detecting an association between disease and a candidate gene. Hum Hered 2002;53:114–129.
40. Witte JS, Gauderman WJ, Thomas DC. Asymptotic bias and efficiency in case-control studies of candidate genes and gene-environment interactions: basic family designs. Am J Epidemiol 1999;149:693–705.
41. Devlin B, Roeder K, Bacanu SA. Unbiased methods for population-based association studies. Genet Epidemiol 2001;21:273–284.
42. Bacanu SA, Devlin B, Roeder K. Association studies for quantitative traits in structured populations. Genet Epidemiol 2002;22:78–93.
43. Pritchard JK, Rosenberg NA. Use of unlinked genetic markers to detect population strat-ification in association studies. Am J Hum Genet 1999;65:220–228.
44. Rothman KJ. No adjustments are needed for multiple comparisons. Epidemiology 1990;1:43–46.
45. Savitz DA, Olshan AF. Multiple comparisons and related issues in the interpretation of epidemiologic data. Am J Epidemiol 1995;142:904–908.
46. Savitz DA, Olshan AF. Describing data requires no adjustment for multiple comparisons: a reply from Savitz and Olshan. Am J Epidemiol 1998;147:813–814.
47. Manor O, Peritz E. Re: "Multiple comparisons and related issues in the interpretation of epidemiologic data". Am J Epidemiol 1997;145:84–85.
48. Thompson JR. Invited commentary: Re: "Multiple comparisons and related issues in the interpretation of epidemiologic data". Am J Epidemiol 1998;147:801–806.
49. Marcus PM, Vineis P, Rothman N. NAT2 slow acetylation and bladder cancer risk: a meta-analysis of 22 case-control studies conducted in the general population. Pharmaco-genetics 2000;10:115–122.

50. Browner WS, Newman TB. Are all significant P values created equal? The analogy between diagnostic tests and clinical research. JAMA 1987;257:2459–2463.
51. Wacholder S, Chanock S, García-Closas M, Elghormli L, Rothman N. Assessing the probability of false positive reports in associative studies. In preparation, 2003.
52. Risch N, Merikangas K. Genetic analysis of complex diseases—response. Science 1997;275:1329–1330.
53. Greenland S, Robins JM. Empirical-Bayes adjustments for multiple comparisons are sometimes useful. Epidemiology 1991;2:244–251.
54. Greenland S, Poole C. Empirical-Bayes and semi-Bayes approaches to occupational and environmental hazard surveillance. Arch Environ Health 1994;49:9–16.
55. Steenland K, Bray I, Greenland S, et al. Empirical Bayes adjustments for multiple results in hypothesis-generating or surveillance studies. Cancer Epidemiol Biomarkers Prev 2000;9:895–903.
56. Greenland, S. Methods for epidemiologic analyses of multiple exposures: a review and comparative study of maximum-likelihood, preliminary-testing, and empirical-Bayes regression. Stat Med 1993;12:717–736.
57. Witte JS, Greenland S, Haile RW, et al. Hierarchical regression analysis applied to a study of multiple dietary exposures and breast cancer. Epidemiology 1994;5:612–621.
58. Witte JS, Greenland S. Simulation study of hierarchical regression. Stat Med 1996; 15:1161–1170.
59. Greenland S. Principles of multilevel modelling. Int J Epidemiol 2000;29:158–167.
60. Witte JS, Greenland S, Kim LL. Software for hierarchical modeling of epidemiologic data. Epidemiology 1998;9:563–566.
61. Witte JS, Greenland S, Kim LL, et al. Multilevel modeling in epidemiology with GLIM-MIX. Epidemiology 2000;11:684–688.
62. Kruglyak L. Prospects for whole-genome linkage disequilibrium mapping of common disease genes. Nat Genet 1999;22:139–144.
63. Fallin D, Cohen A, Essioux L, et al. Genetic analysis of case/control data using estimated haplotype frequencies: application to APOE locus variation and Alzheimer's disease. Genome Res 2001;11:143–151.
64. Johnson GC, Esposito L, Barratt BJ, et al. Haplotype tagging for the identification of common disease genes. Nat Genet 2001;29:233–237.
65. Risch N, Teng J. The relative power of family-based and case-control designs for linkage disequilibrium studies of complex human diseases I. DNA pooling. Genome Res 1998;8:1273–1288.
66. Pfeiffer RM, Rutter JL, Gail MH, Struewing J, Gastwirth JL. Efficiency of DNA pooling to estimate joint allele frequencies and measure linkage disequilibrium. Genet Epidemiol 2002;22:94–102.
67. Barcellos LF, Klitz W, Field LL, et al. Association mapping of disease loci, by use of a pooled DNA genomic screen. Am J Hum Genet 1997;61:734–747.

Appendix

The relationship between proportion of false-positive findings, α-level, study power, and prior probability of the alternative hypothesis can be easily shown by Bayes Theorem.

Imagine a two-sided test where:

H_0: Genotype X is not associated with disease risk
H_a: Genotype X is associated with disease risk

By Bayes Theorem,

Prob (H_0 True | Reject H_0) = Prob (Reject H_0|H_0 is True) • Prob (H_0 is True) /
[Prob (Reject H_0|H_0 is True) • Prob (H_0 is True) +
Prob (Reject H_0|H_a is True) • Prob (H_a is True)]

Or equivalently,

Proportion of false + = α-level • (1 − Prior Prob) /
(α-level • (1 − Prior Prob) + Power • Prior Prob)

9

Applications of human genome epidemiology to environmental health

*Samir N. Kelada, David L. Eaton, Sophia S. Wang,
Nathaniel R. Rothman, and Muin J. Khoury*

Research exploring the role of genetics in determining susceptibility to environmentally induced disease has grown considerably over the last few decades. Many recent epidemiologic investigations have examined associations between polymorphic genes that code for enzymes involved in xenobiotic biotransformation (i.e., metabolism) and disease and have generated interesting findings. These results imply that genetic variability may affect the response to exposure to environmental health hazards. However, without the use of exposure assessment methods traditionally employed in environmental health science research, these studies have not been able to investigate and characterize gene–environment interactions with environmental health hazards. In addition, many of the results from gene–disease association studies have not been replicated in subsequent studies, casting doubt on their validity and leaving the environmental health community with uncertain results with which to proceed.

In this chapter, we assess the integration of genetics into environmental health research using the same exposure leading to disease paradigm traditionally used by environmental health scientists, adding genetics to the existing paradigm as a potential modifier of dose or effect of the initial exposure. To identify gaps in current knowledge, we classify examples of gene–environment interaction into one of two categories on the basis of evidence from laboratory and epidemiologic data. Finally, we describe the benefits of applying this model to future research efforts, and we discuss issues to consider for investigators wishing to pursue this type of endeavor.

Environmental Exposures and Human Genetic Variation

Much of the impetus for this area of research has come from the field of pharmacogenetics, which is primarily concerned with the study of genetic variation in drug efficacy and toxicity. It has been recognized for many decades that individual differences in response to pharmacologic treatment, exhibited as drug toxicity or a lack of therapeutic effect, are often due to genetic differences that result in altered rates of biotransformation (metabolism). Notable examples include nerve damage among individuals homozygous for some variants of the *N*-acetyltransferase 2 gene ("slow acetylators") given isoniazid as an antituberculosis therapy, hemolytic anemia among glucose 6-phosphate dehydrogenase-deficient patients given aminoquinoline antimalarial drugs, and varied rates of biotransformation of debrisoquine, an antihypertensive drug, due to genetic variation at the *CYP2D6* locus (1).

The process of biotransformation, i.e., the enzymatic alteration of foreign or xenobiotic compounds, is conventionally divided into two phases. Phase I enzymes introduce new (or modify existing) functional groups (e.g., -OH, -SH, -NH$_3$) to xenobiotics, and are primarily catalyzed by the cytochrome P450 enzymes (CYPs), although numerous other oxidases, reductases, and dehydrogenases may also participate. These intermediates are then conjugated with endogenous ligands during Phase II, increasing the hydrophilic nature of the compound, facilitating excretion. Enzymes involved in Phase II include the *N*-acetyltransferases (NATs), Glutathione S-Transferases (GSTs), Uridine Diphosphate (UDP) Glucuronosyltransferases, epoxide hydrolases, and methyltransferases. Phase I and II reactions are catalyzed by enzymes collectively known as xenobiotic metabolism enzymes (XMEs). Although most tissues have some XME activity, XMEs are most abundant in the liver. A balance between Phase I and II enzymes is generally necessary to promote the efficient detoxification and elimination of xenobiotics, thereby protecting the body from injury caused by exposure (2).

Mutations in the genes encoding these enzymes and other proteins result from stochastic genetic processes and may accumulate in the population depending on selective pressures. If the frequency of a specific genetic variant reaches 1% or more in the population, it is referred to as a polymorphism. A polymorphism may have no effect (i.e., is "silent"), or it may be considered functional if it results in altered catalytic function, stability, and/or level of expression of the resulting protein. Functional polymorphisms in XMEs include: (*1*) point mutations in coding regions of genes resulting in amino acid substitutions, which may alter catalytic activity, enzyme stability, and/or substrate specificity; (*2*) duplicated or multiduplicated genes, resulting in higher enzyme levels; (*3*) completely or partially deleted genes, resulting in no gene product; and (*4*) splice site variants that result in truncated or alternatively spliced protein products (3). Mutations in the regulatory regions of genes may affect the amount of protein expression as well, and mutations in other noncoding regions may affect mRNA stability or mRNA splicing. Most research in genetics in environmental health has focused on these types of functional genetic variation. Nevertheless, even presumed "nonfunctional" polymorphisms might be useful in some cases either because of a subtle function that has yet to be found or because such polymorphisms may be in linkage disequilibrium with functional polymorphisms in the same region, thus serving as markers for phenotypic effects.

About 90% of all DNA polymorphisms occur as single nucleotide polymorphisms (SNPs), i.e., single base pair substitutions (the first type of functional polymorphism) (4). More than 1,255,000 SNPs have been identified and catalogued as a result of multiple research efforts (see the SNPs Consortium Web site listed in Appendix 9.1). There are estimated to be three to four SNPs in the average gene and roughly 120,000 common coding region SNPs, of which ~40% are expected to be functional (5). These estimates do not include polymorphisms outside the coding region of genes, and thus the total number of SNPs affecting protein function can be expected to be greater.

Functional polymorphisms in XMEs can affect the balance of metabolic intermediates produced during biotransformation, and some of these intermediates can bind and induce structural changes in DNA or binding other critical macromolecules, such as sulfhydryl-containing proteins. Similarly, polymorphisms in DNA repair enzymes can affect an individual's ability to repair DNA damage induced by some exposures, such as ultraviolet radiation. The interindividual differences in these and other components of the human genome that relate to environmental exposures have therefore been predicted to modify environmental disease risk (6). In addition to polymorphisms, age, sex, hormones, and behavioral factors, such as cigarette smoking, alcohol consumption, and nutritional status, can influence the expression of Phase I and II biotransformation genes (7) and thus are also important in understanding environmental disease risk.

One can contrast the role of polymorphisms in XMEs and other components of the environmental response system with genetic variants that are highly penetrant (i.e., that almost invariably lead to disease) but have low population frequency. The interest and focus here is on the role of common genetic variants that alter the effect of exposures that may lead to disease states, or their precursors, and hence are of lower penetrance. Though the individual risk associated with these polymorphisms is often low, they potentially have greater public health relevance (i.e., population-attributable risk) because of their high population frequency (8).

A comprehensive effort to identify genetic polymorphisms in genes involved in environmentally induced disease, known as the Environmental Genome Project (EGP), was initiated by the National Institute of Environmental Health Sciences (NIEHS) in 1998 (9). In addition to the identification of polymorphisms, the EGP aims to characterize the function of these polymorphisms and supports epidemiologic studies of gene–environment interactions as well. Like the Human Genome Project, the EGP has devoted substantial resources to the ethical, legal, and social issues related to this project.

Examples of Genetic Effect Modifiers

The working hypothesis that has typically been employed is that for the majority of genetic polymorphisms that alter responses to chemical hazards, the genetic difference does not result in a qualitatively different response, but rather induces a shift in the dose–response relationship. Thus a genetic variant in an XME that decreases the catalytic efficiency of an enzyme that detoxifies a particular drug might make the standard dose of that drug toxic. This concept extends not only to the acute effects of drugs, but also potentially to chronic response to nondrug chemicals found in the workplace and general environment. Below we describe several examples of "gene–environment interactions" that illustrate the potential public health implications, as well as difficulties in interpretation, of this type of research.

The relationship between aromatic amine exposure, N-acetylation polymorphism (NAT2), and bladder cancer is a classic illustration of the principle of dose–effect modification of an environmental exposure by genetic polymorphism. An initial study by Lower et al. (10) suggested that the effect of exposure to aromatic amines (bladder cancer), by occupation (e.g., dye industry) or smoking, differed by NAT2 phenotype. A preponderance of slow acetylators existed among exposed persons, and subsequent studies have confirmed these results (11,12).

Recently, Marcus and colleagues conducted a case-series meta-analysis of 16 studies of the NAT2 × smoking interaction in bladder cancer (13). Across all studies, they calculated an odds ratio (OR) of 1.3 (95% confidence interval [CI]: 1.0, 1.6) for smokers who are slow acetylators compared with smokers who are rapid acetylators, verifying that smokers who are slow acetylators have a modestly increased risk (13). Limiting the study selection to European studies with large sample sizes (number of cases = 150), the OR was 1.7 (95% CI: 1.2, 2.3). Different patterns of tobacco use and tobacco type may account for some of these differences. In addition, using estimates of the prevalence of smoking and NAT2 genotype, they predicted bladder cancer risk for smokers and nonsmokers by acetylator status, designating never-smoker rapid acetylators as the reference category. Nonsmoking slow acetylators were predicted to have no increase in risk (OR = 1.10), ever-smoking rapid acetylators have about two times the risk (OR = 1.95), and ever-smokers who are slow acetylators have about threefold higher risk (OR = 3.21). Marcus et al. also estimated that the population-attributable risk of the gene–environment interaction was 35% for slow acetylators who had ever smoked and 13% for rapid acetylators who had ever smoked.

In the laboratory setting, complementary experiments can be designed to gain understanding of the biologic basis of the observed effect. This ultimately contributes to the argument of causality. Primary human cell lines, transient and stable transfection assays in cell lines, and transgenic animal models have frequently been used to investigate these questions. With respect to aromatic amines, NAT2, and bladder cancer, in vitro and in vivo studies have demonstrated that polymorphic N-acetylation of some aromatic amines can result in the bioactivation of these procarcinogens in the bladder (14–16). After N-oxidation of aromatic amines such as 4-aminobiphenyl or 2-naphthylamine by CYP1A2 in the liver, O-acetylation of the resulting hydroxylamine by NAT2 can produce unstable acetoxy esters that decompose to form highly electrophilic aryl nitrenium ion species. In addition, the formation of the acetoxy ester, a proximate carcinogen, can proceed through N-acetylation and

N-oxidation reactions that yield *N*-hydroxy-*N*-acetyl aromatic amines, which then form the acetoxy ester through *N, O*-acetyltransfer catalyzed by NAT2. In slow acetylators, initial acetylation in the liver is less efficient, and hence biotransformation of the aromatic amine is more likely to proceed through the CYP1A2 route. Subsequently, the hydroxylated aromatic amine can be further bioactivated in the bladder, either enzymatically or nonenzymatically, potentially leading to DNA binding and point mutations. This is considered a likely mechanism of initiation of bladder carcinogenesis (17–19). Thus, after the early findings by Lower et al. (10), the concerted efforts of epidemiologic and toxicologic studies have quantitatively evaluated this gene–environment interaction and elucidated a probable mechanism.

Recent research exploring genetic modifiers of other common exposures with significant public health importance have begun to yield interesting findings. In addition to gene–environment interactions that link exposures, polymorphisms, and disease states, associations of particular exposures with biomarkers of exposure or effect and polymorphisms have been evaluated. A nonexhaustive list of these exposures and biomarkers or diseases with their potential genetic effect modifiers is given in Table 9.1. (Please see Appendix 9.2 for additional information about the genes.) The evidence for these relationships has been classified according to the whether the associations were proposed from basic scientific laboratory evidence (classified as 2) or from laboratory evidence with suggestive epidemiologic data in some studies (classified as 1). The purpose of using this classification system is to identify gaps in knowledge about the exposure–disease association and effect modification that merit further investigation. The sources of these potential modifiers come from several different fields, including biochemistry, genetics, physiology, pharmacology, and pharmaceutics.

Table 9.1 shows several different types of exposures, including exposures to industrially produced compounds and byproducts (e.g., butadiene and dioxin), substances in the diet (e.g., alcohol and aflatoxin B_1), and both voluntary and involuntary examples of exposure (e.g., tobacco smoke and environmental tobacco smoke). As would be expected, some genes appear to be associated with several different exposures. This can be partially attributed to the relatively nonspecific roles of their gene products in biotransformation of exogenous substrates. It is also likely that once genotyping methods for a particular gene have been developed and streamlined, its role in several pathways will be explored. Few examples in Table 9.1, however, have clear and consistent evidence demonstrating effect modification by polymorphisms, which may be attributable to small sample size or other study design issues.

An example of the evolving knowledge of effect modification by polymorphisms is that of exposure to aflatoxin B_1, a mycotoxin found in some foodstuffs, and an established risk for hepatocellular carcinoma (HCC), especially when combined with hepatitis virus exposure (20). The biotransformation of aflatoxin B_1 proceeds through a CYP450-mediated oxidation and then through a glutathione S-transferase, epoxide hydrolase, and/or glucuronosyl transferase catalyzed reactions to yield excretable

Table 9.1 Proposed Genetic Effect Modifiers of Common Exposures

Exposure	Outcome	Gene(s)	Classification[a]	References
Arsenic	Arsenic metabolites in urine	GSTM1	2	56,57
		GSTT1	2	
		methyl-transferase	2	
Beryllium	Chronic beryllium disease	HLA-DP β_1	1	58–60
Lead	Blood lead level	ALAD	1	61–63
	Bone lead level	ALAD	1	64,65
		VDR	1	66,67
Mercury	Atypical porphyrin profiles	CPOX	2	68,69
		UROD	2	70,71
Alcohol	Esophageal cancer	ALDH2	1	72–76
Aflatoxin B$_1$	Aflatoxin-albumin adducts	CYP1A2	2	77
		CYP3A4	2	78
	Hepatocellular carcinoma	GSTM1	1	22,23
		EPHX1	1	
Heterocyclic amines	Colon cancer	NAT2	1	79–82
	Breast cancer	NAT2	1	83
		SULT1A1	1	84
Aromatic amines (dye industry)	Bladder cancer	NAT2	1	11,12
Halomethanes	Metabolite levels in blood	GSTT1	2	85,86
Benzene	Hematotoxicity	CYP2E1	1	87,88
		NQO1	1	
	Sister chromatid exchange in lymphocytes	GSTT1	1	89
Halogenated solvents (e.g., TCE)	Renal cell carcinoma	GSTT1	1	90,91
Organochlorine compounds (e.g., PCBs, TCDD)	Immunotoxicity	CYP1A1	2	92
		CYP1A2	2	93,94
		AHR	2	92
Organophosphate pesticides (OPs)	Chromosomal aberrations	PON1	1	95
		GSTM1	1	
		GSTT1	1	
		CYP3A4	2	96,97
Butadiene	Sister chromatid exchange in lymphocytes	GSTT1	1	98–100
Lipopolysaccharide (endotoxin)	Decreased forced expiratory volume (FEV$_1$)	TLR4	1	31
Hay dust	TNFα production in hypersensitivity pneumonitis	TNFα	1	101
Ozone	Influx of inflammatory cells in the lung	TLR4	2	102

(continued)

150

Table 9.1 Proposed Genetic Effect Modifiers of Common Exposures (*continued*)

Exposure	Outcome	Gene(s)	Classification[a]	References
Airborne polycyclic aromatic hydrocarbons (PAHs)	PAH metabolites in urine, DNA adducts, or measures of genotoxicity	CYP1A1	1	103–110
		GSTM1	1	
		NAT2	1	
		GSTP1	1	
		EPHX1	1	
	Lung cancer	GSTM1	1	111
Nitro-PAHs	Genotoxic effects in respiratory tract	NAT2	2	112,113
Ultraviolet light (UV)	Basal cell carcinoma	XP-D	1	114
Ionizing radiation (IR)	DNA damage in lymphocytes	XP-D	2	115–117
		XP-F	2	
		XRCC1	2	
	Prolonged cell cycle delay	APE1	1	118
Tobacco smoke	Lung cancer	CYP1A1	1	119–121
		GSTM1	1	121–123
		NAT1	1	80,124
		NAT2	1	
		EPHX1	1	125
		XRCC1	1	126
	Bladder cancer	CYP1A2	1	92
		NAT2	1	13,127
		GSTM1	1	128
	Bronchogenic carcinoma	AHR	2	92
	Emphysema, Chronic	EPHX1	1	129
	Obstructive Pulmonary Disease (COPD)	GSTM1	1	
Environmental tobacco smoke (ETS)	Lung cancer	GSTM1	1	130

[a]Classification 1 = Associations with laboratory evidence and suggestive epidemiologic data; 2 = Associations proposed from basic scientific laboratory evidence.

metabolites (21). For exposed persons, having *GSTM1* and *EPHX1* genotypes conferring a lack of enzyme and less active enzyme, respectively, was shown to result in increased HCC risk (22,23). Similarly, functional polymorphisms in CYP1A2 and CYP3A4, both of which catalyze the Phase I metabolism (epoxidation) of aflatoxin B_1, would be expected to modify HCC risk in exposed persons as well, though epidemiologic data for this have not yet been gathered. Biomarker studies of urinary aflatoxin metabolites and aflatoxin-albumin adducts in peripheral blood have validated their use as indicators of HCC risk at the group level, and polymorphisms in *GSTM1* and *EPHX1* yielded higher levels of adducts (24). Thus, in the case of aflatoxin, exposure-specific, validated biomarkers can be used in lieu of clinical disease measures to estimate the effect modification by specific polymorphisms. Even for this example, however, only a few studies exist and they have limited statistical power; hence, the magnitude of the modifying effect of genetic polymorphism remains highly uncertain. Future efforts to determine the predictive value of biomarkers of other exposures will facilitate the analysis of the effects of common polymorphisms in modifying the effects of those exposures.

Contradictory findings are often found in the literature. Similar issues have been encountered in pharmacogenetic studies. Evans and Relling (25) have commented that the use of different endpoints in assessing response to drugs, the heterogeneous nature of diseases studied, and the polygenic nature of many drug effects all contribute to the study-to-study variation often observed. These same factors will also be important in the types of studies discussed here. Additionally, some investigators have been concerned that population stratification, or bias in estimates of association between a polymorphism and disease due to confounding of a true risk factor with ethnicity, strongly influences study results and may also contribute to study-to-study variation. However, Wacholder et al. (26) have shown that well designed case-control and cohort studies of cancer are free of significant bias due to population stratification.

The examples of gene–environment interaction presented thus far have been fairly simple. More realistically, chronic disease risk is a function of multiple genes in multiple biologic pathways interacting with each other and with cumulative environmental factors over a lifetime. Taylor et al. (27) provided evidence for a three-way interaction between *NAT2, NAT1*, and smoking that modifies bladder cancer risk such that individuals who smoke and have *NAT2* slow acetylator alleles in combination with the high activity *NAT1*10* allele (homozygotes or heterozygotes) have heightened bladder cancer risk (27). Contrasting findings, however, have been reported more recently (28).

Advantages of Incorporating Genetic Polymorphisms into Health Effects Studies

The addition of genetic polymorphisms affords several noteworthy opportunities to health effects studies of exposures to environmental toxicants and toxins. Stratification of a studied health outcome or biomarker by relevant genotype (or phenotype) may allow for detection of different levels of risk among subgroups of exposed persons (29). Collectively, the studies on aromatic amine exposure, *NAT2* genotype, and bladder cancer demonstrate this point. Investigations that assess bladder cancer risk associated with exposure to aromatic amines alone would observe a magnitude of effect that represents the average risk for rapid and slow acetylators combined. This estimate would not suggest that aromatic amines are as etiologically significant, i.e., are potent carcinogens, for particular subpopulations as a stratified analysis would indicate. This has been referred to as effect dilution (30). Effect dilution may be especially important for common exposures—to dietary constituents or air pollution, for example—whose association to a disease outcome is often weak.

Second, evidence of effect modification by genotype yields insights into the potential biologic processes of toxicity or carcinogenicity, as substrates or targets of candidate gene products are identified as potential causative agents (29). The effect of lipopolysaccaride (LPS), also known as endotoxin, a component of particulate matter in rural areas, on lung function parameters may turn out to be a modern example of this. Arbour et al. have shown that response to LPS, measured by decrease in forced expiratory volume in the first second (FEV$_1$), differed by *TLR4* genotype (31). *TLR4* codes for the toll-like receptor that binds LPS and initiates a signal transduction pathway that leads to inflammation of the lung. Their data suggest that individuals with the variant *TLR4* genotype may be resistant to LPS-induced lung inflammation but may be more susceptible to a systemic inflammatory response. These findings may aid in answering the difficult question of what component(s) of particulate matter is (are) responsible for the range of health effects observed, particularly in rural areas where LPS levels are appreciable.

Finally, enhanced understanding of pathologic mechanism gained by the concerted efforts of epidemiologic and toxicologic studies may allow for the development of drugs or dietary interventions that prevent disease onset or progression. As an example, Oltipraz (OPZ, 5-(2-pyrazinyl)-4-methyl-1,2-dithiole-3-thione) is a drug that induces Phase II XMEs, notably the GSTs (32). Early evidence showed that OPZ can protect against the hepatocarcinogenic effects of aflatoxin B$_1$ in rats, and subsequent efforts have demonstrated that administration of OPZ to humans significantly enhanced excretion of a Phase II product, aflatoxin-mercapturic acid (33). Interestingly, there is also evidence that OPZ may act by competitively inhibiting CYP1A2, thereby preventing the activation of aflatoxin (34). In total, the understanding of aflatoxin biotransformation pathways from animal models and in vitro human tissue studies led to the hypothesis-based epidemiologic studies and ultimately contributed to the development of a chemoprevention strategy for aflatoxin-induced HCC.

Additionally, studies on the health effects of exposure to regulated environmental contaminants that incorporate genetic susceptibilities will enlarge the body of knowledge pertaining to the range of human variability in response to these contaminants. For example, the Centers for Disease Control and Prevention (CDC) National Report on Human Exposure to Environmental Chemicals (35), reports body burden among National Health and Nutrition Examination Survey (NHANES) subjects for 27 chemicals. Studies developed to look at the effect of these chemicals should include genes that might confer susceptibility. In this way, the risk assessment process may be improved by using refined estimates of human variability instead of the default assumptions conventionally used (i.e., uncertainty factor of 10), potentially improving public health protection and the regulation of industry through redefinition of acceptable exposure levels. This advantage has been touted for some time, but no clear example yet exists of how this can be done, especially in the face of numerous ethical, legal, and social issues around the use of genetic information. Still, the promise holds, and the potential continues to grow as more functional polymorphisms are discovered and their role in effect modification is deduced.

Considerations for Human Genome Epidemiology Studies

Finally, for environmental health scientists interested in pursuing health effects research that incorporates genetic effect modifiers, we list considerations for health studies that include genetic polymorphisms. Many of these considerations are explored in depth in other chapters in this book. This discussion also assumes that the investigator(s) already have chosen the study design. Case-control and cohort studies are most often used to evaluate gene–environment interaction, and their benefits and drawbacks have been compared and contrasted (36,37).

Exposure Assessment

Exposure assessment is of paramount importance in studies of gene–environment interaction. Typically, efforts aim to characterize the type, duration, intensity, and timing of exposure. Exposure misclassification is a major concern, since it can bias the estimate of the effect of exposures as well as the estimate of the joint genotype–exposure effect (38). New methods, such as biomonitoring approaches (39) and geographic information systems (40–42), can be used to achieve more precise exposure assessments.

Candidate Gene Selection

The selection of candidate genes is one of the first methodologic issues encountered. Generally, one can investigate the role of a gene whose product is hypothesized to be involved in the biotransformation, cell signal transduction, repair, or disease process relevant to a specific exposure. Sources of toxicologic or other biomedical data that can be used to identify candidate genes include previously published literature (PubMed), the Agency for Toxic Substances and Disease Registry's Tox Profiles, the National Library of Medicine ToxNet, the National Institute for Occupational Safety and Health's Registry of Toxic Effects of Chemical Substances (RTECS), the National Toxicology Program Reports on Carcinogens, On-line Mendelian Inheritance in Man (OMIM), and the HuGE NET Database. (Please see Appendix 9.1 for selected Web site addresses.)

Once candidate genes have been selected, sources of genetic information can be used to identify important polymorphisms in candidate gene(s). These sources include Web sites for specific gene families (e.g., CYPs, NATs), OMIM, NIEHS Environmental Genome Project Database, the National Cancer Institute Cancer Genome Anatomy Project (CGAP), and polymorphism databases (e.g., the SNPs consortium and the National Center for Biotechnology Information dbSNPs database). (See Appendix 9.1 for a listing of relevant URLs.) Focusing on polymorphisms with known functional effects is, of course, advantageous.

Efforts to study complex gene–environment interactions are tempered by the difficulty in obtaining adequate sample size (29). Two primary factors to consider are the prevalence of the polymorphism in the population and the magnitude of effect modification. As Caporaso has pointed out (43), there is a trade-off between the prevalence of a polymorphism and the magnitude of effect that may be detected. On the one hand, common variants are less likely to exhibit a strong effect; on the other hand, there is more statistical power in studying these polymorphisms because they are more common. Furthermore, the population-attributable risk of common variants will be greater, even if the penetrance is modest.

More recently, investigators have expanded their study design to include analysis of multiple polymorphisms in single genes that co-segregate (i.e., haplotypes). Haplotype analysis is advantageous in that more information about variation in a gene is captured by this approach relative to single SNPs. Inferring haplotypes from genotype data requires the use of specific algorithms (e.g., ref. 44), and methods are evolving to include adjustment for covariates in the analysis (45).

Selection of a Method to Obtain Samples for Genotyping

Collection of DNA samples from the study population is an area of technologic evolution. Besides venous blood samples, from which DNA can be extracted, buccal cell collection brushes (46) or mouth washes (47,48) have been employed and offer increased convenience to the study participant, but DNA yield can be substantially lower.

Informed Consent

Informed consent for genetic testing is also an important consideration. Beskow et al. (49) recently described the major issues to consider in obtaining informed consent and developed a general template for researchers to utilize (see http://www.cdc.gov/genomics/info/reports/policy/consentarticle.htm; more information can be found at http://www.cdc.gov/genomics/info/perspectives/infmcnst. htm). In addition, the Department of Health and Human Services (DHHS) provides information about human subjects protection, and templates for informed consent protocols can be accessed at the DHHS Web site.

Selection of a Genotyping Method

Many different methods can be used to genotype subjects. Choosing an appropriate method and utilizing quality control procedures are critical, because even minor genotype misclassification can substantially bias study results (38,50). The choice of method depends on both the type of polymorphism to be analyzed and the type of sample obtained. DNA sequence analysis is considered the gold standard, but it is time consuming and expensive. Polymerase chain reaction (PCR) methods are ideal for rapid genotyping of large samples. Restriction fragment length polymorphism analysis can be used if the polymorphism of interest is known to result in the addition or deletion of a restriction site. More recent, high-throughput approaches include 5′-nuclease-based fluoresence assays (Taqman), matrix-assisted laser desorption/ionization—time of flight mass spectrometry analysis (MALDI-TOF), and DNA microarrays (51).

Data Analysis

Khoury and Botto (52) have advocated that, in the context of a case-control study where exposure and genotype are dichotomized, the conventional 2×2 table analysis of exposure and disease be expanded to include genotype, yielding a 2×4 table. In this manner, the raw exposure and genotype data are displayed in such a way that relative risk estimates for each factor alone and their joint effect can be easily generated. Attributable fractions also can be computed from these data. Regression models of interactions can also be employed (53,54). Though not discussed here, issues regarding multiple comparisons and false-positive findings are also important to consider, and the reader is referred to De Roos et al. (55) for guidance.

Conclusions

The role of genetic variations as determinants of health is being explored in many areas of public health research. In environmental health, recently gathered epidemiologic and toxicologic data suggest that the health effects of many different types of exposures can be modified by genetic polymorphisms, although the effect modification may be weak and the power of many studies is inadequate to demonstrate an effect. Current and future efforts to identify new polymorphisms in genes involved in environmental response will broaden the scope of potential genetic effect modifiers. Determining the effect of these polymorphisms (phenotype) will then be of paramount importance.

Though the individual risk associated with a polymorphism may be relatively low, the population-attributable risk may be large, and thus this area of research merits investigation. As newly identified and previously known polymorphisms are incorporated into epidemiologic research, gene–environment interactions can be detected and quantified. Through toxicologic studies, the mechanisms of these interactions can be elucidated. Correlations between biomarkers of exposure and effect with disease outcomes will facilitate the process of identification of polymorphisms that act as effect modifiers. As with any scientific endeavor, intriguing results in this area of research need to be replicated in different studies and populations to confirm the role of a polymorphism as an effect modifier.

Although many "gene–environment" interaction studies on human populations have been completed in the past decade, the number of examples demonstrating important and consistent positive relationships is remarkably small. It now appears that the "one gene—one risk factor" approach to understanding the etiology of environmentally related chronic diseases is not likely to yield high rewards. Nevertheless, it remains clear that most chronic diseases of public health importance arise from a complex and often poorly understood combination of genetic and environmental factors. New tools for high-throughput genotyping of hundreds or thousands of genetic variants in a sample, coupled with very large-scale population-based studies that utilize sensitive biomarkers and comprehensive exposure-assessment strategies are likely to be needed to begin to unravel the complex multi-gene–environment interactions responsible for most chronic diseases of public health importance. This will require new paradigms for inter-disciplinary collaborative research that involve very large-scale studies, as well as new bioinformatics tools to help scientists make sense of the dizzying array of complex data that will come from such studies. Finally, increasing interest and discussion have been generated about the development of an integrated database that links new findings on exposures, etiologic pathways, relevant genes, poly-morphisms in these genes, and their function (55). This database would serve to guide the design of new studies as well as data analysis and interpretation of results (55).

In summary, the ability to detect different levels of risk within the population and greater understanding of etiologic mechanisms are the primary benefits of incorpo-rating genetics into the existing environmental health research framework. The in-sights gained by employing this framework should ultimately allow for the devel-opment of new disease prevention strategies. The use of this information in risk assessments may also be a viable area of development. Whether the use of this in-formation in disease prevention efforts targeted to genetically susceptible individu-als is acceptable is an ethical question that is beginning to be addressed and neces-sitates considerable attention in the future.

Acknowledgments
Portions of this chapter were published as a review article in *Environmental Health Perspectives,* supported in part by NIEMS grants P50ES07033 and 5T32ES07032.

References

1. Weber WW. Pharmacogenetics. New York: Oxford University Press, 1997.
2. Parkinson A. Chapter 6. Biotransformation, in, Toxicology: the Basic Science of Poi-sons, Louis J. Casarett JD, ed. New York: Macmillan.
3. Ingelman-Sundberg M, Oscarson M, et al. Polymorphic human cytochrome P450 en-zymes: an opportunity for individualized drug treatment. Trends Pharmacol Sci 1999;20:342–349.
4. Brookes AJ. The essence of SNPs. Gene 1999;234:177–186.
5. Cargill M, Altshuler D, Ireland J, et al. Characterization of single-nucleotide polymor-phisms in coding regions of human genes. Nat Genet 1999;22:231–238.
6. Perera FP. Environment and cancer: who are susceptible? Science 1997;278:1068–1073.
7. Levy RH. Metabolic drug interactions. Philadelphia: Lippincott Williams & Wilkins, 2000.
8. Caporaso N, Goldstein A. Cancer genes: single and susceptibility: exposing the differ-ence. Pharmacogenetics 1995;5:59–63.
9. Olden K, Wilson S. Environmental health and genomics: visions and implications. Nat Rev Genet 2000;1:149–153.
10. Lower GM, Nilsson T, Nelson CE, et al. N-acetyltransferase phenotype and risk in uri-nary bladder cancer: approaches in molecular epidemiology. Preliminary results in Swe-den and Denmark. Environ Health Perspect 1979;29:71–79.

11. Cartwright RA, Glashan RW, Rogers HJ, et al. Role of N-acetyltransferase phenotypes in bladder carcinogenesis: a pharmacogenetic epidemiological approach to bladder cancer. Lancet 1982;2:842–845.
12. Hanke J, Krajewska B. Acetylation phenotypes and bladder cancer. J Occup Med 1990;32:917–918.
13. Marcus PM, Hayes RB, Vineis P, et al. Cigarette smoking, N-acetyltransferase 2 acetylation status, and bladder cancer risk: a case-series meta-analysis of a gene–environment interaction. Cancer Epidemiol Biomarkers Prev 2000;9:461–467.
14. Hein DW, Doll MA, Rustan TD, et al. Metabolic activation and deactivation of arylamine carcinogens by recombinant human NAT1 and polymorphic NAT2 acetyltransferases. Carcinogenesis 1993;14:1633–1638.
15. Mattano SS, Land S, King CM, et al. Purification and biochemical characterization of hepatic arylamine N-acetyltransferase from rapid and slow acetylator mice: identity with arylhydroxamic acid N,O-acyltransferase and N-hydroxyarylamine O–acetyltransferase. Mol Pharmacol 1989;35:599–609.
16. Trinidad A, Hein DW, Rustan TD, et al. Purification of hepatic polymorphic arylamine N-acetyltransferase from homozygous rapid acetylator inbred hamster: identity with polymorphic N-hydroxyarylamine-O-acetyltransferase. Cancer Res 1990;50: 7942–7949.
17. Autrup H. Genetic polymorphisms in human xenobiotica metabolizing enzymes as susceptibility factors in toxic response. Mutat Res 2000;464:65–76.
18. Williams JA. Single nucleotide polymorphisms, metabolic activation and environmental carcinogenesis: why molecular epidemiologists should think about enzyme expression. Carcinogenesis 2001;22:209–214.
19. Colvin ME, Hatch FT, Felton JS. Chemical and biological factors affecting mutagen potency. Mutat Res 1998;400:479–492.
20. Ross RK, Yuan JM, Yu MC, et al. Urinary aflatoxin biomarkers and risk of hepatocellular carcinoma. Lancet 1992;339:943–946.
21. Eaton DL, Groopman JD. The Toxicology of Aflatoxins: Human health, Veterinary, and Agricultural Significance. San Diego: Academic Press, 1994.
22. London WT, Evans AA, Buetow K, et al. Molecular and genetic epidemiology of hepatocellular carcinoma: studies in China and Senegal. Princess Takamatsu Symp 1995;25:51–60.
23. McGlynn KA, Rosvold EA, Lustbader ED, et al. Susceptibility to hepatocellular carcinoma is associated with genetic variation in the enzymatic detoxification of aflatoxin B1. Proc Natl Acad Sci U S A 1995;92:2384–2387.
24. Wild CP, Turner PC. Exposure biomarkers in chemoprevention studies of liver cancer. IARC Sci Publ 2001;154:215–222.
25. Evans WE, Relling MV. Pharmacogenomics: translating functional genomics into rational therapeutics. Science 1999;286:487–491.
26. Wacholder S, Rothman N, Caporaso N. Population stratification in epidemiologic studies of common genetic variants and cancer: quantification of biase. J Natl Cancer Inst 2000;92:1151–1158.
27. Taylor JA, Umbach DM, Stephens E, et al. The role of N-acetylation polymorphisms in smoking-associated bladder cancer: evidence of a gene-gene-exposure three-way interaction. Cancer Res 1998;58:3603–3610.
28. Cascorbi I, Roots I, Brockmoller J. Association of NAT1 and NAT2 polymorphisms to urinary bladder cancer: significantly reduced risk in subjects with NAT1*10. Cancer Res 2001;61:5051–5056.
29. Rothman N, Wacholder, S, Caporaso, NE, Garcia-Closas, M, Buetow, K, Fraumeni Jr, J.F. The use of common genetic polymorphisms to enhance the epidemiologic study of

environmental carcinogens. BBA–Reviews on Cancer, 1471 (2) (2001) pp. C1–C10 2000;1471:C1–C10.

30. Khoury M, Beaty T, Cohen B. Fundamentals of Genetic Epidemiology. New York: Oxford University Press, 1993.

31. Arbour NC, Lorenz E, Schutte BC, et al. TLR4 mutations are associated with endotoxin hyporesponsiveness in humans. Nat Genet 2000;25:187–191.

32. Carr BA, Franklin MR. Drug-metabolizing enzyme induction by 2,2′-dipyridyl, 1,7-phenanthroline, 7,8-benzoquinoline and oltipraz in mouse. Xenobiotica 1998;28:949–956.

33. Kensler TW, Curphey TJ, Maxiutenko Y, et al. Chemoprotection by organosulfur inducers of phase 2 enzymes: dithiolethiones and dithiins. Drug Metabol Drug Interact 2000;17:3–22.

34. Langouet S, Coles B, Morel F, et al. Inhibition of CYP1A2 and CYP3A4 by oltipraz results in reduction of aflatoxin B1 metabolism in human hepatocytes in primary culture. Cancer Res 1995;55:5574–5579.

35. National Report on Human Exposure to Environmental Chemicals. Atlanta: Centers for Disease Control and Prevention, National Center for Environmental Health, 2001.

36. Caporaso N, Rothman N, Wacholder S. Case-control studies of common alleles and environmental factors. J Natl Cancer Inst Monogr 1999;25–30.

37. Langholz B, Rothman N, Wacholder S, Thomas DC. Cohort studies for characterizing measured genes. J Natl Cancer Inst Monogr 1999;39–42.

38. Rothman N, Garcia-Closas, M., Setwart, W.T., Lubin J. Chapter 9. The impact of misclassification in case-control studies of gene-environment interactions. IARC Scientific Publications. Vol. 148. Lyon: IARC, 1999:89–96.

39. Rothman N, Stewart WF, Schulte PA. Incorporating biomarkers into cancer epidemiology: a matrix of biomarker and study design categories. Cancer Epidemiol Biomarkers Prev 1995;4:301–311.

40. Rushton G, Lolonis P. Exploratory spatial analysis of birth defect rates in an urban population. Stat Med 1996;15:717–726.

41. Kulldorff M, Feuer EJ, Miller BA, et al. Breast cancer clusters in the northeast United States: a geographic analysis. Am J Epidemiol 1997;146:161–170.

42. Ward MH, Nuckols JR, Weigel SJ, et al. Identifying populations potentially exposed to agricultural pesticides using remote sensing and a Geographic Information System. Environ Health Perspect 2000;108:5–12.

43. Caporaso N. Chapter 6. (1999). Selection of Candidate Genes. IARC Sci Publ. vol. 148. Lyon: IARC Press, pp. 23–36.

44. Terwilliger JD, Ott J. Handbook of Human Genetic Linkage. Baltimore: Johns Hopkins University Press, 1994.

45. Schaid DJ, Rowland CM, Tines DE, et al. Score tests for association between traits and haplotypes when linkage phase is ambiguous. Am J Hum Genet 2002;70:425–434.

46. Walker AH, Najarian D, White DL, et al. Collection of genomic DNA by buccal swabs for polymerase chain reaction-based biomarker assays. Environ Health Perspect 1999;107:517–520.

47. Heath EM, Morken NW, Campbell KA, et al. Use of buccal cells collected in mouthwash as a source of DNA for clinical testing. Arch Pathol Lab Med 2001;125:127–133.

48. Garcia-Closas M, Egan KM, Abruzzo J, et al. Collection of genomic DNA from adults in epidemiological studies by buccal cytobrush and mouthwash. Cancer Epidemiol Biomarkers Prev 2001;10:687–696.

49. Beskow LM, Burke W, Merz JF, et al. Informed consent for population-based research involving genetics. JAMA 2001;286:2315–2321.

50. Garcia-Closas M, Rothman N, Lubin J. Misclassification in case-control studies of gene-environment interactions: assessment of bias and sample size. Cancer Epidemiol Biomarkers Prev 1999;8:1043–1050.

51. Shi MM. Enabling large-scale pharmacogenetic studies by high-throughput mutation detection and genotyping technologies. Clin Chem 2001;47:164–172.
52. Botto LD, Khoury MJ. Commentary: facing the challenge of gene-environment interaction: the two-by-four table and beyond. Am J Epidemiol 2001;153:1016–1020.
53. Neter J, Kutner MH, Nachtsheim CJ, et al. Applied Linear Statistical Models. Chicago: Irwin, 1996.
54. Breslow N, Day N. Statistical Methods in Cancer Research, Volume 1: The Analysis of Case-Control Studies. Vol. 32. Lyon: IARC, 1980.
55. De Roos A, Smith MT, Chanock S, et al. Mechanistic Considerations in the Molecular Epidemiology of Cancer. In: Buffler PA. BM, Rice JM., Boffetta P., ed. IARC Scientific Publications. Lyon: IARC, In press.
56. Chiou HY, Hsueh YM, Hsieh LL, et al. Arsenic methylation capacity, body retention, and null genotypes of glutathione S-transferase M1 and T1 among current arsenic-exposed residents in Taiwan. Mutat Res 1997;386:197–207.
57. Vahter M. Genetic polymorphism in the biotransformation of inorganic arsenic and its role in toxicity. Toxicol Lett 2000;112–113:209–217.
58. Richeldi L, Sorrentino R, Saltini C. HLA-DPB1 glutamate 69: a genetic marker of beryllium disease. Science 1993;262:242–244.
59. Saltini C, Amicosante M, Franchi A, et al. Immunogenetic basis of environmental lung disease: lessons from the berylliosis model. Eur Respir J 1998;12:1463–1475.
60. Richeldi L, Kreiss K, Mroz MM, et al. Interaction of genetic and exposure factors in the prevalence of berylliosis. Am J Ind Med 1997;32:337–340.
61. Wetmur JG. Influence of the common human delta-aminolevulinate dehydratase polymorphism on lead body burden. Environ Health Perspect 1994;102:215–219.
62. Schwartz BS, Lee BK, Stewart W, et al. Associations of delta-aminolevulinic acid dehydratase genotype with plant, exposure duration, and blood lead and zinc protoporphyrin levels in Korean lead workers. Am J Epidemiol 1995;142:738–745.
63. Kelada SN, Shelton E, Kaufmann RB, et al. Delta-aminolevulinic acid dehydratase genotype and lead toxicity: a HuGE review. Am J Epidemiol 2001;154:1–13.
64. Schwartz BS, Lee BK, Stewart W, et al. delta-Aminolevulinic acid dehydratase genotype modifies four hour urinary lead excretion after oral administration of dimercaptosuccinic acid. Occup Environ Med 1997;54:241–246.
65. Fleming DE, Chettle DR, Wetmur JG, et al. Effect of the delta-aminolevulinate dehydratase polymorphism on the accumulation of lead in bone and blood in lead smelter workers. Environ Res 1998;77:49–61.
66. Schwartz BS, Stewart WF, Kelsey KT, et al. Associations of tibial lead levels with BsmI polymorphisms in the vitamin D receptor in former organolead manufacturing workers. Environ Health Perspect 2000;108:199–203.
67. Schwartz BS, Lee BK, Lee GS, et al. Associations of blood lead, dimercaptosuccinic acid-chelatable lead, and tibia lead with polymorphisms in the vitamin D receptor and [delta]-aminolevulinic acid dehydratase genes. Environ Health Perspect 2000;108:949–954.
68. Rosipal R, Lamoril J, Puy H, et al. Systematic analysis of coproporphyrinogen oxidase gene defects in hereditary coproporphyria and mutation update. Hum Mutat 1999;13:44–53.
69. Grandchamp B, Lamoril J, Puy H. Molecular abnormalities of coproporphyrinogen oxidase in patients with hereditary coproporphyria. J Bioenerg Biomembr 1995;27:215–219.
70. Mendez M, Sorkin L, Rossetti MV, et al. Familial porphyria cutanea tarda: characterization of seven novel uroporphyrinogen decarboxylase mutations and frequency of common hemochromatosis alleles. Am J Hum Genet 1998;63:1363–1375.

71. Moran-Jimenez MJ, Ged C, Romana M, et al. Uroporphyrinogen decarboxylase: complete human gene sequence and molecular study of three families with hepatoerythropoietic porphyria. Am J Hum Genet 1996;58:712–721.

72. Hori H, Kawano T, Endo M, et al. Genetic polymorphisms of tobacco- and alcohol-related metabolizing enzymes and human esophageal squamous cell carcinoma susceptibility. J Clin Gastroenterol 1997;25:568–575.

73. Chao YC, Wang LS, Hsieh TY, et al. Chinese alcoholic patients with esophageal cancer are genetically different from alcoholics with acute pancreatitis and liver cirrhosis. Am J Gastroenterol 2000;95:2958–2964.

74. Tanabe H, Ohhira M, Ohtsubo T, et al. Genetic polymorphism of aldehyde dehydrogenase 2 in patients with upper aerodigestive tract cancer. Alcohol Clin Exp Res 1999;23:17S–20S.

75. Yokoyama A, Ohmori T, Muramatsu T, et al. Cancer screening of upper aerodigestive tract in Japanese alcoholics with reference to drinking and smoking habits and aldehyde dehydrogenase-2 genotype. Int J Cancer 1996;68:313–316.

76. Yokoyama A, Muramatsu T, Omori T, et al. Alcohol and aldehyde dehydrogenase gene polymorphisms influence susceptibility to esophageal cancer in Japanese alcoholics. Alcohol Clin Exp Res 1999;23:1705–1710.

77. Eaton DL, Gallagher EP, Bammler TK, et al. Role of cytochrome P4501A2 in chemical carcinogenesis: implications for human variability in expression and enzyme activity. Pharmacogenetics 1995;5:259–274.

78. Gallagher EP, Kunze KL, Stapleton PL, et al. The kinetics of aflatoxin B1 oxidation by human cDNA-expressed and human liver microsomal cytochromes P450 1A2 and 3A4. Toxicol Appl Pharmacol 1996;141:595–606.

79. Lang NP, Chu DZ, Hunter CF, et al. Role of aromatic amine acetyltransferase in human colorectal cancer. Arch Surg 1986;121:1259–1261.

80. Hein DW, Doll MA, Fretland AJ, et al. Molecular genetics and epidemiology of the NAT1 and NAT2 acetylation polymorphisms. Cancer Epidemiol Biomarkers Prev 2000;9:29–42.

81. Gil JP, Lechner MC. Increased frequency of wild-type arylamine-N-acetyltransferase allele NAT2*4 homozygotes in Portuguese patients with colorectal cancer. Carcinogenesis 1998;19:37–41.

82. Brockton N, Little J, Sharp L, et al. N-acetyltransferase polymorphisms and colorectal cancer: a HuGE review. Am J Epidemiol 2000;151:846–861.

83. Deitz AC, Zheng W, Leff MA, et al. N-Acetyltransferase-2 genetic polymorphism, well-done meat intake, and breast cancer risk among postmenopausal women. Cancer Epidemiol Biomarkers Prev 2000;9:905–910.

84. Zheng W, Xie D, Cerhan JR, et al. Sulfotransferase 1A1 polymorphism, endogenous estrogen exposure, well-done meat intake, and breast cancer risk. Cancer Epidemiol Biomarkers Prev 2001;10:89–94.

85. Landi S, Hanley NM, Warren SH, et al. Induction of genetic damage in human lymphocytes and mutations in Salmonella by trihalomethanes: role of red blood cells and GSTT1-1 polymorphism. Mutagenesis 1999;14:479–482.

86. Pegram RA, Andersen ME, Warren SH, et al. Glutathione S-transferase-mediated mutagenicity of trihalomethanes in Salmonella typhimurium: contrasting results with bromodichloromethane off chloroform. Toxicol Appl Pharmacol 1997;144:183–138.

87. Rothman N, Smith MT, Hayes RB, et al. Benzene poisoning, a risk factor for hematological malignancy, is associated with the NQO1 609C → T mutation and rapid fractional excretion of chlorzoxazone. Cancer Res 1997;57:2839–2842.

88. Ross D, Traver RD, Siegel D, Kuehl BL, Misra V, Rauth AM. A polymorphism in NAD(P)H:quinone oxidoreductase (NQO1): relationship of a homozygous mutation at position 609 of the NQO1 cDNA to NQO1 activity. Br J Cancer 1996;74:995–996.

89. Xu X, Wiencke JK, Niu T, et al. Benzene exposure, glutathione S-transferase theta homozygous deletion, and sister chromatid exchanges. Am J Ind Med 1998;33:157–163.

90. Bruning T, Lammert M, Kempkes M, et al. Influence of polymorphisms of GSTM1 and GSTT1 for risk of renal cell cancer in workers with long-term high occupational exposure to trichloroethene. Arch Toxicol 1997;71:596–599.

91. Sweeney C, Farrow DC, Schwartz SM, et al. Glutathione S-transferase M1, T1, and P1 polymorphisms as risk factors for renal cell carcinoma: a case-control study. Cancer Epidemiol Biomarkers Prev 2000;9:449–454.

92. Nebert DW, McKinnon RA, Puga A. Human drug-metabolizing enzyme polymorphisms: effects on risk of toxicity and cancer. DNA Cell Biol 1996;15:273–280.

93. Stresser DM, Kupfer D. Human cytochrome P450-catalyzed conversion of the pro-estrogenic pesticide methoxychlor into an estrogen. Role of CYP2C19 and CYP1A2 in O-demethylation. Drug Metab Dispos 1998;26:868–874.

94. Landi MT, Sinha R, Lang NP, et al. Chapter 16. Human cytochrome P4501A2. IARC Sci Publ 1999:173–195.

95. Au WW, Sierra-Torres CH, Cajas-Salazar N, et al. Cytogenetic effects from exposure to mixed pesticides and the influence from genetic susceptibility. Environ Health Perspect 1999;107:501–505.

96. Eaton DL. Biotransformation enzyme polymorphism and pesticide susceptibility. Neurotoxicology 2000;21:101–111.

97. Sams C, Mason HJ, Rawbone R. Evidence for the activation of organophosphate pesticides by cytochromes P450 3A4 and 2D6 in human liver microsomes. Toxicol Lett 2000;116:217–221.

98. Kelsey KT, Wiencke JK, Ward J, et al. Sister-chromatid exchanges, glutathione S-transferase theta deletion and cytogenetic sensitivity to diepoxybutane in lymphocytes from butadiene monomer production workers. Mutat Res 1995;335:267–273.

99. Norppa H, Hirvonen A, Jarventaus H, et al. Role of GSTT1 and GSTM1 genotypes in determining individual sensitivity to sister chromatid exchange induction by diepoxybutane in cultured human lymphocytes. Carcinogenesis 1995;16:1261–1264.

100. Wiencke JK, Pemble S, Ketterer B, et al. Gene deletion of glutathione S–transferase theta: correlation with induced genetic damage and potential role in endogenous mutagenesis. Cancer Epidemiol Biomarkers Prev 1995;4:253–259.

101. Schaaf BM, Seitzer U, Pravica V, et al. Tumor necrosis factor-alpha -308 promoter gene polymorphism and increased tumor necrosis factor serum bioactivity in farmer's lung patients. Am J Respir Crit Care Med 2001;163:379–382.

102. Kleeberger SR, Reddy S, Zhang LY, et al. Genetic susceptibility to ozone-induced lung hyperpermeability: role of toll-like receptor 4. Am J Respir Cell Mol Biol 2000;22:620–627.

103. Wu MT, Huang SL, Ho CK, Yeh YF, Christiani DC. Cytochrome P450 1A1 MspI polymorphism and urinary 1-hydroxypyrene concentrations in coke-oven workers. Cancer Epidemiol Biomarkers Prev 1998;7:823–829.

104. Nielsen PS, de Pater N, Okkels H, Autrup H. Environmental air pollution and DNA adducts in Copenhagen bus drivers—effect of GSTM1 and NAT2 genotypes on adduct levels. Carcinogenesis 1996;17:1021–1027.

105. Merlo F, Andreassen A, Weston A, et al. Urinary excretion of 1-hydroxypyrene as a marker for exposure to urban air levels of polycyclic aromatic hydrocarbons. Cancer Epidemiol Biomarkers Prev 1998;7:147–155.

106. Knudsen LE, Norppa H, Gamborg MO, et al. Chromosomal aberrations in humans induced by urban air pollution: influence of DNA repair and polymorphisms of glutathione S-transferase M1 and N-acetyltransferase 2. Cancer Epidemiol Biomarkers Prev 1999; 8:303–310.

107. Binkova B, Lewtas J, Miskova I, et al. Biomarker studies in northern Bohemia. Environ Health Perspect 1996;104:591–597.

108. Whyatt RM, Perera FP, Jedrychowski W, et al. Association between polycyclic aromatic hydrocarbon-DNA adduct levels in maternal and newborn white blood cells and glutathione S-transferase P1 and CYP1A1 polymorphisms. Cancer Epidemiol Biomarkers Prev 2000;9:207–212.

109. Viezzer C, Norppa H, Clonfero E, et al. Influence of GSTM1, GSTT1, GSTP1, and EPHX gene polymorphisms on DNA adduct level and HPRT mutant frequency in coke-oven workers. Mutat Res 1999;431:259–269.

110. Motykiewicz G, Michalska J, Pendzich J, et al. A molecular epidemiology study in women from Upper Silesia, Poland. Toxicol Lett 1998;96–97:195–202.

111. Lan Q, He X, Costa DJ, et al. Indoor coal combustion emissions, GSTM1 and GSTT1 genotypes, and lung cancer risk: a case-control study in Xuan Wei, China. Cancer Epidemiol Biomarkers Prev 2000;9:605–608.

112. Adamiak W, Jadczyk P, Kucharczyk J. Application of Salmonella strains with altered nitroreductase and O-acetyltransferase activities to the evaluation of the mutagenicity of airborne particles. Acta Microbiol Pol 1999;48:131–140.

113. Watanabe T, Kaji H, Takashima M, et al. Metabolic activation of 2- and 3-nitrodibenzopyranone isomers and related compounds by rat liver S9 and the effect of S9 on the mutational specificity of nitrodibenzopyranones. Mutat Res 1997;388:67–78.

114. Dybdahl M, Vogel U, Frentz G, et al. Polymorphisms in the DNA repair gene XPD: correlations with risk and age at onset of basal cell carcinoma. Cancer Epidemiol Biomarkers Prev 1999;8:77–81.

115. Lunn RM, Helzlsouer KJ, Parshad R, et al. XPD polymorphisms: effects on DNA repair proficiency. Carcinogenesis 2000;21:551–555.

116. Fan F, Liu C, Tavare S, Arnheim N. Polymorphisms in the human DNA repair gene XPF. Mutat Res 1999;406:115–120.

117. Duell EJ, Wiencke JK, Cheng TJ, et al. Polymorphisms in the DNA repair genes XRCC1 and ERCC2 and biomarkers of DNA damage in human blood mononuclear cells. Carcinogenesis 2000;21:965–971.

118. Hu JJ, Smith TR, Miller MS, Mohrenweiser HW, et al. Amino acid substitution variants of APE1 and XRCC1 genes associated with ionizing radiation sensitivity. Carcinogenesis 2001;22:917–922.

119. Xu X, Kelsey KT, Wiencke JK, Wain JC, Christiani DC. Cytochrome P450 CYP1A1 MspI polymorphism and lung cancer susceptibility. Cancer Epidemiol Biomarkers Prev 1996;5:687–692.

120. Houlston RS. CYP1A1 polymorphisms and lung cancer risk: a meta-analysis. Pharmacogenetics 2000;10:105–114.

121. Bartsch H, Nair U, Risch A, Rojas M, Wikman H, Alexandrov K. Genetic polymorphism of CYP genes, alone or in combination, as a risk modifier of tobacco-related cancers. Cancer Epidemiol Biomarkers Prev 2000;9:3–28.

122. Houlston RS. Glutathione S-transferase M1 status and lung cancer risk: a meta-analysis. Cancer Epidemiol Biomarkers Prev 1999;8:675–682.

123. McWilliams JE, Sanderson BJ, Harris EL, Richert-Boe KE, Henner WD. Glutathione S-transferase M1 (GSTM1) deficiency and lung cancer risk. Cancer Epidemiol Biomarkers Prev 1995;4:589–594.

124. Bouchardy C, Mitrunen K, Wikman H, et al. N-acetyltransferase NAT1 and NAT2 genotypes and lung cancer risk. Pharmacogenetics 1998;8:291–298.

125. Benhamou S, Reinikainen M, Bouchardy C, Dayer P, Hirvonen A. Association between lung cancer and microsomal epoxide hydrolase genotypes. Cancer Res 1998;58: 5291–5293.

126. Ratnasinghe D, Yao SX, Tangrea JA, et al. Polymorphisms of the DNA repair gene XRCC1 and lung cancer risk. Cancer Epidemiol Biomarkers Prev 2001;10:119–123.

127. Marcus PM, Vineis P, Rothman N. NAT2 slow acetylation and bladder cancer risk: a meta-analysis of 22 case-control studies conducted in the general population. Pharmacogenetics 2000;10:115–122.

128. Engel LS, Taioli E, Pfeiffer R, et al. Pooled analysis and meta-analysis of GSTM1 and bladder cancer: a HuGE Review. Am J Epidemiol 2002;156:95–109.

129. Koyama H, Geddes DM. Genes, oxidative stress, and the risk of chronic obstructive pulmonary disease. Thorax 1998;53:S10–S104.

130. Bennett WP, Alavanja MC, Blomeke B, et al. Environmental tobacco smoke, genetic susceptibility, and risk of lung cancer in never-smoking women. J Natl Cancer Inst 1999;91:2009–2014.

131. Spurr NK, Gough AC, Stevenson K, et al. Msp-1 polymorphism detected with a cDNA probe for the P-450 I family on chromosome 15. Nucleic Acids Res 1987;15:5901.

132. Persson I, Johansson I, Ingelman-Sundberg M. In vitro kinetics of two human CYP1A1 variant enzymes suggested to be associated with interindividual differences in cancer susceptibility. Biochem Biophys Res Commun 1997;231:227–230.

133. Sachse C, Brockmoller J, Bauer S, et al. Functional significance of a C—>A polymorphism in intron 1 of the cytochrome P450 CYP1A2 gene tested with caffeine. Br J Clin Pharmacol 1999;47:445–449.

134. Chida M, Yokoi T, Fukui T, et al. Detection of three genetic polymorphisms in the 5'-flanking region and intron 1 of human CYP1A2 in the Japanese population. Jpn J Cancer Res 1999;90:899–902.

135. Marchand LL, Wilkinson GR, Wilkens LR. Genetic and dietary predictors of CYP2E1 activity: a phenotyping study in Hawaii Japanese using chlorzoxazone. Cancer Epidemiol Biomarkers Prev 1999;8:495–500.

136. Hayashi S, Watanabe J, Kawajiri K. Genetic polymorphisms in the 5'-flanking region change transcriptional regulation of the human cytochrome P450IIE1 gene. J Biochem (Tokyo) 1991;110:559–565.

137. Rebbeck TR, Jaffe JM, Walker AH, et al. Modification of clinical presentation of prostate tumors by a novel genetic variant in CYP3A4. J Natl Cancer Inst 1998;90:1225–1229.

138. Walker AH, Jaffe JM, Gunasegaram S, et al. Characterization of an allelic variant in the nifedipine-specific element of CYP3A4: ethnic distribution and implications for prostate cancer risk. Mutations in brief no. 191. Hum Mutat 1998;12:289. http://www.interscience.wiley.com/jpages/1059-7794/pdf/mutation/191.pdf

139. Smart J, Daly AK. Variation in induced CYP1A1 levels: relationship to CYP1A1, Ah receptor and GSTM1 polymorphisms. Pharmacogenetics 2000;10:11–24.

140. Hassett C, Aicher L, Sidhu JS, Omiecinski CJ. Human microsomal epoxide hydrolase: genetic polymorphism and functional expression in vitro of amino acid variants. Hum Mol Genet 1994;3:421–428.

141. Moran JL, Siegel D, Ross D. A potential mechanism underlying the increased susceptibility of individuals with a polymorphism in NAD(P)H:quinone oxidoreductase 1 (NQO1) to benzene toxicity. Proc Natl Acad Sci U S A 1999;96:8150–8155.

142. Raftogianis RB, Wood TC, Otterness DM, Van Loon JA, Weinshilboum RM. Phenol sulfotransferase pharmacogenetics in humans: association of common SULT1A1 alleles with TS PST phenotype. Biochem Biophys Res Commun 1997;239:298–304.

143. Seidegard J, Vorachek WR, Pero RW, et al. Hereditary differences in the expression of the human glutathione transferase active on trans-stilbene oxide are due to a gene deletion. Proc Natl Acad Sci U S A 1988;85:7293–7297.

144. Ali-Osman F, Akande O, Antoun G, et al. Molecular cloning, characterization, and expression in Escherichia coli of full-length cDNAs of three human glutathione S-trans-

ferase Pi gene variants. Evidence for differential catalytic activity of the encoded proteins. J Biol Chem 1997;272:10004–10012.

145. Pemble S, Schroeder KR, Spencer SR, et al. Human glutathione S-transferase theta (GSTT1): cDNA cloning and the characterization of a genetic polymorphism. Biochem J 1994;300:271–276.

146. Wiebel FA, Dommermuth A, Thier R. The hereditary transmission of the glutathione transferase hGSTT1-1 conjugator phenotype in a large family. Pharmacogenetics 1999; 9:251–256.

147. Costa LG, Cole TB, Jarvik GP, Furlong CE. Functional genomics of the paraoxonase (PON1) polymorphisms: effects on pesticide sensitivity, cardiovascular disease, and drug metabolism. Annu Rev Med 2003;54:371–392.

148. Cooper GS, Umbach DM. Are vitamin D receptor polymorphisms associated with bone mineral density? A meta-analysis. J Bone Miner Res 1996;11:1841–1849.

149. Shen MR, Jones IM, Mohrenweiser H. Nonconservative amino acid substitution variants exist at polymorphic frequency in DNA repair genes in healthy humans. Cancer Res 1998;58:604–608.

150. Hadi MZ, Coleman MA, Fidelis K, Mohrenweiser HW, Wilson ID. Functional characterization of Ape1 variants identified in the human population. Nucleic Acids Res 2000;28:3871–3879.

151. Abraham LJ, Kroeger KM. Impact of the -308 TNF promoter polymorphism on the transcriptional regulation of the TNF gene: relevance to disease. J Leukoc Biol 1999;66: 562–566.

Appendix 9.1. List of Websites

Environmental Health Websites

Agency for Toxic Substances and Disease Registry's Tox Profiles
http://www.atsdr.cdc.gov/toxpro2.html

National Library of Medicine's ToxNet http://toxnet.nlm.nih.gov/

PubMed http://www4.ncbi.nlm.nih.gov/PubMed/

National Institute of Environmental Health Sciences Environmental Genome Project http://www.niehs.nih.gov/envgenom/home.htm

National Toxicology Program Report on Carcinogens
http://ntp-server.niehs.nih.gov/NewHomeRoc/AboutRoC.html

National Institute for Occupational Safety and Health's (NIOSH's) Registry of Toxic Effects of Chemical Substances (RTECS)
http://www.cdc.gov/niosh/rtecs.html

Gene Families

Cytochrome P450s http://www.imm.ki.se/CYPalleles/

N-Acetyl Transferases
http://www.louisville.edu/medschool/pharmacology/NAT.html

Genetic Information Websites

On-line Mendelian Inheritance in Man (OMIM)
http://www.ncbi.nlm.nih.gov/Omim

Human Genome Epidemiology (HuGE) Net
   http://www.cdc.gov/genomics/hugenet/
Cancer Genome Anatomy Project (CGAP) http://cgap.nci.nih.gov/
PubMed http://www4.ncbi.nlm.nih.gov/PubMed/
SNPs Consortium http://snp.cshl.org/
National Center for Biotechnology Information (NCBI) dbSNPs
   http://www.ncbi.nlm.nih.gov/SNP/

Informed Consent
   http://www.cdc.gov/genomics/info/reports/policy/consentarticle.htm
   http://www.cdc.gov/genomics/info/perspectives/infmcnst.htm
   Department of Health and Human Services
      http://ohrp.osophs.dhhs.gov/polasur.htm#INF

Appendix 9.2

Genes and Polymorphisms with Relevance to Environmental Health

Gene	Gene Product	Polymorphism	Effect of Polymorphism	References
CYP1A1	Aryl hydrocarbon hydroxylase	T3801C (m1) A2455G (m2)	Unknown None	131 132
CYP1A2	Arylamine hydroxylase	C-164A	Decreased inducibility	133,134
CYP2E1	Ethanol-indudible P450	5' flanking repeat region	Increased activity after ethanol exposure	135,136
CYP3A4	Steroid-inducible P450	5' promoter A → G mutation	Unknown, perhaps expression levels	137,138
AHR	Aryl hydrocarbon receptor	G1721A	CYP1A1 inducibility?	139
EPHX1	Epoxide hydrolase	Tyr113His His139Arg	Altered protein stability?	140
NQO1	NAD(P)H: quinone oxido-reductase 1	C609T	Altered enzyme induction	88,141
NAT1	N-Acetyl transferase 1	Many alleles	Rapid vs. slow acetylation	80
NAT2	N-Acetyl transferase 2	Many alleles	Rapid vs. slow acetylation	80
SULT1A1	Sulfotransferase	Arg213His	Low activity and low thermal stability	142
GSTM1	Glutathione S-Transferase-μ	Deleted (null) allele(s)	No enzyme produced	143

Genes and Polymorphisms with Relevance to Environmental Health (*continued*)

Gene	Gene Product	Polymorphism	Effect of Polymorphism	References
GSTP1	Glutathione S-Transferase-π	Ile104Val Ala113Val	Altered activity and substrate affinity	144
GSTT1	Glutathione S-Transferase-θ	Deleted (null) allele(s)	No enzyme produced	145,146
PON1	Paraoxonase	Arg192Gln Met55Leu	Change in activity and substrate specificity:	147
		Promoter point mutations	Change in enzyme expression levels	
VDR	Vitamin D receptor	RFLP in 3' UTR; multiple point mutations	3' UTR = unknown; known effects for some point mutations	148
HLA-DP β_1	Antigen recognition protein	Lys69Glu	Change in $CD4^+$ recognition	58
XPD(ERCC2) XPF	Nucleotide excision repair (NER) enzyme system	Lys751Gln	Improved function	114
			Amino acid sequence change	116
XRCC1	Base excision repair (BER)	Arg399Gln	Unknown	149
APE1	Apurinic/ apyrimidinic endonuclease 1	Asp148Glu	Reduced endonuclease activity	150
ALAD	δ-Amino-levulinic Acid Dehydratase	G177C	Alleles 1 and 2, 2 allele yields a more electronegative protein	61
TLR4	Type I trans- membrane protein	A896G D299G	Unknown Altered cell signal transduction after LPS exposure	31
TNFα	Cytokine	G-308A	Altered transcriptional regulation	151

RFLP, restriction fragment length polymorphism; UTR, untranslated region.

10

Reporting and review of human genome epidemiology studies

Julian Little

The recent completion of the human genome sequence (1,2,16) and advances in technologies for genomic analysis are generating tremendous opportunities for epidemiologic studies to evaluate the role of genetic variants in the etiology of human disease (3). The basis of this evaluation will be identification of the allelic variants of human genes, description of the frequency of these variants in different populations, identification of diseases influenced by these variants and assessment of the magnitude of the associated risk, and identification of gene–environment and gene–gene interactions. The process of identifying DNA variation that may be associated with disease is under way through the cataloguing and mapping of single nucleotide polymorphisms (SNPs) throughout the genome. The analysis of genotype data on SNPs may aid in the identification of DNA alterations that result in or contribute to disease states.

Not surprisingly, the number of published human genome epidemiologic studies has increased rapidly (4). Therefore, integration of evidence will become increasingly important as a means of dealing with potentially unmanageable amounts of information. Heterogeneity between studies can be assessed, and when this occurs, attempts can be made to explain it. So far, few gene–disease associations have been replicated (5–7). This is also true for gene–environment and gene–gene interaction (8,9). It is important to determine how far methodologic issues may account for differences between studies. This requires that the studies are adequately reported and appraised. Investigation of heterogeneity between studies can lead to the formulation of new hypotheses.

In this chapter, we consider the reporting and systematic review of human genome epidemiologic studies. Systematic reviews differ from traditional reviews in that systematic reviews are supported by evidence that is integrated in explicitly defined stages (see below). Meta-analyses form a subset of systematic reviews in which quantitative methods are obtained to obtain an overall measure of effect across different studies or to detect and explain heterogeneity among studies. Pooled analysis of data on individual subjects from multiple studies has many features in common with systematic reviews but involves obtaining and re-analyzing the primary data, as distinct from aggregating published information (10).

Critical appraisal and integration of evidence require that the evidence be adequately reported. Brief checklists or guidelines for reporting gene–disease associations have been proposed (11,12). In this chapter, we present a more detailed overview of issues in the critical appraisal of studies of genotype prevalence, gene–disease associations, and gene–environment interactions, based in large part on the deliberations of an expert panel workshop convened by the Centers for Disease Control and Prevention and the National Institutes of Health in January 2001 (4,13). A checklist intended to guide investigators in preparation of manuscripts, to guide those who need to appraise manuscripts and published papers, and to be useful to journal editors and readers is presented in Table 10.1. It should not be regarded as an exhaustive list of points that have to be presented in all journal articles. Addressing all of the considerations, for example in studies of rare conditions in clinical settings, may not always be feasible.

Reporting and Appraisal of Single Studies

Hypothesis Specification

Associations between several genes and a disease can be tested according to a priori hypotheses based, for example, on a documented biologic mechanism of these genes in determining the disease. For example, the associations between a number of gene variants whose products are thought to influence the metabolism of folate and related nutrients and colorectal neoplasia have been investigated, because of the roles of folate in methylation and DNA synthesis (see Chapter 20). It is becoming usual practice in human genome epidemiology studies to initiate a study to test hypotheses that are current at that time and to establish a resource to test additional hypotheses proposed later on the basis of knowledge external to the resource. These are all a priori hypotheses. Hypothesis-testing is important to distinguish from hypothesis-generation.

In gene–disease association studies and studies of the prevalence of allelic variants, it has been suggested that data on genotypes should be presented, because it is the genotype that determines risk (13). A point to consider in appraising studies is the choice of categories. In a two-allele system, for example, justification would be sought for the decision to consider heterozygotes separately, include them in the reference category with homozygotes for the common variant, or group them with homozygotes for the rarer variant(s). This is more complex for multi-allelic systems.

In studies of gene–environment and gene–gene interactions, many hypotheses of interaction can potentially be tested. The distinction between a priori hypotheses

Table 10.1 Proposed Checklist for Reporting and Appraising Studies of Genotype Prevalence, Gene–Disease Associations, and Gene–Environment Interactions

Item To Be Specified	DETAILS BY TYPE OF STUDY		
	Genotype Prevalence	Gene–Disease Associations	Genotype–Environment Interaction
1. Purpose of study	✓	Detect associations or estimate magnitude of association	Describe joint effects; test specific hypotheses about interaction
2. Analytical validity of genotyping			
Types of samples used	✓	For cases and for controls	For cases and for controls
Timing of sample collection and analysis, by study group[a]	E.g., ethnic group	E.g., cases vs. controls	E.g., cases vs. controls
Success rate in extracting DNA, by study group[a]	E.g., ethnic group	E.g., cases vs. controls	E.g., cases vs. controls
Definition of the genotype(s) investigated; when there are multiple alleles, those tested for should be specified	✓	✓	✓
Genotyping method used (reference; for PCR methods—primer sequences,[a] thermocyle profile,[a] number of cycles[a])	✓	✓	✓
Percentage of potentially eligible subjects for whom valid genotypic data were obtained, by study group	E.g., ethnic group	E.g., cases vs. controls	E.g., cases vs. controls
If pooling was used, strategy for pooling of specimens from cases and controls		✓	
Quality control measures[a]	✓	Including blinding of laboratory staff	Including blinding of laboratory staff
Samples from each group of subjects compared (e.g., cases and controls) included in each batch analyzed[a]			✓
3. Assessment of exposures			
Methods of assessing exposure documented			✓
Reproducibility and validity of exposure documented			✓
Categories or exposure scale justified			✓

(continued)

4. Selection of study subjects

Geographical area from which subjects were recruited	✓	✓	✓
The recruitment period	✓	✓	✓
Recruitment methods for subjects whose genotypes were determined, such as random population-based sampling, blood donors, hospitalized subjects with reasons for hospitalization	✓	✓	
Definition of cases and method of ascertainment		✓	✓
Number of cases recruited from families and methods used to account for related subjects		✓	✓
Exclusion criteria for cases and controls		✓	✓
Recruitment rates		For cases and controls	For cases and controls
Mean age (±SD) or age range of study subjects, and the distribution by sex	Where possible, by sex, age, and ethnic group	For cases and controls	For cases and controls
If the subjects were controls from a case-control study, information on the disease under investigation and any matching criteria such as age, gender, and/or risk factor levels	✓	✓	
Ethnic group of study subjects	✓		
Similarity of sociodemographic (or other) characteristics of subjects for whom valid genotypic data were obtained with characteristics of subjects for whom such data were not obtained[a]		✓	✓
Steps taken to ensure that controls are noncases[a]		✓	✓

5. Confounding, including population stratification

Design		✓	✓
If other than a case-family control design, matching for ethnicity, or adjustment for ethnicity in analysis		✓	✓
Potential correlates of the genotype identified and taken into consideration in design or analysis		✓	✓

Table 10.1 Proposed Checklist for Reporting and Appraising Studies of Genotype Prevalence, Gene–Disease Associations, and Gene–Environment Interactions (*continued*)

Item To Be Specified	DETAILS BY TYPE OF STUDY		
	Genotype Prevalence	Gene–Disease Associations	Genotype–Environment Interaction
6. Statistical issues			
Distinguish clearly *a priori* hypotheses and hypotheses generated		✓	✓
If haplotypes used, specify how these were constructed	✓	✓	✓
Number of subjects included in the analysis, by cell numbers where possible	✓	✓	✓
Method of analysis, with reference, and software used to do this		✓	✓
Confidence intervals	Of genotype frequency	Of measures of association with the genotype	✓
For interaction analysis, 2 × K presentation used, or choice of stratified analysis justified			✓
For interaction analysis, *P* value for interaction calculated and choice of Wald test or likelihood ratio test specified and justified			✓
For interaction analysis, null interactions listed			✓
Assessment of goodness-of-fit of the model used[a]		✓	✓

[a]Additional information recorded (ideally in Web-based methods register).

and hypothesis generation is, again, important. Even in the simplest case of a dichotomous genotype and dichotomous exposure, genotype and environment can interact in six ways (14). Many more can be defined if more categories are introduced. For instance, Taioli et al. (15) have proposed a model in which an effect of the genotype is apparent at low environmental exposures but is not apparent at high exposures. Once multiple categories of dose are defined for the environmental variable, many different dose–response models can be tested in the data. Clearly, model specification becomes more difficult as more environmental factors (and levels of exposure) and more genes (and alleles) are included.

Design

In appraising studies, it is important to consider design as this affects the biases that may occur and generalizability (Chapters 6 and 8). Most studies of gene–disease associations and gene–environment and gene–gene interactions for late-onset diseases have used the case-control design. Much of the discussion therefore focuses on this design. However, DNA samples are being collected in a number of ongoing cohort studies. Compared with case-control studies, cohort studies have a number of advantages, including the capacity to examine age-at-onset distributions and multiple-disease outcomes (17–19). The use of case-cohort and nested case-control analysis of archived samples that are suitable for genotypic analysis potentially can minimize the disadvantages of the cost of genotyping an entire cohort. A major advantage of the case-cohort design for studies in which use of expensive assays is planned is that the same comparison group can be used for several different disease outcomes. Therefore, this design is likely to be used increasingly. Because the detection of gene–environment and gene–gene interaction is particularly challenging, novel study designs, most notably the case-only design and multistage designs, have been proposed (Chapter 6). Concern about the possible impact of population stratification has stimulated the development of family-based case-control designs; these are discussed briefly in the section on population stratification.

Issues that are particularly important in the appraisal of studies of genotype prevalence, gene–disease associations, and gene–environment and gene–gene interactions include the analytical validity of genotyping, selection of subjects, confounding (especially as a result of population stratification), statistical power, and multiple statistical comparisons. In addition, exposure assessment is an important issue in the appraisal of studies of gene–environment interaction. Because many methodologic issues are common to the three types of study, these are discussed in parallel.

Assessment of Genotypes

The definition of the genotype(s) investigated should be clearly presented. The validity of grouping genotypes on the basis of putative functional effects depends on the availability and quality of functional studies of gene variants, and information on functional effects is likely to change over time. For multi-allelic systems, genotypes have been grouped according to functional effects in some investigations. For example, grouping according to inferred rapidity of acetylation has been done for the *NAT2* polymorphisms (8).

True functional variants are important to distinguish from markers associated with a disease only because they are in linkage disequilibrium with a functional variant. Typing several polymorphisms throughout a candidate gene may be useful in order to construct haplotypes, which could then be tested for association with the phenotype of interest. The increasing availability of mapped SNP markers (20–24) offers the opportunity for such an approach and presents methodologic challenges (see below).

Other factors affecting the analytical validity of genotyping, including the types of samples and timing of collection, the method used for genotyping, and quality control procedures are summarized in Table 10.1. These issues are discussed in Chapter 5 and Little et al. (13).

Assessment of Exposures

Not surprisingly, exposure assessment is important in studies of gene–environment interaction. Points that need to be considered are the method of exposure assessment, and its validity and reproducibility. Exposure misclassification can bias the estimation of an interaction effect, the magnitude of which depends on the prevalence of the misclassified exposure and on the interaction model (Chapter 8, 25). If interaction is defined as lack of fit to a multiplicative model, a test for interaction will be conservative (26). In theory, case-control studies are more vulnerable to differential misclassification than are cohort studies (and the related case-cohort and nested case-control designs). However, provided that the extent of misclassification of exposure does not vary by genotype, differential misclassification between cases and controls is not a serious problem for the detection of departures from a multiplicative gene–environment joint effect (26).

Selection of Subjects

Evaluation of potential selection bias requires consideration of study design and fieldwork. It is important to distinguish studies that aim to detect an association from those that aim to estimate the magnitude of an association. In the former situation, cases may be "overselected" from multiplex families to increase the power to detect an association; presenting the measure of association as an estimate of population association would be inappropriate. In the latter situation, the principles underlying study design are essentially the same as for the investigation of the magnitude of association with environmental risk factors, including the minimization of the potential for selection bias emphasized in many epidemiologic textbooks (27–30). In a number of studies, the selection of cases has not been well described (31). In a review of type 1 diabetes and *HLA-DQ* polymorphisms, the authors noted that many studies were based on convenience samples of cases in which persons with type 2 diabetes who used insulin in their treatment regimen had been included (32). In several studies of cancer, prevalent cases have been included to varying extents (33). In these studies, bias would occur if the genotype affected survival or if genotypes were assayed by a phenotypic test that was influenced by disease progression or treatment.

A recurrent problem in case-control studies of gene–disease associations with unrelated controls has been that the controls were not selected from the same source population as the case-subjects (8,9,31,32). The potential problem of selecting controls who do not represent the population from which case-subjects arise is illustrated by the divergence in odds ratios for the association between colorectal cancer and the *GSTT1* null genotype (34), when the different control groups were analyzed (9). In regard to genotype prevalence, many early studies were based on convenience samples and not infrequently, little information was given about how the samples were selected (8,9,31,35).

Population Stratification

Concern has been raised about the possible effects of population stratification on the results of population-based case-control studies (36–41). Population stratification includes differences between groups in ethnic origin and can arise because of differences between groups of similar ethnic origin but between which there has been limited admixture, such as in isolated populations. For example, a population might comprise the descendants of waves of immigrants from the same source but differ generally because of founder effects. The differences may then be apparent because insufficient time has elapsed for mixture between the groups. In an exploration of the possible degree of bias from population stratification in U.S. studies of cancer among non-Hispanic Americans of European descent, this bias was considered unlikely to be substantial when epidemiologic principles of study design, conduct, and analysis were rigorously applied (42). A similar conclusion was reached with the use of data from case-unrelated control studies of non-Hispanic U.S. whites with hypertension or type 2 diabetes, and Polish subjects with type 2 diabetes (43). Variations in the frequency of certain genotypes in African Americans appear to be much wider than those observed in persons of European origin and therefore the possibility of stratification may be higher (44). Evidence was weak for an effect of population stratification in data from a case-unrelated control study of hypertension in African Americans, but this was no longer apparent when the study was restricted to persons with U.S.-born parents and grandparents (43).

Concern about the possible effects of population stratification has stimulated development of family-based case-control designs, which essentially eliminate potential confounding from this source (45,46). The most commonly used examples of such designs involve the use of siblings or parents as controls. Sibling controls are derived from the same gene pool as cases. However, selection bias could result because a sibling may not be available for every case–bias would arise if determinants of availability (e.g., sibship size) were associated with genotype. In addition, compared with a study in which unrelated controls were used, a study using an equivalent number of sibling controls has less statistical power because of overmatching on genotype (47). This loss of power generally does not occur for case-parental control studies (46), which have been advocated for the identification of modest gene–disease associations (48). However, the need to obtain samples from parents is a practical problem limiting the applicability of the design for diseases of late onset. Clearly, the study design is appropriate to consider in assessing the possible impact of population stratification.

Another approach proposed to minimize the potential problem of population stratification when unrelated controls are used is to measure and adjust for genetic markers of ethnicity that are not linked to the disease under investigation (49–52). This would be expected to control for ethnic variation in disease risk attributable to genetic factors. However, residual confounding from other sources of ethnic variation in disease risk would be a potential issue. A single measure is unlikely to capture the important sources of ethnic variation (53). In appraising case-unrelated control studies, or cohort studies, points to consider are the adequacy of matching for ethnicity or adjusting for it in analysis.

Confounding from Other Sources

Confounding of a gene–disease association, and of gene-environment and gene-gene interactions, potentially could result from linkage disequilibrium. Linkage disequilibrium depends on population history and on the genetic make-up of the founders of that population (7,54). Linkage disequilibrium varies between populations (54) and may in part account for the variable results of studies of gene–disease associations (7). In a correctly designed association study, except for allelic associations that extend for a short genomic region from the locus under investigation, the comparison of groups of individuals defined by genotype could be equivalent to a randomized comparison (26). However, so far data on linkage disequilibrium for SNPs show that the extent of linkage disequilibrium varies by region of the genome, and that its variation at all distances is great (54). Moreover, studies of microsatellite polymorphisms have shown linkage disequilibrium between a few loci that are separated by many megabases (≥ 1 cM) (55).

In studies of gene–environment interaction, confounding of exposures is a potential problem. The principles regarding the control of confounding are the same as those for studying the relation between exposure and disease. In practice, the use of biomarkers of exposure may need care in interpretation, because the genotype may influence the presence or level of the biomarker. Rothman (56) noted that an extraneous risk factor is a confounder only if its effect becomes mixed with the effect under study. For example, an exposure may cause altered physiology, which in turn causes disease. A biomarker of the altered physiology is a risk factor for the disease and is unrelated to exposure because it results from exposure. It is not confounding because the effect of the exposure is mediated through the effect of the altered physiology, and therefore no effects are mixed. However, decisions about whether a biomarker represents an intermediate factor in etiology or is a potential confounder are difficult when uncertainties exist about the mechanism of effect of exposure. This would also apply to genes.

In case-cohort studies, controls are a random sample of the cohort, and the effect of age, which is the key time variable, is controlled for in the analysis only. In more traditional nested case-control designs, controls are selected to match the cases on a temporal factor, such as age, and the main comparisons are within the time-matched sets (57). In appraising case-cohort studies, the method of age adjustment and, in appraising nested case-control studies, details of the matching on age or other temporal factors are important to consider.

Statistical Issues

In appraising studies, the main statistical issues are study power, multiple testing, and method of analysis.

Power. A small study size is a limitation of many studies testing a priori hypotheses about gene–disease associations (e.g., references 9 and 58). This problem is exacerbated in studies of gene–environment and gene–gene interactions. To test for departures from multiplicative effects, it has been noted that study size should be at least four times larger than needed to detect only the main effects of the individual factors (59). In studies of modest gene–environment interactions, the sample size requirement is of the order of a thousand cases or more. When nondifferential misclassification of exposure is taken into account, many thousands of cases may be needed (25). The biggest problem facing the field of gene–environment interaction is that almost no published studies have these sample sizes. A possible solution is pooled analysis (see below).

Multiple Testing. One proposed research strategy is large-scale testing by genome-wide association mapping (48,60–62). This strategy is hypothesis-generating rather than hypothesis-based and thus may require additional safeguards against type 1 error. For example, Risch and Merikangas (48) suggested specifying a higher significance level. However, increasing the significance level will increase the number of subjects required to have adequate statistical power, although this may not make studies unfeasible (48).

In the analysis of gene–environment interactions, a large number of potential interactions could be tested for in a typical data set. Current data sets often already have several dozen genotypes determined, and many dozen, or even hundreds, of different environmental variables may be determined for each person in the data set (e.g., a typical food–frequency questionnaire will measure intake of more than 100 foods and permit estimation of more than 50 nutrients). Moreover, it is important to know whether there was an a priori choice of categories or scale used to quantify the amount of exposure, because this will give insight as to whether multiple testing is an issue for interpretation. In addition, the interaction model must be specified (see earlier discussion of the many models of gene–environment interaction). An approach of assessing interaction of every genotype with every environmental variable under every possible interaction model would generate a large number of false-positive results. Increasing the significance level is unlikely to solve the multiple comparisons problem in this context. The limited power to detect even established interactions at the $p < 0.05$ level in most studies (because of modest effects and limited sample sizes) means that adjusting for multiple comparisons would be almost equivalent to never declaring statistical significance for "true" interactions. In other words, reducing the nominal p value would mitigate the false-positive problem by creating a potentially unacceptably high false-negative rate.

Methods of Analysis. Well-established methods exist for describing the prevalence of exposure and for measuring associations (27,28). These can be applied to describing genotype prevalence and assessing gene–disease associations. In regard to trend-tests for gene–disease associations, even in the case of a single gene with two alleles, a decision is needed about whether to treat genotype as a trichotomous variable in which heterozygotes are categorized separately (i.e., assuming codominance) or to combine them with one of the two groups of homozygotes (i.e., assuming a dominant or recessive model). A problem in the choice of such models is the lack of functional information. There can be substantial loss of statistical power when a test suitable for one mode of inheritance is used where another mode is the true one (63).

Methodologic issues relating to haplotype analysis are still under development. In particular, in studies based on unrelated persons, haplotypes can be estimated only probabilistically on the basis of allele frequencies. If external estimates of haplotype frequency in the population are applied, inference may be affected by the quality and availability of the data on haplotype frequencies in the relevant population. As more SNP loci are identified, the number of possible haplotypes can become huge, in turn raising the issues of multiple comparisons and sparse data for many haplotypes (60,64). A potential limitation of the approach of constructing haplotypes is that the effect of a true functional variant might be diluted when haplotypes rather than loci are the units of analysis.

The methods of assessing gene–environment and gene–gene interaction are less established. Three common methods have been used to assess the statistical significance of gene–environment interactions, when defined as departures from multiplicative effects. First, an interaction term is introduced into a logistic model, and the Wald p value for the coefficient of the interaction term is reported. In the case of multiple ordered categories of the environmental variable entered as an ordered categorical variable, the interaction term tests whether the linear trend in the environmental variable is significantly different between the dichotomous categories of genotype. Second, a cross-product "dummy" term is introduced into the logistic model for each combination of genotype and environment category (omitting the combination for the reference category). The p value for interaction is then given as the difference in the log-likelihood between this model and the model containing the main effect estimates for the genotype and environment variables. When both genotype and exposure are dichotomous, then these two tests are equivalent.

However, when there is more than one category, they test different models. In this situation, a point to appraise is whether the model of interaction was specified a priori. A potential problem with the likelihood ratio test for interaction is that it does not directly test for trend. In situations in which the data depart from an ordered trend, the likelihood ratio test may give a significant result because the cross-product terms improve the fit of the model to the data. Therefore, assessing gene–environment interaction solely by screening for level of significance of a formal test for interaction should be avoided. Third, estimates of environmental effects are compared between genotype strata. However, the finding of a significant effect in one or more strata but no significance in at least one other stratum does not constitute statistical evidence of interaction. Often such a pattern has been observed when inadequate power exists in one of the strata. Whether a formal test of statistical interaction has been performed to assess the strength of the evidence for interaction should be considered.

Analytic methods to test for gene-environment and gene-gene interactions are still under development. For example, the application of hierarchical models is being explored (65,66). Little work has been done on testing for departures from additive models of genetic and environmental effects (26,67).

Systematic Review

The stages involved in systematic review are (*1*) specification of the issue for which integrated evidence is needed, (*2*) identification of studies, (*3*) critical appraisal of studies, (*4*) abstraction of data, and (*5*) synthesis.

Specification of the Issue

Typically, the need exists to specify the allelic variant, then consider one or more of questions relating to its frequency (at an early stage in research), its variation in frequency (as data accumulate), its relations with specific diseases, and whether it modifies the effect of exposures that are etiologically important (and vice versa).

Identification of Studies

A comprehensive search is one of the key differences between a systematic review and a traditional review (68). Typically, the strategy used to identify relevant papers for a systematic review involves specifying the search terms, the time period of publication, the databases searched, and software used to do this (69). Because problems may exist with the indexing of papers, hand-searches of the reference lists of relevant papers identified from the original search and of key journals are common practices. Thus, for example, in a review of the association between glutathione S-transferase polymorphisms and colorectal cancer, Medline and EMBASE were searched using the MeSH heading "glutathione transferase" and the textwords "GST" and "glutathione S transferase" for papers published between 1993 and 1998 (9). The CDC Office of Genomics and Disease Prevention

Medical Literature Search was also searched, and reference lists in published articles were hand-searched.

A further issue is the possible inclusion of unpublished sources, including abstracts, technical reports, and non-English journals (70) that may not be identified by electronic searches, as a means of minimizing the potential impact of publication bias (see below). However, this material should be treated with caution because it may not be peer-reviewed and may be subject to modification and revision. In addition, information on study methods may be insufficient to assess study quality.

Several instances have occurred of sequential or multiple publications of analyses of the same or overlapping datasets. For example, in studies of *CYP1A1* polymorphisms and breast cancer, substantial overlap between the studies of Ambrosone at al. (71) and Moysich et al. (72), and between that of Taioli et al. (73) and Taioli et al. (74), is likely. An aid to identifying this problem is to organize evidence tables (see below) first by geographic area and then by study period within a specified area. If the reports clearly relate to the same or overlapping datasets, then a consistent method of dealing with this should be adopted, such as including data only from the largest or most recent publication. Under these circumstances, details of the methodology may be described in greater detail in an earlier publication. If so, the reference to the earlier publication should be given with the reference to the publication from which the data were abstracted in the evidence tables.

Critical Appraisal

Issues in the appraisal of single studies have been discussed above. A number of reports have been published about the rating of the quality of analytical observational studies. Several relate to case-control studies (27,29,75–79). Some (75,76) are part of a series of articles documenting the deficiencies of epidemiologic research; they have been challenged on the grounds of technical errors, failure to distinguish important from unimportant biases, and ignoring the need to weight the totality of the evidence about a relation (80,81). Other issues include possible overemphasis of the potential problems of case-control studies in comparison with cohort studies (78) and difficulty in assessing differences between methods applied in the case and control groups, or between different exposure (prognostic) groups (79,82).

Several authors have proposed quantitative quality scoring systems for critical appraisal (82). Other schemes have been developed for meta-analyses in which an attempt has been made to assess the importance of study quality in accounting for heterogeneity of results between studies (83–85). This type of assessment also has been considered for pooled analysis (86,87). Certain features of the assessment schemes are specific to the disease or the exposure under consideration, and each aspect of the study is given equal weight. Thus, summation of points might result in worse quality scores for a study with several minor flaws than for a study with one major flaw. Although empirical studies on a large number of primary investigations might suggest an overall relation between a specific aspect of study design and the reported results, this relation is ecologic and may not be true for a specific investigation. Therefore, specific noncausal factors, which might affect the interpretation of a single investigation, are difficult to isolate. Jüni et al. (88) observed that the use of scores to identify clinical trials of high quality is problematic and recommended that relevant methodologic aspects should be assessed individually and their influence on the magnitude of the effect of the intervention explored. Similar caution in consideration of studies of gene–disease associations is likely to be justified. As in clinical trials, multidimensional domains may be more appropriate to consider than a single grade in the integration of evidence from observational studies.

Little or no empirical evaluation exists of the quality scoring of association studies. However, many users of data on genotype prevalence and gene–disease associations need a robust means of grading evidence. This approach has been proposed by the Scottish Intercollegiate Guidelines Network (89). In this approach, studies of gene–disease association in which all or most of the criteria specified as appropriate to a research question are satisfied would be graded as "+ +." Criteria that have not been fulfilled would not affect the grade if the conclusions of the study were considered *very unlikely* to be affected by their omission. Studies in which some of the criteria have been fulfilled, and criteria that were not fulfilled considered *unlikely* to alter the conclusions would be graded as "+." Studies in which few or no criteria were fulfilled and the conclusions of the study considered *likely* or *very likely* to be altered by multiple omissions in required criteria for an acceptable study, would be graded as "−."

Abstraction of Data

Specific forms are often used for this purpose, for example, that used for the Human Genome Epidemiology Network's e-journal reviews (90). The form should be piloted to ensure a consistent approach to data abstraction. Ideally, this would be done by two independent reviewers and discrepancies would be resolved, but resources may not permit this (91,92). Typically, such forms include reference details, information about study eligibility, study methods, and study results.

Synthesis of the Evidence

The first steps include describing the volume of evidence and preparing evidence tables that summarize the basic characteristics of the studies, factors relating to study quality, measures of association (with indicators of precision), and the reference. On this basis, consideration is given to combining results. The simplest way of combining results is counting the number of studies showing positive, negative, and inverse associations. However, this approach is very limited as no account is taken of study quality or of the magnitude of the association. Other approaches take account of these issues.

Hierarchy of Evidence. In many schemes of qualitative synthesis of evidence, a hierarchy exists whereby certain study designs are considered inherently superior to others. In general, analytical epidemiologic designs are stronger than ecologic designs and studies of case series or reports. Although cohort studies may be less subject to bias than case-control studies, important issues exist about quality of follow-up and case ascertainment. Therefore, it seems more rigorous to weight the evidence from specific studies of these types on the basis of a full critical appraisal rather than solely on the basis of general design.

Quantitative Synthesis. There are two types of quantitative synthesis of evidence: (1) meta-analysis of the results of studies and (2) pooled analysis of data on individual subjects obtained in several studies. The validity of meta-analysis of observational studies has been debated (69,93). On the one hand, meta-analysis may indicate a "spurious precision" and either meta-analysis of observational studies should be abandoned altogether (94) or possible sources of heterogeneity between studies should be considered (95). On the other hand, meta-analysis can help clarify whether an association exists and indicate the quantitative relation between the dependent and independent variables (96). The indication of the quantitative relation, although potentially biased, may be valuable in considering public health effects of interventions based on knowledge of the genetic factor or its interactions.

Pooled analysis requires *data* on individual subjects. This approach offers many advantages over the meta-analysis of the *results* of studies, including standardization of definitions of cases and variables, better control of confounding, and consistent determination of subgroup effects (10,86). For example, this approach has been used successfully to study the effect of chemokine and chemokine receptor alleles on HIV-1 disease progression (97). Nevertheless, pooling approaches require much greater resources (98). Interestingly, the results of meta-analyses of the glutathione S-transferase M1 polymorphism and cancer of the lung (99) and bladder (67) were similar. Pooled analysis is preferred to meta-analysis of the results of studies when a high degree of accuracy of the measures of effect is required. However, stratification by original study may still be important, to allow for and elucidate causes of heterogeneity among the data sets being pooled.

Interpretation

The main issues appear to be consideration of possible publication bias and application of guidelines for causal inference.

Publication Bias. Publication bias is the selective publication of studies on the basis of the magnitude and direction of their findings (100). Research with statistically significant results has long been accepted to be more likely to be submitted and published than work with null or nonsignificant results (101), and this has led to a preponderance of false-positive results in the literature (102). Therefore, publication bias is a potentially serious problem for the integration of evidence on

Box 10.1 Guidelines for Causal Inference (modified from Hill [120] and Surgeon General [119])

- Consistency
- Strength
- Dose–response[a]
- Biologic plausibility (including analogy[b])[a]
- Temporality
- Experimental support[a]
- Coherence

[a]Additional considerations specified by Hill (120)
[b]Analogy is a variant of biologic plausibility (29, 121)

gene–disease associations (6,7), especially in relation to gene–environment and gene–gene interactions. In addition to the larger number of potential comparisons implicit in the concept of multiple interacting variables, authors face the problem that large tables of gene–environment interaction estimates are very cumbersome and difficult to assemble in publishable format. This inevitably increases the potential for publication bias.

In other fields, quantitative and qualitative methods of detecting publication bias have been used, such as the fail-safe technique where the number of new studies averaging a null result needed to bring the overall effect to nonsignificance is calculated (69,103). Then a judgment can be made as to whether it is realistic to assume that such a number of studies have been unpublished in the field of investigation. If the assumption were realistic, then the validity of conclusions based on published evidence would be doubtful. Other quantitative and qualitative methods have been reviewed by Sutton et al. (92) and by Thornton and Lee (104). In general, all the methods have limitations. Therefore, it seems appropriate to account for the possibility that the evidence base may be skewed toward positive results in drawing conclusions about causal relations.

Another potential method of identifying publication bias is to search research registers such as Computer Retrieval of Information on Scientific Projects (CRISP) (105) and the *Directory of On-going Studies in Cancer Prevention* (106). Administering research registers on studies of genotype prevalence and gene–disease associations is challenging, because data for each additional allele genotyped would need to be added to the database. It is even more difficult for studies of gene–environment and gene–gene interactions, because of the diversity of joint effects which can be investigated.

Causal Inference. Well-established guidelines exist for causal inference (Box 10.1). However, in practice, only limited subsets of these tend to be used (107). For example, in cancer epidemiology, the guidelines most often applied are consistency, strength, dose–response, and biologic plausibility.

Consistency In relation to consistency of gene–disease associations and gene–environment and gene–gene interactions, differences between studies in distributions of subjects by age and sex are sources of heterogeneity. For example, hormonal alterations can affect ligand binding, enzyme activity, gene expression, and the metabolic pathways influenced by gene expression. In particular, some inconsistency between the results of gene–disease association studies may be secondary to variation among studies in the prevalence of interacting environmental factors that have not been assessed. Testing a priori hypotheses about differences in gene–disease associations and genotype frequencies between studies that may arise from these sources would be appropriate.

In relation to interactions, heterogeneity may occur if the allele under study is associated with disease due to linkage disequilibrium with a gene that is truly causal. Such a "marker" allele may behave differently in populations with different genetic backgrounds resulting from differences in the extent of the linkage disequilibrium, even if the "causal gene" has the same effect in the different populations. Differences between populations in allele prevalence may result in differences between studies in the statistical power to detect both the main effect of the genotype and gene–environment interactions. Similarly, the prevalence of exposure or variability of exposure may influence whether an interaction exists or is detectable.

Strength As noted by Rothman (56), the strength of an association is not a biologically consistent feature but rather a characteristic that depends on the relative prevalence of other causes. In studies of the general population, the associations between disease and biomarkers of susceptibility are not likely to be strong. In particular, many of the genetic variants so far identified as influencing susceptibility to common diseases are associated with a low relative and absolute risk (108). Therefore, exclusion of noncausal explanations for associations is crucial. In this situation, an interaction between a gene and exposure (or another gene) would be expected.

Dose–response In the context of gene-disease associations, the value of considering dose–response relations will depend on information about the functional effect(s) of the relevant gene. As already noted, in the particular instance of gene-environment interaction, when multiple categories of dose are defined for the exposure, then many different dose-response models can be tested in the data, and tests for interaction can be applied to the trends across strata. Consequently, false-positive results are likely to be a problem.

Biologic plausibility This is a particularly important issue in the evaluation of gene–disease associations, gene–gene, and gene–environment interactions. For example, in investigations of associations with genetic polymorphisms of carcinogen metabolism and DNA repair, many genotypes have been assessed without data on their functional significance. Investigations confined solely to genotypes potentially would lead to numerous false-positive associations. Consideration of biologic plausibility involves determining (*1*) whether a known function of the gene product can be linked to the observed phenotype; (*2*) whether the gene is expressed in the tissue of interest; and (*3*) temporal relations, including the time window of gene-expression in relation to age-specific gene–disease relations. Thus, the gene should be in the disease pathway and/or involved in the mechanism that is responsible for the development of the disease. If not, then the effect of the gene may be indirect. In studies of cancer in young persons, maternally mediated effects of the maternal genotype and parental imprinting also may be relevant to consider. As an example of the need for careful interpretation, N-acetyltransferases (NATs) have been considered to be important in detoxification. However, NAT has been observed to catalyse O-acetylation (109). O-acetylation is thought to be an activating step.

Specificity Although specificity has been included as a criterion of causation, it may be inappropriate in relation to the effects of complex exposures that may influence several outcomes such as tobacco smoking, or genetic variants that may influence the metabolism of a variety of exposures, such as cytochrome P450 gene variants. For example, *CYP1A1* gene variants have been investigated in relation to a variety of types of cancer (33), macular degeneration (110), Parkinson's disease (111), endometriosis (112), primary dysmenorrhea (2), and orofacial clefts (113).

Temporality Although a correct time relation is specified in many methodologic texts, it seems to be seldom used in causal inference (107). In the situation of gene–disease associations, the disease could influence the result of a phenotypic assay of the genotype under investigation. This should not be a problem with PCR methods. If data were available on the time window of gene–expression, it would be relevant to consider this in relation to age-specificity of gene–disease relations. As a perhaps extreme example, if an association existed between a type of cancer in infants and the *CYP1A1* or *CYP1A2* genotype of the index child, this probably would be indirect (e.g., reflecting an effect of maternal genotype) because the enzymes coded by these genes are not expressed in the fetal liver (114,115).

Experimental support In the context of gene–disease associations, experimental support is most likely to be derived from studies of gene expression in knockout or other experimental animals, from in vitro data on gene function, or from experimental interventions based on clinical trials of interventions aimed at normalizing the function or levels of a product regulated by the gene. For example, initially transgenic mouse models appeared to support a role for certain genes in the etiology of orofacial clefts (116). It is now apparent that clefts often occur in knockout and insertion experiments, and that gene expression at a critical time and in a tissue relevant to development of the lip and palate should also be taken into account. An example of in vitro investigation on gene function is an investigation of the effect of the *MTHFR* C677T polymorphism on folic acid deficiency-induced uracil incorporation into human lymphocyte DNA (117). In regard to trials of interventions aimed at normalizing a gene product, trials of the drug CPX are under way (118). CPX acts by binding to the mutant channel protein, helping it to mature and gain access to the plasma membrane, and it is thought that repair of the defect in trafficking to the membrane helps suppress the high level of synthesis and secretion of IL-8 that is involved in pathogenesis.

Coherence This criterion has been defined as being satisfied when an association being consistent with the state of knowledge of the natural history and biology of the disease (119). In practice, this criterion has been little used, perhaps because it has been considered equivalent to biologic plausibility (107). Elwood (29) defines an association as being coherent "if it fits the general features of the distribution of both the exposure and the outcome under assessment." He notes that the concept holds only if a high proportion of the outcome is caused by the exposure, and if the frequency of the outcome is fairly high in those exposed. An additional constraint on the use of this criterion arises when information about the distribution of the relevant exposure and outcome is inadequate. Information about the distribution of many biomarkers is limited. In the situation of gene–disease associations, the "exposure" would be the genotype being investigated.

Conclusions

There has been a tremendous increase in the number of published human genome epidemiologic studies, and this increase is set to continue. So far, few gene–disease associations, gene–environment, or gene–gene interactions have been replicated. This may in part be due to methodologic issues. Methodologic issues that are particularly important include the assessment of genotypes, selection of subjects, confounding, statistical power and multiple statistical testing. In the assessment of gene–environment interaction, assessment of exposure is also an important issue. It is hoped that the checklist presented in this chapter will be useful to investigators preparing manuscripts, to those who need to appraise manuscripts and published papers, and to journal editors and readers. In regard to the integration of evidence, established principles of systematic review should be applied. Meta-analysis and pooled analysis can help address concerns about statistical power and provide a formal means of investigating possible heterogeneity between studies. Pooled analysis is labor intensive. It is preferred to meta-analysis when a high degree of accuracy of the measures of effect is required. In interpreting this evidence, the potential for publication bias is an important consideration. To address the problem of publication bias, a register of research is needed that would include negative findings. In terms of specifying hypotheses to be tested and interpretation of the biologic plausibility of study findings, inter-disciplinary collaboration in this fast-expanding field is crucial.

Acknowledgments
Much of this chapter is the result of discussions at an expert panel workshop convened by the Centers for Disease Control and Prevention and the National Institutes of Health in January 2001. We thank the following contributors for comments: Linda Bradley, Molly S Bray, Daniel Burns, Mindy Clyne, Gwen W. Collman, Janice Dorman, Darrell L. Ellsworth, James Hanson, Robert A. Hiatt, David J. Hunter, Muin J. Khoury, Joseph Lau, Thomas R O'Brien, Nathaniel Rothman, Donna Stroup, Emanuela Taioli, Duncan Thomas, Harri Vainio, Sholom Wacholder, Clarice Weinberg, and Paula Yoon.

References

1. McPherson JD, Marra M, Hillier L et al. A physical map of the human genome. Nature 2001;409:934–941.
2. Venter JC, Adams MD, Myers EW et al. The sequence of the human genome. Science 2001;291:1304–1351.
3. Shpilberg O, Dorman JS, Ferrell RE et al. The next stage: molecular epidemiology. J Clin Epidemiol 1997;50:633–638.
4. Khoury MJ. Commentary: epidemiology and the continuum from genetic research to genetic testing. Am J Epidemiol 2002;156:297–299.
5. Dunning AM, Healey CS, Pharoah PD et al. A systematic review of genetic polymorphisms and breast cancer risk. Cancer Epidemiol Biomarkers Prev 1999;8:843–854.
6. Ioannidis JP, Ntzani EE, Trikalinos TA et al. Replication validity of genetic association studies. Nat Genet 2001;29:306–309.
7. Hirschhorn JN, Lohmueller K, Byrne E et al. A comprehensive review of genetic association studies. Genet Med 2002;4:45–61.
8. Brockton N, Little J, Sharp L et al. N-acetyltransferase polymorphisms and colorectal cancer: a HuGE review. Am J Epidemiol 2000;151:846–861.
9. Cotton SC, Sharp L, Little J et al. Glutathione S-transferase polymorphisms and colorectal cancer: a HuGE review. Am J Epidemiol 2000;151:7–32.
10. Ioannidis JP, Rosenberg PS, Goedert JJ et al. Commentary: meta-analysis of individual participants' data in genetic epidemiology. Am J Epidemiol 2002;156:204–210.
11. Weiss ST. Association studies in asthma genetics. Am J Respir Crit Care Med 2001;164:2014–2015.
12. Cooper DN, Nussbaum RL, Krawczak M. Proposed guidelines for papers describing DNA polymorphism-disease associations. Hum Genet 2002;110:207–208.
13. Little J, Bradley L, Bray MS et al. Reporting, appraising, and integrating data on genotype prevalence and gene-disease associations. Am J Epidemiol 2002;156:300–310.
14. Khoury MJ, Adams MJ Jr, Flanders WD. An epidemiologic approach to ecogenetics. Am J Hum Genet 1988;42:89–95.
15. Taioli E, Zocchetti C, Garte S. Models of interaction between metabolic genes and environmental exposure in cancer susceptibility. Environ Health Perspect 1998;106:67–70.
16. National Human Genome Research Institute, National Institutes of Health, Department of Health and Human Services and Office of Science, U.S. Department of Energy. International Consortium Completes Human Genome Project. Accessed May 15, 2003, from http://www.genome.gov/11006929.

17. Dean M, Carrington M, Winkler C et al. Genetic restriction of HIV-1 infection and progression to AIDS by a deletion allele of the CKR5 structural gene. Hemophilia Growth and Development Study, Multicenter AIDS Cohort Study, Multicenter Hemophilia Cohort Study, San Francisco City Cohort, ALIVE Study. Science 1996;273:1856–1862.
18. Michael NL, Chang G, Louie LG et al. The role of viral phenotype and CCR-5 gene defects in HIV-1 transmission and disease progression. Nat Med 1997;3:338–340.
19. Langholz B, Rothman N, Wacholder S et al. Cohort studies for characterizing measured genes. J Natl Cancer Inst Monogr 1999;39–42.
20. Sachidanandam R, Weissman D, Schmidt SC et al. A map of human genome sequence variation containing 1.42 million single nucleotide polymorphisms. Nature 2001;409: 928–933.
21. Reich DE, Cargill M, Bolk S et al. Linkage disequilibrium in the human genome. Nature 2001;411:199–204.
22. Altshuler D, Pollara VJ, Cowles CR et al. An SNP map of the human genome generated by reduced representation shotgun sequencing. Nature 2000;407:513–516.
23. Gray IC, Campbell DA, Spurr NK. Single nucleotide polymorphisms as tools in human genetics. Hum Mol Genet 2000;9:2403–2408.
24. Porter CJ, Talbot CC, Cuticchia AJ. Central mutation databases—a review. Hum Mutat 2000;15:36–44.
25. Garcia-Closas M, Rothman N, Lubin J. Misclassification in case-control studies of gene-environment interactions: assessment of bias and sample size. Cancer Epidemiol Biomarkers Prev 1999;8:1043–1050.
26. Clayton D, McKeigue PM. Epidemiological methods for studying genes and environmental factors in complex diseases. Lancet 2001;358:1356–1360.
27. Breslow N, Day N. Statistical methods in cancer research. Volume 1. The analysis of case-control studies. 1980. Lyon, IARC.
28. Kelsey JL, Whittemore AS, Evans AS et al. Methods in observational epidemiology. Oxford: Oxford University Press, 1996.
29. Elwood M. Critical appraisal of epidemiological studies and clinical trials. Oxford: Oxford University Press, 1998.
30. dos Santos Silva I. Cancer epidemiology: Principles and methods. Lyon: IARC, 1999.
31. Botto LD, Yang Q. 5,10-Methylenetetrahydrofolate reductase gene variants and congenital anomalies: a HuGE review. Am J Epidemiol 2000;151:862–877.
32. Dorman JS, Bunker CH. HLA-DQ locus of the human leukocyte antigen complex and type 1 diabetes mellitus: a HuGE review. Epidemiol Rev 2000;22:218–227.
33. d'Errico A, Malats N, Vineis P, et al. Review of studies of selected metabolic polymorphisms and cancer. In: Vineis P, Malats N, Lang M et al., eds. Metabolic polymorphisms and susceptibility to cancer. IARC Scientific Publications No. 148. Lyon: IARC, 1999: 323–393.
34. Chenevix-Trench G, Young J, Coggan M et al. Glutathione S-transferase M1 and T1 polymorphisms: susceptibility to colon cancer and age of onset. Carcinogenesis 1995;16:1655–1657.
35. Wang SS, Fernhoff PM, Hannon WH et al. Medium chain acyl-CoA dehydrogenase deficiency human genome epidemiology review. Genet Med 1999;1:332–339.
36. Knowler WC, Williams RC, Pettitt DJ et al. Gm3;5,13,14 and type 2 diabetes mellitus: an association in American Indians with genetic admixture. Am J Hum Genet 1988;43:520–526.
37. Gelernter J, Goldman D, Risch N. The A1 allele at the D2 dopamine receptor gene and alcoholism. A reappraisal. JAMA 1993;269:1673–1677.
38. Khoury M, Beaty TH, Cohen BL. Fundamentals of genetic epidemiology. New York: Oxford University Press, 1993.

39. Caporaso N, Rothman N, Wacholder S. Case-control studies of common alleles and environmental factors. J Natl Cancer Inst Monogr 1999;25–30.
40. Thomas DC, Witte JS. Point: population stratification: a problem for case-control studies of candidate-gene associations? Cancer Epidemiol Biomarkers Prev 2002;11:505–512.
41. Wacholder S, Rothman N, Caporaso N. Counterpoint: bias from population stratification is not a major threat to the validity of conclusions from epidemiological studies of common polymorphisms and cancer. Cancer Epidemiol Biomarkers Prev 2002;11: 513–520.
42. Wacholder S, Rothman N, Caporaso N. Population stratification in epidemiologic studies of common genetic variants and cancer: quantification of bias. J Natl Cancer Inst 2000;92:1151–1158.
43. Ardlie KG, Lunetta KL, Seielstad M. Testing for population subdivision and association in four case-control studies. Am J Hum Genet 2002;71:304–311.
44. Garte S. The role of ethnicity in cancer susceptibility gene polymorphisms: the example of CYP1A1. Carcinogenesis 1998;19:1329–1332.
45. Teng J, Risch N. The relative power of family-based and case-control designs for linkage disequilibrium studies of complex human diseases. II. Individual genotyping. Genome Res 1999;9:234–241.
46. Witte JS, Gauderman WJ, Thomas DC. Asymptotic bias and efficiency in case-control studies of candidate genes and gene-environment interactions: basic family designs. Am J Epidemiol 1999;149:693–705.
47. Gauderman WJ, Witte JS, Thomas DC. Family-based association studies. J Natl Cancer Inst Monogr 1999;31–37.
48. Risch N, Merikangas K. The future of genetic studies of complex human diseases. Science 1996;273:1516–1517.
49. Devlin B, Roeder K. Genomic control for association studies. Biometrics 1999;55: 997–1004.
50. Pritchard JK, Stephens M, Rosenberg NA et al. Association mapping in structured populations. Am J Hum Genet 2000;67:170–181.
51. Reich DE, Goldstein DB. Detecting association in a case-control study while correcting for population stratification. Genet Epidemiol 2001;20:4–16.
52. Satten GA, Flanders WD, Yang Q. Accounting for unmeasured population substructure in case-control studies of genetic association using a novel latent-class model. Am J Hum Genet 2001;68:466–477.
53. Lin SS, Kelsey JL. Use of race and ethnicity in epidemiologic research: concepts, methodological issues, and suggestions for research. Epidemiol Rev 2000;22: 187–202.
54. Ardlie KG, Kruglyak L, Seielstad M. Patterns of linkage disequilibrium in the human genome. Nat Rev Genet 2002;3:299–309.
55. Pritchard JK, Przeworski M. Linkage disequilibrium in humans: models and data. Am J Hum Genet 2001;69:1–14.
56. Rothman KJ. Modern Epidemiology. Boston/Toronto: Little, Brown and Company, 1986.
57. Wacholder S. Practical considerations in choosing between the case-cohort and nested case-control designs. Epidemiology 1991;2:155–158.
58. Boffetta P, Pearce N. Epidemiological studies on genetic polymorphisms: study design issues and measures of occurrence and association. In: Vineis P, Malats N, Lang M et al., eds. Metabolic polymorphisms and susceptibility to cancer. IARC Scientific Publications No. 148. Lyon: IARC, 1999:97–108.
59. Smith PG, Day NE. The design of case-control studies: the influence of confounding and interaction effects. Int J Epidemiol 1984;13:356–365.

60. Schork NJ, Fallin D, Lanchbury JS. Single nucleotide polymorphisms and the future of genetic epidemiology. Clin Genet 2000;58:250–264.

61. Morton NE, Collins A. Tests and estimates of allelic association in complex inheritance. Proc Natl Acad Sci U S A 1998;95:11389–11393.

62. Risch N, Teng J. The relative power of family-based and case-control designs for linkage disequilibrium studies of complex human diseases I. DNA pooling. Genome Res 1998;8:1273–1288.

63. Freidlin B, Zheng G, Li Z et al. Trend tests for case-control studies of genetic markers: power, sample size and robustness. Hum Hered 2002;53:146–152.

64. Fallin D, Cohen A, Essioux L et al. Genetic analysis of case/control data using estimated haplotype frequencies: application to APOE locus variation and Alzheimer's disease. Genome Res 2001;11:143–151.

65. Aragaki CC, Greenland S, Probst-Hensch N et al. Hierarchical modeling of gene-environment interactions: estimating NAT2 genotype-specific dietary effects on adenomatous polyps. Cancer Epidemiol Biomarkers Prev 1997;6:307–314.

66. Witte JS. Genetic analysis with hierarchical models. Genet Epidemiol 1997;14:1137–1142.

67. Engel LS, Taioli E, Pfeiffer R, et al. Pooled analysis and meta-analysis of glutathione S-transferase M1 and bladder cancer: a HuGE review. Am J Epidemiol 2002;156:95–109.

68. Oxman AD. The Cochrane Collaboration Handbook: preparing and maintaining systematic reviews. 1992, Oxford: Cochrane Collaboration.

69. Stroup DF, Berlin JA, Morton SC, et al. Meta-analysis of observational studies in epidemiology: a proposal for reporting. Meta-analysis Of Observational Studies in Epidemiology (MOOSE) group. JAMA 2000;283:2008–2012.

70. Gregoire G, Derderian F, Le Lorier J. Selecting the language of the publications included in a meta-analysis: is there a Tower of Babel bias? J Clin Epidemiol 1995;48:159–163.

71. Ambrosone CB, Freudenheim JL, Graham S, et al. Cytochrome P4501A1 and glutathione S-transferase (M1) genetic polymorphisms and postmenopausal breast cancer risk. Cancer Res 1995;55:3483–3485.

72. Moysich KB, Shields PG, Freudenheim JL, et al. Polychlorinated biphenyls, cytochrome P4501A1 polymorphism, and postmenopausal breast cancer risk. Cancer Epidemiol Biomarkers Prev 1999;8:41–44.

73. Taioli E, Trachman J, Chen X et al. A CYP1A1 restriction fragment length polymorphism is associated with breast cancer in African-American women. Cancer Res 1995;55:3757–3758.

74. Taioli E, Bradlow HL, Garbers SV et al. Role of estradiol metabolism and CYP1A1 polymorphisms in breast cancer risk. Cancer Detect Prev 1999;23:232–237.

75. Feinstein AR. Methodologic problems and standards in case-control research. J Chronic Dis 1979;32:35–41.

76. Horwitz RI, Feinstein AR. Methodologic standards and contradictory results in case-control research. Am J Med 1979;66:556–564.

77. Kopec JA, Esdaile JM. Bias in case-control studies. A review. J Epidemiol Community Health 1990;44:179–186.

78. Crombie IK. The Pocket Guide to Critical Appraisal. London: BMJ Publishing Group, 1996.

79. Liddle J, Williamson M, Irwig L. Method for evaluating research and guideline evidence (MERGE). Sydney: NSW Health Department, 1996.

80. Savitz DA, Greenland S, Stolley PD et al. Scientific standards of criticism: a reaction to "Scientific standards in epidemiologic studies of the menace of daily life," by A.R. Feinstein. Epidemiology 1990;1:78–83.

81. Weiss NS. Scientific standards in epidemiologic studies. Epidemiology 1990;1:85–86.
82. Dixon RA, Munro JF, Silcocks PB. The evidence based medicine workbook. Critical appraisal for clinical problem solving. Oxford: Butterworth-Heinemann, 1997.
83. Longnecker MP, Berlin JA, Orza MJ, et al. A meta-analysis of alcohol consumption in relation to risk of breast cancer. JAMA 1988;260:652–656.
84. Longnecker MP, Orza MJ, Adams ME, et al. A meta-analysis of alcoholic beverage consumption in relation to risk of colorectal cancer. Cancer Causes Control 1990; 1:59–68.
85. Berlin JA, Colditz GA. A meta-analysis of physical activity in the prevention of coronary heart disease. Am J Epidemiol 1990;132:612–628.
86. Friedenreich CM. Methods for pooled analyses of epidemiologic studies. Epidemiology 1993;4:295–302.
87. Friedenreich CM, Brant RF, Riboli E. Influence of methodologic factors in a pooled analysis of 13 case-control studies of colorectal cancer and dietary fiber. Epidemiology 1994;5:66–79.
88. Jüni P, Witschi A, Bloch R, et al. The hazards of scoring the quality of clinical trials for meta-analysis. JAMA 1999;282:1054–1060.
89. SIGN. SIGN 50: A Guideline Developer's Handbook. Edinburgh, UK: Scottish Intercollegiate Guidelines Network, 2001.
90. HuGE. Human Genome Epidemiology Network e-journal club. Accessed October 7, 2002, from http://www.cdc.gov/genomics/hugenet/ejournal.htm
91. Deville WL, Buntinx F, Bouter LM et al. Conducting systematic reviews of diagnostic studies: didactic guidelines. BMC Med Res Methodol 2002;2:9.
92. Sutton AJ, Abrams KR, Jones DR et al. Systematic reviews of trials and other studies. Health Technol Assess 1998;2:1–276.
93. Blettner M, Sauerbrei W, Schlehofer B et al. Traditional reviews, meta-analyses and pooled analyses in epidemiology. Int J Epidemiol 1999;28:1–9.
94. Shapiro S. Meta-analysis/Shmeta-analysis. Am J Epidemiol 1994;140:771–778.
95. Egger M, Schneider M, Davey Smith G. Spurious precision? Meta-analysis of observational studies. BMJ 1998;316:140–144.
96. Doll R. The use of meta-analysis in epidemiology: diet and cancers of the breast and colon. Nutr Rev 1994;52:233–237.
97. Ioannidis JP, Rosenberg PS, Goedert JJ et al. Effects of CCR5-Delta32, CCR2-64I, and SDF-1 3'A alleles on HIV-1 disease progression: An international meta-analysis of individual-patient data. Ann Intern Med 2001;135:782–795.
98. Steinberg KK, Smith SJ, Stroup DF, et al. Comparison of effect estimates from a meta-analysis of summary data from published studies and from a meta-analysis using individual patient data for ovarian cancer studies. Am J Epidemiol 1997;145:917–925.
99. Benhamou S, Lee WJ, Alexandrie AK, et al. Meta- and pooled analyses of the effects of glutathione S-transferase M1 polymorphisms and smoking on lung cancer risk. Carcinogenesis 2002;23:1343–1350.
100. Stroup DF, Thacker SB (2000). Meta-analysis in epidemiology, in Encyclopedia of Epidemiologic Methods. Gail MH, Benichou J, eds. Chichester, New York: Wiley & Sons Publishers, pp. 557–570.
101. Easterbrook PJ, Berlin JA, Gopalan R et al. Publication bias in clinical research. Lancet 1991;337:867–872.
102. Begg CB, Berlin JA. Publication bias and dissemination of clinical research. J Natl Cancer Inst 1989;81:107–115.
103. Rosenthal R . The file drawer problem and tolerance for null results. Psychological Bulletin 1979;86:638–641.
104. Thornton A, Lee P. Publication bias in meta-analysis: its causes and consequences. J Clin Epidemiol 2000;53:207–216.

105. CRISP. Computer Retrieval of Information on Scientific Projects. Accessed October 7, 2002, from http://crisp.cit.nih.gov.
106. Sankaranarayannan, R., Becker, N., and Démaret, E. Directory of on-going research in cancer prevention (http://www-dep.iarc.fr/direct/prevent.htm). 2000. Lyon, IARC.
107. Weed DL, Gorelic LS. The practice of causal inference in cancer epidemiology. Cancer Epidemiol Biomarkers Prev 1996;5:303–311.
108. Caporaso N. Selection of candidate genes for population studies. IARC Sci Publ 1999;23–36.
109. Hein DW. Acetylator genotype and arylamine-induced carcinogenesis. Biochim Biophys Acta 1988;948:37–66.
110. Kimura K, Isashiki Y, Sonoda S, et al. Genetic association of manganese superoxide dismutase with exudative age-related macular degeneration. Am J Ophthalmol 2000;130: 769–773.
111. Chan DK, Mellick GD, Buchanan DD, et al. Lack of association between CYP1A1 polymorphism and Parkinson's disease in a Chinese population. J Neural Transm 2002; 109:35–39.
112. Hadfield RM, Manek S, Weeks DE, et al. Linkage and association studies of the relationship between endometriosis and genes encoding the detoxification enzymes GSTM1, GSTT1 and CYP1A1. Mol Hum Reprod 2001;7:1073–1078.
113. van Rooij IA, Wegerif MJ, Roelofs HM, et al. Smoking, genetic polymorphisms in biotransformation enzymes, and nonsyndromic oral clefting: a gene-environment interaction. Epidemiology 2001;12:502–507.
114. Cresteil T. Onset of xenobiotic metabolism in children: toxicological implications. Food Addit Contam 1998;15(suppl):45–51.
115. Sonnier M, Cresteil T. Delayed ontogenesis of CYP1A2 in the human liver. Eur J Biochem 1998;251:893–898.
116. Schutte BC, Murray JC. The many faces and factors of orofacial clefts. Hum Mol Genet 1999;8:1853–1859.
117. Crott JW, Mashiyama ST, Ames BN, et al. Methylenetetrahydrofolate reductase C677T polymorphism does not alter folic acid deficiency-induced uracil incorporation into primary human lymphocyte DNA in vitro. Carcinogenesis 2001;22:1019–1025.
118. Eidelman O, Zhang J, Srivastava M et al. Cystic fibrosis and the use of pharmacogenomics to determine surrogate endpoints for drug discovery. Am J Pharmacogenomics 2001;1:223–238.
119. Surgeon General (Advisory Committee). Smoking and health. Washington DC: US Department of Health, Education and Welfare, 1964.
120. Hill AB. The environment and disease: association or causation? Proceedings of the Royal Society of Medicine 1965;58:295–300.
121. Schlesselman JJ. "Proof" of cause and effect in epidemiologic studies: criteria for judgment. Prev Med 1987;16:195–210.

III

METHODS AND APPROACHES II: ASSESSING GENETIC TESTS FOR DISEASE PREVENTION

11

Epidemiologic approach to genetic tests: population-based data for preventive medicine

Marta Gwinn and Muin J. Khoury

Sequencing the human genome ahead of schedule has raised expectations for quick translation of the data into tools for medical practice. When the initial sequence was published in February 2001, Francis Collins and Victor McKusick wrote that "genetic prediction of individual risks of disease and responsiveness to drugs will reach the medical mainstream in the next decade or so (1)." The idea that genetic tests could offer patients personal estimates of risk and interventions on the basis of their genotypes has captured the imagination of scientists and the public.

Genetic Tests

Until now, genetic tests have been used mostly to aid the diagnosis of rare hereditary disorders. In November 2000, when we reviewed the list of tests in GeneTests, a Web-accessible database that serves as the main directory of U.S. clinical and research laboratories offering genetic testing (2–3), we found that fewer than 5% of tests available for clinical use applied to common, adult-onset diseases (4). Most of these were tests for variants of single genes associated with disease susceptibility in high-risk families (e.g., *BRCA1* for breast cancer). A recent review of entries in the online version of Mendelian Inheritance in Man (5) suggests that this situation is unlikely to change soon: although the discovery of disease-associated gene variants is accelerating rapidly, the number of identified "susceptibility genes" remains small (6).

Genetic tests that predict future risk for disease in asymptomatic people, thereby suggesting specific strategies for prevention or early detection, are the starting point for models of individualized preventive medicine. An example that helps illustrate the expectations, limitations, and future potential of predictive genetic tests is Francis Collins' "hypothetical case in 2010 (7)," in which a 23-year-old man named

John undergoes DNA testing for genes related to several common chronic diseases. The genetic test report includes relative risks (range: 0.3–6) as well as lifetime risks (range: 7%–30%) for each of these diseases, predicted on the basis of John's genotype for one to three genes related to each condition. John's physician recommends that he stop smoking, undergo regular colonoscopy beginning at age 45 years, and take lipid-lowering medications.

In this example, not only the patient but also the relative and lifetime risks are hypothetical. Where will we obtain the data needed to interpret and act on the results of genetic tests? Despite the media's tendency to depict genetic tests as definitive, no test can predict with certainty the behavior of a complex biologic system (in this case, John) over a lifetime. Furthermore, risk cannot be predicted solely on the basis of individual information; it must be estimated by analysis of the characteristics, experience, and outcomes of a group of people "similar" to the individual of interest. Information about the family is used to assess risk of classic, Mendelian disorders, and the ability to predict disease based on inheritance is the foundation for the clinical specialty of genetic counseling. However, estimating the risk for complex disorders (without a clear pattern of inheritance) requires genetic information from larger population samples. Information on prevalence of gene variants, genotype–phenotype correlations, and gene–gene and gene–environment interactions must be collected systematically by epidemiologic studies conducted in populations resembling those to which inferences will be drawn (8). These populations are likely to be much more diverse than the genetically homogeneous groups in which susceptibility genes are usually first identified.

Epidemiologic Approach

Epidemiologic studies that collect genetic information depend on access to valid, reproducible, economical tests for the genetic variants of interest. New technology has made such tests available for use in large-scale, population-based studies (9). Choosing among analytic methods involves practical considerations, such as availability of collaborators, as well as characteristics of the variant itself. Before a test can be used in epidemiologic research, its analytic validity must be established. *Analytic validity* refers to the sensitivity, specificity, and predictive value of the test in relation to genotype; these characteristics are measured by comparing the test result with a gold standard in a set of well-described samples, only some of which contain the genetic variant.

The final report of the National Institutes of Health-Department of Energy Task Force on Genetic Testing (10) distinguished analytic validity from *clinical validity*, which they defined as the sensitivity, specificity, and predictive value of a test in relation to a particular phenotype. In contrast to analytic validity, which must be determined before a study begins, clinical validity is defined by epidemiologic studies that measure gene–disease associations. A third parameter defined by the Task Force, *clinical utility*, refers to the net value of the information gained from a genetic test in changing disease outcomes. Clinical utility is best assessed in clinical trials or by synthesis of observational data (e.g., as in cost-effectiveness analysis). Table 11.1 summarizes these measures of validity and utility.

Classic epidemiologic study designs include cross-sectional, cohort, and case-control studies. Each design is useful for addressing particular aspects of genetic variation or gene–environment interaction in relation to disease outcomes (11). Cross-sectional studies can be used to estimate the prevalence of gene variants, although variants associated with poor survival may be underrepresented. Cohort studies are unique in providing direct estimates of absolute risk and relative risk in people with different genotypes. "Experimental" cohort studies (randomized, controlled trials) are ideal for evaluating the effects of gene–environment interactions or specific interventions (see Chapter 15). Retrospective cohort studies are attractive because genetic information is invariant and can be measured long after the study has ended; however, they are subject to the usual biases of observational studies.

Case-control studies can generally measure gene–disease associations more quickly, efficiently, and at lower cost than cohort studies. Case-control studies yield odds ratios, which approximate the relative risk of disease as long as the disease is rare or controls are sampled randomly (independent of disease status or genotype) from the source population (12). The defining characteristic of a population based case-control study is the set of a priori criteria—applied to selection of controls as well as cases—that specifies the study's source population (e.g., by geographic area and ethnicity). Studies comparing genotypes of patients in a clinical case series with those of an undefined convenience sample of control subjects (or worse, "control specimens") are numerous in the published literature, but they provide little basis for risk estimation. From a public health perspective, a population-based estimate of relative risk is critical because it provides the basis for estimating attributable fraction—the proportion of cases that would not occur in the absence of a particular exposure (or genotype) in the population. To examine the role of epidemiologic studies in eliciting genetic factors in common diseases, we consider the example of colorectal cancer (CRC).

Table 11.1 Analytic Validity, Clinical Validity, and Clinical Utility of a Genetic Test

Characteristic	How Is It Defined?	How Is It Determined?	How Is It Used?
Analytic validity	Sensitivity, specificity, and predictive value in relation to genotype	Laboratory analyses comparing test result with gold standard	Validating test before clinical or research use
Clinical validity	Sensitivity, specificity, and predictive value in relation to phenotype	Population-based, epidemiologic studies (cohort, case-control)	Predicting risk, screening, making diagnosis
Clinical utility	Benefits and risks accruing from both positive and negative tests	Clinical trials; synthesis of observational data	Assessing added value of testing in preventing disease outcomes

Box 11.1 Example: Colorectal Cancer
___

Like other cancers, colorectal cancer (CRC) is a "genetic disease" caused by mutations that disable normal regulation of cell growth and differentiation. An estimated 945,000 new cases of CRC occur annually worldwide (13). Epidemiologic studies have identified many exposures associated with increased risk for CRC, including smoking, overweight, inactivity, and dietary factors. Since familial clustering of CRC was first reported more than 100 years ago, clinical research has identified several high-risk syndromes, including familial adenomatous polyposis (FAP) and hereditary nonpolyposis CRC (HNPCC).

Table 11.2 compares estimates of the absolute (lifetime) risk and relative risk for CRC in people who have FAP (14), pedigrees consistent with HNPCC (15–16), or at least one first-degree relative with CRC (17), with risks in people in none of these groups, who are considered at "average risk." Without intervention, people with FAP are virtually certain to develop CRC (absolute risk approaching 1), usually by their mid-30s; however, FAP is very rare (prevalence approximately 1/8000) and thus accounts for a very small share of CRC cases in the U.S. population (attributable fraction <1%) (14). FAP is diagnosed clinically on endoscopic and histologic criteria. Although it is an autosomal dominant disorder, about one third of cases result from new mutations; thus genetic testing is useful mainly for counseling an affected person's family members, offering increased surveillance if positive, and reassurance otherwise (18).

The diagnosis of HNPCC is based on a pedigree consistent with autosomal dominant inheritance of CRC (as well as cancers at other sites in some high-risk families) (16). Unlike FAP, HNPCC cannot be diagnosed on the basis of clinical characteristics, and thus far, genotypic information has not become part of the case definition. Since 1994, mutations in several DNA mismatch-repair (MMR) genes have been found in HNPCC families, with MSH2 and MLH1 mutations by far most often implicated.(16) Identifying a mutation within an HNPCC family can be useful for testing and counseling family members. Although absolute risk for CRC is approximately 80% in HNPCC family members with inherited susceptibility (16), data are insufficient to estimate the absolute and relative risks for CRC based on MMR genotype. The population prevalence of genetic variants associated with HNPCC is unknown, although one study analyzing data from Scotland, Finland, and the United States arrived at an estimate of 1/3000 for MSH2 and MLH1 variants combined (15). Population-based data on frequency of HNPCC among CRC cases is also scarce, although recent studies suggest that it may be lower than previously estimated from case series, perhaps as low as 1% (19).

Family history can capture shared, unmeasured genetic risk, as well as the potential influences of shared diet, behavior, and other nongenetic factors (20). Accumulated evidence from epidemiologic studies of CRC suggests that having at least one first-degree relative with CRC increases the relative risk for colon cancer approximately twofold. However, because the average risk for CRC is only about 4%, the absolute risk of CRC in this group remains <10% (17); thus, family history offers poor predictive value as a "genetic screening test" for CRC.

Genetic information obtained thus far from population-based, epidemiologic studies is relevant to only about 10% of CRC cases in the population. Although epidemiologic studies consistently identify CRC in a first-degree relative as one of the strongest risk factors for CRC, other factors (e.g., dietary habits, physical activity) that are less strongly associated but more prevalent in the population have higher attributable fractions (21).

Table 11.2 Risk Factors for Colorectal Cancer: Estimated Absolute Risks, Relative Risks, and Attributable Fractions

Risk factor	AVERAGE	MODERATE	HIGH	
	No 1° relative[a]	*Any 1° relative*[a]	*HNPCC*	*FAP*[b]
Prevalence	9/10	1/10	1/3,000[c]	1/8,000
Absolute risk	0.04	0.055	0.80[d]	~1
Relative risk		1.7	~20	~30
Attributable fraction		0.07	unk	~0.004

[a]Fuchs CS, Giovannucci EL, Colditz GA, et al. A prospective study of family history and the risk of colorectal cancer. New Engl J Med 1994;331:1669–1674.

[b]Bodmer W. Familial adenomatous polyposis (FAP) and its gene, APC. Cytogenet Cell Genet 1999;86:99–104.

[c]Dunlop MG, Farrington SM, Nicholl I, et al. Population carrier frequency of hMSH2 and hMLH1 mutations. Brit J Cancer 2000;83:1643–1645.

[d]Lynch HT, Lynch JF. Hereditary nonpolyposis colorectal cancer. Semin Surg Oncol 2000;18:305–313.

Estimating Individual Risk from Population-Based Data

Epidemiologic studies of gene–disease associations—particularly those reporting results in terms of predicted risk—have lately come under fire as an unwarranted extension of the individual risk paradigm (22), in which relations among disease risk factors are teased out at the individual level. The exclusive focus on individuals can be criticized on both practical and philosophic grounds, for failing to prompt effective public health interventions while "blaming the victim (22–23)." The "privatization of risk (24)" also appears to defy the concept of risk as an aggregate measure and to ignore the reality that individual risk factors generally make poor screening tools (25). A recent commentary on the potential impact of genetics on preventive medicine echoes these concerns (26), arguing that because most genetic tests have low predictive value, low clinical sensitivity, and little potential for stimulating tailored intervention, they will have limited value in preventing common complex diseases.

Viewing the potential contribution of a single test in isolation reflects a time-honored perspective in public health screening, as well as the traditional use of clinical genetic tests. Mass screening programs are generally delivered to whole populations without regard to prior information (such as family history or race/ethnicity), both to maximize sensitivity and to achieve social goals, such as fairness and program efficiency. On the other hand, clinical geneticists have used tests mostly for diagnosis of hereditary disorders resulting from single gene variants with very high penetrance. In this setting, a single genetic test may be definitive, although DNA sequencing is revealing increasingly diverse genotype–phenotype relations, even in classic "single gene disorders" like cystic fibrosis (27).

Most common chronic diseases arise from interactions among multiple genes, environmental exposures, and behaviors; thus, genetic tests are most likely to be useful when combined with results of other information to uncover interactions associated with markedly elevated risks. This concept can be demonstrated by using basic principles from either epidemiology (28–29) or genetics (30). Real examples are still scarce, however, partly because they are likely to be complicated, involving multiple genes and multiple environmental exposures.

This state of affairs is familiar to most medical practitioners, who are used to considering the results of multiple clinical tests in the context of other, often incomplete, information about an individual patient, including family history, lifestyle, and physical examination. The basis for integrating and interpreting this information is experience—whether clinical experience of an individual physician, expert consensus, or data gathered systematically from scientific studies, such as clinical trials. During the last decade, the methods developed by clinical epidemiologists for critical analysis, synthesis, and application of accumulated experience have become the foundation for "evidence-based medicine (31)." In this medical paradigm, diagnosis is a Bayesian process: as each test result is added to the body of evidence, some possible diagnoses become more likely, while others are less likely or altogether ruled out (31, pp. 121–140).

Results of genetic tests can be integrated into the same framework, as long as the association of genotype with disease outcome (genotype–phenotype correlation) has been well described (32a). When the goal is to predict future disease, rather than to make a diagnosis, a genotype becomes part of the evidence that can be used to make a probabilistic estimate of risk (32b). The underlying relationship between a susceptibility genotype (defined by variant alleles at one or more genetic loci) and disease outcome, and its reflection in measures of test performance and risk, can be summarized in the familiar framework of a 2×2 table (Appendix).

As observed by critics of individualized preventive medicine, few risk factors for common chronic diseases have sufficient predictive ability to serve as screening tools (25); in this respect, common polymorphisms associated with disease susceptibility are unlikely to be different. Most risk estimates useful to individuals will be obtained only by considering the joint effects of many factors; however, although technical advances have made large-scale genotyping feasible in epidemiologic studies, the ability to assimilate, synthesize, and interpret the data has not yet fully caught up. The sheer number of variables potentially available for analysis tests the limits of conventional methods.

Challenges for "Genomic" Epidemiology

Despite their independent origins, genetic and epidemiologic methods for investigating causes and predicting risks for human diseases share many concerns common to observational sciences. During the last 50 years, the synthesis of genetic and epidemiologic methods has been accelerated by growth in statistical and computing techniques and by development of molecular methods for measuring environmental exposures, biological processes, and genetic traits (9,33).

Much recent development in genetic epidemiology and statistical genetics has focused on methods for identifying disease susceptibility genes in families; however, describing the distribution of genetic traits in populations and evaluating the role of genetic factors in disease occurrence requires larger studies of unrelated people. Epidemiologic studies of genotype prevalence, gene–disease association, and gene–environment interaction are subject to the usual sources of bias, including confounding and misclassification. Confounding is analogous to "population stratification" in studies of gene–disease association (34); misclassification of genotype occurs as a function of analytic validity. Also, type I errors are of concern when multiple gene–disease associations are tested, and type II errors are likely when results are analyzed for small subgroups defined by genotype or gene–environment interactions.

The nature of genomic data poses additional challenges for epidemiologic analysis. For example, genetic variants at different loci cannot be assumed to occur independently, even when they are found on different chromosomes (35). Furthermore, although the genetic sequence of an individual remains the same, gene expression naturally varies tremendously among tissues, in response to environmental stimuli, and with age, reflecting cumulative effects over the course of a lifetime. Thus, the interactions of gene products with each other and with other factors in their milieu reflect an underlying complex "genetic architecture (36)," in which health and disease are "defined by the same continuum of biological traits (36, p. 217)." New models are needed for analyzing these relations and using them for prediction and intervention.

Most common chronic diseases result from multiple gene–environment interactions over a long period of time, involving invariant features (e.g., genotype), "context-dependent features" (e.g., diet), and chance processes (36). Prediction from "first principles" (genotype) is thus an unrealistic goal. One strategy for capturing the effects of multiple factors pursues data more proximate to the disease outcome, such as acquired (somatic) mutations, gene expression, protein markers, or intermediate states or conditions that recapitulate prior gene–environment interactions. Predictions based on observations made closer to the outcome are likely to be more accurate. To examine the potential of this approach, we revisit the example of CRC (Box 11.2).

New Opportunities and Challenges for Public Health

Pathogenesis at the molecular level is far better understood for CRC than for other chronic diseases, including most other cancers. Ready access to a premalignant lesion—the adenomatous polyp—has afforded researchers a rare window for dissecting the process of tumorigenesis, and public health a ready opportunity for prevention. However, new technology coupled with new analytic methods may be opening other windows onto diseases for which preventive medicine and public health have had little to offer until now. For example, collaborating researchers from several federal agencies, academic medical centers, and industry recently reported a method for using proteomic patterns in serum to identify ovarian cancer (40). They developed and tested a new algorithm for discriminating serum protein profiles in patients with ovarian cancer from those in controls. The algorithm used data-driven

Box 11.2 Example Revisited: Colorectal Cancer

The ability to examine DNA sequence information in clinical and epidemiologic studies of CRC reveals that traditional categories for classifying CRC cases and their causes are not as distinct as they once seemed. On the other hand, insights at the molecular level may suggest more useful models for sorting out pathogenetic mechanisms of CRC, along with more specific targets for prevention, diagnosis, and treatment. For example, we now know that inherited variation in the APC gene gives rise to several different cancer syndromes with involvement at various extracolonic sites (e.g., Gardner syndrome, Turcot syndrome); a form of attenuated FAP in which far fewer polyps are found; and a susceptibility polymorphism in Ashkenazi Jews associated with only modestly increased risk for CRC (37).

HNPCC for now remains a diagnosis based on pedigree because genotype–phenotype correlation is not well enough understood to establish sensitive and specific diagnostic criteria on the basis of genotype. Predictive tests for HNPCC (other than family history) would thus currently seem to be out of reach. Even DNA-based diagnosis remains problematic. Most tumors in affected persons exhibit microsatellite instability (MSI), which has been proposed as an initial screening test before sequencing *MSH2* and *MLH1*; however, MSI testing itself is a costly and complex procedure that is not entirely sensitive or specific for cancer in HNPCC families (16).

The molecular pathways that give rise to FAP and HNPCC are also important in sporadic CRC; thus events leading to loss of functional APC and MMR gene products can begin either with inherited or acquired mutations. In 1999, John Potter reviewed the evidence implicating these and other pathways in the pathogenesis of CRC, along with recognized or postulated gene–environment and gene–gene interactions (37). He pointed out that epidemiologic studies that examine agents affecting only one or some of the pathways could be expected to find weak or inconsistent associations. However, he also predicted that as these pathways became better understood, population subgroups of similar susceptibility could be recognized for more specific preventive interventions, and that early molecular changes could serve as screening markers.

A recent study reported the feasibility of examining fecal DNA for APC mutations that occur early in the pathogenesis of CRC, suggesting future potential for new, noninvasive approaches to screening (38). Although highly specific, the fecal DNA analysis was only 57% sensitive, positive in 26 of 46 patients with neoplasia (9/18 adenomas, 11/28 carcinomas). A commentary accompanying this article (39) suggested that while this study had "drawn back a curtain to reveal a tantalizing possibility, . . . there are other curtains and other possibilities. The basis of the next molecular screening test for CRC may not be a mutant gene but an abnormal protein that the new science of proteomics may find (39, p. 304)."

methods (cluster analysis and "genetic algorithms," in which principles of natural selection were used to select key measurements for analysis) to distinguish patterns generated by mass spectrometry. The algorithm successfully classified all 50 cancer and 50 noncancer serum samples in a "training set," and all 50 cancer samples in a "masked set"; however, 3 of 66 noncancer samples were incorrectly classified (specificity 95%). The authors concluded, "These findings justify a prospective population-based assessment of proteomic pattern technology as a screening tool for all stages of ovarian cancer in high-risk and general populations (40, p. 572)."

New technologies are likely to continue to improve the prospects for early intervention in disease processes and prevention of morbidity, but epidemiologic studies are the key to their potential, beginning at the population level for translation into tools for individualized preventive medicine. However, evaluating the contribution of genomic data to the evidence base for prevention only begins to address the issue of clinical utility, which presents a larger array of complex issues. These include the probable timing and severity of disease outcomes, availability and effectiveness of interventions, and costs of alternative tests, interventions, and treatments (41). Furthermore, the needs and preferences of tested individuals and their family members must be taken into account.

Even if we recognize the potential of genomics for preventive medicine, what does it mean for public health? Developments in the science and technology of genomics have prompted widespread use of the term "paradigm shift (42)" to describe the future of biologic research and clinical medicine. Epidemiology, the basic science of public health, also faces challenges. Future studies of the distribution and causes of disease in human populations will be incomplete if they do not consider the potential contribution of genetic variation. Amid this sea of changes, the mission of public health remains the same: to prevent morbidity and mortality by using science-based approaches that serve the interests of the total population, with special responsibility for underserved communities. However, new understanding of gene–environment interactions in disease etiology and progression may suggest interventions that require rethinking the "one size fits all" paradigm for public health interventions.

As more genetic tests are developed and marketed for use in public health and health-care settings, it will be important to evaluate the value they add to existing interventions. Public health policies, backed by strong epidemiologic research, must provide a balance to intense commercial pressures, which have identified high-technology screening tests, including those based on genomics, as a new opportunity for direct marketing to individuals (43–44). Ultimately, the public will benefit only when genetic tests are used appropriately, interventions are tailored to those at risk, and access is assured. Thus public health institutions clearly have a role in helping realize the potential of genetic information to prevent disease and improve health, by developing appropriate research and policies, and by helping educate health-care providers and the public.

References

1. Collins FS, McKusick VA. Implications of the Human Genome Project for medical science. JAMA 2001;285:540–544.

2. Pagon RA, Tarczy-Hornoch P, Baskin PK, et al. GeneTests-GeneClinics: Genetic testing information for a growing audience. Hum Mutat 2002;19:501–509.
3. GeneTests. Web site. Accessed April 26, 2002, from http://www.genetests.org/.
4. Yoon PW, Chen B, Faucett A, et al. Public health impact of genetic tests at the end of the 20th century. Genet Med 2001;3:405–410.
5. Online Mendelian Inheritance in Man. Web site. Accessed April 26, 2002, from http://www3.ncbi.nlm.nih.gov/Omim/.
6. Peltonen L, McCusick VA. Genomics and medicine: dissecting human disease in the postgenomic era. Science 2001;291:1224–1229.
7. Collins FS. Shattuck lecture—medical and societal consequences of the Human Genome Project. N Engl J Med 1999;341:28–37.
8. Khoury MJ, Little J. Human genome epidemiologic reviews: the beginning of something HuGE. Am J Epidemiol 2000;151:2–3.
9. Ellsworth DL, Manolio TA. The emerging importance of genetics in epidemiologic research. I. Basic concepts in human genetics and laboratory technology. Ann Epidemiol 1999;9:1–16.
10. Task Force on Genetic Testing. Ensuring the safety and effectiveness of genetic tests. *In* Holtzmann NA, Watson MS (eds) (1997). Promoting safe and effective genetic testing in the United States: final report of the Task Force on Genetic Testing. Web site. Accessed April 26, 2002, from http://www.nhgri.nih.gov/ELSI/TFGT_final/.
11. Whittemore AS, Nelson LM. Study design in genetic epidemiology: theoretical and practical considerations. Monogr Natl Cancer Inst 1999;26:61–69.
12. Rothman KJ, Greenland S (1998). Case-control studies, in Modern Epidemiology, second edition, Rothman KJ (ed.). Philadelphia:Lippincott Williams & Wilkins, pp. 93–114.
13. Ferlay J, Bray F, Pisani P, et al. Globocan 2000: Cancer Incidence, Mortality and Prevalence Worldwide. International Agency for Research on Cancer, World Health Organization. Lyon, France: IARC Press, 2001.
14. Bodmer W. Familial adenomatous polyposis (FAP) and its gene, APC. Cytogenet Cell Genet 1999;86:99–104.
15. Dunlop MG, Farrington SM, Nicholl I, et al. Population carrier frequency of hMSH2 and hMLH1 mutations. Brit J Cancer 2000;83:1643–1645.
16. Lynch HT, Lynch JF. Hereditary nonpolyposis colorectal cancer. Semin Surg Oncol 2000;18:305–313.
17. Fuchs CS, Giovannucci EL, Colditz GA, et al. A prospective study of family history and the risk of colorectal cancer. New Engl J Med 1994;331:1669–1674.
18. Cole TRP, Sleightholme HV. ABC of colorectal cancer: the role of clinical genetics in management. BMJ 2000;943–946.
19. Samowitz WS, Curtin K, Lin HH, et al. The colon cancer burden of genetically defined hereditary nonpolyposis colon cancer. Gastroenterology 2001;121:830–836.
20. Yoon P, Peterson-Oehlke KL, Scheuner MT, et al. Can family history be used as a tool for public health and preventive medicine? Genet Med 2002;4:304–310.
21. Tomeo CA, Colditz GA, Willett WC, et al. Harvard report on cancer prevention. Volume 3: Prevention of colon cancer in the United States. Cancer Causes and Control 1999;10:167–180.
22. Rockhill B, Kawachi I, Colditz GA. Individual risk prediction and population-wide disease prevention. Epidemiol Rev 2000;22:176–180.
23. Pearce N. Traditional epidemiology, modern epidemiology, and public health. Am J Public Health. 1996;86:678–683.
24. Rockhill B. The privatization of risk. Am J Public Health 2001;91:365–368.
25. Wald NJ, Hackshaw AK, Frost CD. When can a risk factor be used as a worthwhile screening test? BMJ 1999;319:1562–1565.

26. Holtzman NA, Marteau TM. Will genetics revolutionize medicine? New Engl J Med 2000;343:141–144.

27. Wang X, Moylan B, Leopold DA, et al. Mutation in the gene responsible for cystic fibrosis and predisposition to chronic rhinosinusitis in the general population. JAMA 2000;284:1814–1819.

28. Khoury MJ. Will genetics revolutionize medicine? Letter. New Engl J Med 2000;343:1497.

29. Khoury MJ, Wagener DK. An epidemiological evaluation of the use of genetics to improve the predictive value of disease risk factors. Am J Hum Genet 1995;56:835–844. [Erratum, Am J Hum Genet 1996;58:253.]

30. Pharoah PDP, Antoniou A, Bobrow M, Zimmern RL, Easton DF, Ponder BAJ. Polygenic susceptibility to breast cancer and implications for prevention. Nat Genet 2002;31:33–36.

31. The Evidence-Based Medicine Working Group. Users' guides to the medical literature: a manual for evidence-based clinical practice. Guyatt G, Rennie D (eds.) Chicago: American Medical Association, 2002.

32a. Sotos JG, Rienhoff Y Jr. Will genetics revolutionize medicine? Letter. New Engl J Med 2000;343:1496.

32b. Yang Q, Khoury MJ, Botto L, Friedman JM, Flanders WD. Improving the prediction of complex diseases by testing for multiple disease-susceptibility genes. Am J Hum Genet 2003;72:636–649.

33. Beaty TH, Khoury MJ. Interface of genetics and epidemiology. Epidemiol Rev 2000;22:120–125.

34. Satten GA, Flanders WD, Yang Q. Accounting for unmeasured population substructure in case-control studies of genetic association using a novel latent-class model. Am J Hum Genet 2001;68:466–477.

35. Zerba KE, Ferrell RE, Sing CF. Genetic structure of five susceptibility gene regions for coronary artery disease: disequilibria within and among regions. Hum Genet 1998;103:346–354.

36. Sing CF, Haviland MB, Reilly SL. Genetic architecture of common multifactorial diseases. Ciba Foundation Symposium 1996;197:211–229.

37. Potter JD. Colorectal cancer: molecules and populations. J Natl Cancer Inst 1999;91:916–932.

38. Traverso G, Shuber A, Levin B, et al. Detection of APC mutations in fecal DNA from patients with colorectal tumors. N Engl J Med 2002;346:311–320.

39. Schwartz RS. A needle in a haystack of genes. N Engl J Med 2002;346:302–304.

40. Petricoin EF, Ardekani AM, Hitt BA, et al. Use of proteomic patterns in serum to identify ovarian cancer. Lancet 2002;359:572–577.

41. Evans JP, Skrzynia C, Burke W. The complexities of predictive genetic testing. BMJ 2001;322:1052–1056.

42. Subramanian G, Adams MD, Venter JC, Broder S. Implications of the human genome for understanding human biology and medicine. JAMA 2001;286:2296–307.

43. Lee TH, Brennan TA. Direct-to-consumer marketing of high-technology screening tests. N Engl J Med 2002;346:529–531.

44. Personalized medicine: opportunities and challenges for commercialization. Web site. Accessed April 26, 2002, http://www.lifesciencesinfo.com/email/2707/.

Appendix

Joint Probability Distribution of Susceptibility Genotype and Disease Outcome in a Hypothetical Population

		DISEASE		
		Yes	*No*	*Total*
Genotype	Present	$P(GD)$	$P(G\bar{D})$	$P(G)$
	Absent	$P(\bar{G}D)$	$P(\bar{G}\bar{D})$	$P(\bar{G})$
	Total	$P(D)$	$P(\bar{D})$	1

Probability of genotype $= P(G)$
Probability of disease $= P(D)$
Population odds of disease $= P(D) / P(\bar{D})$
Joint probability of G and D $= P(GD)$
Sensitivity $= P(GD) / P(D) = P(G|D)$
Specificity $= P(\bar{G}|\bar{D})$, $1 -$ specificity $= P(G|\bar{D})$
Positive predictive value (ppv) of genotype $= P(D|G)$
Negative predictive value (npv) of genotype $= P(\bar{D}|\bar{G})$

The improved ability to predict disease using genotype can be summarized:

$$P(GD)/P(G\bar{D}) = [P(G|D) / P(G|\bar{D})] [P(D)/P(\bar{D})]$$

or

odds of disease in presence of genotype $=$ [sensitivity/(1 $-$ specificity)](population disease odds)

The factor by which disease prediction can be improved in persons with the genotype, relative to the whole population, is [sensitivity/(1 $-$ specificity)], known as the *likelihood ratio*.[a]

Likewise, the relative risk of disease in persons with the genotype can be expressed:

$$[P(GD)/P(G)]/[P(\bar{G}D)/P(\bar{G})] = P(D|G) / P(D|\bar{G}) = ppv / (1 - npv)$$

[a]Reference 31.

12

Genetic tests in populations: an evidence-based approach

Paolo Vineis

This chapter describes the main epidemiologic aspects that need to be considered before any genetic test is offered to a population. In an "evidence-based approach," we base testing on knowledge of the epidemiologic characteristics of the population and on an evaluation of the impact of testing rather than only on "mechanistic" evidence (i.e., a gene predisposes to a diseases and therefore testing may be useful). The introduction of a genetic test implies a complex series of measurements that involve:

1. The estimate of the population prevalence of the "high-risk" genotype;
2. The estimate of the penetrance of the genetic trait;
3. Sufficiently good evidence that after the carrier of a mutation or a variant (polymorphism) is identified an effective preventive or therapeutic treatment will exist for the carrier, and
4. Sufficiently good evidence that the harm associated with testing is lower than the benefit of testing (where harm must be interpreted widely, including psychological and indirect harm, e.g., insurance implications).

Prevalence

Prevalence is the frequency of the relevant gene variant in the target population. Low prevalence has several consequences for testing. First, given sensitivity and specificity of a test, the predictive value is proportional to prevalence: low prevalence means a low proportion of true positives to false positives. Let us consider an example. Alison Dunning and colleagues (personal communication), using 864 DNA samples, determined the N372H genotype of *BRCA1* by allele specific oligonucleotide (ASO) and by Taqman (two common genotyping techniques) assays. If the two results did not agree, they additionally used a forced RsaI digest. Thus, they developed a "Consensus genotype" of the same result by at least two methods for each sample. Subsequently, they typed a random subset of DNA samples with the RsaI forced digest and typed 96 samples with the Invader Cleavase technology. These latter results were also related back to the original "Consensus genotype." For each method they assessed the "Sensitivity" [=(total number of samples for which a result was called)/(total number of samples in evaluation)] and "Specificity" [=(total number of samples called as Consensus)/total number of samples called by method)]. The results obtained are given in Table 12.1.

The best performance is for TaqMan, with sensitivity and specificity of 96% to 98%. With such levels of accuracy of the test, when the prevalence of the variant is, for example, 50%, the positive predictive value is 98% (i.e., of every 100 persons testing positive, 98 are true positive, and 2 are false positive); if prevalence is 5%, the predictive value becomes 74%, i.e., of every 100 persons testing positive, 74 have the variant, but 26 are false positives.

Second, low prevalence is also important for the population impact of the benefits of testing, as measured by the Number Needed to Screen (see below).

Penetrance

Penetrance is defined as the proportion of the mutation carriers who develop the phenotypic manifestations. Even in the case of diseases caused by highly penetrant mutations, e.g., Huntington disease; and phenylketonuria (PKU), penetrance is not 100%. Penetrance depends on at least six aspects: (*1*) the importance of the function of the protein encoded by the gene (e.g., in crucial metabolic pathways, as in PKU, or in key regulatory aspects of the cell cycle as affected by the *BRCA1* gene); (*2*) the functional importance of the mutation (e.g., a deletion vs. a mild loss of function from a point mutation); (*3*) the interaction with other genes, including modifying genes; (*4*) the onset of somatic mutations; (*5*) the interaction with the environment; (*6*) the existence of alternative pathways that can substitute for the loss of function.

An approximately inverse relationship exists between the frequency of a variant and its penetrance. The more penetrant (i.e., deleterious) a mutation, the less frequently we expect to find it in the population—although it may be concentrated in particular groups or families because of a founder effect or segregation. We will see how penetrance affects the estimation of benefits and harms of testing in a population.

Table 12.1 Results

Method	Sensitivity	(%)	Specificity	(%)
ASO	836/864	97	753/836	90
Taqman	826/864	96	812/826	98
RsaI digest	125/173	72	103/125	82
Invader	62/92	67	45/62	73

How to Measure the Population Impact of Testing: Number Needed to Screen

Let us consider a low-penetrance genetic trait that occurs frequently in the general population (with a prevalence of 13.8%, i.e., a polymorphism) and whose identification can lead to a 58% reduction in the risk for disease. If the absolute lifetime risk for disease in the carriers of the polymorphism is 1.4% or 14 per 1000, then a reduction of 58% gives 6 per 1000, i.e., an absolute reduction of 8 per 1000. The inverse of 8 per 1000 (1/0.008) is 125, i.e., the number of carriers we need to treat (NNT) to obtain one success. However, because the prevalence of the trait is 13.8%, we in fact need to screen 125/0.138, i.e., 905.8 people, to prevent one case of disease. This is the NNS. In contrast, imagine we have a highly penetrant mutation that is rare in the general population (0.16%) but common in some families (50% prevalence). Suppose the lifetime risk for disease is 37%, and detection and intervention reduce the relative risk by 58%. With such figures, we have an absolute risk reduction of 21.5% (i.e., 58% of 37%) to 15.5%, and a NNT (1/0.215) of 4.5. If we decide to screen families, then we have to screen 4.5/0.5 (9 people) to prevent 1 case of disease; if we screen the entire population, however, the figure becomes 4.5/0.0016, i.e., 2812.5 people. The example shows that a reasonable NNS is attained only by screening for highly penetrant mutations in high-risk families, not for such mutations in the general population or for low-penetrant polymorphisms.

Two Examples: *BRCA1* and *GSTM1*

Two examples of the implications of the points considered above are given in Table 12.2.

Table 12.2 addresses NNS with real figures. Imagine we want to screen high-risk families for a highly penetrant gene (*BRCA1*). In this example, we assume that the cumulative (lifetime) risk for breast cancer is approximately 80% in the mutation carriers, and the prevalence of the mutations in high-risk families is about 50%. Let us hypothetically suppose that tamoxifen or Raloxifene halves the risk (this figure is based on a 45% reduction in risk, reported by the Breast Cancer Prevention Trial (BCPT) and a 76% reduction based on the Raloxifene trial; however, two studies on tamoxifen do not report any benefit). This means that we have to treat 2.5 family members (carriers of the mutation) to prevent one cancer. This requires us to screen five members to achieve the same result. However, if we want to screen the general population, things change dramatically. Now the cumulative risk in mutation carriers is 40%, with an absolute risk reduction by tamoxifen of 20%, which means a NNT of five within those carrying the mutations. However, because the mutation carriers in the general population are only 0.2%, the NNS is as large as 2500 to prevent one cancer. This makes *BRCA1* an unrealistic marker for use in the general population for screening purposes.

Table 12.2 Calculation of the Number Needed to Screen in the Case of a Screening for a Low Penetrant Gene (*GSTM1* in smokers) and a Highly Penetrant Gene (*BRCA1*), Respectively, in the General Population or in Families

Measures	LUNG CANCER		BREAST CANCER	
	Smokers GSTM1 Null	*Smokers GSTM1 Wild*	*BRCA1 General Population*	*BRCA1 Families*
Relative risk	1.34 (1.21–1.48)[a]	1.0	5	10
Cumulative risk	13%	10%	40%[b]	80%
Risk reduction	50%[c]	50%[c]	50%(Tamoxifen or Raloxifene)[d]	50%
Cumulative risk after intervention	6.5%	5%	20%	40%
Absolute risk reduction	6.5%	5%	20%	40%
NNT in mutation carriers	15	20	5	2.5
Prevalence	50%	50%	0.2%[e]	50%
NNS in whole target population	30	40	2500	5
NNS in all smokers		35		

NNT, number needed to treat; NNS, number needed to screen.
[a]From Vineis et al, 1999 (5); the OR for *GSTM1* in smokers was 1.22 (0.96–1.54).
[b]From Hopper et al, 1999 (4).
[c]Theoretical maximum reduction in risk of lung cancer because of chemopreventive agent.
[d]Theoretical benefit, based on the BCPT trial with a 45% benefit, and the Raloxifene trial with a 76% benefit. The figures concerning tamoxifen and Raloxifene (two drugs with side effects) are provisional and need confirmation.
[e]Coughlin et al, 1999 (2).

Consider now a low-penetrance gene, *GSTM1* null. We might consider screening smokers for the *GSTM1* genotype and addressing prevention/chemoprevention only to them. What would be the advantage? According to a meta-analysis, the relative risk for lung cancer associated with the *GSTM1* genotype is 1.34. Therefore, if the cumulative risk for lung cancer in smokers is 10%, it will be approximately 13% among the null *GSTM1* carriers. Suppose that the chemopreventive intervention has a 50% efficacy. This leads to a cumulative risk of 6.5% among smokers who are *GSTM1* null, with a NNT of 15 (1/6.5%) to prevent 1 cancer. However, because the carriers of the null genotype are 50% in the population, we need to screen 30 individuals to prevent one cancer. Now we repeat the same calculations including the carriers of the wild (normal) genotype, ending up with a NNS of 40 (Table 12.2). Without screening the population for *GSTM1*, we would have a NNS of 35 (the average of the previous two). Clearly, little advantage exists in screening for a low-penetrance gene if the NNS decreases from 35 to 30.

Risks and Benefit

The issue of risks can be exemplified by the epidemiologic measure of the number needed to harm (NNH), and of its ratio to NNS.

Figure 12.1 shows a crucial relation in clinical epidemiology, i.e., usually the harm associated with a treatment is not related to the frequency of the outcome, whereas the benefit increases with an increasing prevalence of the outcome. For ex-

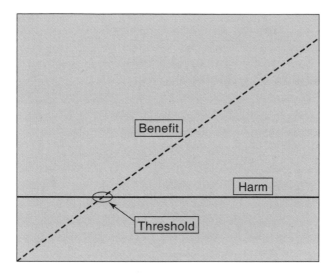

Figure 12.1 Relation between benefits and harms for genetic testing. While benefits tend to increase with the penetrance of the gene mutation (see text), harms to the screenees tend to be constant (on the abscissa: penetrance of the gene).

ample, if we treat with statins a population with a very high risk of myocardial infarction, the effectiveness of treatment will be much higher than if we treat people with lower levels of cholesterol and low risk for infarction; however, the probability of side effects of statin administration will not change. Thus, the threshold for treatment depends on the frequency of the outcome we want to prevent. The same concepts apply to genetic testing. Testing a population for a low-penetrance gene means that the NNS is at the low (left) extreme of Figure 12.1, and the benefit does not overcome the negative implications, which are legal and ethical.

Combination of Genotypes

One could object to the reasoning above by saying that several genes contribute to defining the individual risk on the basis of the individual combination of different "at-risk" genotypes (e.g., *CYP1A1* plus *GSTT1*, etc.). However, if the genes are not in linkage disequilibrium, i.e., the corresponding genes are not adjacent in the same chromosome, the probability of a given combination of genotypes is low, being the product of the individual probabilities. In fact, the frequencies of combinations observed in a large data set, (a pooled analysis of the studies on metabolic polymorphisms, (3)), are comparable with the frequencies expected under the hypothesis of no linkage (Table 12.3). All 25 possible double combinations for the different genes were examined for Caucasians and Asians separately, and some examples of these comparisons are shown in Table 12.3. In no case were any significant deviations observed from expected allele combinations, suggesting that for these alleles no

Table 12.3 Examples of Frequencies of Observed Genotype Combinations Compared to Expected Frequencies Assuming Non-linkage

Genotype Combination	Race	No.	Observed	Expected
GSTM1*0/*0 + GSTT1*0/*0	Caucasian	5532	0.104	0.105
GSTM1*0/*0 + GSTT1*0/*0	Asian	407	0.246	0.248
GSTM1*0/*0 + CYP1A1*1/*2A	Caucasian	3192	0.0573	0. 0558
GSTM1*0/*0 + CYP1A1*1/*2A	Asian	509	0.132	0.144
GSTM1*0/*0 + CYP1A1*2A/*2A	Caucasian	3192	0.0025	0.0028
GSTM1*0/*0 + CYP1A1*1/*2B	Caucasian	3192	0.0326	0.0341
GSTM1*0/*0 + CYP1A1*1/*2B	Asian	509	0.165	0.175
GSTM1*0/*0 + CYP1A1*2B/*2B	Asian	509	0.0275	0.0245
GSTM1*0/*0 + CYP2E1*1/*5A	Asian	283	0.209	0.194
GSTM1*0/*0 + NAT2*5/*5	Caucasian	3266	0.122	0.116
GSTM1*0/*0 + NAT2*6/*6	Caucasian	3069	0.0401	0.0370
GSTT1*0/*0 + CYP1A1*1/*2A	Caucasian	2502	0.0164	0.0207
GSTT1*0/*0 + CYP2E1*1/*6	Caucasian	395	0.0253	0.0201
CYP1A1*1/*2A + NAT2*5/*5	Caucasian	1335	0.0217	0.0230
CYP1A1*1/*2B + (NAT2*4/*6 & NAT2*5/*6 & NAT2*6/*7)[a]	Caucasian	1151	0.0278	0.0276
CYP2E1*1/*6 + (NAT2*4/*5 & NAT2*5/*6 & NAT2*5/*7)[b]	Caucasian	409	0.0416	0.0491

[a]Refers to all NAT2*6 heterozygotes.
[b]Refers to all NAT2*5 heterozygotes.
Source: From Garte et al. (3).

linkage exists between any of the polymorphic alleles at these loci. For the combination of different susceptibility alleles, we can apply the same reasoning of the previous paragraph, i.e., common low-penetrance polymorphisms can combine to generate rare combinations (for example, GSTT1 null plus CYP1A1 variants) associated with high risk of one or more diseases in the presence of exposures. With such combinations, we fall into the situation we have already depicted through BRCA1, i.e., high penetrance but frequencies of less than 1% in the population. In these circumstances, gene combinations might not overcome the difficulties associated with population testing.

Combinations can be responsible for a continuous distribution of susceptibility (Figure 12.2). The example in the figure refers to adverse reactions to ionizing radiation: a small group of patients (e.g., those heterozygous for mutations in the AT gene) has an exceptionally high risk for adverse reactions, but the other people are normally distributed, probably for the existence of several genes implicated in the process, such as DNA repair genes.

Discrimination

One of the concerns about genetic testing is that it could be used to discriminate against people by, for example, insurers or employers (see Chapter 4). Such discrimination might even be associated with the ethnic group; the frequencies of some alleles differ considerably by ethnic group, for example, CYP1A1 in Asians (Table 12.4). The database of control subjects analyzed for metabolic gene polymorphisms

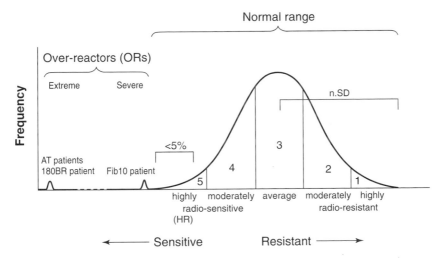

Figure 12.2 Distribution of the adverse effects of radiation in the population; from Burnet et al. (1).

(Garte et al., 2001) consisted of 15,843 subjects. Of these, 12,525 (79.1%) were Caucasians, 2136 (13.5%) were Asians, 936 (5.9%) were African Americans, 60 (0.4%) were Africans, 186 (0.16%) were of uncertain or other ethnicity, and 75 had no ethnic information in the database.

Genetic Testing or Reduction of Exposures?

The diseases that dominate today in Western societies are multifactorial diseases. This means, for example, that lung cancer is not entirely attributable to smoking but to causal complexes of which smoking is a part. Conversely, smoking also contributes to cardiovascular and other chronic diseases. The elimination of one exposure like smoking would eliminate a large proportion of chronic diseases in the Western world (and the same can hold true for, for example, some dietary agents and environmental pollutants). Genetic traits, on the contrary, may relate differently to different diseases: subjects with the *NAT-2* slow genotype have an increased risk for bladder cancer but a decreased risk for colon cancer. This is not likely to be an unusual situation. In contrast, few exposures are causal for one disease and protective for another. In addition, although the relation "one exposure, many diseases" is true, for low-penetrance genes the opposite seems to apply: "one disease, many genotypes" (Table 12.5). If we consider *GSTM1* and lung cancer, because the "null" genotype is present in 50% of the general population, the risk attributable to *GSTM1* is approximately (0.34/1.34)/2, i.e., 12.6%. However, the same person may be at high risk because of *GSTM1* and at low risk because of other polymorphisms, so the sum of the effects of single polymorphisms is unpredictable and can be greater or lower than 12.6%. In general, in the population we expect to find a whole distribution of risks, with few individuals carrying sev-

Table 12.4 Metabolic Gene Allele Frequencies in Different Ethnic Groups

Gene	Race	No.[a]	Heterozygous	Homozygous	Allele
CYP1A1*1	Caucasians	3814	0.190 (0.13–0.27)[b]	0.795 (0.71–0.87)	0.890
	Asians	626	0.460 (0.43–0.49)	0.395 (0.34–0.41)	0.625
	Africans	445	0.465	0.432	0.664
CYP1A1*2A	Caucasians	3814	0.105 (0.054–0.16)	0.005 (0–0.015)	0.058
	Asians	626	0.272 (0.23–0.31)	0.0128 (0–0.056)	0.149
	Africans	445	0.333	0.0517	0.218
CYP1A1*2B	Caucasians	3814	0.064 (0.025–0.12)	0.0001 (0–0.0057)	0.032
	Asians	626	0.331 (0.32–0.44)	0.0463 (0.031–0.058)	0.212
	Africans	445	0.036	0.00	0.018
CYP1A1*2C	Caucasians	3814	0.033 (0–0.095)	0.0021 (0–0.012)	0.0186
	Asians	626	0.027 (0–0.032)	0.0016 (0–0.012)	0. 0152
	Africans	445	0.0135	0	0.00675
CYP1A1*3	Caucasians	735	0	0	0
	Africans	464	0.177	0.0043	0.0927
CYP2E1*5A	Caucasians	854	0.0480 (0.034–0.095)	0.0012 (0–0.005)	0.0252
	Asians	286	0.367 (0.35–0.38)	0.0594 (0.054–0.63)	0.243
CYP2E1*5B	Caucasians	854	0.0105 (0–0.05)	0.00	0.00525
	Asians	286	0.0210 (0.006–0.045)	0.00	0.0105
CYP2E1*6	Caucasians	854	0.102 (0.08–0.12)	0.0023 (0–0.0032)	0.0533
	Asians	286	0.126 (0.071–0.093)	0.00	0.0630
EPHX*3	Caucasians	685	0.398	0.117	0.316
EPHX*4	Caucasians	686	0.353	0.038	0.215
GSTM1*0	Caucasians	10514		0.531 (0.42–0.60)	
	Asians	1511		0.529 (0.42–0.54)	
	Africans	479		0.267 (0.16–0.36)	
GSTT1*0	Caucasians	5577		0.197 (0.13–0.26)	
	Asians	575		0.470 (0.35–0.52)	
GSTM1*0	Caucasians	5532		0.104	
+T1*0	Asians	407		0.246	
GSTP1*1	Caucasians	1137	0.493	0.438	0.685
GSTP1*2	Caucasians	1138	0.442	0.0413	0.262
GSTP1*3	Caucasians	878	0.126	0.0057	0.0687
NAT2*5	Caucasians	3847	0.482 (0.42–0.55)	0.219 (0.13–0.32)	0.46
NAT2*6	Caucasians	3618	0.430 (0.35–0.56)	0.070 (0.032–0.11)	0.285
NAT2*7	Caucasians	3129	0.055 (0.028–0.099)	0.0013 (0–0.071)	0.029

[a]No., number of subjects tested.
[b]Numbers in parentheses give the range of values for individual studies used.
Source: From Garte et al. (2001).

eral high-risk polymorphisms and a majority with a balance of "bad" and "good" genotypes (Fig. 12.2).

Conclusions

In an "evidence-based approach" to genetic testing knowledge of the main features of the test and of epidemiologic characteristics of the target population should be

Table 12.5 An Illustration of the Principle of One Exposure–Many Diseases, One Disease–Many Low-Penetrance Genes

Exposure	Disease	Proportion Attributable
Tobacco smoke	Lung cancer	90%
	Bladder cancer	70% men
		30% women
	Larynx cancer	90%
	Coronary heart disease	12.5%
	Chronic bronchitis	80%

Disease	Low-Penetrance Genes	Odds Ratio[a]
Lung cancer	*CYP1A1* MspI (Asians)	1.73
	CYP1A1 MspI (Caucasians)	1.04
	CYP1A1 Exon 7 (Asians)	2.25
	CYP1A1 Exon 7 (Caucasians)	1.30
	CYP2D6	1.26
	GSTM1	1.34
Bladder cancer	*NAT-2* slow	1.37
	GSTM1	1.57
Colon cancer	*NAT-2* rapid	1.19

[a]Meta-analysis from Vineis et al. (1999).

provided. The introduction of a genetic test implies a complex series of measurements that involve the estimate of the population prevalence of the "high-risk" genotype; the estimate of the penetrance of the genetic trait; sufficiently good evidence that after the carrier of a mutation or a variant (polymorphism) is identified an effective preventive or therapeutic treatment will exist for him or her; and sufficiently good evidence that the harm associated with testing is lower than the benefit of testing. In the absence of such knowledge, we risk to cause more harm than benefit, e.g., by a large proportion of false positives to true positives, a modest or marginal number of successes per unit of population screened, and discrimination based on ethnic group.

Acknowledgments

I am grateful to Alison Dunning (Cambridge) and Sy Garte (Milano) for some of their data that I quote in this chapter. This paper was made possible by a grant from the Compagnia di San Paolo to the ISI Foundation (Torino) and a grant of the World Cancer Research Fund.

References

1. Burnet NG, Johansen J, Turesson I, et al. Describing patients' normal tissue reactions concerning the possibility of individualising radiotherapy dose prescriptions based on potential predictive assays of normal tissue radiosensitivity. Int J Cancer 1998;79:606–613.
2. Coughlin SS, Khoury MJ, Steinberg KK. BRCA1 and BRCA2 gene mutations and risk of breast cancer. Public health perspectives. Am J Prev Med 1999;16(2):91–98.

3. Garte S, Gaspari L, et al. Metabolic gene frequencies in control populations. Cancer Epidemiol Biomarkers Prev 2001;10:1239–1248.
4. Hopper JL, Southey MC, Dite GS, et al. Population-based estimate of the average age-specific cumulative risk of breast cancer for a defined set of protein-truncating mutations in BRCA1 and BRCA2. Australian Breast Cancer Family Study. Cancer Epidemiol Biomarkers Prev 1999;8(9):741–747.
5. Vineis P, Malats N, Lang M, et al. Metabolic polymorphisms and susceptibility to cancer. IARC Scientific Publication No. 148, International Agency for Research on Cancer, Lyon, 1999.
6. Vineis P, Schulte P, McMichael AJ. Misconceptions about the use of genetic tests in populations. Lancet 2001;357(9257):709–712.

13

ACCE: a model process for evaluating data on emerging genetic tests

James E. Haddow and Glenn E. Palomaki

In this chapter, we outline a process for collecting, evaluating, interpreting, and reporting data about DNA (and related) testing for disorders with a genetic component. The process follows a format that allows policymakers to access up-to-date and reliable information for decision-making. An important by-product of this process is the identification of gaps in knowledge. This approach builds upon a methodology described by Wald and Cuckle (1) and terminology introduced by a committee established by the Secretary of Health and Human Services (the Secretary's Advisory Committee on Genetic Testing) (2). We then describe the components of that extended methodology, along with several examples that include applications for both screening and diagnostic purposes.

ACCE: A Model System for Assessing DNA Testing

Figure 13.1 displays the ACCE model system. Before beginning the evaluation, the clinical disorder shown at the center of the target needs to be carefully defined, along with the setting in which the testing is to be performed (e.g., population screening). Next, an in-depth assessment of the four ACCE components (*A*nalytic validity; *C*linical validity, *C*linical utility, and *E*thical, legal, and social issues) can begin. Each of these components is labeled in Figure 13.1, along with its key elements. Systematic implementation is achieved by identifying a standardized set of specific questions aimed at fully characterizing the clinical disorder and each of the four ACCE components.

The *clinical disorder* needs to be carefully defined at the outset. Particular attention is paid to this first activity because successfully assessing the other components depends upon a clear definition. Problems have arisen in the past when a "disorder" has been described in terms of the test being used to identify it, rather than

Figure 13.1 A graphical representation of the ACCE model system. Organized beginning at the center, the figure represents the organization of information that is necessary to collect, evaluate, interpret and present as part of a comprehensive review of DNA testing. PPV, positive predictive value; NPV, negative predictive value.

in terms of clinical manifestations. One such example is "hypercholesterolemia" being characterized as the condition sought by cholesterol measurement. Hypercholesterolemia is a laboratory finding rather than a clinical disorder. Cholesterol testing is performed to classify risk for coronary artery disease, and most people with elevated serum cholesterol measurements do not suffer from that disorder.

Analytic validity defines a test's ability to accurately and reliably measure the genotype of interest. This aspect of evaluation focuses on the laboratory DNA test. The four main elements of analytic validity include analytic sensitivity (or the analytic detection rate), analytic specificity (or $1 -$ the analytic false-positive rate), laboratory quality control, and assay robustness.

Clinical validity defines a test's ability to detect or predict the associated disorder (phenotype). An important distinction exists between analytic and clinical validity. For example, a given DNA test may reliably detect a mutation, but if that mutation is associated with a serious medical problem only once in every 100 times, the clinical predictive value will be poor. Clinical validity builds upon the analysis of analytic validity by assessing five more elements (Fig. 13.1). These include clinical sensitivity (or the clinical detection rate), clinical specificity (or $1 -$ the clinical false-positive rate), prevalence of the clinical disorder, positive and negative predictive values, and penetrance (including gene and environmental modifiers).

Clinical utility defines the risks and benefits associated with a test's introduction into practice. Specifically, clinical utility focuses on the health outcomes (both positive and negative) associated with testing. The natural history of the clinical disorder needs to be understood, so that such considerations as optimal age for testing might be taken into account. It is necessary to determine the availability and effectiveness of interventions aimed at avoiding adverse clinical consequences (if no effective interventions are available, for example, testing may not be warranted). Quality assurance assesses procedures in place for controlling preanalytic, analytic, and postanalytic factors that could influence the risks and benefits of testing. Pilot trials assess the performance of testing under real-world conditions, including the psychologic and social risks and benefits of testing. Health risks define adverse consequences of testing or interventions in individuals with either positive or negative test results. Economic evaluation helps define and compare the financial costs and benefits of testing. Facilities assess the capacity of resources to manage all aspects of the service. Education assesses the quality and availability of validated informational materials and expertise. Monitoring and evaluation assess a program's ability to maintain surveillance over its activities and make adjustments.

Ethical, legal, and social implications (ELSI) surrounding the testing process for the clinical disorder refer to two types of concerns: those inherent in any medical technology and those particularly germane to testing for diseases that have a genetic component. The latter concerns include implications for relatives of the person undergoing testing, the possibility of insurance discrimination, and stigmatization based on genotype (disease risk) rather than on phenotype (actual disease). The precise nature of these risks, however, depends to a great degree on the preceding three components. Thus, ELSI concerns are represented in Figure 13.1 by a penetrating pie slice.

The following is a more in-depth review of information sought by targeted questions directed at each of the preceding categories.

Disorder/Setting

Question 1: What Is the Specific Clinical Disorder to Be Studied?
As previously stated, the medical disorder needs to be defined in terms of its clinical characteristics, rather than the laboratory test being used to detect it. Furthermore, in order to qualify for a full-scale evaluation by the ACCE process, the medical disorder needs to be considered 'important' in terms of morbidity and/or mortality. There are times when the definition of a medically important clinical disorder is not as straightforward as might initially be thought. One such example is adult-onset hemochromatosis.

Question 2: What Are the Clinical Findings Defining This Disorder?

Answering this question offers the opportunity to gather relevant information concerning the extent of morbidity and mortality, and to itemize specific manifestations of the disorder which might either be prevented or remedied. In some instances, it might be necessary to document the frequency and severity of nonspecific manifestations such as fatigue and right upper quadrant pain in adult-onset hemochromatosis in comparison to the general population to determine attributable risk. This is not necessary when the manifestations are highly specific, such as malabsorption occurring as a result of pancreatic insufficiency in cystic fibrosis.

Question 3: What Is the Clinical Setting in Which the Test Is to Be Performed?

Understanding the nature of the setting is important at the outset, since positive and negative predictive values for the same test may differ greatly when a diagnostic versus screening setting is chosen. The target of testing (and subsequent intervention), and the composition of the testing panel can also vary by setting.

Question 4: What DNA Test(s) Are Associated with This Disorder?

Once the medical disorder has been defined according to its clinical characteristics, appropriate DNA tests can be identified. For example, if the disorder were venous thrombosis, then DNA testing for both the factor V Leiden and prothrombin genes would be indicated.

Question 5: Are Preliminary Screening Questions Employed?

A preliminary question might be used as part of a formal screening strategy to identify individuals who might benefit most from laboratory testing. Many genetic disorders vary in prevalence depending on race (e.g., sickle cell disease) or ethnicity (e.g., Tay-Sachs disease). Often, however, a question of this type is not recognized as a screening test. It is important that such questions be carefully evaluated.

Question 6: Is It a Stand-Alone Test or Is It One of a Series of Tests?

This question is aimed at the evaluation of screening tests but may also be appropriate in some diagnostic settings. For example, prenatal DNA (or biochemical) screening for Tay-Sachs disease is preceded by a question regarding ethnicity, and testing is usually offered only to couples where at least one partner is of Ashkenazi Jewish heritage. Tay-Sachs disease does occur in other ethnic/racial groups, but at a much lower prevalence. The initial screening question is important in that it sets an upper limit on the overall impact of the screening strategy on birth prevalence.

Question 7: If It Is Part of a Series of Screening Tests, Are All Tests Performed in All Instances (Parallel) or Are Some Tests Performed Only on the Basis of Other Results (Series)?

In general, when multiple screening tests are offered, they are performed in series. That is, only individuals with positive results in the first test are offered subsequent testing. For example, in some newborn screening programs for cystic fibrosis, the first screening test is a measurement of immunoreactive trypsinogen (IRT) on the dried blood spot. Only those with elevated IRT measurements have subsequent DNA testing performed. The DNA test may be diagnostic in some individuals, but many will require further testing (e.g., a sweat test or expanded DNA panel).

Analytic Validity

Question 8: Is the DNA Test Qualitative or Quantitative?

In general, DNA test results are qualitative, in that a mutation is either present or absent, as is true when testing for breast cancer (*BRCA1/2* mutations). In some instances, DNA test results are, however, quantitative, as with quantifying the number of CGG repeats when testing for fragile X. Laboratories may categorize these fragile X test results as "normal," "pre-mutation," or "full mutation," but the laboratory measurement is a continuous variable and needs to be treated as such throughout the process of exploring analytic validity.

Question 9: How Often Is the Test Positive When a Mutation Is Present?

This question deals directly with analytic sensitivity. At this stage, it is not necessary to have information about phenotype or penetrance. Analytic sensitivity focuses only on the proportion of positive test results when a mutation that is being tested for is present. If the mutation is present and the test is positive, the result is classified as a true positive. If the mutation is present and the test is negative, the result is classified as a false negative. The analytic sensitivity is the number of true positive results divided by the sum of the true positives and false negatives. External proficiency testing can be a major source of information about analytic sensitivity. Other sources include published method comparisons and validation studies. Confidence intervals for analytic sensitivity should be included along with the point estimates. Only samples with a known genotype can be used to directly compute analytic sensitivity.

Question 10: How Often Is the Test Negative When the Mutation Is Not Present?

This question addresses analytic specificity and focuses on results in samples that do not contain the mutation(s) being tested for. For DNA testing, the analytic specificity ought to approach 100%. In other words, false positive results are expected to be relatively uncommon. It is necessary to obtain test results on hundreds of samples without the mutation to confidently estimate the analytic specificity. If data are available for 30 samples without mutations and test results on all of those samples are negative (i.e., no false positives), the point estimate for analytic specificity is 100%. However, the 95% confidence interval derived using the binomial distribution indicates that the analytic specificity could be as low as 88%. Because of the time and costs associated with testing an adequate number of samples, confident estimates may be difficult to obtain.

Question 11: Is An Internal QC Program Defined and Externally Monitored?

Internal quality control (QC) is a set of laboratory procedures designed to ensure that the test method is working properly. Such procedures can document that high quality reagents are used, equipment is properly calibrated and maintained, and good laboratory practices are being applied at every level. Regulatory and professional organizations provide specific regulations and recommendations concerning internal quality control programs, and certifying organizations provide oversight of internal quality control programs as part of the inspection process. If testing is being performed in a laboratory that is not certified for clinical testing (e.g., a research laboratory), specific documentation of performance may need to be sought from that laboratory.

Question 12: Have Repeated Measurements Been Made on Specimens?

As part of ongoing quality control, laboratories routinely include positive (and negative) control samples in their assay runs. The same control sample will be included repeatedly in assays over months or even years, as one measure of assay performance. In the course of a year, test results from these control samples may include occasional failures, or the results may be incorrect. The failure rate of control samples is important, because these samples are the only ones tested for which the mutation status is known. Although failure rates for control samples are documented as part of a clinical laboratory's internal quality program, they are not usually reported. It may be possible to request this information from laboratories as part of the ACCE review process.

Question 13: What Is the Within- and Between-Laboratory Precision?

This question is only applicable when the DNA test results can be interpreted as a continuous variable, as is described in Question 8 for fragile X testing. In such instances, repeated measurements on a single sample within a laboratory can provide an estimate of within- and between-run variability. The process can be extended to multiple laboratories with differing methodologies to estimate between-laboratory and between-method variability, as well.

Question 14: If Appropriate, How Is Confirmatory Testing Performed to Resolve False Positive Results in a Timely Manner?

Confirmatory testing is usually limited to situations where a positive test result is found and is especially important when an assay is known to yield occasional false positive results (i.e., the analytic specificity is less than 100%). Confirmatory testing can range from re-assaying the same sample to obtaining a new sample and utilizing another testing technology. When possible, the effect of confirmatory testing on analytic specificity should be calculated.

Question 15: What Range of Patient Specimens Has Been Tested?

Important pre-test variables, including the type and condition of patient specimens, should be evaluated to determine whether they adversely affect the quantity or quality of the DNA and subsequent test results. Time and temperature are important considerations. Blood samples obtained via phlebotomy provide large amounts of high quality DNA, and diagnostic testing often utilizes this source. For screening purposes, however, other sample types (e.g., buccal scraping or dried blood spots) may also be used. It is important to evaluate analytic performance for these different specimen types, separately.

Question 16: How Often Does the Test Fail to Give a Useable Result?

In the pre-analytic phase, it may be determined that the sample is not suitable for testing, because specific clinical criteria are not met. These events are usually not considered a laboratory or methodologic 'failure'. True assay failures occur for reasons such as improperly processed samples, problems with reagents, or equipment malfunction. Optimally, method-specific failure rates should be obtained under routine operating conditions. Although it is required that most of these data be maintained as part of a quality assurance program, failure rates and other information on assay robustness are often not published, and this may represent a gap in knowledge. The ACCE process includes an attempt to learn about test failure rates.

Question 17: How Similar Are Results Obtained in Multiple Laboratories Using the Same, or Different, Technology?

For an analytic methodology to be considered robust, evidence should be presented that it can be successfully implemented in multiple laboratories. It is also important to document that different methodologies will provide equivalent results. Both of these characteristics can be assessed by examining external proficiency testing results or published method comparisons.

Clinical Validity

Question 18: How Often Is the Test Positive When the Disorder Is Present?

This question deals directly with clinical sensitivity and focuses on the proportion of positive test results found among individuals who have (or will develop) the disorder of interest. If the test is positive among those individuals, it is a true positive. If the test is negative among those individuals, it is a false negative. Clinical sensitivity is defined as the true positive results divided by the sum of the true-positive and false-negative results. Clinical performance is aimed at correctly identifying a phenotype. This is in contrast to analytic performance, where the goal is to correctly identify a genotype. A clinical false negative result is usually not due to laboratory error. Rather, it indicates that the disorder can be caused not only by the mutation(s) being tested for, but also by other mutations or causal agents, as well.

Question 19: How Often Is the Test Negative When a Disorder Is Not Present?

This question deals with clinical specificity and is the proportion of individuals with a negative test result among those who do not have (or will not develop) the disorder. If the test is negative among those individuals, it is a true negative. If the test is positive among those individuals, it is a false positive. Clinical specificity is defined as the true negative results divided by the sum of the true negative and false positive results. There are two main ways in which false positive results will occur. The test may be positive because of analytic error, or it may correctly identify the genotype, but the gene may not be penetrant. For example, genotyping will correctly identify individuals as being homozygous for the C282Y mutation, but some homozygous individuals do not have, and will not develop, the phenotype of iron overload.

Question 20: Are There Methods to Resolve Clinical False Positive Results in a Timely Manner?

A clinical false positive result can occur when the laboratory has correctly identified the mutation(s) being sought, but the individual does not have (or will not develop) the disorder. Unlike analytic false positive findings that can often be resolved by confirmatory testing, clinical false positives are more difficult to resolve. Clinical false positives are relatively uncommon when the penetrance (the relationship between genotype and phenotype) is high (e.g. cystic fibrosis). However, when penetrance is low, as is the case for most DNA testing currently available, clinical false positives are more common. The practical consequence of this is that more individuals with positive test results will, necessarily, require treatment as though they had, or would develop, the disorder. Preliminary evidence suggests that between 50% and 75% of males identified as being homozygous for the C282Y mutation in the *HFE* gene will not develop serious clinical manifestations of hemochromatosis in their lifetime.

Question 21: What Is the Prevalence of the Disorder in This Setting?

In order to determine the positive and negative predictive values of the DNA test being used, it is necessary to estimate the prevalence of the specific medical disorder in the population to be tested. If the test is to be used in a diagnostic setting in which the indications for testing are limited to individuals with specific clinical complaints associated with the disorder, the prevalence is expected to be high. On the other hand, were the same test to be used in a population-based screening setting of healthy individuals, the prevalence would be much lower. Factors such as race, ethnicity, and gender may also need to be explored.

Question 22: Has the Test Been Adequately Validated on All Populations to Which It May Be Offered?

If the test is to be offered to various racial/ethnic groups, evidence should be collected to demonstrate that reliable clinical performance estimates are available for those groups. This would include clinical sensitivity, clinical specificity, and prevalence. For example, African Americans are known to have a lower prevalence of cystic fibrosis than Caucasians. In addition, the clinical sensitivity for the recommended panel of 25 mutations is considerably lower for African Americans, because about half of the cases will have at least one mutation not included in the panel. Clinical specificity, however, is likely to be similar for all racial/ethnic groups, as is analytic performance. The clinical performance of prenatal screening for cystic fibrosis in African Americans is, therefore, clearly different from that for Caucasians.

Question 23: What Are the Positive and Negative Predictive Values?

The positive and negative clinical predictive values are well-defined epidemiological terms, calculated by using analytic sensitivity, analytic specificity, clinical sensitivity, clinical specificity, and prevalence.

Question 24: What Are the Genotype/Phenotype Relationships?

One reason for addressing the question of genotype/phenotype relationships is to provide an estimate of penetrance. Penetrance can be viewed as the positive predictive value, in that it is the proportion of individuals with a positive test result that actually has (or will develop) the disorder. Another feature of the genotype/phenotype relationship also deserves consideration. Sometimes, different mutations in the same gene cause distinctly different disease phenotypes. One series of mutations in the muscular dystrophy gene causes the Becker form of the disorder, while other mutations in the same gene cause the Duchenne form. This genotype/phenotype relationship is highly reliable and is used by some clinicians for prognosis and counseling purposes.

Question 25: What Are the Genetic, Environmental, or Other Modifiers?

The phenotypic expression of a medical disorder with a given genotype may differ, depending on the presence or absence of specific genetic or environmental factors. These factors may influence whether any clinical manifestations are present, or they may influence the specific types of manifestations that appear. One example of a genetic modifier is the effect of the 5T/7T/9T variant on the *CFTR* mutation, R117H.

Clinical Utility

Question 26: What Is the Natural History of the Disorder?

Rational development of recommendations for implementing DNA testing requires an understanding of the disorder's natural history. If the clinical disorder for which DNA testing is proposed does not have serious health consequences, testing may not be warranted. If the disorder is serious, then the typical age of onset can be useful in determining the optimal time for either screening or early diagnostic testing. Another aspect to the natural history that would be important to policymakers is a more in-depth documentation of the type, frequency and severity of clinical manifestations. These can be readily obtained for some disorders (such as cystic fibrosis). Other disorders, however, are more difficult to characterize. For example, HFE-related hemochromatosis is associated with a number of nonspecific clinical manifestations, making it difficult to determine what proportion of those manifestations is attributable specifically to the disorder in question.

Question 27: What Is the Impact of a Positive (or Negative) Test Result on Patient Care, and Does the Individual Understand the Meaning of His or Her Test Result?

Understanding the implications of a positive test is important, not only to policymakers, but also to those delivering health care, as well as to the patients, themselves. Policymakers will be interested not only in potential health benefits, but also in the utilization of health care resources. The health care provider carries the responsibility for interpreting the test results to the patients (and sometimes to the family, as well). The provider is also responsible for overseeing the subsequent testing and treatment. From the patient's perspective, the testing and treatment that follows a positive test may carry health and social risks, as well as financial burdens. Although less obvious, there can also be an impact on the patient, when the test result is negative. For testing to have the potential to benefit, individuals need to understand what their results mean. The four critical points for achieving this understanding include (*1*) the nature and severity of the condition for which they have been tested, (*2*) what their result was, (*3*) the chances that they do or do not have the condition, and (*4*) what action is indicated.

Question 28: If Applicable, Are Diagnostic Tests Available?

When DNA testing is being used for diagnostic purposes, this question is not applicable. In instances where DNA testing is being used for screening purposes, however, diagnostic testing for those with positive screening test results will be available for some, but not all, medical disorders. The purpose of such testing would be to reduce the number of clinical false positives. A second purpose of this testing might be to identify the extent to which the clinical disorder had progressed at the time of detection.

Question 29: Is There an Effective Remedy, Acceptable Action, or Other Measurable Benefit?

This is a critical question to be answered when policymakers are attempting to shape recommendations for introducing new DNA tests. If the disorder of interest cannot be either treated or avoided, then it is unlikely that justification can be made for routinely identifying it. Having an effective intervention to prevent or avoid the morbidity or mortality associated with the disorder (possibly including risk-reducing behavior) is particularly important when decisions are being made about using a test for population screening. In that setting, large numbers of people will be tested who have not sought medical attention on account of symptoms of the disorder. If a subgroup of that population is suddenly informed that they have a problem for which there is no solution, anxiety will be raised without the opportunity for resolution. In contrast, when diagnostic testing of high risk family members is undertaken (e.g., presymptomatic testing for Huntington disease), some candidates may decide against testing, while others may decide that testing will be helpful in either family planning or in providing psychological benefit by removing uncertainty.

Question 30: Is There General Access to that Remedy or Action?

Individuals with positive test results need to have access to diagnostic testing and treatment, or, in some cases, expert consultation and counseling. Barriers may be financial, or relate to travel or time constraints. Policymakers might recommend against routine DNA testing for a given disorder when the region being considered lacks necessary support services for follow-through.

Question 31: Is the Test Being Offered to a Socially Vulnerable Population?

The ACCE process evaluates whether there is any information indicating that members of the population to be offered testing need to be considered especially vulnerable. Two well recognized examples of vulnerable populations include prisoners and individuals with mental retardation. Any DNA testing process needs to be examined to determine whether its results may have an adverse influence on people who might be considered socially vulnerable. It may also be important to consider whether there are barriers to offering the test to individuals in this category. Pregnant women are considered a medically vulnerable population, and a case can be made for their being psychosocially vulnerable, as well. There is often a time-related urgency involved in medical decisions during pregnancy, which can fore-shorten time to carefully consider options. In addition, data have shown that women's motivation to do whatever they can to protect the health of their fetus can lead to overestimates or misunderstandings of the benefits of prenatal testing.

Question 32: What Quality Assurance Measures Are in Place?

Clinical molecular genetic testing laboratories are required to follow good labora-tory practice guidelines and subscribe to external quality assessment programs, if they exist. For DNA testing that is done routinely in clinical laboratories, these qual-ity assurance measures and practice guidelines are generally available, and infor-mation can be collected, summarized and assessed by the ACCE process. For DNA testing that is performed in research laboratories, it will be more difficult to collect data to assess whether quality assurance measures are in place and externally monitored.

Question 33: What Are the Results of Pilot Trials?

Pilot trials study the transition of testing from the research phase to routine care, under controlled conditions. An important characteristic of such trials is that it is possible to document the study subjects' responses to the testing process. Pilot tri-als also subject the DNA testing process to the day-to-day pressures of clinical test-ing. In addition, a pilot trial requires that adequate information concerning analytic validity, clinical validity, and clinical utility be available, in order to gain Institu-tional Review Board approval and for use in educating health-care providers and study subjects. Pilot trials provide information that cannot be collected in any other way. These include acceptance rates at various stages of the testing process, pat-terns of decision making, and economic information. If results from several pilot trials are available, an analysis across studies can provide additional information about the robustness of these estimates, and whether the results might be widely applicable.

Question 34: What Health Risks Can Be Identified for Follow-up Testing and/or Intervention?

Health risks, as well as economic costs, need to be considered when balancing the pros and cons of introducing a new DNA test. Health risks might include quantifi-able morbidity/mortality associated with subsequent procedures for diagnosis or treatment. They might also include less tangible risks, such as anxiety and labeling. Following a report that a *BRCA1/2* mutation is present in a woman, the diagnostic test might be mammography, which carries few negative health risk implications. One of the treatment options (bilateral mastectomy), however, carries more important implications, because it is associated with the well-defined risks of a major surgical procedure. Anxiety and sense of well-being also need to be considered.

Question 35: What Are the Financial Costs Associated with Testing?

A formal economic analysis of any proposed DNA test will involve listing the steps in the testing process, along with the cost estimates for each. These steps should go beyond laboratory testing, extending from patient education and informed consent to diagnostic and follow-up services. Analyses can be from several perspectives (e.g., the public, the health care system, or the patient) and may be based on actual costs or charges.

Question 36: What Are the Economic Benefits Associated with Actions Resulting from Testing?

Assessment of benefits that result from testing is usually limited to financial benefits, such as reduced health care costs. Some of the more recent analyses also examine nonfinancial benefits, such as increased longevity and quality of life. If clinical test performance varies by population subgroups, separate economic analyses should be available for policy-makers.

Question 37: What Facilities/Personnel Are Available or Easily Put in Place?

This question presupposes that policymakers have already determined that a given DNA testing process is worthwhile. As part of policy evaluation prior to making final recommendations, the facilities and personnel necessary for implementation then need to be evaluated. At the same time, available resources need to be identified and a determination made as to whether they are sufficient, or whether additional or new resources might be needed. At this stage of the process, this information is likely to be found from reports of pilot trials and can be supplemented by consultation with experts in the field. It might also be worthwhile to examine resources needed for other testing programs that might be similar to the one proposed.

Question 38: What Is the Availability of Validated Educational Materials, and Have They Been Shown to Be Effective in Achieving Understanding?

Not all written materials are effective in communicating information so that it is understood and retained. Prior to the introduction of any new clinical test, educational materials will need to be developed and validated. The ACCE process calls for verifying that these types of materials have been developed in pilot programs, or from other expert sources. Several published methods exist to evaluate both provider and patient educational materials in medical settings. The validation process usually involves providing materials to the target audience and measuring level of satisfaction and increase in knowledge. Revisions are then made, based on insights gained.

Depending on the target audience, the materials may need to be translated or the reading level adjusted. In order for a new DNA testing process to be effective, validated educational materials should be readily available. Prior to introduction, these materials should be assessed in well-designed studies with sufficient power to detect effectiveness in those with lower levels of education. At the outset, individuals in this category generally have a lower level of understanding of the principles of screening. It is impractical for each laboratory or program to attempt to develop its own materials.

Question 39: Are There Informed Consent Requirements and Is There Evidence That Those Offered the Test Make Informed Choices?

Policymakers need to be aware of existing regulations and guidelines that deal with properly informing individuals prior to testing, in either the diagnostic or screening setting. Generic regulations and guidelines are available, and there may also be guidelines specific to the DNA test under consideration. The General Medical Council in the United Kingdom has identified five critical points that individuals should understand for making an informed choice: *(1)* the purpose of screening, *(2)* the likelihood of positive and negative findings alongside the possibility of false positive and false negative findings, *(3)* the uncertainties and risks attached to the screening process, *(4)* any significant medical social, or financial implications of screening, and *(5)* follow-up plans including the availability of counseling.

Question 40: What Methods Exist for Long-Term Monitoring?

It is important to document the efficacy of any given DNA testing process, following its introduction into medical care. An ongoing estimate of analytic validity can be documented, if the test is included in an external proficiency testing program. This monitoring is especially important in the early years following introduction, a time when inexperienced laboratories will begin providing services. Components of clinical validity and utility can be collected by individual laboratories for monitoring purposes on an ongoing basis. For example, a DNA test may be used in both a screening and diagnostic setting, and a laboratory might collect and store information separately from these two settings for future analysis. Although this may seem relatively simple, some clinical laboratories do not currently collect and store this information. The ACCE process is aimed at identifying gaps of this type, so that policymakers can decide whether specific recommendations are warranted.

Question 41: What Guidelines Have Been Developed for Evaluating Program Performance?

Some DNA tests, especially if done in a screening setting, will be performed in the context of a program. A program implies centralized administration, usually managed by the laboratory. Besides laboratory testing, additional programmatic services could include education, follow-up testing, and genetic counseling. Methods for evaluating such programs have been published, and some policies have included recommendations for specific data collection. When guidelines of this type are developed, they need to consider the feasibility and cost of data collection, as well as the program characteristics to be quantified.

Ethical, Legal, and Social Implications

Question 42: What Is Known About Stigmatization, Discrimination, Health Disparities, Privacy/Confidentiality and Personal/ Family/Societal Issues?

It is important to document the social and ethical issues that might occur when DNA testing is being introduced into routine care. Pilot trials will be one source of specific information, but generic insights that have been gained over the past decade on these subjects are also likely to be applicable. In reviewing the pilot trials and other sources of information, the ACCE review might identify gaps where more focused data collection would be helpful. Policymakers may use this information for recommending testing strategies or specific safeguards that minimize the opportunity for these issues to arise.

Question 43: Are There Legal Issues Regarding Consent, Ownership of Data and/or Samples, Patents, Licensing, Proprietary Testing, Obligation to Disclose, or Reporting Requirements?

These topics are appropriate to consider for any kind of laboratory testing, and generic regulations and guidelines are available which cover such topics as consent and ownership. All of these topics, however, also need to be examined in the specific context of the DNA test being evaluated, along with the setting in which the test is being used. Expert legal review of information gathered in the ACCE process might lead to highlighting issues that could benefit from the development of new policy, or the modification of existing policy. Certain items, such as patents and licensing, can have economic implications, and proprietary testing could have an impact on access.

Question 44: What Safeguards Have Been Described and Are These Safeguards in Place and Effective?

The ACCE process is aimed at identifying important problems in the testing process. Examples of such problems could be unnecessary anxiety, incorrect or outdated information, or incorrect laboratory results. By this point in the ACCE process, safeguards to avoid these problems (such as a well-defined consent process) will have been identified, but this question offers the opportunity for a more formal review to evaluate effectiveness and identify potential gaps.

Methodology to Select Appropriate Topics for ACCE Review

The ACCE review process is sufficiently comprehensive that it is appropriate for use only in circumstances where there are important medical, social, legal or economic implications for testing. Characteristics of a DNA test that can be used for determining eligibility include, for example, whether it is presently in widespread use or whether its use is controversial. Characteristics of the associated disorder can also influence eligibility. These might include the prevalence, the seriousness, the penetrance, and the availability of potentially effective interventions or treatments. These characteristics place emphasis on determining whether, how often, and for what application the test is being used. If a currently used test is determined to be controversial as a result of inappropriate commercial promotion, for example, it can be placed in line for immediate evaluation for the purpose of avoiding or minimizing adverse consequences. This underscores the need for having an efficient, responsive system in place for all aspects of the evaluative process.

Resources Required for an ACCE Review

Many areas of expertise are needed in order to perform an ACCE review. They include clinical medicine, clinical genetics, molecular genetics, epidemiology, biostatistics, law, economics and the social sciences. While one person could have expertise in more than one of these areas, a complete ACCE review would inevitably require several individuals working together. Secretarial and library support are essential. A method for coordinating the activities of the experts needs to be in place, especially when members of this group are in separate institutions. Collection, evaluation and interpretation of the relevant literature can be time consuming, including the time necessary for consensus to be reached on both the approach to presenting the information and on conclusions. Both conference calls and face-to-face meetings can help facilitate this process. After a given ACCE review has reached final draft form, it is necessary to have a larger group of experts in various disciplines review the document for completeness, accuracy and acceptability. New findings that are identified during this process may be reported separately in the peer-reviewed literature, even before the entire document has been completed. It is especially important to have this flexibility, in that individual findings may have the potential to improve medical practice, in and of themselves. The final document needs to be available both in hardcopy and electronic form. A genetic test brief should also be prepared and made available with links to the final report. The aim of this brief is to provide reliable summary data for policymakers and public health administrators. This format is also more understandable to the public than detailed reports. It is unlikely that an ACCE review will be published in its entirety in a peer-reviewed journal, given the size and wide range of topics. An important component of the process will be to periodically identify areas that require updating and ensure that such updates are distributed appropriately.

Acknowledgments
This work was funded by a cooperative agreement with the U.S. Department of Health and Human Services, Centers for Disease Control and Prevention, Office of Genomics and Disease Prevention (UR3/CCU119356); the ACCE project. We thank Linda A. Bradley, Ph.D., Clinigene Laboratories, Hauppauge, New York and Carolyn Sue Richards, Ph.D., Baylor College of Medicine, Houston, Texas, for their review and comments. These two individuals are also members of the Core Group for the ACCE project. We also thank Theresa Marteau, Ph.D., Kings College, London, for her helpful comments about information and communication.

References

1. Wald N, Cuckle H. Reporting the assessment of screening and diagnostic tests. Br J Obstet Gynaecol 1989;96:389–396.
2. Enhancing the Oversight of Genetic Tests: Recommendations of the SACGT, http://www4.od.nih.gov/oba/sacgt/gtdocuments.html
3. Hanson EH, Imperatore G, Burke W. HFE gene and hereditary hemochromatosis. Am J Epidem 2001;154:193–206.
4. Preconception and prenatal carrier screening for cystic fibrosis: clinical and laboratory guidelines. American College of Obstetricians and Gynecologists, American College of Medical Genetics, 2001, Washington, DC
5. Grody WW, Cutting GR, Klinger KW, Richards CS, Watson MS, Desnick RJ. Laboratory standards and guidelines for population-based cystic fibrosis carrier screening. Genet Med 2001;3:149–154.

14

The interface between epidemiology and pharmacogenomics

David L. Veenstra

One of the goals of genomic medicine is to individualize drug therapy based on patient-level genetic information, an application referred to as pharmacogenomics. Genetic variation can affect both the safety and efficacy of drugs when mutations occur in proteins that are drug targets (e.g., receptors) that are involved in drug transport mechanisms (e.g., ion channels), or that are drug-metabolizing enzymes (e.g., cytochrome P450). Based on terminology from pharmaceutics, pharmacogenomic therapies can be broadly categorized into (*1*) those based on variation in drug targets (pharmacodynamic strategies) and (*2*) those based on variation in drug metabolism enzymes (pharmacokinetic strategies).

Pharmacogenomics, more traditionally referred to as pharmacogenetics, has a relatively long history. Variable drug response was first noted in the 1940s and 1950s in response to drugs such as isoniazid, succinlcholine, and primaquine (1). The causes of these variations were later found to be genetic variations in the enzymes N-acetyl transferase, pseudocholinesterase, and glucose-6-phosphate dehydrogenase, respectively. Based on these initial observations, Arno Motulsky proposed in 1957 that the inheritance of acquired traits could explain individual differences in drug efficacy and adverse drug reactions (ADRs) (2). Genetic variation in a major drug metabolizing enzyme, cytochrome P450 *CYP2D6*, was identified based on a patient's severe hypotensive response to debrisoquin (3). Family studies showed that the metabolism of debrisoquin is under monogenic control, and the poor metabolizers were homozygous for a recessive allele of *CYP2D6*. Over the past several decades, additional polymorphisms have been identified and characterized in many of the drug-metabolizing enzymes (4). This work serves as a foundation for the broader field of pharmacogenomics, which has rapidly expanded to include drug response (efficacy) in addition to drug toxicity, and the identification of novel polymorphisms via single nucleotide polymorphism (SNP) discovery programs (5).

Although research in this field has spanned several decades, pharmacogenomic tests are only just beginning to be used in clinical practice. Despite this slow start, pharmacogenomic testing will soon become widely available for the general population. For example, home-testing for cytochrome P450 genotype is now available on the internet (6). Thus, a better understanding of the implications of genetic testing to guide drug therapy is imperative.

Pharmacokinetic Strategies

Many of the first applications of pharmacogenomics will likely be in the pharmacokinetic area because of the extensive basic research conducted in this area. The anticoagulant warfarin and the antileukemic agent 6-mercaptopurine (6-MP) are two of the most commonly presented examples of pharmacokinetic-based testing in the literature. Warfarin exhibits large variability in drug response, primarily because of disease, diet, and drug interactions. However, part of the variability has been attributed to polymorphisms of the enzyme that metabolizes warfarin, the cytochrome P450 enzyme CYP2C9. Individuals deficient in CYP2C9 activity may be at higher risk for severe bleeding episodes and require lower starting doses or more frequent monitoring (7). Genetic information may thus assist clinicians in initiating and monitoring warfarin dosing.

Polymorphisms of the enzyme thiopurine S-methyltransferase (TPMT) play an important role in the metabolism of 6-MP, which is used for treatment of acute lymphoblastic leukemia (ALL) in children. TPMT is responsible for the inactivation of 6-MP, and TPMT deficiency is associated with severe hematopoietic toxicity when deficient patients are treated with standard doses of 6-MP (8). In general, drugs with a narrow therapeutic index (i.e., high toxicity) are good candidates for evaluation of variations in metabolism; thus, the majority of applications of pharmacokinetic strategies will likely be found in oncology (9).

Pharmacodynamic Strategies

A recent example of pharmacodynamic testing in the literature is a study that found an association between a variant allele of the enzyme CETP and clinical response to pravastatin (10). Interestingly, drug response as measured by coronary vessel intraluminal diameter was correlated with CETP genotype but not with lipid levels. This study has significant implications for the management of tens of thousands of patients in the United States currently receiving statin therapy, though additional studies are needed to confirm the findings.

The use of genetic testing to identify viral or tumor cell genotype represents the most widespread application of pharmacodynamic strategies. Although if is the viral or tumor genome that is tested and not the patient's genome, many of the clinical and economic implications are similar. For example, treatment of hepatitis C with interferon and ribivirin (combination therapy) can be guided by viral genotype.

Patients with the more virulent viral genotype (genotype 1) respond significantly better to 48 weeks of treatment than to 24 weeks of treatment, while patients with non-genotype 1 respond similarly to 24 or 48 weeks of therapy (11).

The evaluation of viral genotype is also used to individualize human immunodeficiency virus (HIV) treatment cocktails. The Community Programs for Clinical Research on AIDS (CPCRA) trial 046 showed that among patients who failed highly active anti-retroviral therapy (HAART), viral suppression with a subsequent regimen was significantly more common for patients randomized to receive genetic testing (34%) than for patients treated based on clinical judgment alone (22%) (12). The Antiretroviral Adaptation (VIRADAPT) trial reported similar results (13). It is likely that genetic testing of bacterial genomes also will become a routine part of medical care.

The evaluation of tumor cell lines is often used for guiding chemotherapy in oncology. One of the most widely used genetic tests evaluates the amplification of the proto-oncogene N-Myc, which is present in about one thrid of children with neuroblastoma. Children without amplification have survival rates of 80% to 90% with conventional chemotherapy, while most patients with amplification die. However, with high-dose chemotherapy and bone marrow transplantation, children with N-Myc amplification have survival rates of almost 30% (9). A more recent example is the growth-factor-receptor gene *HER-2/neu*, which is overexpressed in approximately 25% of metastatic breast tumors. Patients with tumors that overexpress HER-2/neu are eligible for treatment with the targeted monoclonal antibody drug Herceptin (9).

Why have some applications found their way into clinical practice relatively quickly, while other potential applications that have been known for years have not yet been developed? The decision to implement pharmacogenomic tests has been driven by an implicit combination of clinical, epidemiologic, and economic factors. To ensure that pharmacogenomic tests are developed and applied in a clinically and economically reasonable fashion, explicit and quantitative evaluations are needed.

Evaluating Pharmacogenomic Tests

Veenstra and colleagues developed a set of criteria, based on a formal cost-effectiveness framework, for evaluating the potential clinical and economic benefits of pharmacogenomic technologies (Table 14.1) (14). The key aspects of these criteria are highlighted in the following paragraphs, and a more in-depth evaluation of the epidemiologic considerations in bringing pharmacogenomics to clinical practice is presented.

Formal cost-effectiveness analysis provides a quantitative framework for evaluating the complex and conflicting factors involved in the evaluation of health care technologies. Most importantly, it helps ensure that all costs, clinical benefits, and patient outcomes have been properly evaluated. The U.S. Panel on Cost-Effectiveness in Healthcare has provided general recommendations for performing such

Table 14.1 Framework for Evaluating the Potential Cost-Effectiveness of
Pharmacogenomic Therapies

Factors	Characteristics Favoring Cost-Effectiveness
Severity of outcome avoided	Severe outcome—high mortality, significant impact on quality of life, or expensive medical care costs
Drug monitoring	Monitoring of drug response currently not practiced or difficult
Genotype–phenotype association	Strong association between gene variant and clinically relevant outcomes
Assay	A rapid and relatively inexpensive assay is available
Polymorphism	Variant allele frequency is relatively high

Source: Reproduced from Veenstra et al. (14).

studies (15), and countries outside the United States such as the United Kingdom, Australia, and Canada have formal requirements for submitting economic evaluations to government agencies (16). Recently, managed care organizations in the United States have begun adopting similar policies (17).

Several types of economic evaluation used in health care include: cost minimization, cost-consequence analysis, cost-benefit analysis, cost-effectiveness analysis, and cost-utility analysis (see also Chapter 17). These methods vary primarily in the way they measure health outcomes, e.g., in monetary terms, medical events, or life expectancy adjusted for a patient's quality of life. Cost-utility analysis has been more accepted in health care than have other types of economic evaluation because it measures benefit in patient-oriented terms (quality of life) and permits comparison between different interventions by standardizing the outcome measure. Once the economic, clinical, and patient outcomes have been valued, the novel technology is compared to current medical practice in an incremental analysis. The incremental cost-effectiveness ratio (ICER) is defined as

$$\text{ICER} = \frac{(C_2 - C_1)}{(E_2 - E_1)}$$

where C_2 and E_2 are the cost and effectiveness of the new intervention being evaluated and C_1 and E_1 are the cost and effectiveness of the standard therapy.

Medical interventions are considered to be cost effective when they produce health benefits at a cost comparable to that of other commonly accepted treatments. A general guide is that interventions that produce 1 quality-adjusted life-year (QALY), equivalent to 1 year of perfect health, for under $50,000 are considered cost-effective, those between $50,000 and $100,000 per QALY are of questionable cost-effectiveness, and those above $100,000 per QALY are not considered cost effective (18).

An example of the use of cost-effectiveness analysis to evaluated pharmacogenomic testing is the study conducted by Weinstein and colleagues, who evaluated primary and secondary HIV resistance testing using cost-effectiveness and decision analysis techniques to model life expectancy, quality of life, and lifetime costs

(19). The authors estimated that secondary resistance testing was cost-effective compared with no testing ($17,900 per QALY). The cost effectiveness of primary testing was highly dependent on the prevalence of resistance in a treatment-naive population: $22,300 per QALY at 20% prevalence and $69,000 per QALY at 4% prevalence. The results support current guidelines for the use of secondary testing, but additional studies are needed to evaluate the effectiveness of primary testing and elucidate the prevalence of HIV resistance in different settings.

From Cost to Cost-Effectiveness

Testing Costs
In addition to the direct cost of a genetic assay, there are induced costs, including direct costs such as additional medical care follow-up and indirect costs such as patient time away from work. These costs are potentially of greater magnitude than the direct cost of purchasing the test. Another consideration should be the impact of knowledge of genetic status on the behavior of patients. For example, patients with genetic variants may exhibit adverse behavior such as avoiding all drug therapies. In contrast, patients with no major genetic variants may adopt a careless attitude with regard to drug compliance and consumption.

Fortunately, these induced costs will be offset to a certain extent. One of the benefits of pharmacogenomic testing is that the information can be used throughout the lifetime of the patient. For example, rather than measuring serum drug levels to infer the metabolic capability of a patient every time a novel drug is introduced to the patient, a genetic test to identify variations in all drug metabolizing enzymes could be used throughout the lifetime of the patient for a variety of medications.

Cost Offsets
Pharmacokinetic strategies will offer one primary benefit that offsets costs: the prevention ADRs. ADRs have been shown to lead to significant patient morbitidy, and mortality, and excess health care costs. Pharmacodynamic strategies, on the other hand, will provide the greatest economic benefit when it can be used to avoid long-term therapy with an ineffective drug or acute treatment with an expensive drug.

Comparator Treatment Strategies
The *incremental* cost effectiveness of pharmacogenomics will depend on the current state of technology for monitoring patients for toxic effects or drug response and individualize their therapy accordingly. Plasma drug levels are often used to monitor toxic drugs, while surrogate markers such as blood pressure for hypertension, lipid levels for hypercholesteremia, and HbA1c for diabetes are used to measure drug response for chronic diseases. When there are readily available, inexpensive, and validated means of monitoring drug response, pharmacogenomics may offer little incremental benefit.

From Efficacy to Effectiveness

Effectiveness of Genetic Tests

The "efficacy" of a genetic test can be distinguished from its "effectiveness." The efficacy of a test can be viewed as the diagnostic ability of the assay—that is, the ability of the test to accurately detect the genetic variation it was designed to identify. Because genetic tests have high sensitivity and specificity ($>90\%$) when direct sequencing or restriction site assays are used, they are often viewed as being highly accurate. From a cost-effectiveness or clinical perspective, however, it is the prognostic significance of the test result that is important (its effectiveness).

The prognostic significance of a test is determined by the degree of association between the identified genetic variation and its physical manifestation (phenotype). The association between genotype and phenotype (gene penetrance) will drive both clinical and economic outcomes. For example, if half of all patients with a gene variant experience a severe side effect from a drug (gene penetrance of 50%), avoiding the use of that drug in all patients with the variant would unnecessarily deprive the other half of the patients (the "false positives") of medication. The issue of "false positives" will be important for almost all applications of pharmacogenomics, and the consequence of labeling patients as having a genetic variation despite the fact that not all of them will have clinically relevant effects must be considered. The phenotype that has been associated with a genetic variant will be as critical as outcomes measured in a randomized clinical trial. In pharmacogenetics, "phenotype" has often referred to drug plasma levels. Abnormal drug levels are not necessarily associated with adverse drug reactions, however.

Clinical Benefits

The potential clinical benefits of individualized drug therapy have driven much of the excitement about pharmacogenomics. Despite the inherent appeal of pharmacogenoimcs, however, the value of its benefits must be carefully considered. As previously stated, avoiding ADRs will be the most direct benefit of pharmacokinetic strategies. A recent analysis by Phillips and colleagues (4) suggests that genetic variations in drug metabolizing enzymes are associated with serious ADRs. The authors found that drugs associated with serious ADRs were significantly more likely to be metabolized by enzymes with genetic variations than were drugs that were not associated with serious ADRs (59% vs. 22%, $p = 0.006$).

Testing costs for pharmacodynamic strategies, on the other hand, will be offset by either (*1*) avoiding unnecessary drug expenditures for patients who are unlikely to respond or (*2*) improving the *effectiveness* of drug treatment strategies.

Patient Quality of Life

As mentioned previously, patients may have dramatically different behavioral reactions to the results of genetic tests, and these changes could lead to important changes in clinical and economic outcomes. Additionally, the impact of knowledge of genetic status on a patient's quality of life must be considered. The effects on a patient's emotional, physical, and social functioning will vary depending on the type of test conducted; most likely, pharmacogenomic testing will lead to less impairment than testing for disease risk. Although initial work in the area of breast cancer (20) suggests that patients benefit from knowledge of their genetic status, there is a critical need for additional studies in this area.

Gene Prevalence

Finally, and perhaps most importantly, the cost effectiveness of pharmacogenomics will be highly dependent on the underlying prevalence of disease—in this case, the frequency of the variant allele (or prevalence of the variant genotype) in the population being tested. For example, if the frequency of a variant allele is 0.5%, on average only one patient with a variant allele would be detected for every 200 patients tested.

Clearly, epidemiologic factors play a central role in determining the usefulness of pharmacogenomic tests. Key components of this role are the degree of gene-"outcome" association, the prevalence of the variant gene(s), and the validity of surrogate markers.

Epidemiologic Considerations

Gene Variant Prevalence

The importance of gene variant prevalence was highlighted by Veenstra and colleagues (14) in an evaluation of *TPMT* genotyping for children with ALL receiving

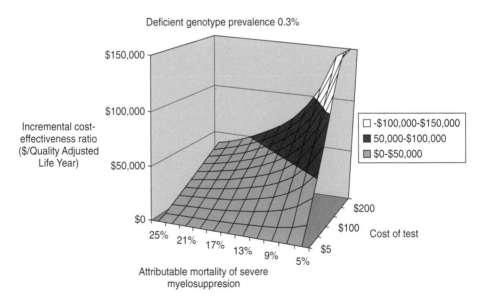

Figure 14.1 Influence of cost of genetic test and severity of clinical outcome on the hypothetical cost-effectiveness of TPMT genotyping with a deficient genotype prevalence of 0.3%. Reproduced from Veenstra DL, Higashi MK, Phillips KA. Assessing the cost-effectiveness of pharmacogenomics. Pharmaceutical-Sciences 2000 (Sept. 14);2(3): Article 29.

6-MP therapy (Figs. 14.1 and 14.2). As can be seen, what might be considered a fairly small variation in genotype prevalence (0.3% vs. 1.0%) leads to a dramatic difference in the cost effectiveness of testing.

What's more, gene variant prevalence will drive drug development decisions within the pharmaceutical and biotechnology industries. Targeting rare genotypes will not be financially viable if companies are not able to recoup investments by charging higher prices for these targeted drugs. Danzon and Towse recently presented a formal evaluation of this issue (22). Their analysis suggests that manufacturers and health-care payers will need "to use economic evaluation to identify the higher value associated with such targeting."

In a recent systematic review of known genetic variants of drug metabolizing enzymes, Phillips and colleagues (4) found incomplete and sometimes inconsistent data on genotype prevalence and variant allele frequency (Table 14.2). Most of the estimates were derived from limited, clinic-based patient populations. Few population-based estimates were available for the U.S. population, and even fewer were available for racial and ethnic subgroups. The recent excitement over collaborative efforts such as the SNP Consortium (23) is warranted in that thousand of genetic variants are being identified. However, these data are based on DNA samples from just 24 unrelated individuals. For researchers to make use of this information in a clinically meaningful and financially sound fashion, population-based prevalence studies will be needed.

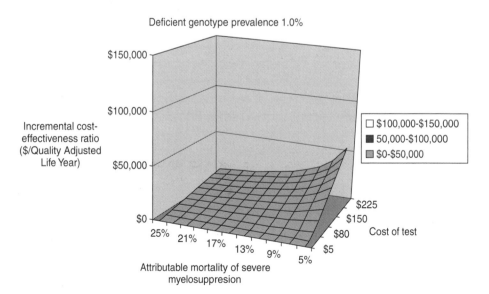

Figure 14.2 Influence of cost of genetic test and severity of clinical outcome on the hypothetical cost effectiveness of TPMT genotyping with a deficient genotype prevalence of 1.0%. Reproduced from Veenstra DL, Higashi MK, Phillips KA. Assessing the cost-effectiveness of pharmacogenomics. Pharmaceutical-Sciences 2000 (Sept. 14);2(3): Article 29.

Table 14.2 Variant Alleles of Enzymes that Metabolize Drugs Associated with Serious Adverse Drug Reactions

Drug-Metabolizing Enzymes	Variant Alleles	Prevalence of Poor Metabolizers
CYP1A2	*1C	12% Caucasian
CYP2C9	*2, *3	2%–6% Caucasian
CYP2C18	*3	No data available
CYP2C19	*2A, *3, *4, *2B, 5A, 5B,6,7,9	2%–6% Caucasian, 15%–17% Chinese, 18%–23% Japanese
CYP2D6	2A, *3A, *3B, *4A,B, *5, *6A, *7, *8, *9, *10, *10A,B, *11, *12, *17, *36, *4C,D,K,4X2,6B,6C *13,14,15,16,18,19,20,38	3%–10% Caucasian, <2% Chinese; Japanese; African American
CYP2E1	*2	No data available
UGT2	B7	No data available
NAT2	*5A, *5b, *5C, *6A, *7A, *7B, *13, *14A, *14B	50%–59% Caucasian, 41% African-American, 20% Chinese, 8%–10% Japanese, 92% Egyptian

Source: Adapted from Phillips et al. (4).

The relationship between genotype prevalence and ethnicity, particularly for drug-metabolizing enzymes, will present some significant challenges for the equitable delivery of health care. For instance, will clinicians offer genetic testing based on a patient's race or ethnicity? Will a patient's race/ethnicity or genotype be incorporated into treatment guidelines or drug formulary development? Finally, patients of different racial/ethnic backgrounds may agree to participate in genetic testing and respond to their results differentially.

Association Studies

Only a few studies have evaluated the association between genes and drug-related clinically relevant outcomes (7,8,10,24–28). In contrast, readily accepted associations such as that between *CYP2C9* genotype and the risk of bleeding from warfarin exposure need to be carefully examined. In a recent and oft-cited study, Aithal and colleagues (29) conducted a retrospective case-control study in patients receiving chronic warfarin therapy. The authors found a significant association between *CYP2C9* genotype and warfarin low-dose requirement. This finding suggests that genotyping patients before initiating warfarin therapy may help avoid overdosing susceptible patients, but perhaps clinicians, who closely monitor warfarin treatment and response, are able to adjust dosing quickly enough to avoid any adverse outcome. Even though an association between low-dose warfarin patients and a higher risk of bleeding was reported, the critical link between genotype and the clinical outcome of greatest interest—bleeding—was not established. In a subsequent study, Taube and colleagues (30) reported a similar association between genotype and dose requirement, but they did not find an association between genotype and International Normalized Ratio (INR, which is a measure of anticoagulation status). Thus, clinicians were left wondering whether genotyping of warfarin patients is a useful clinical tool.

In a recently completed retrospective cohort study of patients in a U.S.-based warfarin anticoagulation clinic, Higashi et al (7) found that *CYP2C9* genotype was significantly associated with warfarin dose, time to stabilization, and serious or life-threatening bleeding events (Figs. 14.3 and 14.4). This study provides the first critical evidence that warfarin patients with variant *CYP2C9* genotype are at higher risk for clinically (and economically) relevant outcomes. Subsequent studies are needed to determine whether clinicians are able to utilize genotype information to improve patient management and outcomes.

Large population-based association studies, (e.g., Iceland study by deCODE Genetics [31]), may be able to address some of these concerns. However, selection bias caused by depletions of susceptibles will likely be a significant problem in any retrospective study design.

Surrogate Markers
Preliminary association studies designed to test initial hypotheses about drug–gene interactions often used limited patient populations and relied on surrogate markers.

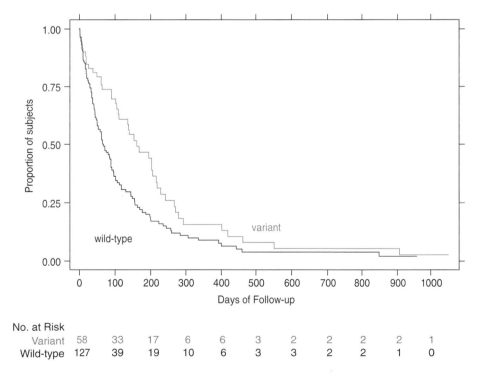

No. at Risk										
Variant	58	33	17	6	6	3	2	2	2	2	1
Wild-type	127	39	19	10	6	3	3	2	2	1	0

Figure 14.3 Kaplan-Meier curves of time to stable dosing. Log-rank test for equality of survivor functions, $X^2 = 8.30$, 1 d.f., $P = 0.004$. Higashi MK, Veenstra DL, Kondo ML, Wittkowsky AK, Srinouanprachanh SL, Farin F, Rettie A. Association between CYP2C9 genetic variants and anticoagulation-related outcomes during warfarin therapy. JAMA 2002;287: 1690–1698.

However, many surrogate markers explain only a portion of drug response, and correlations with clinical outcomes can vary across drug classes (32). As noted previously, pharmacokinetic strategies have traditionally relied on serum drug levels. Current studies have progressed toward surrogate markers, such as intraluminal coronary artery diameter with *CETP* genotyping and anticoagulation status (INR) with warfarin therapy. The risks of relying on surrogate markers are highlighted by both of these examples. In the case of *CETP* genotyping for patients receiving pravastatin, no association would have been observed had the investigators relied on lipid levels as an outcome measure (10). In contrast, the study provides the important suggestion that statin therapy in patients with the B2B2 genotype may lower lipid levels but not reduce atherosclerosis. The study of warfarin patients by Taube and colleagues (30) failed to find an association between INR variability and *CYP2C9* genotype, in contrast to the findings by Higashi et al (7). Thus, two well-validated surrogate markers, lipid levels and INR, were not associated with genotype as might be presumed *a priori*. As medicine moves into the genomics era, pharmacogenomics offers the opportunity to move beyond traditional clinical markers of drug response

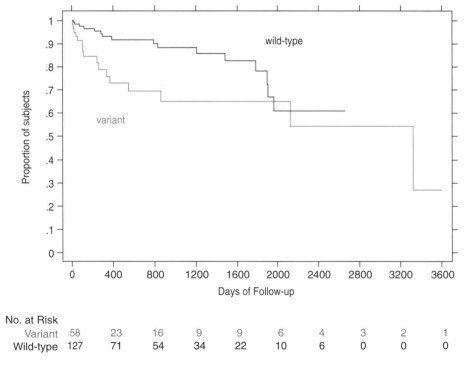

Figure 14.4 Kaplan-Meier curves of time to serious or life-threatening bleeding events. Logrank test for equality of survivor functions, $X^2 = 6.21$, 1 d.f., p = 0.013. Higashi MK, Veenstra DL, Kondo ML, Wittkowsky AK, Srinouanprachanh SL, Farin F, Rettie A. Association between CYP2C9 genetic variants and anticoagulation-related outcomes during warfarin therapy. JAMA 2002;287:1690–1698.

and identify gene and gene expression profiles (33) that are more closely correlated with ultimate clinical outcomes.

Future Studies

The evaluation of pharmacogenomic tests is a complicated, challenging process. The foundations of this process are well-conducted epidemiologic studies designed to evaluate the prevalence of genetic variants in the population and associate them with clinically relevant outcomes. Other important studies can build upon such epidemiologic findings. Randomized controlled trials should be conducted to evaluate the effectiveness of clinicians in responding to genomic information. Economic studies are needed to ensure that health care resources are being used wisely. Legal issues need to be investigated to ensure that patients are treated equitably. Finally, more information should be gathered to evaluate the impact of genetic testing on patients' management of their health and on their quality of life. But until more quality epidemiologic studies can be conducted, it will be difficult to move forward in an informed fashion.

References

1. Mancinelli L, Cronin M, Sadee W, Pharmacogenomics: The Promise of Personalized Medicine. Pharmaceutical-Sciences 2000 (Sept. 14);2(3): article 4 Accessed March 7, 2002, from (http://www.pharmsci.org/scientificjournals/pharmsci/journal/4.html).
2. Motulsky AG. Drug reactions, enzymes and biochemical genetics. JAMA 1957;165: 835–837.
3. Eichelbaum M, Steincke B, Dengler JJ. Defective N-oxidation of sparteine in man: a new pharmacogenetic defect. Eur J Clin Pharmacol 1977;16:183–187.
4. Phillips K, Veenstra DL, Oren E, et al. The potential role of pharmacogenomics in preventing adverse drug reactions: a systematic review. JAMA 2001;286:2270–2279.
5. Evans WE, Relling MV. Pharmacogenomics: translating functional genomics into rational therapeutics. Science 1999;286:487–491.
6. Genelex Acessed March 7, 2002, from http://genelex.com/.
7. Higashi MK, Veenstra DL, Kondo ML, Wittkowsky AK, Srinouanprachanh SL, Farin F, Rettie A. Association between CYP2C9 genetic variants and anticoagulation-related outcomes during warfarin therapy. JAMA 2002;287:1690–1698.
8. Evans WE, Horner M, Chu YQ, Kalwinsky D, Roberts WM. Altered mercaptopurine metabolism, toxic effects, and dosage requirement in a thiopurine methyltransferase-deficient child with acute lymphocytic leukemia. J Pediatr 1991;119:985–989.
9. Flowers C, Veenstra DL. Will pharmacogenomics in oncology be cost-effective? Oncology Economics 2000;1:26–33.
10. Kuivenhoven JA, Jukema JW, Zwinderman AH, et al. The role of a common variant of the cholesteryl ester transfer protein gene in the progression of coronary atherosclerosis. N Engl J Med 1998;338:86–93.
11. McHutchinson JG, Gordon SC, Schiff ER, Shiffman ML, Lee WM, Rustigi VK, Goodman ZD, Ling M, Cort S, Albrecht JK. Interferon Alfa-2b Alone or in Combination with Ribavirin as Initial Treatment for Chronic Hepatitis C. N Engl J Med 1998; 339 (21): 1485–1492.
12. Baxter JD, Mayers DL, Wentworth DN, et al. A randomized study of antiretroviral management based on plasma genotypic antiretroviral resistance testing in patients failing

therapy. CPCRA 046 Study Team for the Terry Beirn Community Programs for Clinical Research on AIDS. AIDS 2000;14:F83–F93.

13. Durant J, Clevenbergh P, Halfon P, et al. Drug-resistance genotyping in HIV-1 therapy: the VIRADAPT randomised controlled trial. Lancet 1999;353:2195–2199.

14. Veenstra DL, Higashi MK, Phillips KA. Assessing the cost-effectiveness of pharma-cogenomics. Pharmaceutical-Sciences 2000 (Sept. 14);2(3): Article 29. (http://www.pharmsci.org/scientificjournals/pharmsci/journal/veenstra/manuscript.htm). Accessed 3/7/02.

15. Weinstein MC, Siegel JE, Gold MR, et al. Recommendations of the Panel on Cost-effectiveness in Health and Medicine. JAMA. 1996;276:1253–1258.

16. Drummond M, Dubois D, Garattini L, et al. Current trends in the use of pharmacoeco-nomics and outcomes research in Europe. Value in Health 1999;2:323–332.

17. Sullivan SD, Lyles A, Luce B, et al. AMCP guidance for submission of clinical and eco-nomic data to support formulary listing in U.S. health plans and pharmacy benefits man-agement organizations. J Manag Care Pharm 2001;7:272–282.

18. Garber AM, Phelps CE. Economic foundations of cost-effectiveness analysis. J Health Econ 1997;1–31.

19. Weinstein MC, Goldie SJ, Losina H, et al (2001): Use of genotypic resistance testing to guide HIV therapy: Clinical impact and cost-effectiveness. Ann Intern Med 134:440–450.

20. Lerman C, Narod S, Schulman K, et al. BRCA1 testing in families with hereditary breast-ovarian cancer. JAMA 1996;275:1885–1892.

21. Lindpainter K. Genetics in drug discovery and development: challenge and promise of individualizing treatment in common complex diseases. Br Med Bull 1999;55:471–491.

22. Danzon P and Towse A. The economics of gene therapy and of pharmacogenetics. Value in Health 2002;5:5–13.

23. The SNP Consortium. Accessed March 7, 2002, from http://snp.well.ox.ac.uk/.

24. Israel E, Drazen JM, Liggett SB, et al. Effect of polymorphism of the B2-adrenergic re-ceptor on response to regular use of albuterol in asthma. Int Arch Allergy Immunol 2001;124:183–186.

25. Tanis BC, van den Bosch MA, Kemmeren JM, Cats VM, Helmerhorst FM, Algra A, van der Graaf Y, Rosendaal FR. Oral contraceptives and the risk of myocardial infarction. N Engl J Med 2001;Dec 20;345(25):1787–1793.

26. Dishy V, Sofowora GG, Xie HG, Kim RB, Byrne DW, Stein CM, Wood AJ. The effect of common polymorphisms of the beta2-adrenergic receptor on agonist-mediated vascu-lar desensitization. N Engl J Med 2001;Oct 4;345(14):1030–1035.

27. Richard F, Helbecque N, Neuman E, Guez D, Levy R, Amouyel P. APOE genotyping and response to drug treatment in Alzheimer's disease. Lancet 1997;349:539.

28. Farlow MR, Lahiri DK, Poirier J, Davignon J, Hui S. Apolipoprotein E genotype and gender influence response to tacrine therapy. Annals NY Acad Sci 1996;802:101–110.

29. Aithal GP, Day CP, Kesteven PJ, Daly AK. Association of polymorphisms in the cy-tochrome P450 CYP2C9 with warfarin dose requirement and risk of bleeding complica-tions. Lancet 1999. 353:717–719.

30. Taube J et al. Influence of cytochrome P-450 CYP2C9 polymorphisms on warfarin sen-sitivity and risk of over-anticoagulation in patients on long-term treatment. Blood 2000;96:1816–1819.

31. deCODE Genetics. Accessed 3/7/02 from http://www.decode.com/company/.

32. Temple R. Are surrogate markers adequate to assess cardiovascular disease drugs? JAMA. 1999;282:790–795.

33. van't Veer LJ, Dai H, van de Vijver MJ, et al. Gene expression profiling predicts clini-cal outcome of breast cancer. Nature 2002;415:530–536.

15

Integrating genetics into randomized controlled trials

John P.A. Ioannidis and Joseph Lau

The new era of human genetics poses significant challenges to randomized controlled trials (RCTs) for exploiting genetic information in the evaluation of preventive and therapeutic interventions. Genetic information may have two different, potentially complementary uses. First, a genetic parameter may be a predictor of outcomes, such as disease susceptibility, disease progression, target organ disease or toxicity. Second, genetic parameters may modify the postulated positive or negative effects of preventive or therapeutic interventions. A better predictive ability and a more detailed understanding of effect modification both lead to increasing our chances for rationally individualizing treatment (1,2) to optimize health outcomes.

Individualized treatment is a goal that has been difficult to attain or even approach until recently. RCTs have typically been designed, conducted, and analyzed with the view of obtaining average answers for the average patient with a given condition. Subgroup analyses (3) within the population of an RCT and predictive modeling (4) have been viewed with skepticism—perhaps justifiably given the high probability of type I error. Multiple comparisons between different subgroups, e.g. those defined by genotype results, may yield different results simply by chance. On the other hand, stringently defined genetic subgroups may have too few subjects to be adequately powered in a clinical trial to show the efficacy of interventions, let alone the differential efficacy in different subgroups. Finally, even simple predictive modeling is far from "simple" in any population, including the population of patients enrolled in a clinical trial.

Genetic information may change the design and conduct of clinical trials, and in their turn, clinical trials of genetic tests themselves may also change our appreciation about the importance and uses of this information. With the explosive development of human genetics, a challenge of clinical trials would be to evaluate whether the availability and use of genetic information is actually effective and improves outcomes in clinical practice. Trials aiming to study such effects face several challenges that are worthwhile understanding and overcoming, whenever possible.

The integration of genetic information in RCTs has potentially important implications for the practice of medicine. Guideline development for clinical practice is increasingly based on RCTs. Genetic issues should be given due weight not only when designing and analyzing RCTs, but also when the evidence from RCTs is being critically appraised for guideline development and clinical practice.

In this chapter, we shall try to examine the advent of various uses of genetic information in RCTs and the unique opportunities as well as problems that are generated in this very challenging interface. The chapter is divided in two parts. The first part deals with general issues of genetic factors in the design and evaluation of RCTs, and the second part deals with the application of RCTs to the evaluation of genetic tests.

Genetic Factors in the Design and Evaluation of Randomized Clinical Trials

Integrating Genetics in the Design of Randomized Clinical Trials

Selection of Study Populations and Eligibility Criteria. Genetic parameters may allow a more accurate definition of various diseases that have a strong genetic component. They may be helpful in defining disease conditions where genetic factors or gene–environment interactions may be equally important or even more important than environmental and acquired parameters. Accurate definition of a condition is essential and may improve the efficiency of a study design when the intervention is likely to work only in patients with that condition. While disease conditions in the past have been defined mostly on the basis of phenotype, definitions based on detailed genotyping information are likely to become more common. The selection of the genetic group to be targeted may be made based on prior evidence from predictive risk models or from biologic rationale. Since the validity of genetic risk factors is often not fully studied and the underlying biology may be poorly understood, one should be prepared for surprises. For example, the PARIS study (5) was designed to include only patients who have the DD genotype for the angiotensin-I converting enzyme deletion/insertion (D/I) allele. The investigators had background evidence that this genotype may be a risk factor for restenosis after coronary stent implantation. Thus they targeted the DD high-risk group in order to assess whether quinapril, a drug that blocked angiotensin-I converting enzyme, would reduce the risk of angiographic restenosis. Coronary stenting was performed in 345 consecutive patients and genotyping showed that 115 of them had the DD genotype. Ninety-one of them were randomized to quinapril versus placebo. Paradoxically, the trial showed that quinapril was associated with a significantly greater reduction in lumen diameter in this selected population.

Stratification and A Priori Adjusted Analyses of the Collected Data. An alternative approach to the one presented above would be to avoid limiting enrollment to subjects with specific genotypes but design a trial with explicit a priori stratification according to genotype. Or, specify a priori that the final analyses will adjust for genotype and the comparison of the treatment effects between subgroups with different genotypes will be a key endpoint of the trial. Stratification has been a contentious and controversial issue in RCTs (6). In theory, randomization in very large RCTs should help obviate concerns about an imbalance between the compared groups. Nevertheless, even if overall imbalance is avoided, it is still conceivable that the absolute magnitude of the treatment effect may be substantially different in high versus low risk subgroups (7), especially if these subgroups differ markedly in their baseline level of risk. Strong risk factors would thus need to be considered for stratification. Imbalance is more likely to occur in smaller RCTs, but in this case stratification during the randomization phase may be more cumbersome. The use of adjusted analyses and subgroup analyses may overcome the lack of stratification by evaluating differences between subgroups at the analysis stage.

With the exception of clear-cut monogenetic diseases where one gene is directly responsible for the disease outcome, most proposed associations between genetic polymorphisms and diseases to date are represented by relatively small effect sizes (relative risks, 0.2 to 5). Such risk factors may be less influential in creating imbalance during randomization or differentiating the baseline risk. Nevertheless, it may be useful to account for them upfront in the study design, when possible. If the genetic risk factors are known to be "strong" ones (i.e., they have large risk ratios associated with them and are not uncommon), at a minimum they should be included in the description of the baseline characteristics, to show that the compared arms are similar in that regard.

There is no absolute rule on what constitutes a "strong" risk factor at the population level. The attributable fraction is one potential approach to quantify the overall importance of a risk factor at the population level. Attributable fraction is given by $PR(RR - 1)/[1 + PR(RR - 1)]$, where PR is the prevalence of the risk factor and RR is the risk ratio associated with it. This means that for a prevalence of 40% and a risk ratio of 1.5, the attributable fraction is 0.167, suggesting that 16.7% of the disease or outcome of interest may be attributed to the risk factor. One might suggest that it may be unnecessary to take seriously into account genetic risk factors that have an attributable fraction associated with them of less than 0.05, while those with higher attributable fractions may require more attention. Other more formal approaches have also been proposed to model at the design stage the expected variability of risk of the population targeted by an RCT based on the prevalence and risk ratios of known risk factors (8).

Clinical trialists may often routinely stratify for parameters that are unlikely to reflect the disease risk, such as gender or clinical site. In such cases, it may be better to stratify or adjust for genetic or other parameters that are stronger determinants of risk.

Integrating Genetics in the Analysis of Randomized Clinical Trials

In most cases, genetic information is incorporated in clinical trials at the analysis stage, most often in the setting of secondary and exploratory analyses. Many times, the post hoc nature of the analyses is unavoidable, since new genetic polymorphisms may be identified or their role may be the object of speculation after the commencement of an RCT or even after an RCT has been completed. Such post hoc research should be viewed with the same caution that should accompany any kind of exploratory research. The essential issue is that findings need cautious replication, since they are largely hypothesis generating. Even though the study population may carry the name of a "randomized trial," in fact the study population may be treated in a manner where the emphasis of the comparisons is not on the randomization process. While nonrandomized designs may tend to agree on average with randomized studies (9–11), overinterpretation may be more common with nonrandomized studies (12).

Disease-Association Studies. The population of an RCT may well be used as a sample for performing disease association studies. The classic study design in this setting is a case-control experiment, and since the study sample is derived from a large population, the terminology "nested case control" design is more accurate. For example, in an early application of this approach, one group of investigators (13) evaluated a sample of 619 of the 12,866 participants of the Multiple Risk Intervention Trial (93 with death from coronary artery disease, 113 with nonfatal myocardial infarction, and 412 matched controls). Patients with and without coronary disease outcomes were compared with regard to the allele frequencies of the apolipoprotein E gene (epsilon 2, epsilon 3, epsilon 4).

Besides main genetic effects, case-control studies may also investigate situations where a specific polymorphism may be considered to be a modifier of the effect of an environmental or other risk factor. For example, a group of investigators (14) performed a nested case-control study using 141 cases of nonfatal myocardial infarction and 270 matched controls from the study population of the Helsinki Heart Study, a primary prevention trial. The selected subjects were genotyped for the 344C/T polymorphism of the gene encoding aldosterone synthase (CYP11B2). The investigators found that the polymorphism was not a strong risk factor for myocardial infarction. However, they suggested that there may be a strong interaction with the effect of smoking. Overall, smokers had a relative risk of myocardial infarction of 2.5 compared with nonsmokers. In the presence of the 344CC genotype, this relative risk became 4.67, while in the presence of 344TT homozygosity, this relative risk dissipated to 1.09. Gene–gene interactions represent more convoluted "second-order" effects. They may provide insights to pathogenetic mechanisms, such as the involvement of specific genes in modulating environmental risk factors. However, given their more complex nature, the risk of false-positive findings is probably higher than for "first-order" relationships.

For randomized trials that possess a robust design and adequate archiving of clinical information and blood or tissue samples, a multitude of nested case-control studies may be performed. Such studies may extend the scientific value of the original randomized research effort. In this regard, RCTs are similar to prospective cohort studies that may also be utilized for nested case-control designs. This creates the need for improved archiving and adequate data banks to be supported during the conduct of clinical trials. At the other end, data banking may be expensive and meaningless if performed without purpose and if case-control studies are performed in a haphazard fashion without some underlying biologic and clinical rationale. Such efforts may likely lead to false-positive, spurious, and clinically misleading findings.

The ethics of performing subsequent case-control studies should also be considered. Ideally, at the time of consent for DNA storage as part of an RCT, subjects need to know what is going to be tested, thus the design of the RCT and its consent form should take this into account. Trying to obtain additional consent at a later stage is often difficult, since the RCT population may be difficult to reassemble, once the trial is completed. On the other hand, consent needs to be generic enough so that the study investigators could have the option of testing new genetic polymorphisms. Striking a balance between protecting the rights of trial participants and satisfying the growing needs of genetic association research may not be straightforward. Procedures should be improved, standardized, and, when possible, simplified, to protect patient confidentiality without hindering research.

Predictive Modeling of Disease Risk—Genetic Predictors of Study Outcomes.

The population of the RCT may be used to evaluate specific polymorphisms as predictors of the outcome of interest. The outcomes of interest may be hard clinical endpoints, such as disease progression or death, or surrogate endpoints such as laboratory or genetic markers.

For example, one group of investigators (15) examined whether polymorphisms of the genes for apolipoprotein B, apo AIV, lipoprotein lipase, and cholesterol ester transfer protein may be associated with a greater change in dense LDL cholesterol in a crossover RCT study population treated with two dietary interventions of 4 weeks each (high saturated fat diet vs. high polyunsaturated fat diet). Of the polymorphisms tested, only the Q360H polymorphism in the apo AIV gene was significantly predictive of the change in dense LDL cholesterol. In cases such as this, randomization is not really any longer the essential feature of the study design. The RCT design simply serves to provide a population with fairly standardized exposures to important parameters, in this case diet, that may be influencing also the outcome of interest. The RCT is treated as a cohort study.

Soft biologic outcomes may sometimes be misleading for use in clinical decision-making (16). However, sometimes the interest of such analyses may be more focused on making pathophysiologic investigations rather than deriving clinical inferences. For example, basic research may be performed on tissue samples from randomized patients. One group of investigators (17) found that the angiotensin II type 1 receptor A1166C polymorphism is a significant predictor determining potassium chloride (KCL)-induced angiotensin II responses in excess segments of the internal mammary artery in both the experimental and the control group of a randomized study comparing an angiotensin-converting enzyme inhibitor versus placebo in patients undergoing bypass surgery. Such information may yield helpful pathophysiologic support for further hypothesis testing.

Effect Modification. Effect modification is probably the most challenging feature in the integration of genetics into RCTs. The old question is "can we find out which patients are likely to benefit more from a specific therapy?" If benefit is measured on an absolute scale (e.g., absolute risk reduction), then for a treatment that achieves a consistent relative risk reduction at all levels of baseline risk, the absolute benefit is likely to vary substantially across patients in different categories of risk. For example, a risk ratio of 0.7 for mortality translates to a 0.3% absolute risk reduction for death when the baseline risk of death is 1%, while it translates to a 6% absolute risk reduction for death when the baseline risk is 20%. Thus predictive modeling with individual patient data, including genetic and other predictors, may provide in essence evidence for effect modification in an absolute risk scale. However, usually effect modification as a term is reserved for cases where different subgroups (of different or even similar baseline risk) show significantly diverse relative responses to treatment. Such subgroups may be defined by genetic parameters. Specific genetic parameters may separate patients who benefit differentially from the same treatment, regardless of whether these parameters also affect the prognosis in the absence of treatment.

Most of the effect modification work to date in genetics has not examined hard clinical endpoints, such as survival, but surrogate laboratory and biologic parameters. The reason is probably that for hard endpoints such as survival, very large trials are required to show main effects, and the sample sizes required to show effect modification are even larger—and thus largely prohibitive. There are several examples of postulated effect modification with surrogate markers, however. For example, one group of investigators (18) found that in a sample of patients enrolled in a RCT of maintenance antiretroviral treatment, the haplotypes of the *CCR5* and *CCR5* promoter genes might be determinants of the magnitude of decrease of plasma human immunodeficiency virus RNA in response to potent antiretroviral therapy. In another example, in the Lipoprotein and Coronary Atherosclerosis Study (19), the investigators detected a strong significant genotype-by-treatment interaction in the relative response of total cholesterol, low-density lipoprotein cholesterol, and apoliporotein B with fluvastatin versus placebo. Patients with the DD genotype had greater reductions in these lipid parameters with fluvastatin than placebo recipients in the same randomized arm. In a study of lisinopril versus placebo in patients after renal transplantation, lisinopril had a beneficial effect on LV mass index reduction, and the effect was more prominent in patients with the DD genotype (8.4% vs. −7.2%) than in the other two genotypes [ID and II] (2.8% vs. −11.4%) (20).

Occasionally, effect modification may be seen predominantly in the control group rather than in the treatment group. For example, a RCT examined the influence of the PvuII polymorphism of the estrogen receptor alpha gene on the response of bone mineral density in postmenopausal women treated with hormone replacement therapy versus placebo (21). Overall, bone mineral density fared better in women given hormonal replacement than in those who got no placebo. Moreover, in the group of patients receiving hormonal replacement therapy, the bone mineral density change was not affected by genotype. Conversely, in the control group, the loss of bone mineral density was larger in the PP and Pp genotypes (6.4% and 5.2%, respectively) than in the pp genotype (2.9%, $p = 0.002$). This information is fairly similar to what can be obtained by predictive modeling in a uniformly untreated cohort of patients. It can be used to select the patients who might *not* have to be treated, especially when the available therapy is potentially toxic or controversial for other reasons.

Predicting Adverse Drug Reactions. A useful potential application of genetics in clinical trials is to identify genetic parameters that can be used to select patients who have the best or worst tolerance for toxic treatments. For example, it might be possible to detect patients who have a worse reaction to specific chemotherapeutic agents. A study (22) based on subjects from the St. Jude's Children's Research Hospital Protocol Total XII addressed whether mercaptopurine therapy intolerance is associated with polymorphisms within the thiopurine S-methyltransferase gene. The drug 6-mercaptopurine causes accumulation of thiopurine nucleotides. The investigators found that dose reductions due to toxicity were ubiquitous in patients homozygous in thiopurine S-methyltransferase enzyme activity deficiency, occurred in 35% of those with heterozygosity, and were very uncommon for wild-type patients (7%), based on phenotyping and confirmed also with genotyping in a subset of patients. Thus such knowledge, if validated, could be used in selecting the starting dose of 6-mercaptopurine for individual patients. In a different field, another group of investigators (23) found that homozygosity for the insertion allele (II) of the angiotensin-I-converting enzyme gene affects the cough threshold in patients treated with an angiotensin-converting enzyme inhibitor, cilazapril, in a crossover placebo-controlled trial. To maximize statistical efficiency in the comparison of interest, the investigators only recruited those subjects with II and DD genotypes (homozygous for the insertion allele or for the deletion allele). The knowledge of an increased susceptibility to cough among II individuals may be used to select whether an angiotensin-converting enzyme inhibitor or an agent from a different drug class should be used in specific individual patients, when several alternative regimens of equal efficacy are available.

Assessment of Generalizability
An interesting frontier where genetic information may have applications in RCTs is the assessment of the generalizability of the trial results. For diseases where ge-

netic factors are strong predictors of the disease outcome or an effect modifier for the impact of a treatment, determination of the genetic profile of the study population may provide insight on whether the results may be generalizable to other patient populations. Several genetic polymorphisms have explicit diversity in their distribution in different racial or ethnic subgroups. This means that effective treatments that depend on the presence of a specific genetic polymorphism may not be generalizable to ethnic or other subgroups where this polymorphism is missing or is encountered in low frequency. In a different approach, if a trial shows surprisingly no efficacy for an intervention, post hoc genetic testing of the study population might lead to hypotheses about specific parameters in the genetic profile of the study populations that might be more amenable to a new treatment. These are hypothesis-generating findings and should always be interpreted with due caution. Nevertheless, genetic information could contribute further in the assessment of the external validity of treatment recommendations derived from the interpretation of randomized evidence.

Table 15.1 summarizes the aspects of RCT design and analysis where genetic information could be used, in a similar, although perhaps more informative fashion, as other more traditional parameters that have been used to date for these purposes.

Caveats in the Use of Genetic Information in Clinical Trials

Several caveats must be pointed out in the use of genetic information in the design, analysis, and interpretation of results of RCTs. Several of these caveats pertain to the use of genetic information in other settings as well, while others may be more specific to RCTs.

Validity of genotyping Genetic information may sometimes suffer from low accuracy, regardless of whether it is performed as part of an RCT or for other purposes. Reasons could include lack of internal validation, lack of blinding in the as-

Table 15.1 Main Uses of Genetic Parameters in the Design and Analysis of Randomized Trials

Primary Design and Analysis	*Other Traditional Parameters Used for Similar Purposes*
Eligibility (inclusion and exclusion) criteria	Age range, disease characteristics, background treatment, contraindications, other laboratory parameters
Stratification and a priori adjustment	Clinical site, strong predictors of disease risk
Effect modification	Demographics, disease characteristics, background treatment
Safety outcomes	Demographics, disease characteristics
Generalizability assessment	Demographics, disease characteristics, background treatment
Secondary Studies	
Nested studies for disease associations	Any epidemiologic parameter of interest
Predictive modeling for disease risk	Composites of several parameters

sessment of the genetic test, a large test failure rate and many gray measurements, and large observer variability (24).

Linkage disequilibrium Genetic markers in linkage disequilibrium may complicate the interpretation of genetic associations or effect modification that is observed in RCTs. The observed relationships may not reflect a true association with direct pathophysiologic consequences, but may result from linkage disequilibrium of the tested genetic marker with some other unknown or unprobed marker that is the one truly responsible for the association or effect modification (25).

Heterogeneity in linkage The strength of linkage between genetic markers may vary in different samples and patient populations. This may result in further heterogeneity in the strength of the detected genetic relationships in RCTs and may lead to a low ability to replicate the findings in other RCTs or generalize their interpretation for clinical use.

Definitions of clinical trial endpoints and outcomes Unclear definitions of outcomes or "moving the goalposts" may generate spurious associations in genetic analyses involving RCTs. This problem occurs for both genetic and nongenetic parameters. One of the great design advantages of RCTs is the fact that outcomes should ideally be specified upfront in a specific and accurate manner. This is in contrast to hypothesis-generating epidemiologic research where outcomes may be (appropriately) manipulated in search of new associations. In some occasions, genetic research in RCT patient populations may examine new outcomes that may or may not be robustly defined. For newly conceived outcomes, a population of subjects derived from an RCT does not offer any clear advantage to a population derived from a well-designed epidemiologic nonrandomized cohort (26).

Surrogate outcomes Surrogate outcomes may give us clear insights about a pathophysiologic process. In some occasions they may clearly replicate the findings of hard clinical endpoints. However, RCTs have often been misled by surrogate endpoints that were not validated with corresponding clinical outcomes. As clinical trials become more linked with molecular medicine, use of biologic markers as endpoints is likely only to increase. Several of these biologic endpoints present problems of validation, replication, reproducibility in measurements, random error, and a complex correlation pattern with each other. Given their potential multiplicity, issues of multiple comparisons should also be considered as a potential problem.

Nonrandomized uses of randomized study populations As we stated above, most of the situations where genetics have interacted with RCT research to date have involved the use of the RCT population or samples thereof in ways that the advantage of randomization is lost. Such research should be seen more in the context of nonrandomized semi-experimental designs rather than randomized experiments and, therefore, inferences should be appropriately more cautious.

Multiple comparisons As discussed above, in genetic studies within RCTs, there may exist a multiplicity of outcomes, and a multiplicity of potential subgroup comparisons. To complicate matters, most diseases with a genetic background are likely to have very complex genetic patterns. Thus, there may be a multitude of potential putative genetic markers to be probed for disease association or effect modification. Some of the mutation sites are polymorphic, i.e., they may be several variations at the same site. Let us consider, for example, a very simple genetic polymorphism where there are only two different alleles, A and a. The number of potential genotypes is 3, i.e. AA, Aa, and aa. The number of potential genetic contrasts is 5: AA vs. others, Aa vs. others, aa vs. others, AA and aa vs. Aa, a allele vs. A allele. For a genetic polymorphism with 3 alleles, the number of potential genotypes is 6 and the number of potential contrasts increases exponentially. In exploratory analyses, all of these contrasts may be analyzed and one or more of them may show statistical significance that may simply reflect type I error. The situation is further compounded by the fact that there may be variations at multiple sites within the same gene. This creates a plethora of possible genetic comparisons. Large-scale testing in genetics (27), although exciting, may further increase the problem of type I error. Multiple comparisons with sparse data for many haplotypes may often lead to spurious results (28,29).

Replication of findings Given the above caveats, replication of findings is essential in genetic epidemiology (30). This applies to all aspects of genetic associations including those derived from randomized studies. Empirical evidence suggests that the findings of subsequent research tends to have a greater likelihood of disagreeing with the results of the original research on a genetic polymorphism, when the first studies are of small sample size, and when more subsequent evidence accumulates. Functional data, evolutionary conservation, and biologic plausibility should also be considered in determining which polymorphisms should be tested first and are likely to be most important, but it is unclear how much they improve the validation potential of genetic association studies.

Randomized Trials Evaluating the Clinical Use and Impact of Genetic Information

Randomized trials are considered the reference standard for evaluating medical technologies. Genetic tests are a rapidly expanding area of biotechnology that is being rapidly introduced into clinical care. However, in most cases, the supporting evidence for the introduction of genetic tests into routine care may be lacking or suboptimal. It is estimated that currently more than 700 genetic tests are already available or in late research development (GeneTests; www.genetests.org). RCTs have hardly ever been performed to document that these tests are warranted and have beneficial consequences when applied in specific clinical setting.

Prerequisites for Randomized Trials

RCTs are likely to be performed for tests that are candidates in possessing some meaningful clinical utility. In order for a genetic test to have clinical utility it must meet several requirements. We discuss these requirements in the context of performing and interpreting RCTs that evaluate the clinical use and impact of genetic tests.

First, accurate and reproducible routine methods must be available for the determination of the genetic trait of interest (31). It is conceivable that highly experimental, novel methods may be used in hypothesis-generating studies of genetic disease association or effect modification. However, when a genetic test reaches the stage of clinical use, it must be standardized and routinely applicable with adequate accuracy and reproducibility. It would be difficult to make inferences about the use of a test in the general population, if the assays used during clinical development cannot yet be applied to the general population. Designing a clinical trial to assess the usefulness of a screening strategy that depends on a nonstandardized test may result in low generalizability of the trial findings.

Second, the trial population must be readily identifiable and usually should be a rather limited/circumscribed group of subjects. Otherwise it is unlikely that the test would be cost-effective, unless the disease is very common in the general population. The frequency of the disease-related genotype(s) or allele(s) in the screened population is an important consideration. For a rare genotype, even a good test with high sensitivity and specificity may have relatively limited positive and/or negative predictive value. Thus, the eligibility criteria should be carefully selected in an RCT appraising a genetic test.

Third, the diagnostic test under study must be acceptable to the target population. Issues related to acceptability include costs, perceived and actual side effects, ease of administration and test accuracy, especially its false-positive rate. There are substantial ethical and social issues involved in genetic testing. These issues are often latent and difficult to eliminate (32–35). Genetic testing may provoke anxiety and sometimes result in psychological harm, insurance and employment discrimination, and worsening personal, family, and social relationships. False-negative results may also have grave consequences, as they may convey a false sense of reassurance to the misled patient and this could result in postponement of diagnosis or of use of indicated therapies in the future. All of these "side-effects" are difficult to measure in a clinical trial or other study design, but they should not be neglected in the interpretation of the results.

Fourth, the genetic test should ideally be a strong determinant of the disease process or a potent effect modifier of the response to available treatment. Nevertheless, genetic markers with modest effects may still be worthwhile as screening targets, if they have a high prevalence in the screened population. In this setting, the attributable fraction associated with them may still be substantial. For weak and rare, silent genetic traits, clinical trials may not be feasible to perform, since they would require the screening of very large number of subjects and a very large sample size of test-positive subjects in order to have adequate power to detect differences in outcomes with different approaches.

Fifth, effective and acceptable prevention or treatment options must be available for subjects where the test is positive, and it should be possible to initiate therapy promptly. Also, genetic effect modification may be more useful to know when there are several alternative preventive or therapeutic regimens and only some of them are affected by the genetic trait. Moreover, given the rapid change in therapeutics in many medical fields, one would have to be cautious about whether long-term trials would yield results that still hold true in a radically modified therapeutic environment by the time they are completed.

Sixth, preventive and therapeutic interventions must be accessible and affordable to the population identified to be at-risk, and they should have a favorable cost-effectiveness ratio. Ideally, they should have both short-term and long-term benefits for major disease outcomes. Long-term benefits may be more important to document, but they are likely to be more difficult and expensive to study with an RCT design. Nevertheless, it is unclear whether observational research can ever supplement and cover the lack of long-term randomized data in this field (11,12).

Other Design Considerations

The design of RCTs to assess specific genetic tests is still at its infancy. Studies of primary prevention screening and early interventions are intuitively the most attractive, given the theoretically anticipated larger gains of primary prevention versus late interventions. However, these studies are also the most challenging, given the need for very large sample size and long-term follow-up. The theoretical promise of preventive medicine may not be justified when tested in real life. The design of such trials poses challenges similar to those faced in the conduct of long-term RCTs in nutritional chemoprevention (e.g., with various antioxidants), have started appearing in the literature during the last decade. Moreover, additional problems may arise. For example, patient preferences may be an important obstacle to randomization. Or, large genetic heterogeneity may make guidance difficult to standardize. Finally, long-term follow-up may be problematic and associated with high rates of loss to follow up or voluntary crossover of subjects into the opposite study arms.

One may discuss some of the issues that arise in trying to implement screening for hereditary breast and ovarian cancer. There is some evidence that for *BRCA1* and *BRCA2* screening, subjects may have strong preferences both in regards to genetic testing and in regards to subsequent interventions (36). A study has found (37) that positive results in *BRCA1* and *BRCA2* screening tended to reinforce the intention towards prophylactic surgery among women who were already leaning towards this intervention; however, women who were reluctant to have surgery upon study entry were still reluctant after testing and counseling. Consent for randomization might be difficult to obtain for testing the comparative merits of different preventive or therapeutic options. Differences between options may be subtle in the short-term, but more clinically meaningful in the long-term, when major events start accruing. However, maintaining a largely asymptomatic trial population under routine follow-up for very lengthy periods of time may be unrealistic.

Decision analysis has been used in order to model some of the decisions that may be involved in genetic testing and the actions derived from the genetic information. The inferences of such models may illustrate some of the problems that may be faced by RCTs in these areas. For example, a decision analysis compared prophylactic mastectomy, bilateral prophylactic oophorectomy, tamoxifen, and no intervention for women with breast cancer and *BRCA1* or *BRCA2* mutations (38). It found that the three interventions increased life expectancy by 0.6–2.1, 0.2–1.8, and 0.4–1.3 years over a horizon of 10 years for the baseline scenario of a 30-year-old woman with early breast cancer. However, the results were substantially sensitive on the penetrance rate of the mutation. The differences between the three interventions would be difficult to study unless one had a very large sample size. Even documenting the superiority of these interventions over no intervention at all in an RCT would still require a large sample size and long-term follow-up. In some cases, decision analysis may help decide whether an RCT is desirable at all in a specific population. For example, a different group of investigators (39) found that *BRCA1* and *BRCA2* screening would not benefit women without a family history or early breast cancer, because the pretest probability is very low and surgical prophylaxis is largely undesirable. Conversely, up to 2 quality-adjusted life years may be gained in women with a family history or early breast or ovarian cancer.

RCTs may also be designed to examine what are the relative merits of genetic testing versus use of some other technology or a combination of various technologies. The same challenges apply here as in the case of testing versus no testing comparisons. Examples include whether screening for familial adenomatous polyposis of the colon should use genetic testing for mutations in the implicated *APC* gene or colonoscopy; or whether screening for familial hemochromatosis should involve genetic analysis of the hemochromatosis *HFE* gene or iron studies. Both questions have been approached with decision analysis modeling that suggests the superiority of genetic testing for both examples (40,41). Questions comparing technologies may be even more difficult to subject to the rigorous standards of randomized evaluation and may require even larger numbers of subjects, since the differences are likely to be smaller than in test versus no test comparisons. While modeling approaches are a useful substitute in this setting, one is left with the wish that actual randomized evidence were available.

Trials of Educational and Counseling Approaches in Genetic Testing

We need to learn more about the proper implementation of genetic testing for different conditions and the value of adjunctive educational and counseling measures.

Modern medical practice in many developed countries is moving away from physician-initiated prescriptions and towards a greater emphasis on patient-initiated choices. Patients have prompt access to vast amounts of medical information through various sources, including in particular the Internet. Such information may be loaded with errors (42). Genetic information may be difficult to comprehend. Health-care consumers may often misunderstand genetic testing, and there may be misconceptions about the actual implications of a genetic test. For example, patients may overestimate the diagnostic ability of a test. Or, they may perceive a positive test as a sign of irreparable "genetic doom." Given this situation, it is important to study the optimal approaches towards enhancing the appreciation and use of genetic testing by health-care consumers. This is a promising field that is suitable for randomized trials.

For example, a randomized trial (43) evaluated pretest education regarding *BRCA1* testing versus education plus counseling versus a waiting-list (control) condition among women at low-to-moderate risk with a family history of breast or ovarian cancer. Both education and counseling led to increases in overall knowledge, but only counseling heightened the perception about the limitations and risks of *BRCA1* testing. Neither intervention changed the intention of women to have *BRCA1* testing and about half of the women eventually gave a blood sample. In another trial (44), written and video information was found to be equally effective in providing information about cystic fibrosis carrier screening and achieved high levels of subject-matter knowledge. This might suggest that information technologies may often substitute effectively face-to-face education and counseling, but this may not hold true in all circumstances and for all genetic tests.

RCTs may also study the setting where a genetic test should be recommended and/or implemented. Genetic tests often have implications that extend beyond the individual and affect also couples or whole families. This may generate differential reactions to genetic testing recommendations, depending on whether information is conveyed to an individual, a couple, or a family. For example, one group of investigators (45) randomized offering counseling and carrier testing for cystic fibrosis either to pregnant women in the first instance (stepwise screening) or to couples upfront (couple screening). The two groups differed significantly in transient and late anxiety levels and in the false reassurance rates among subjects testing negative.

Concluding Comments

Implementation of randomized research in the field of genetics is difficult and challenging, but not unfeasible. A genetic test needs to be evaluated rigorously as does any other diagnostic technology. The cost savings or the wasted expense associated with the use of a genetic test may rival that of any other diagnostic technology, especially when one considers genetic tests that target the general population or large segments thereof. The introduction of genetic tests into clinical practice without some strong supporting evidence is worrisome. While regulatory actions should not strangle this exciting, rapidly expanding field, greater attention should be given towards materializing randomized experiments testing the usefulness of genetic tests. Such research may give us valuable lessons.

References

1. Glasziou PP, Irwig LM. An evidence based approach to individualising treatment. BMJ 1995;311:1356–1359.
2. Ioannidis JP, Lau J. Uncontrolled pearls, controlled evidence, meta-analysis and the individual patient. J Clin Epidemiol 1998;51:709–711.
3. Oxman AD, Guyatt GH. A consumer's guide to subgroup analyses. Ann Intern Med 1992;116:78–84.
4. Altman DG, Royston P. What do we mean by validating a prognostic model? Stat Med 2000;19:453–473.
5. Meurice T, Bauters C, Hermant X, et al. Effect of ACE inhibitors on angiographic restenosis after coronary stenting (PARIS): a randomized, double-blind, placebo-controlled trial. Lancet 2001;357:1321–1324.
6. Meinert CL. Design and conduct of clinical trials: course slides. Baltimore: Johns Hopkins University Center for Clinical Trials, 1994.
7. Ioannidis JP, Lau J. The impact of high-risk patients on the results of clinical trials. J Clin Epidemiol 1997;50:1089–1098.
8. Ioannidis JP, Lau J. Heterogeneity of the baseline risk within clinical trial populations: a proposed evaluation algorithm. Am J Epidemiol 1998;148:1117–1126.
9. Benson K, Hartz AJ. A comparison of observational studies and randomized, controlled trials. N Engl J Med 2000;342:1878–1886.
10. Concato J, Shah N, Horowitz RI. Randomized, controlled trials, observational studies and the hierarchy of research designs. N Engl J Med 2000;342:1887–1892.
11. Ioannidis JPA, Haidich A-B, Lau J. Any casualties in the clash between randomised and observational evidence? BMJ 2001;322:879–880.
12. Ioannidis JP, Haidich AB, Pappa M, et al. Comparison of evidence of treatment effects in randomized and non-randomized studies. JAMA 2001;286:821–830.
13. Eichner JE, Kuller LH, Orchard TJ, et al. Relation of apolipoprotein E phenotype to myocardial infarction and mortality from coronary artery disease. Am J Cardiol 1993;71: 160–165.
14. Hautanen A, Toivanen P, Manttari M, et al. Joint effects of an aldosterone synthase (CYP11B2) gene polymorphism and classic risk factors on risk of myocardial infarction. Circulation 2000;100:2213–2218.
15. Wallace AJ, Humphries SE, Fisher RM, Mann JI, Chisholm A, Sutherland WH. Genetic factors associated with response to LDL subfraction to change in the nature of dietary fat. Atherosclerosis 2000;149:387–394.
16. Fleming TR, DeMets DL. Surrogate endpoints in clinical trials: are we being misled? Ann Intern Med 1996;125:605–613.
17. van Geel PP, Pinto YM, Voors AA, et al. angiotensin II type 1 receptor A1166C gene polymorphism is associated with an increased response to angiotensin II in human arteries. Hypertension 2000;35:717–721.
18. O'Brien TR, McDermott DH, Ioannidis JP, et al. Effect of chemokine receptor gene polymorphisms on the response to potent antiretroviral therapy. AIDS 2000;14:821–826.
19. Marian AJ, Safari F, Ferlic L, et al. Interactions between angiotensin-I converting enzyme insertion/deletion polymorphism and response of plasma lipids and coronary

arterosclerosis to treatment with fluvastatin: the lipoprotein and coronary atherosclerosis study. J am Coll Cardiol 2000;35:89–95.

20. Hernandez D, Lacalzada J, Salido E, et al. Regression of left ventricular hypertrophy by lisinopril after renal transplantation: role of ACE gene polymorphism. Kidney Int 2000;58:889–897.

21. Salmen T, Heikkinen AM, Mahonen A, et al. Early postmenopausal bone loss is associated with PvuII estrogen receptor gene polymorphism in Finnish women: effect of hormone replacement therapy. J Bone Miner Res 2000;15:315–321.

22. Relling MV, Hancock ML, Rivera GK, et al. Mercaptopurine therapy intolerance and heterozygosity at the thiopurine S-methyltransferase gene locus. J Natl Cancer Inst 1999; 91:2001–2008.

23. Takahashi T, Yamaguchi E, Furuya K, Kawakami Y. The ACE gene polymorphism and cough threshold for capsaicin after cilazapril usage. Respir Med 2001;95:130–135.

24. Bogardus ST, Jr, Concato J. Feinstein, AR. Clinical epidemiological quality in molecular genetic research. The need for methodological standards. JAMA 1999;281:1919–1926.

25. Reich DE, Cargill M, Bolk S., et al. Linkage disequilibrium in the human genome. Nature 20001;411:199–204.

26. Langholz B, Rothman N, Wacholder S, Thomas DC. Cohort studies for characterizing measured genes. J Natl Cancer Inst Monogr 1999;26:39–42.

27. Risch N, Merikangas K. The future of genetic studies of complex human diseases. Science 1996;273:1516–1517.

28. Schork NJ, Fallin D, Lanchbury JS. Single nucleotide polymorphisms and the future of genetic epidemiology. Clin Genet 2000;58:250–264.

29. Fallin D, Cohen A, Essioux L, et al. Genetic analysis of case/control data using estimated haplotype frequencies: application to APOE locus variation and Alzheimer's disease. Genome Research 2001;11:143–151.

30. Ioannidis JPA, Ntzani E, Trikalinos TA, Contopoulos-Ioannidis JPA. Replication validity of genetic association studies. Nat Genet 2001;29:306–369.

31. Holtzman NA, Watson MS. Promoting Safe and Effective Use of Genetic Testing in the United States: Final Report of the Task Force on Genetic Testing. Baltimore: Johns Hopkins University Press, 1998.

32. Billings P, Kohn MA, deCuevas M et al. Discrimination as a consequence of genetic testing Am J Hum Genet 1992;50:472–482.

33. Rothenberg KH. Genetic information and health insurance: state legislative approaches. J Law Med Ethics 1995;23:312–319.

34. Lapham EV, Kozma C, Weiss JO. Genetic discrimination: perspectives of consumers. Science 1996;274:621–624.

35. Khoury MJ, Thrasher JF, Burke W, Gettig EA, Fridinger F, Jackson R. Challenges in communication genetics: a public health approach. Genet Med 2000; 2:198–201.

36. Weber BL, Giusti RM, Liu ET. Developing strategies for intervention and prevention in hereditary breast cancer. J Natl Cancer Inst Monogr 1995;(17):99–102.

37. Miron A, Schildkraut JM, Rimer BK, et al. Testing for hereditary breast and ovarian cancer in the southeastern United States. Ann Surg 2000;231:624–634.

38. Schrag D, Kuntz KM, Garber JE, et al. Life expectancy gains from cancer prevention strategies for women with breast cancer and BRCA1 or BRCA2 mutations. JAMA 2000;283:617–624.

39. Tengs TO, Winter EP, Paddock S, et al. Testing for BRCA1 and BRCA2 breast-ovarian cancer susceptibility genes: a decision analysis. Med Dec Making 1998;18:365–375.

40. Bapat B, Noorani H, Cohen Z, et al. Cost comparison of predictive genetic testing vs. conventional clinical screening for familial adenomatous polyposis. Gut 1999;44:698–703.

41. El-Seray HB, Inadoni JM, Kowdley KV. Screening for hereditary hemochromatosis in siblings and children of affected patients. A cost-effectiveness analysis. Ann Intern Med 2000;132:261–269.
42. Jadad AR, Gagliardi A. Rating health information on the Internet: navigating to knowledge or to Babel? JAMA 1998;279:611–614.
43. Lerman C, Biessecker B, Benkendorf JL, et al. Controlled trial of pretest education approaches to enhance informed decision-making for BRA1 gene testing. J Natl Cancer Inst 1997;89:148–157.
44. Clayton EW, Hannig VL, Pfotenhauer JP, et al. Teaching about cystic fibrosis carrier screening by using written and video information. Am J Hum Genet 1995;57:171–181.
45. Miedzybrodzka ZH, Hall MH, Mollison J, et al. Antenatal screening for carriers of cystic fibrosis: randomised trial of stepwise v. couple screening. BMJ 1995;310:353–357.

16

Developing guidelines for the clinical use of genetic tests: a U.S. perspective

Linda Pinsky, David Atkins, Scott Ramsey, and Wylie Burke

Medical information is gathered primarily for the purpose of improving health outcomes. With growing information about the genetic contributors to human health, many new genetic tests will be available to clinicians for this purpose (1,2). Genetic tests may be used to diagnose a medical condition, to direct management in symptomatic patients, or to predict future health risk in asymptomatic patients in order to guide preventive care. As with other new medical tests, they must be assessed carefully to determine their appropriate use (3).

Clinical practice guidelines represent an effort to define the appropriate use of medical interventions (4,5). In the process of developing guidelines, panels or expert groups examine evidence to determine the effectiveness and applicability of the new test under consideration (6–9). They must look at the quality of evidence supporting test use to ask whether benefits outweigh potential harms and, if so, whether the testing process and interventions based on tests results are feasible and cost-effective. For genetic testing, this evaluative process must consider the added value/harm to the patient of genetic testing over conventional approaches. Embedded in the examination are issues of patient preference, individual and societal ethics and values, and laws regarding the use of medical information. In this chapter, we consider the evidence needed for practice guidelines for genetic testing and the strategies available to clinicians and policymakers to accomplish this task.

Evaluation of Genetic Tests

Test Properties

Many genetic tests use DNA-based technology, but any laboratory test used primarily to identify an inherited condition is considered a genetic test (10). In creating clinical practice guidelines for genetic tests, three basic questions must be answered: (*1*) Does the test accurately identify the genetic variant of interest (analytic validity)? (2) Does identifying that variant accurately predict the presence or risk of having the related clinical condition? (clinical validity) and (*3*) Does identifying the clinical condition (either disease or risk for developing disease) improve the patient's health outcome? (clinical utility).

In addressing these basic questions, the evaluation must include a definition of (*1*) the population to which a given test would be applied in clinical practice; (2) the testing procedures and the interventions to be based on test results; (*3*) the outcome of interest, e.g., the health improvement resulting from use of the test and associated interventions; and (*4*) the alternative strategy to which test use is being compared. In the case of genetic tests, the comparison is generally with existing strategies to accomplish the same health goal that do not utilize genetic testing (for example, reduction in breast cancer mortality through testing for *BRCA1/2* mutations and subsequent interventions would be compared to a strategy of identifying inherited risk through analysis of the family history, or simply through regular mammography in all eligible women). Usually the comparison asks whether the addition of a genetic test to identify people with increased risk will enhance efforts to prevent disease or other adverse health consequences, such as drug complications. In research assessing these questions, the same diagnostic methods and outcome assessments must be applied to all study participants, not just those identified by the test of interest. Further, the population studied must have a sufficient spectrum of disease or risk to represent those for whom the test will be used. Often, these requirements can be met only through studies specifically designed to assess clinical use of a test, after epidemiological or clinical studies have indicated a potential clinical value for the test.

Genetic tests can be used for diagnosis (e.g., confirming the presence of a condition suspected based on clinical information) or for screening. These two uses have different implications for assessment. Diagnostic testing is applied to relatively narrow populations (e.g., family members, persons with specific clinical findings, etc.) to help guide clinical management in order to improve short- or long-term health outcomes. Screening, in contrast, refers to testing in broader, relatively unselected populations to identify conditions that may not yet be causing health problems, for which there are interventions that might prevent future problems. Most individuals undergoing a screening test do not have the condition of interest; of those that do, only a proportion may suffer health problems because of it. As a result, assessment of screening tests must consider the potential harms of the screening test (false–positive tests, downstream testing) and of interventions that might be instituted as a result of screening (e.g., prophylactic mastectomy in women with *BRCA1* mutations) as well as their potential benefits. In order to justify screening, early detection and intervention should produce greater net benefits (benefits minus any harms of screening) than intervention at the time of symptoms. It is also important to examine the acceptability of the intervention in asymptomatic persons, who may be less motivated to comply with interventions than individuals who were identified on the basis of symptoms or risk status, such as having affected relatives.

Analytic validity refers to the accuracy with which a particular genetic alteration (for example, a DNA sequence variant) can be identified by a given laboratory test. One way to report analytic validity is in terms of the test's sensitivity and specificity *for the genetic variant* in question. Most genetic variants can be tested by a variety of protocols, and a number of technical issues arise in evaluating analytic validity. These include the assay chosen, the reliability of the assay, the degree to which reliability varies from laboratory to laboratory, and the complexity of test interpretation. Thus, an oligonucleotide probe for a single nucleotide sequence variant is a simpler test than a linkage analysis. In the latter, accuracy of the test is based on the accuracy with which samples and medical history are collected from family members, as well as on technical aspects of the testing process in the laboratory.

In addition, the same genetic trait may be assessed by using different laboratory methods. For example, some studies of factor V Leiden, a genetic variant associated with increased risk of venous thromboembolism, utilize a DNA-based assay while others utilize a functional assay, with differences in sensitivity and specificity (11). An adequate description of the analytic validity of a test requires systematic collection of data for this purpose, with the use of defined populations with known genotypes (Table 16.1). Few studies report systematic evaluation of analytic validity, however, and this lack represents an important limitation in the available evidence about genetic tests (12). These concerns find a parallel in the assessment of nongenetic tests, e.g., the accuracy with which different analytic measures of clot degradation products, termed D-dimers, predict the presence of deep venous thrombosis (13).

Clinical validity describes the accuracy with which a test predicts a particular clinical outcome. When a test is used diagnostically, clinical validity measures the association of the test with the current existence of that disorder. When a test is used to identify genetic susceptibility, as in genetic screening, clinical validity measures the accuracy with which it predicts a future clinical outcome. Clinical validity can be expressed as the positive and negative predictive value of the test for the occurrence of disease within a defined population. This measurement of clinical validity incorporates the prevalence of the disease in the population of interest.

Clinical validity is often uncertain for new genetic tests. Mutations associated with disease susceptibility are often first defined in "high risk" families, characterized by multiple affected family members, and these studies may overestimate risk. Initial estimates of an 85% lifetime risk of breast cancer associated with *BRCA1* and *BRCA2* mutations came from such families (14), while later estimates from population-based studies have provide lower and more variable risk estimates, ranging form 26% to 74% (15–21). (See also Chapter 26.) These observations suggest that other factors, both genetic and nongenetic, may modify risk. As a result, studies in families selected for multiple cancer cases will overestimate the risk conferred by the mutations (22). Even population-based studies may overestimate risk if they are based on families selected through an index case with cancer; like studies based on high-risk families, this methodology may also select for other cancer risk fac-

Table 16.1 Presenting Data about Genetic Tests

Type of Study	Parameters to Be Assessed
All studies	*Study population* How were subjects selected? Is information provided concerning age, gender, racial or ethnic origin? If a control population is included, was it selected from the same population as the cases; were matching criteria used? What inclusion and exclusion criteria were specified? *Laboratory assay* What was the source of samples? What variant was measured? What laboratory method was used?
Analytic validity	*Reference standards* Did study include samples with known genotypes, with and without the variant being assayed? What was the source of reference standards? What criteria were used to define genotypes? *Laboratory performance* How was reproducibility of assay assessed? Was reliability of assay assessed in a routine clinical laboratory setting?
Clinical validity	*Study design* Were data collected prospectively or retrospectively? Did study include measurement of potential modifying factors? Was there an independent, blind comparison with a reference standard? Did the subject group include an appropriate spectrum of patients representative of those to whom the test will be applied in clinical practice? Did the results of the test being evaluated influence the decision to perform the reference standard? *Clinical and other end points* What case definition was used? What end points were measured? Was interpretation of end points blinded? Were negative results verified? Are likelihood ratios for the test results presented or data necessary for their calculation provided?
Clinical utility	*Intervention* What interventions were used? What were the criteria for use of the intervention? *Study design* Were data collected prospectively or retrospectively? Was an experimental study design used? If so, was a randomization method used? Was intervention blinded? *Clinical and other end points* What outcomes were measured? Was interpretation of end-points blinded? Were negative results verified? Will the reproducibility of the test result and its interpretation be satisfactory in my setting? Are the results applicable to my patient? Will the results change my management? Are the interventions acceptable to my patients? Will my patients be better off as a result of the test?

tors (23). Similarly, initial reports of gene–disease associations in epidemiologic studies also often report stronger associations than subsequent follow-up studies (24,25). This variation in clinical validity could be due to complex gene–gene and gene–environment interactions. The result is that the predictive value of many genetic tests is likely to remain imprecise even when all appropriate population-based studies have been completed. Additionally, different methodologic approaches may influence estimates of clinical validity.

Uncertainty about clinical validity is an important consideration in guideline development, particularly when policy-makers are considering whether the risk information provided by the test provides a sufficient rationale for the use of interventions with putative but unproven benefits. In evaluating evidence of clinical validity, several variables are important, including the size and selection criteria for the study population, the type of test used (and its analytic validity), study design, and end points measured (Table 16.1). Careful definition of the measures used to determine the clinical outcome is needed. Studies will provide more convincing results when assessment of clinical outcome incorporates the contribution of nongenetic risk factors. Key methodologic considerations include the comparability of case and control populations (where a case/control design is used), whether the interpretation of clinical end points was blinded, and whether negative results, e.g., absence of disease, were verified.

Many genetic variants affect more than one clinical outcome; for example, factor V Leiden is associated with both venous thromboembolism and pregnancy loss (26), and genetic variants associated with cancer risk may increase risk for two or more different cancers (27). Multiple disease end points need to be considered when studying such variants; clinical validity may be high for one end point and low for another. Guidelines must specifically state which outcome is addressed in their recommendations.

Evaluation of Interventions: The Analytic Framework

Evidence-based guidelines for genetic testing need to address a central question: can health outcomes be improved by testing a target population to identify those at increased genetic susceptibility (for a screening test) or with a previously undiagnosed genetic condition (for diagnosis)? Only rarely do studies provide a direct answer to this overarching question. In the absence of such direct studies, an *analytic framework* provides a way of organizing the component questions that need to be answered, and organizing the evidence relevant to those questions to inform a guideline (5,6,8). An analytic framework (Fig. 16.1) defines a designated population, an initial evaluation action (screening or diagnosis), a resulting state (increased risk or the presence of disease), possible treatment actions (interventions), and final health outcome (morbidity and mortality). The framework makes explicit the considerations of both the benefits and potential harms for each action in the testing/ treatment pathway. The former includes decreased morbidity, including improved quality of life, and decreased mortality. The latter include unnecessary interventions

Evidence of overall benefit of screening/diagnostic approach

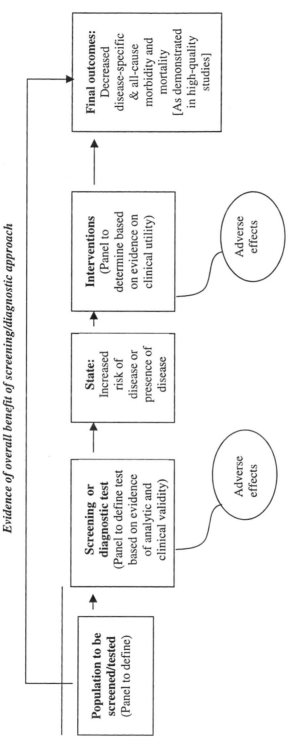

Figure 16.1 Analytic framework (task for guidelines panel).

occurring as a result of false-positive results, adverse effects of interventions, and personal or psychosocial consequences of testing or intervention, such as the loss of access to insurance or employment. Harms also include other negative effects on quality of life that are more difficult to quantify, such as anxiety or other nontangible effects of labeling. Overall, the framework provides an outline for asking: how do net benefits compare to net harms and costs? In an analytic framework, the population to be tested narrows with each step on the framework, but guidelines developed from the framework may also address the needs of those with negative test results at each step in the framework.

The analytic framework considers all potential interventions available to the population at risk. Interventions commonly available for those identified with increased genetic susceptibility or genetic disease may include information/education, risk reduction through lifestyle modification or exposure avoidance, more intensive screening than usual for early detection of disease, earlier intervention than usual when symptoms or positive lab results occur, or interventions not used for those at average risk. Each of these may vary with respect to the likelihood of reducing future health risk and the potential for harm. This estimate of benefits and harms represents an assessment of clinical utility: the likelihood that the test will lead to an improved health outcome. Standardized approaches are available to assess the quality of evidence concerning the outcomes of interventions (28). Genetic testing must be compared to alternative approaches to achieving the same health benefit; these include other methods of risk assessment or diagnosis; or the using of the same treatment or preventive treatment for all patients in the same clinical category, in the absence of a specific genetic diagnosis or risk assessment.

Knowledge about optimal prevention or treatment strategies associated with genetic risk is usually accumulated incrementally. People known to be genetically susceptible are initially offered interventions based on clinical reasoning or extrapolation from data in other populations—for example, early initiation of mammography screening in women with *BRCA1/2* mutations because they are at risk for early-onset breast cancer. The degree of risks and benefits of these interventions can be quantified accurately only after systematic observation, in the form of well-designed randomized controlled trials (RCTs), cohort or case-control studies. Of these study types, the RCT presents the most reliable form of data due to a decreased possibility for confounding. A high quality RCT, through randomization, decreases the potential confounding that can be seen in cohort studies as the latter allows participants choice of receiving the treatment; this choice may reflect the individuals' characteristics and introduces confounding factors. Randomization, if sample size is adequate, assures equal distribution of both known and unsuspected confounding factors. Blinding of participants, providers, and investigators to what intervention each participant is receiving decreases the possibility for placebo effects and observational bias. Because most genetic risk factors, even for "common" genetic diseases, occur in a small percentage of the population—for example, factor V Leiden is present in 1% to 5% of the population (29) and *BRCA1/2* mutations are estimated to have a prevalence of 1/500 or lower (20)—sample sizes of genetically suscepti-ble subjects are often small. Definitive understanding of clinical utility may be de-pendent on reliable methods for the pooled analysis of different studies.

Assessments of clinical utility need to consider factors that may affect internal validity (are these results true for the population studied?) and external validity (are these results likely to be true for the population of interest?). Internal validity is pri-marily a function of study design, execution and analysis. Although well-done ran-domized trials provide the highest internal validity, prospective intervention trials enrolling patients identified by genetic testing are rare at present. More commonly, interventions may have been assessed through prospective or retrospective obser-vations in populations identified by clinical signs or symptoms or by affected rela-tives. Important issues affecting internal validity of such studies include whether the outcomes were measured in reliable and unbiased manner, whether all important outcomes were included, and whether adequate control groups were identified. Ex-ternal validity is equally important: are results of such studies applicable to patients who are identified by genetic testing, rather than clinical symptoms? Issues affect-ing external validity include representativeness of the study population (are the sub-jects similar to those who would be eligible for this test?); the type of laboratory assay (is this comparable to assays in clinical practice?); and intervention (is it com-parable in scope and quality to what a patient is likely to be offered, or accept, in a typical practice?) (Table 16.1). When screening is contemplated based on obser-vational data from a clinically selected population, critical questions are raised con-cerning the external validity of the data. Is the underlying physiology of the disease in the studied population similar to that in the population to be screened? Is the risk similar? Is the intervention likely to be as acceptable or compliance comparable in an asymptomatic population? Are the observed benefits likely to be comparable in a population identified by broader screening? As with clinical validity studies, the comparability of the cases and controls is an important issue, when a case/control design is used. If an interventional design is used, elements of the study methodol-ogy such as randomization strategy, blinding, and intention-to-treat analysis are im-portant in determining the quality of the study. The methods for measuring clinical and other end-points should be specified, as well as whether negative results were verified.

Despite recognition of the need to assess clinical utility, genetic tests have been offered for clinical use before outcome studies are available to determine the ben-efits of interventions for those with genetic susceptibility (32–34). Usually, this has occurred because the risk identified by the test is deemed high enough to justify pursuing the potential benefit provided by available interventions. Often the evalu-ation process that is part of guideline development needs to determine whether or not a test should be used in clinical practice when limited data are available, and if so, what uses are appropriate. This evaluation process can also identify the evidence needed to improve practice guidelines. Two emerging genetic tests offer examples of this assessment process.

Factor V Leiden and Venous Thromboembolism. Factor V Leiden is a variant of the factor V gene, which codes for a protein involved in clot formation. It is present in approximately 5% of European Americans, 2% of Hispanic Americans, and 1% of African and Asian Americans (29), and is estimated to increase risk of venous thromboembolism (VT) two- to eightfold (35,36). Most people with factor V Leiden are heterozygous, that is, they carry one copy of factor V Leiden and one copy of a normal factor V gene. A small minority are homozygous; these individuals have a much higher risk of VT. (See also Chapter 19.)

Other established risk factors for VT, such as pregnancy, surgery, and use of oral contraceptives, interact to produce higher risks in people with factor V Leiden (35,37–39). In addition, interactions occur between factor V Leiden and other genetic risk factors. For example, individuals with both factor V Leiden and the 20210A variant in the prothrombin gene have a substantially higher risk of VT than people with either trait alone (36,40,41). Further, in a series of families with antithrombin III deficiency ascertained on the basis of early VT, the high risk of VT was due to the additional presence of factor V Leiden or the prothrombin variant in more than half of the families (42). Thus, when VT occurs in the presence of factor V Leiden it often—if not always—represents the effect of interactions with one or more other genetic and nongenetic risk factors (35).

Partly as a result of the complex mix of risk factors in VT, the implications of factor V Leiden for clinical management are unclear. Several rationales for factor V Leiden screening have been proposed. Carriers of the variant might benefit from avoiding oral contraception (34,37) or postmenopausal estrogen (43). Anticoagulant prophylaxis might reduce the risk of VT in the setting of transient risk factors such as surgery, pregnancy, or prolonged bed rest, or might prevent pregnancy loss (34,35,41). However, these interventions all pose potential risks as well as benefits, and none has been systematically studied in people with factor V Leiden. Further, people with heterozygous factor V Leiden who have had a VT do not appear to have a higher risk of recurrence than other VT patients, unless other genetic or metabolic risk factors are present (44,45). As a result, they do not appear to benefit from prolonged anticoagulant therapy or anticoagulant prophylaxis.

How might a guideline deal with this incomplete collection of evidence in an attempt to provide useful clinical guidance regarding the appropriate use of testing for factor V Leiden? First, the numerous gaps in the evidence argue against widespread use of testing in asymptomatic populations. Important issues of harms versus benefits need to be addressed if population screening is to be contemplated, and prospective studies are needed to provide reliable answers. Cost-effectiveness studies may suggest that even under optimistic assumptions, broad screening is cost ineffective (46). Thus, the process of considering guidelines can be a powerful tool for advocating for the necessary studies and focussing on the areas of greatest potential.

At the same time, a guidelines panel can assemble what data are available to address narrower clinical areas, where the information from testing has greater potential to help guide clinical decisions even when data are imperfect. Such recommendations should clearly identify the evidence supporting the recommendations, including its limitations and where recommendations are based on expert opinion or extrapolation from limited evidence. It should also identify information that is lacking, discuss possible alternative approaches to practice, and consider the potential effects of new evidence on practice recommendations.

For factor V Leiden testing, the guidelines might address the management of those already known to have the mutation. The higher risk of thromboembolism on oral contraceptives might be a basis for advising women with factor V Leiden to seek other methods of contraception (46). Similarly, testing might be considered for female relatives of a patient with factor V Leiden who are considering oral contraception. This approach is based on higher efficiency of such testing, due to the higher a priori risk in a family member. Similarly, the higher risk of harms from hormone therapy (with no higher benefit) can support a recommendation that women with factor V Leiden avoid hormone replacement therapy (43). Some women with factor V Leiden have been treated with low-dose anticoagulation during pregnancy (39), presumably on the basis of similar reasoning, although this extrapolation is harder to justify due to the potential harms of anticoagulant therapy. Finally, the occurrence of VT in a young adult with a positive family history of VT, in the absence of other causes, may indicate the presence of homozygous factor V Leiden or a combination of factor V Leiden with a second predisposing factor. Testing of a male in this situation would be done only to identify these rare circumstances, and if heterozygous factor V Leiden were found, it would have medical implications only for female relatives considering pregnancy or therapy with estrogens.

Although these considerations might inform a cohesive practice guideline regarding factor V Leiden testing in women with a family history of a relative with factor V Leiden, or in a person with VT occurring at a young age in the absence of other risk factors, important uncertainties would need to be acknowledged. In particular, advice about contraception must take into account competing issues. The risks of VT with oral contraception is higher in women with factor V Leiden, but the absolute risk remains low (37). In addition, the lower effectiveness of most alternative methods of contraception, the risks of pregnancy itself (including risk of VT), and the acceptability of the alternative birth control methods all need to be considered. For some patients, the higher risk may tip the balance against oral contraceptives, whereas in others these other considerations may be more important. A practice guideline for the clinician should note the lack of high-quality evidence associated with the decision to base contraceptive choice on factor V Leiden status. Similarly, for pregnant women with factor V Leiden, the lack of definitive evidence about the benefits and risks of anticoagulation for women with factor V Leiden during pregnancy should be noted. In both these situations, recommendations for shared decision making would be appropriate. In considering anticoagulation during pregnancy, this process would include a review of the risks of clotting if untreated and of hemorrhage or other serious bleeding complications if treated. Considerable weight would be give to the patient's preferences in considering these risks. The guideline might also acknowledge the bias toward action that has existed historically in medical practice, in this case the current practice of continuing an existing but untested obstetrical intervention because it may protect a pregnancy at risk

Pharmacogenetic Testing. Pharmacogenetic testing represents an application of genetic testing that is predicted to play an important role in the medical practice of the future (2). (See also Chapter 14) Genetic variation in drug metabolism is likely to account for much of the variation in susceptibility to the adverse and therapeutic effects of drugs seen in clinical practice. As information about the genetic contributors to drug response accumulates, it may be possible to individualize drug therapy for many conditions, to achieve either more effective treatment or decreased side effects or both, through tailoring prescriptions to an individual's genetic profile for drug metabolism.

An example that highlights this potential is the use of testing for *CYP2C9* polymorphisms to estimate appropriate warfarin dosing. Warfarin is an anticoagulant prescribed for over one million patients in the United States annually (47). It requires close monitoring to prevent both bothersome and potentially fatal side effects of bleeding. In a study by Higashi et al. (48), a clinic population was studied for the relationship between *CYP2C9* genotype and anticoagulation status. The authors found a correlation between the presence of *CYP2C9* *2 and *3 alleles (as compared with the more common *1 allele) and the rate of above-range anticoagulation levels, the maintenance anticoagulation dose, the time to stable warfarin dose, and the number of bleeding events on warfarin. Their findings indicate a potential benefit from testing for these genetic variants prior to prescribing warfarin. However, the authors note the need for more data before such testing is considered for use in clinical practice. They recommend a randomized controlled trial to ensure accurate measurement of outcomes after testing, in particular to determine whether warfarin dosing is in fact more accurate and bleeding complications are actually reduced with genetic testing, as predicted by their study. They also note the need for cost-effectiveness studies, comparing this approach with the empiric dosing and monitoring which is the current standard of care for patients on warfarin.

The study identifies the essential factors in achieving benefit from pharmocogenetic testing: (*1*) the test measures a response to a medication in common clinical use, (*2*) the test has been used in a defined and representative study population and clinical setting, and (*3*) clinical outcomes link the identified genetic variant to health outcomes. Observations of this kind justify a randomized clinical trial to assess the feasibility of this genetic testing strategy. Guidelines formulated prior to the results of a trial might note the potential biases of observational studies—such as population selection bias—and might use the current data to advocate for the trial rather than for the immediate introduction of testing. However, the knowledge that there are unidentified patients within treatment populations who are at increased risk for bleeding might also influence current practice. For example, this observation could lead to a more careful monitoring of treatment in patients who achieve a higher than expected level of anticoagulation on a standard initial dose or take a longer than usual to achieve a stable warfarin maintenance dose.

Cost-Effectiveness and Other Social Outcomes

Cost-effectiveness has not universally been incorporated into clinical practice guidelines, because guidelines have traditionally focused on individual patient care, rather than on population health (49,i). Economic analyses are constructed to address the allocation of limited resources across the health system, and thus attempt to define the most efficient way of financing health care to improve the health of the population. To accomplish this task, any expenditure for a new test must be compared relative to the other choices available, for example, other types of diagnostic or screening tests or not testing at all. The benefit of the approach chosen over the other choices is compared in terms of the cost of that choice relative to the outcome benefit achieved.

Costs to be considered in the evaluation of a genetic tests would include any genetic counseling or education about the genetic implications of the test, before or after testing, and the costs of any tests or interventions that result from the initial genetic tests; and the limitations of the testing program resulting from false positive and false negative test results would also need to be considered. For many new genetic tests, the pretest counseling will likely be less detailed and time-intensive than the current counseling model used in medical genetics—as is proposed for factor V Leiden testing (34)—and accurate estimates of the costs of testing will not be possible until a standard of care is established.

A testing strategy may be effective but be cost-ineffective; that is, the cost is deemed too high for the amount of health benefit derived. The threshold for defining an intervention as cost-effective may be either an absolute cost or a cost comparable to that of an already accepted intervention. Conflicts may occur between decisions driven by the goals of cost-effectiveness, which are to maximize return on a given investment, and those of equity, as determined from an individual patient's perspective. For example, if two screening programs have the same cost—one offering an expensive screening test to half the population, and the other offering a cheaper test to the whole population—and the expensive test reduces net mortality more, the expensive test would be more cost effective. However unless there were a factor like age or gender that provided a medical justification for offering the test to half rather than all the population, an ethical argument could be made for a policy using the less cost-effective approach, because of concerns about equal access to medical testing. Additionally, a cost-effective approach does not necessarily translate into a less expensive approach. One approach may be more cost-effective than another, in that it provides more health benefits per extra dollar, but still have net costs that are substantially higher and as a result could be unacceptable to some societies or health-care systems.

Genetic tests have social implications beyond cost-effectiveness. Studies that are limited to narrowly defined medical outcomes—e.g., whether or not a particular clinical diagnosis is present or a particular outcome of treatment occurs—will be inadequate to evaluate the full implications of testing. Some social factors, such as access to testing and treatment, are critical determinants of medical outcomes. Others may determine the potential for a testing program to cause harm or its cost effectiveness. Unfortunately, the outcomes that have generated the greatest concern about genetic testing, such as insurance or employment discrimination, stigmatization, and long-term psychological harms from testing, are difficult to study (50–52). To fully evaluate them, separate studies designed to provide adequate measurements of these outcomes might be necessary. Similarly, evaluation of inequities in access to genetic services may require studies designed to examine this issue.

The appropriate application of genetic testing also requires information about the effects of different pretest and post-test counseling procedures and about post-test behaviors. The goal of some genetic tests may be to guide choices about health behaviors, but knowledge about genetic risk does not necessarily lead to risk-reducing behaviors and could, contrariwise, produce a fatalistic attitude that reduces motivation to change (53). To the extent that such effects occur, they demonstrate the potential for interaction between social outcomes of testing and their clinical utility and cost-effectiveness.

Developing Guidelines

The quality of available evidence (Table 16.1) and difficulties in completely assessing the balance of benefits and harms (Table 16.2) are challenges in the assessment of all diagnostic and screening tests. Several factors, however, pose unique challenges for genetic testing. In genetic testing, "labeling" effects of a positive test result may be more pronounced than for other medical tests, because genetic information is typically presented as highly predictive (54). Thus, the potential psychological harms, of false-positive and true-positive tests, may be greater. Second, the marginal benefits of genetic information may be hard to quantify. They depend on the alternatives that might be pursued in the absence of screening, and on the prevalence of the genotype in the target population, its associated risk, available interventions and their applicability, compliance with offered interventions, and harms from interventions. Assessing the cost-effectiveness of genetic testing is an added

Table 16.2 Problems in Weighing Benefits Versus Harms of an Intervention
- Evidence of benefits often indirect or projected based on multiple assumptions
- Need to quantify *incremental* benefit
- Interventions possible without screening
- Effect of compliance
- Nonmedical harms hard to quantify
- Effect of individual preferences

challenge. Cost-effectiveness analysis must incorporate estimates of the costs of testing, counseling, and follow-up; the costs of intervention; the savings due to averted health-care costs; and averted disease treatment (for those found not to be at risk) and compare that result to estimated health gains. These in turn need to incorporate psychological benefits and harms in the equation.

This research agenda is challenging, and data are likely to accumulate slowly and incrementally. Both the warfarin and factor V Leiden examples demonstrate the importance of presenting data available about genetic tests—with its limitations—in formats that allow the specific clinical outcomes of tests to be evaluated. This represents the first step in guideline development. Accurate reporting of the properties of genetic tests, by using standardized methodologies serves an important clinical purpose. Reports summarizing what is known and not known about a genetic test allow policymakers, health-care providers, and patients to evaluate new testing opportunities with the assurance that their decisions about test use will be based on the available evidence. In-depth summaries of evidence can also guide future research. To facilitate this process, authors should be encouraged to present methods and results in standardized ways, as outlined in Table 16.1 (12,55).

The second step in guideline development is a consideration of test use in a specific clinical practice setting. In this step, the guideline must consider the value of testing for the specific population being served, its cost and acceptability, and its priority relative to other health-care services. Tests suitable in some settings may not be suitable in others. Objective information concerning the current state of knowledge about a test is an essential, but not sufficient, component of this process.

Physicians contemplating the use of a guideline must ask themselves the following questions: Will this guideline help me give better care to my patients? Will following it result in my patients living longer and/or being healthier while they are alive? Is it clear for which patients the guideline is helpful and for which patients it is not? Will it help me choose the most effective and cost-efficient approach and avoid harmful or unnecessarily expensive treatments? Does it represent the best evidence available on the subject and will it be updated when better evidence is available? Criteria for evaluating a practice are outlined in Table 16.3. In providing answers to these questions, guidelines should use a process that is transparent to the user, revealing how data were gathered and analyzed, the perspective from which data were interpreted; and how recently the guidelines have been updated.

Physicians must also be aware of the perspective and potential biases of the panel that constructs the guidelines, i.e., the constitution of the panel, how panel members were chosen, how the guideline process was supported, and whether the panel represented group(s) with potential advocacy bias or self-interest. Not all guidelines can be relied upon to provide disinterested guidance, based on a careful review of evidence and a well-delineated reasoning process. Additionally, different panels make different needs based or political decisions about the composition of the panel. For example, a panel on newborn screening may choose to include consumer representatives such as parents of a child affected with a genetic disease, to gain the

Table 16.3 Assessing the Quality of Guidelines

- Explicit disclosure of panel's members, their expertise and perspective (e.g., primary care, population-based, high-risk case finding, group with potential advocacy bias or self-interest)
- Financial sponsors of the guidelines clearly stated
- Construction of analytic framework
- Description of how evidence was found and filtered and how the references were used
- Critical review of literature informing the framework and explicit grading of the quality of evidence
- Explicit labeling of gaps in evidence
- Explicit discussion of other reasonable conclusions that could be reached from this data
- Explicit discussion of the evidence, when available, that would override opinion
- Explicit labeling of expert opinion, reasons for its inclusion, line of reasoning, and strength of extrapolation from other data used in formation of the expert opinion
- Explicit discussion of benefits and harms considered in arriving at the opinion and relative weight placed on specific benefits and harms in different scenarios to provide means of incorporating patient preference
- Discussion of cost-effectiveness
- Proposed plan and explicit date for upgrading the guidelines

perspective of those affected by the panel decision and enhance the credibility of the resulting guideline with the public. Other guideline panels, with different goals, constraints and political environments, may seek different compositions. There is no single correct composition for a guideline panel: the compelling standard is that users should know who was involved in the guideline formation. Similarly, transparency of the panel's underlying philosophy is important. This difference can be seen by contrasting the philosophy of American Cancer Society guideline processes with those of the U.S. Preventive Services Task Force. The former will adopt a screening intervention that theoretically is of benefit but lack supporting outcome data. The underlying assumption is that one should endorse an opportunity with a strong possibility of doing good until there is evidence that it causes harm (56). The U.S. Preventive Services Task Force's stand when making recommendations is that "first do no harm" must be the organizing principle, and that an intervention should not be adopted until there is high-quality outcome evidence of benefit (28). As a result, these two groups may analyze existing data similarly but reach disparate recommendations. The purpose of evaluating underlying assumptions of this kind is not to suggest that there is a single correct approach to guideline development but rather to allow the health-care provider using the guideline readily to determine the philosophy under which the guideline was created.

Summary

New techniques to organize and disseminate evidence about genetic tests are an important step in support of the rational and prudent use of this new technology. To be effective, evaluation methods must be as complete and unbiased as possible. The most important use of such evidence summaries will be in the development of clinical practice guideline. Clinicians, guideline panels, and other policymakers must consider genetic tests in light of alternative approaches to care; that is, the health benefits and other outcomes of a genetic test must be evaluated compared to the outcomes that could be obtained in the absence of genetic testing. To make this comparison, adequate evidence concerning the analytic validity, clinical validity and clinical utility of genetic tests is needed, based on studies with appropriate control populations and, ideally, measurement of social outcomes.

Geneticists and other clinicians using genetic tests face the need, not uncommon in medicine, to formulate guidelines prior to the availability of the desired evidence. Factors to be considered in formulating and evaluating these guidelines have been reviewed for factor V Leiden and pharmacogenetic testing related to warfarin use. These factors can be summarized as providing transparency to the guidelines formation process. More specifically, guidelines panels should make explicit the evidence that was considered in the formulation of the guideline, how the quality of the evidence was evaluated, and where the panel has relied on expert opinion in formulating its recommendations. When expert opinion has been used, the compelling clinical argument that justifies this approach should be articulated, with an explicit discussion of the line of reasoning used to inform the guideline and of alternative opinions that might be reached from the same evidence. Most importantly, guidelines based on incomplete evidence should advocate for the appropriate studies to improve the evidence and anticipate a revision of the guideline when new evidence is available.

References

1. Collins FS. Shattuck lecture—Medical and societal consequences of the Human Genome Project. N Engl J Med 1999;341:28–37.
2. Roses AD. Pharmacogenetics and the practice of medicine. Nature 2000;405:857–865.
3. Eisenberg J. Ten lessons for evidence-based technology assessment. JAMA 1999; 282:1865–1869.
4. Mowatt G, Bower DJ, Brebner JA, et al. When and how to assess fast-changing technologies: a comparative study of medical applications of four generic technologies. Health Technol Assess. 1997;1(14):i–vi, 1–149.
5. Pinsky LE, Deyo RA (1999). Clinical Guidelines: A Strategy for Translating Evidence into Clinical Practice, in Evidence-based Clinical Practice; Concepts and Approaches, Geyman J, Deyo RA, Ramsey S (eds). Woburn: Butterworth Heinemann.
6. Berg A, Atkins D, Tierney W. Clinical Practice Guidelines in Practice and Education. J Gen Intern Med. 1997;12 (suppl):25–33.
7. Institute of Medicine. Guidelines for Clinical Practice: from Development to Use. National Academy Press, Washington, DC 1992.
8. Woolf SH. Evidence-based medicine and practice guidelines: an overview. Cancer Control 2000;7:362–367.
9. Miller J, Petrie J. Development of practice guidelines. Lancet 2000;355:82–83.
10. Holtzman NA, Watson MS. Promoting safe and effective genetic testing in the United States. Final report of the Task Force on Genetic Testing. J Child Fam Nurs 1999;2: 388–390.

11. Grody WW, Griffin JH, Taylor AK, Korf BR, Heit JA. American College of Medical Genetics consensus statement on factor V Leiden mutation testing. Genet Med 2001; 3:139–148.

12. Burke W, Atkins D, Gwinn M, Guttmacher A, Haddow J, Lau J et al. Genetic test evaluation: information needs of clinicians, policy makers and the public. Am J Epidemiol 2002;156:311–318.

13. Becker D, Philbrick J, Bachhuber T, Humphries J. D-dimer testing and acute venous thromboembolism. Archives of Internal Medicine 1996;156:939–946.

14. Ford D, Easton DF, Stratton M, Narod S, Goldgar D, Devilee P, et al. Genetic heterogeneity and penetrance analysis of the *BRCA1* and *BRCA2* genes in breast cancer families. Am J Hum Genet 1998;62:676–689.

15. Warner E, Foulkes W, Goodwin P, Meschino W, Blondal J Paterson C, et al. Prevalence and penetrance of *BRCA1* and *BRCA2* gene mutations in unselected Ashkenazi Jewish women with breast cancer. J Natl Cancer Inst 1999;91:1241–1247.

16. Hopper JL, Southey MC, Dite GS, Jolley DJ, Giles GG, McCredie MR, et al. Population-based estimate of the average age-specific cumulative risk of breast cancer for a defined set of protein-truncating mutations in *BRCA1* and *BRCA2*. Australian Breast Cancer Family Study. Cancer Epidemiol Biomarkers Prev 1999;8(9):741–747.

17. Struewing JP, Hartge P, Wacholder S, Baker SM, Berlin M, McAdams M et al. The risk of cancer associated with specific mutations of *BRCA1* and *BRCA2* among Ashkenazi Jews. N Engl J Med 1997;336:1401–1408.

18. Satagopan JM, Offit K, Foulkes W, Robson ME, Wacholder S, Eng GM, et al. The lifetime risks of breast cancer in Ashkenazi Jewish carriers of *BRCA1* and *BRCA2* mutations. Cancer Epidemiol Biomarkers Prev 2001;10:467–473.

19. Risch HA, McLaughlin JR, Cole DEC, Rosen B, Bradley L, Kwan E, et al. Prevalence and penetrance of germline *BRCA1* and *BRCA2* mutations in a population series of 649 women with ovarian cancer. Am J Hum Genet 2001;68:700–710.

20. Anglian Breast Cancer Study Group. Prevalence and penetrance of *BRCA1* and *BRCA2* mutations in a population-based series of breast cancer cases. Br. J Cancer 2000;83: 1301–1308.

21. Thorlacius S, Struewing JP, Hartge P, Olafsdottir GH, Sigvaldason H, Tryggvadottir L, al. Population-based study of risk of breast cancer in carriers of *BRCA2* mutation. Lancet 1998;352:1337–1339.

22. Burke W, Press NA, Pinsky LE. BRCA1 and BRCA2: a small part of the puzzle. J Natl Cancer Inst 1999;91:904–905.

23. Begg C. On the use of familial aggregation in population-based case probands for calculating penetrance. J Natl Cancer Inst 2002;94:1221–1226.

24. Ioannidas JPA, Ntzani EE, Trikalinos TA, Contopoulos-Ioannidis DG. Replication validity of genetic association studies. Nat Genet 2001;29:306–309.

25. Hirschhorn JN, Lohmeuller K, Byrne E, Hirschhorn K. A comprehensive review of genetic association studies. Genet Med 2002;4:45–61.

26. Kupferminc MJ, Eldor A, Steinman N, et al. Increased frequency of genetic thrombophilia in women with complications of pregnancy. N Engl J Med 1999;340:9–13.

27. Ford D, Easton DF, Bishop DT, Narod SA, Goldgar DE. Risks of cancer in BRCA1-mutation carriers. Breast Cancer Linkage Consortium. Lancet 1994;343:692–695.

28. US Preventive Services Task Force. Guide to Clinical Preventive Services. Baltimore: Williams & Wilkins, 1996.

29. Ridker PM, Miletich JP, Hennekens CH, et al. Ethnic distribution of factor V Leiden in 4047 men and women. Implications for venous thromboembolism screening. JAMA 1997;277:1305–1307.

30. Renkonen-Sinisalo L, Aarnio M, et al. Surveillance improves survival of colorectal cancer in patients with hereditary nonpolyposis colorectal cancer. Cancer Detect Prev 2000;24:137–142.
31. Burke W, Petersen G, Lynch P, et al. Recommendations for follow-up care of individuals with an inherited predisposition to cancer I. hereditary nonpolyposis colon cancer. JAMA 1997;277:915–919.
32. Laken SJ, Petersen GM, Gruber SB, et al. Familial colorectal cancer in Ashkenazim due to a hypermutable tract in APC. Nat Genet 1997;17:79–83.
33. Burke W, Daly M, Garber J, et al. Recommendations for follow-up care of individuals with an inherited predisposition to cancer II.BRCA1 and BRCA2. JAMA 1997;277: 997–1003.
34. Grody WW, Griffin JH, Taylor AK, Korf BR, Heit JA. American College of Medical Genetics consensus statement on factor V Leiden mutation testing. Genet Med 2001; 3:139–148.
35. Rosendaal FR. Venous thrombosis: a multicausal disease. Lancet 1999;353:1167–1173.
36. Meyer G, Emmerich J, Helley D, et al. Factors V leiden and II 20210A in patients with symptomatic pulmonary embolism and deep vein thrombosis. Am J Med 2001;110:12–15.
37. Bloemenkamp KW, Rosendaal FR, Helmerhorst FM, Buller HR, Vandenbroucke JP. Enhancement by factor V Leiden mutation of risk of deep-vein thrombosis associated with oral contraceptives containing a third-generation progestagen. Lancet 1995;346: 1593–1596.
38. Simioni P, Sanson BJ, Prandoni P, et al. Incidence of venous thromboembolism in families with inherited thrombophilia. Thromb Haemost 1999;81:198–202.
39. Middeldorp S, Meinardi JR, Koopman MM, et al. A prospective study of asymptomaic carriers of the factor V Leiden mutation to determine incidence of venous thromboembolism. Ann Intern Med 2001;135:322–327.
40. De Stefano V, Martinelli I, Mannucci PM, et al. The risk of recurrent deep venous thrombosis among heterozygous carriers of both factor V Leiden and the G20210A prothrombin mutation. N Engl J Med 1999;341:801–806.
41. Martinelli I, Bucciarelli P, Margaglione M, et al. The risk of venous thromboembolism in family members with mutations in the genes of factor V or prothrombin or both. Br J Haematol 2000;111:1223–1229.
42. van Boven HH, Vandenbroucke JP, Briet E, et al. Gene-gene and gene-environment interactions determine risk of thrombosis in families with inherited antithrombin deficiency. Blood 1999;94:2590–2594.
43. Rosendaal FR, Vessy M, Rumley A, et al. Hormonal replacement therapy, prothrombotic mutations and the risk of venous thrombosis. Br J Haematol 2002;116:851–854.
44. Meinardi JR, Middeldorp S, de Kam PJ, et al. The incidence of recurrent venous thromboembolism in carriers of factor V Leiden is related to concomitant thrombophilic disorders. Br J Haematol 2002;116:625–631.
45. Mienardi JR, Middeldorp S, de Kam PJ, et al. Risk of venous thromboembolism in carriers of factor V Leiden with a concomitant thrombophilic defect: a retrospective analysis. Blood Coagul Fibrinolysis 2001;12:713–720.
46. Lidegaard O, Bygdeman M, Milsom I, et al. Oral contraceptives and thrombosis: from risk estimates to health impact. Acta Obstet Gynecol Scand 1999;78:142–149.
47. Hirsch J, Dalen DE, Anderson D, et al. Oral anticoagulants: mechanism of action, clinical effectiveness, and optimal therapeutic range. Chest. 2001;119:8–21.
48. Higashi MK, Veenstra DL, Kondo LM, et al. Association between CYP2C9 genetic variants and anticoagulation-related outcomes during warfarin therapy. JAMA. 2002;287: 1690–1698.

49. Ramsey S. Economic analyses and clinical practice guidelines: why not a match made in heaven. J Gen Intern Med 2002;17:235–237.
50. Saha S, Hoerger TJ, Pignone MP, et al. The art and science of incorporating cost effectiveness into evidence-based recommendations for clinical preventive services. Am J Prev Med. 2001 Apr;20(3 suppl):36–43.
51. Ubel PA, DeKay ML, Baron J, et al. Cost-effectiveness analysis in a setting of budget constraints—is it equitable? N Engl J Med 1996 May 2;334(18):1174–1177.
52. Rothenberg KH. Genetic information and health insurance: state legislative approaches. J Law Med Ethics 1995;23:312–319.
53. Billings PR, Kohn MA, de Cuevas M, et al. Discrimination as a consequence of genetic testing. Am J Hum Genet 1992;50:476–482.
54. Lapham EV, Kozma C, Weiss JO. Genetic Discrimination: perspectives of consumers. Science 1996;274:621–624.
55. Marteau TM, Lerman C. Genetic risk and behavioural change. BMJ 2001;322:1056–1059.
56. Khoury MJ, Thrasher JF, Burke W, et al. Challenges in communicating genetics: A public health approach. Genet Med 2000;2:198–202.
57. Little J, Bradley L, Bray MS, et al. Reporting, appraising, and integrating data on genotype prevalence and gene-disease associations. Am J Epidemiol 2002;156:300–310.
58. Condor B, "Complex Picture" Chicago Tribune. 20 March 1994 quoted in Malm HM. Medical screening and the value of early detection: when unwarranted faith leads to unethical recommendations. Hastings Center Report 29.1999:26–37.

17

Using human genome epidemiologic evidence in developing genetics services: the U.K. experience

Brenda J. Wilson, Jeremy M. Grimshaw, and Neva E. Haites

The U.K. National Health Service

The United Kingdom's National Health Service (NHS) came into being in 1948 with the aim of removing financial barriers to health care for the entire population. It was naively assumed that the cost of the service would fall as the population became progressively healthier, and this error was soon recognized. However, the NHS continues to provide reasonably comprehensive primary care, hospital and specialist services with universal population coverage, and is still largely free at the point of use (although co-payment exists in some areas, e.g., for prescription drugs). Access to hospital and specialist services is constrained by means of the referral system and, at any given time, a patient can be registered to receive care from only one family practitioner. Except in cases of emergency, access to specialist services is available only by direct referral from that practitioner or another hospital specialist.

The devolution of government authority from the Westminster parliament to the Scottish parliament and the Welsh and Northern Irish legislative assemblies has meant a formal transfer of responsibility for NHS policy to these bodies. However, the NHS has always been administered separately for the four countries in the United Kingdom, with matters pertaining to the population of England the responsibility of the Secretary of State for Health, and for those to the other three countries the responsibility of their respective representatives in the U.K. cabinet, namely the Secretaries of State for Scotland, Wales, and Northern Ireland. Both before and after devolution, an executive department of health in each country has been responsible for the implementation and oversight of services for its own resident population. Thus, from the start of the NHS, differences in some administrative arrangements, bodies, and to some extent, culture, were evident in the four U.K. countries. Nevertheless, the core NHS values, policies, and objectives have remained essentially the same across the United Kingdom as a whole. In addition, whether set at a U.K. or country level (1,2), policies have generally been presented as a framework within which local (district) community, primary care, and hospital services are planned and evaluated, taking into account analyses of local health needs (3–5).

Specialist Genetics Services within the NHS

The current organization of specialist genetics services within the NHS reflects their historical evolution from academic departments to regional centres, each of which serves a number of smaller districts and usually has integrated clinical and laboratory services (6). The latter are often arranged as consortia in order to achieve economies of scale, share scarce expertise, and ensure consistent quality standards (7). Currently, there are 25 regional centres, each of which typically serves a population of 2 million to 6 million people, with an average of one or two consultant geneticists per million population (8,9). Genetics teams now routinely include laboratory scientists, doctors, specialist nurses, and genetic counsellors, with data handling and genetic record facilities in single functional units. Regional services often extend to include clinics in smaller district hospitals and sometimes patients' homes (9,10). Until the last few years, U.K. genetics services were concerned in the main with the diagnosis of fairly uncommon congenital and inherited disorders, and counseling generally concerned reproductive issues (8).

For many years, genetic counseling has been provided mainly by physicians. The introduction of genetic counselors with a science or nursing background has been a fairly recent development (11–13). This meant that clinical services had limited capacity to deal with the increasing demand which resulted from the rapid expansion of molecular genetics knowledge (8). This led to consideration of new ways of dealing with patient demand, including arguments for an increased role for primary care physicians (8,14–16), the development of guidelines (17) and computer support aids for family physicians (18,19), the introduction of intermediate levels of specialist advice (8,9), and an enhanced role for nurses (8,11,12,20). Surveys, qualitative studies, and statements by professional bodies suggest that U.K. primary care physicians identify a role for themselves which reflects traditional general practice activities (18,21), for example taking a family history, making appropriate referrals to specialist services, and providing emotional support, but they identify significant barriers to an expanded role (15,22), including inadequate knowledge or confidence (23), practical (24), and ethical and legal issues (18,24–29).

Evidence-Based Health Services

There has been a gradual, if still incomplete, shift in NHS professional and managerial culture towards evidence-based practice and services, and this reflects the emphasis on improving the quality of health care which pervades government policymaking (30). This can be seen in practice in the implementation of increasingly formal appraisal mechanisms of effectiveness and costs (31) as well as in numerous local initiatives (32,33). In this context, evidence is considered primarily to relate to effectiveness and efficiency (34). *Effectiveness* refers to the ability of an intervention, technology, or service to achieve what it intends to achieve, in practice, and efficiency (or cost-effectiveness) indicates how well it makes use of scarce resources in meeting these goals. (With this definition, effectiveness differs from efficacy, which is the performance of the intervention or technology under ideal circumstances, e.g. in a highly selected or motivated group of patients.) The most efficient intervention, therefore, is one which achieves the greatest effect per unit of resource use. In evaluating health-care interventions and technologies, comparisons of efficiency may be considered at two levels. The first is assessment of technical efficiency, where different interventions designed to meet the same specific goal (e.g., different interventions to reduce mortality from breast cancer in *BRCA1* or *BRCA2* carriers) are compared against each other. The second is assessment of allocative efficiency, where overall distribution of resources within the health-care system across different programs are examined to determine whether shifting resources from one program to another would lead to better overall health outcomes for the population being served; an example would be comparing renal replacement programs with prenatal care.

If these concepts are applied to genetics, the effectiveness of a test may be considered to be approximately equivalent to clinical utility (the balance of benefits and harms to an individual patient which flow from the decision to take the test). However, the concept of effectiveness may also be applied to a service, and may be thought of as its ability to meet its goals in relation to a particular population. Extending evaluation to this level requires consideration of two other issues: *equity*, which may be another goal of the service, and the success of the *implementation* in practice of potentially effective interventions. In the later part of the chapter, we will explore the interplay of these issues as they relate to a genetics service within the NHS, by using population hemoglobinopathy screening as a case study.

Equity is the concept of "fairness," which has been a consistent principle at the heart of the NHS from its inception in 1948 (35). It is generally interpreted as "equality of access (36)," and has led to an underpinning policy goal of minimizing geographical, socioeconomic, or other barriers which might lead to inequalities in access to, or (sometimes) use of, services. The two goals of maximizing efficiency and avoiding inequity are not totally compatible, and lead to difficult trade-offs (36,37).

The transfer of research-based evidence into practice (*implementation*) is unpredictable and can be a slow and haphazard process (38,39). Implementation of interventions may be inappropriately lengthy (denying patients proven benefits and policymakers information which could influence their decision-making), or premature, before effectiveness is established (leading to the exposure of patients to potentially ineffective or harmful interventions, and inefficient use of resources). Whether an effective intervention is effectively implemented in practice (or whether an apparently ineffective intervention is not implemented or discontinued) depends on overcoming structural, organizational, professional, or other issues (see Table 17.1).

Until the early 1980s, the views of medical professionals defined what was considered necessary or appropriate in relation to services, treatments, and interventions (40). Increasing political control over the NHS policy agenda has fundamentally challenged this clinical dominance at a central level (30), although clinicians still largely retain control at the operational clinical level (41). The central agenda emphasizes the importance of an evidence-based model of development, which has its most recent examples in initiatives promoting clinical effectiveness (42), the establishment of bodies such as the National Screening Committee (43), the NHS Research & Development Programme (44), and the National Institute for Clinical Excellence (45). However, the NHS is a multilevel system and does not work in a completely top-down manner. Therefore, although central policy decisions may appear to drive the development and configuration of new services (such as in genetics), they may have in reality been driven themselves by the need to remove perceived inequities in the provision of services interventions across the country. Thus, the standard of care or provision has often been driven by largely independent decisions made at regional, district, or hospital levels, which may or may not be evidence-based. For example, many regional clinics across the United Kingdom incorporated counseling and testing for familial breast cancer into existing service models in advance of any robust evidence of benefit. The increasingly obvious variations in provision across the country (46), accompanied by calls for the NHS to invest more resources in clinical genetics, prompted central reviews (47–49), with subsequent recommendations for national service models and quality standards. To some extent, this reflects a model of policymaking referred to as "extemporaneous (50)," in which the decisions which are made result from the combined influence

Table 17.1 Factors Influencing the Effective Implementation of Evidence-Based Practice

Level	Examples of Factors
Health-care system	National policies
	Remuneration systems
	National human resources
Provider organization	Equipment
	Case mix
Professional peer group	Habits, norms
Individual professional	Knowledge
	Attitudes
	Skills
Professional–patient interaction	Information overload within consultation
	Acts of omission
Patient/community	Inappropriate expectations/demands (too high or too low)

of the various stakeholders and the force of their arguments, not on an explicit and formalized assessment of empirical data (the so-called "evidentiary" model).

However, it would be unfair to suggest that U.K. clinical geneticists disregard the need for an evidence base. For example, regional clinics throughout the United Kingdom worked to develop cancer genetics referral and management criteria which represented their interpretation of "best practice" and which more recently have been disseminated by the U.K. Cancer Family Study Group as representing expert consensus (17) (Table 17.2). Generally similar guidelines are used throughout Europe

Table 17.2 U.K. Cancer Family Study Group Breast Cancer Referral and Mammography Guidelines

Family History of Breast Cancer	*Expected Breast Cancer Cases 40–50y (Population Risk = 1%)*	*Lifetime Risk (Population Risk 20–80y = 9%)*	*Risk Group*[a]	*Early Mammography*[b]	*Specialist Genetics Clinic*[c]
1° Relative[d]					
>40y	Max 1 in 50	Max 1 in 8	Low	No	No
<40y	1 in 30–1 in 50	1 in 12–1 in 6	Low/moderate	Yes	No[e]
Female <30 or male affected at any age	Max 1 in 25	Max 1 in 6			
2° Relatives[d]					
Average age 50–60	1 in 40	1 in 8	Low	No	No
Average age 40–49	1 in 25	1 in 6–1 in 4	Moderate	Yes	No[e]
Average age 30–39	1 in 14	1 in 4–1 in 3	High	Yes	Yes
3° Relatives[d]					
Average age 50–60	1 in 15	1 in 4	Moderate	Yes	Yes
Average age 40–50	1 in 11	1 in 3	High	Yes	Yes
					Yes
Breast and other cancers					
≥1 relative with breast cancer ≥50y plus ≥1 relative with ovarian cancer at any age *or* 1 relative with both	Usually >1 in 25	Usually >1 in 6	Moderate/high		
≥1 relative with breast cancer <40y plus ≥1 relative with childhood malignancy	—	—	May be high	Avoid mammograms pending genetics review	Yes

[a]Low risk, <2 × population lifetime risk; moderate, 2–3 × population lifetime risk; high, >3 × population lifetime risk.
[b]Early screening mammography should start not younger than 35 years of age in the moderate risk group; the risk of cancer and potential benefits of screening are most likely to be seen in the 40 to 50 year age group.
[c]Formal risk assessment and possible genetic mutation analysis
[d]"Relative" includes first degree relative and their first degree relatives. A relative with clearly bilateral breast cancer can be viewed as two relatives for simplicity. A male relative with breast cancer counts as a young female (<40).
[e]Ethnic origin may make mutation searching easier, for example Ashkenazi Jewish ancestry might mean genetic testing would be more helpful even with a less striking family history.
Source: Ref. 17.

(51) and have the effect that very few women meet criteria for *BRCA1* or *BRCA2* testing, although many more at moderate and high risk are offered appropriate risk reduction interventions (clinical breast examination, screening mammography, chemoprevention trials, and prophylactic mastectomy). As the rationale for these interventions is based on largely observational data (52–56), most U.K. centers collaborate with their European and North American counterparts in pooling data on practice and outcomes in an effort to improve the evidence base. An example of the outcome of such collaboration is the emerging consensus that surveillance activities alone do not appear to make a major impact on mortality in women who carry *BRCA1* or *BRCA2* mutations (57).

Case Study: Population Hemogobinopathy Screening

Prenatal and neonatal hemoglobinopathy (HbP) screening presents a complicated set of issues regarding evidence and how it is used. The issues relate to what is considered "effectiveness" in relation to screening (particularly prenatal), variations in the effective implementation of screening interventions, and how these affect goals of equity and efficiency. Underlying these considerations is the influence of geographic variation in gene prevalence across the United Kingdom, the association of genetic risk with ethnicity, and of ethnicity with the relative acceptability of different interventions.

Clinical Features

The HbPs (sickle cell disorders and thalassemias) (58) are caused by alterations in the globin protein chains which are responsible for oxygen transport by erythrocytes and they are amongst the commonest genetic conditions in northwest Europe. They mainly affect populations of black, Asian, and Mediterranean origin, with an autosomal recessive pattern of inheritance, although the clinical picture is modified by genetic and environmental factors.

The sickle cell (SCD) disorders include sickle cell anemia (SCA) and other related conditions of varying severity. Undiagnosed individuals with SCA usually present with painful crisis, pneumococcal infection, acute chest syndrome, splenic sequestration, stroke, and acute anemia (59–63), and it may lead to death in early infancy.

The thalassemias are characterized by failure to synthesize parts of the globin chain, leading to poor production and high rates of destruction of red cells. Like SCD, there are different forms, but the most common types compatible with survival, β-thalassemia major and intermedia and HbE β-thalassemia (64), are associated with progressive hemolytic anemia (65). This causes death by late childhood or early teens if untreated (66) by regular blood transfusion, which itself requires chelation therapy (67–69) to prevent iron overload and organ damage (70,71).

It is estimated that there are approximately 10,000 people in the United Kingdom with SCDs and 600 with β-thalassemia major (72). The birth rate in some migrant communities is generally higher than in the authochthonous white population, leading to rising overall prevalence of carriers and increasing incidence of annual homozygote conceptions and births. The limited available data suggests that the lifetime health service costs are approximately £50–£80,000 (US$80–$127,000) for SCD and £120,000 (US$190,000) for β-thalassemia (1997 prices) (73–75).

Early detection of infants with sickle cell disorders allows penicillin prophylaxis and comprehensive care to be instituted before the age of 3 months, and these interventions have been shown to reduce morbidity and mortality (76). The early detection of major thalassemia syndromes has not been shown to be of benefit (74).

Guidelines recommending screening for HbP in the United Kingdom have been in existence since 1988 (77–79). Since 1994, all screening policy in the United Kingdom has been based on the recommendations of the NSC (43,80), which makes explicit use of modified WHO criteria (81,82) in formulating its advice. These criteria (Table 17.3) suggest that for the health problem under consideration, important, acceptable screening tests and effective, acceptable interventions should be available, and that the program, once implemented, should be cost-effective and sustainable. Advocates of HbP prenatal and neonatal screening have argued for a co-ordinated national public health approach in order to improve screening standards, data collection, and professional and community education. In 2000, the U.K. government, informed by the NSC, appeared to agree with this conclusion and announced plans for a national screening program (83).

It is highly likely that the HbPs originally arose through spontaneous mutations, with a selection advantage arising through the protection they confer against malaria, which may explain their high prevalence in areas of high malaria prevalence (84,85). Uneven settlement of inward migrants from countries in these regions has led to wide variation in the distribution of the genes across U.K. populations (86). It is argued that this uneven distribution of high-risk populations has led to geographic variation in the expertise of health care providers and in their awareness of the need for culturally sensitive services: as specialists in lower prevalence areas have fewer cases to deal with, they have less opportunity to acquire expertise, and there is no perceived need or demand for dedicated services (87,88). Threefold variation in the use of prenatal diagnosis (PND) in the United Kingdom has been reported: this has been attributed to variations to delays on antenatal screening, lack of own-language counselling, and poor follow-up of at-risk couples rather than variations in attitudes to prenatal testing among different ethnic groups (86). If the problems lie with the quality of service provision, this variation may be viewed as geographic inequity, as there are few areas with no at-risk population.

However, variation in expertise does not explain all of the variation in service provision. There is also evidence to suggest that some health districts with similar proportions of ethnic groups have implemented different combinations of antenatal and neonatal screening strategies, with no explicit policies existing in some districts. Before the recent decision to implement a national screening policy, many investigators and commentators suggested that HbP screening provision in the United King-

Table 17.3 The U.K. National Screening Committee Criteria for Appraising the Viability, Effectiveness, and Appropriateness of a Screening Program

The condition	1. The condition should be an important health problem.
	2. The epidemiology and natural history of the condition, including development from latent to declared disease, should be adequately understood, and there should be a detectable risk factor or disease marker and a latent period or early symptomatic stage.
	3. All the cost-effective primary prevention interventions should have been implemented as far as possible.
The test	4. There should be a simple, safe, precise and validated screening test.
	5. The distribution of test values in the target population should be known and a suitable cut-off level defined and agreed on.
	6. The test should be acceptable to the population.
	7. There should be an agreed policy on the further diagnostic investigation of individuals with a positive test result and on the choices available to those individuals.
The treatment	8. There should be an effective treatment or intervention for patients identified through early detection, with evidence of early treatment leading to better outcomes than late treatment.
	9. There should be agreed evidence-based policies covering which individuals should be offered treatment and the appropriate treatment to be offered.
	10. Clinical management of the condition and patient outcomes should be optimized by all health care providers prior to participation in a screening program.
The screening program	11. There must be evidence from high quality randomized controlled trials that the screening program is effective in reducing mortality or morbidity.
	12. Where screening is aimed solely at providing information to allow the person being screened to make an "informed choice" (e.g., Down's syndrome, cystic fibrosis carrier screening), there must be evidence from high-quality trials that the test accurately measures risk. The information that is provided about the test and its outcome must be of value and readily understood by the individual being screened.
	13. There should be evidence that the complete screening program (test, diagnostic procedures, treatment/intervention) is clinically, socially, and ethically acceptable to health professionals and the public.
	14. The benefit from the screening program should outweigh the physical and psychological harm (caused by the test, diagnostic procedures, and treatment).
	15. The opportunity cost of the screening program (including testing, diagnosis, treatment, administration, training, and quality assurance) should be economically balanced in relation to expenditure on medical care as a whole (i.e., value for money).
	16. There must be a plan for managing and monitoring the screening program and an agreed set of quality assurance standards.
	17. Adequate staffing and facilities for testing, diagnosis, treatment and program management should be made available prior to the commencement of the screening program.

Table 17.3 The U.K. National Screening Committee Criteria for Appraising the Viability, Effectiveness, and Appropriateness of a Screening Program (*continued*)

18. All other options for managing the condition should have been considered (e.g. improving treatment, providing other services), to ensure that no more cost effective intervention could be introduced or current interventions increased within the resources available.
19. Evidence-based information, explaining the consequences of testing, investigation, and treatment, should be made available to potential participants to assist them in making an informed choice.
20. Public pressure for widening the eligibility criteria for reducing the screening interval, and for increasing the sensitivity of the testing process, should be anticipated. Decisions about these parameters should be scientifically justifiable to the public.

Source: Ref. 82.

dom, taken as a whole, was failing to achieve the potential outcomes suggested by programs in high-prevalence areas (89–93). For example, overall utilisation of antenatal diagnosis for thalassemias in the United Kingdom lay around 50%, compared with over 75% in Mediterranean countries (77,94,95), and uptake of testing for sickle cell disorders was 13%, compared with the 50% uptake in other high-risk populations (87). Uptake of testing is an indirect and imperfect measure of whether an antenatal screening program is achieving an objective of ensuring fully informed reproductive choice. However, despite it being the explicit goal of such programs (78,79,82,83,96), there is no standard method of measuring it, and it has not been used as an outcome measure in evaluations (73,97).

Universal and Selective Screening Strategies

A consequence of the variable prevalence of HbP is the variable cost-effectiveness of different screening strategies. *Universal* screening is offered to all eligible individuals or couples, whereas *selective* (or targeted) screening uses predetermined criteria to select those at higher risk. The arguments for and against these two approaches in the UK have been the subject of considerable analysis and discussion (36,73,74). In general, selective screening is justified on the basis of lower costs than universal approaches in areas of low gene prevalence. However, it carries the risk of missing potentially affected individuals, is dependent on the appropriate application of valid selection criteria (usually based on ethnic group), and incurs administrative costs related to the selection process (73). Compared with selective approaches, universal screening is likely to miss fewer cases, but it usually creates a higher workload and potentially increases the burden of adverse screening consequences, such as loss of a healthy fetus from a prenatal diagnostic procedure (although it is argued that these would be very rare [74]). Therefore, arguments regarding the relative merits of universal or selective screening strategies tend to revolve around efficiency issues: the incremental costs and benefits of identifying each extra case through a universal approach that would have been missed in a selective approach. Inevitably, the balance is influenced by the frequency of the mutations, and, in the case of prenatal screening, the likelihood of a decision to terminate or continue the pregnancy (73). This makes it difficult to draw easily generalizable conclusions based on gene prevalence in different communities: the severity of the condition and the cultural and religious acceptability of pregnancy termination are both heavy influences on this decision. In an examination of the interplay of equity and efficiency in relation to universal and selective screening strategies, Sassi and colleagues (36) point out that the incidence of disease in the (potentially) nonscreened population is extremely low. Although their analysis is confined to neonatal screening, they point out that a universal policy in the United Kingdom would incur a marginal cost per life-year gained (over a selective approach) in the otherwise unscreened population around 20 times that of other NHS-funded health-care interventions. Their conclusion is that the implementation of universal screening in almost all of the United Kingdom would imply that a very high value is placed on equity of access in this situation, or that decisions are being made on inadequate or misleading economic data. The value placed on equity of access is reflected in the comments of Zeuner and colleagues (74), who suggest that the use of economic criteria to determine whether to implement a universal or selective program leads to geographically based inequities of access. This, of course, assumes that the selection criterion is not applied perfectly and that some at-risk individuals or couples are missed (see below).

Acceptability

The available evidence suggests that the acceptability of antenatal screening and subsequent interventions varies by ethnic group. If offered PND, almost all from a Cypriot background are likely to accept it, around 60% from an Indian background, with lower uptake in Pakistani Muslim women, and very low uptake in Bangladeshi families (87). Before the introduction of first trimester PND in 1982, the acceptance in Indian and Pakistani Muslim populations was lower than it is now (73). Termination of pregnancy (TOP) is generally more acceptable for β-thalassemia major than SCD (96,98,99). African-Caribbean populations are less likely to request termination than African populations (96,98). However, it is difficult to use data on uptake of antenatal screening to determine whether the services are "acceptable." For example, U.K. data indicate that PND uptake in at-risk ethnic groups in Britain is lower than in those same groups in their country of origin; some argue that this means PND and TOP are generally acceptable in these ethnic groups (95,100,101) and that U.K. programs have been failing to deliver education and counselling in effective or appropriate ways (73). However, these differences might also be interpreted as revealing the "true" preferences of parents with an affected pregnancy in a country where there is greater availability of health services or more tolerance of disability at a societal level.

Implementation Factors

The NSC concluded that their screening criteria were generally met by antenatal screening for both SCD and thalassemias (the goal being reproductive choice) and by neonatal screening for SCD (the goal being improved quality of life and survival for affected infants) (1). A target has been set to implement "effective and appropriate" screening for thalassemia and SCD by 2004 (1). A policy of universal neonatal screening for SCD will be implemented across the UK (102); universal antenatal screening for HbPs will begin in high-prevalence areas, while research to inform its implementation in lower-prevalence areas is conducted (103). The NHS faces a number of organizational issues in trying to ensure a high-quality prenatal screening program. The first challenge will be to organise screening and PND within the first trimester of pregnancy; the second, assuming that a selective program is implemented in lower-prevalence areas, will be the development and application of a valid selection criterion to identify at-risk pregnancies (104).

First Trimester PND

The timing of screening in pregnancy influences acceptance rates, but in 1990–1994 the goal of first trimester PND, by chorionic villus sampling (CVS) was achieved in less than half of at-risk pregnancies in the United Kingdom as a whole, a large proportion of high risk pregnancies being identified only in the second trimester, or after the birth of an affected child (87,88). First trimester PND is difficult to achieve if the initial screening is not performed until the first antenatal clinic visit at the hospital, which may often be in the second trimester, especially for women from particular ethnic groups at high risk. In the United Kingdom, pregnant women usually see their general practitioner (GP) or community midwife first, sometimes as early as 6 weeks' gestation, but these professionals often lack awareness and training in relation to HbP (105–107).

The challenge of achieving first trimester PND for at risk couples will therefore depend on improving the knowledge and awareness of providers and the at-risk population, and on organizing the service to minimize delays. It is will be easier to achieve these goals in areas of high prevalence of the conditions, where families, parents, and providers are likely to have higher levels of knowledge and awareness than in lower-prevalence areas. The Department of Health response has been to commission research on the best approach to achieving these goals in medium- and low-prevalence areas (108).

Ethnicity as a Selection Criterion

Ethnic group or ancestry is essentially used as a proxy indicator for risk of carrying a relevant mutation. However, the concept of ethnicity is imprecise because it is a social construct, not biologically determined (109–112). It is also not a perfect predictor of risk, because some at-risk groups frequently marry outside their ethnic group (e.g., Cypriots, and Italians) (78). Therefore, although selective screening is intended to target parents of non–northern European origin, who have a higher risk of carrying a mutation than northern Europeans, individuals may self-identify with the "wrong" group (from a screening perspective). This will lead to offering screening to those who do not really meet the selection criterion (so using resources unnecessarily), and failing to offer it to people who ought to receive it (so potentially missing cases). This problem is compounded by the practice of leaving the identification of ethnic group to the subjective judgment of individual health care providers, who may or may not have explicit guidance and may or may not involve the at-risk individual or couple in making the judgment (113). For example, in an audit of pregnancies affected by major β-thalassemia, Modell and colleagues (88) found that individual health-care providers had failed to offer screening in a number of cases, documenting in the case notes that the couples "did not want" screening because they were Muslim. Prejudging attitudes in this way leads to removal of reproductive choice (the supposed goal of prenatal genetic screening) and failure to identify at-risk individuals or pregnancies. This therefore also undermines the policy goal if programs are implemented on the basis of judgments of cost-effectiveness.

The NHS Haemoglobinopathy Screening Programme has commissioned research on the feasibility of using a standard question on ethnicity to identify high-risk populations within areas of medium to low prevalence (108).

Other Issues
There are other issues on which research has been commissioned, including the implications of identifying fetuses and infants heterozygous for disease (114) (for which no policy decisions have yet been taken regarding the provision of genetic counseling), the nature and quality of communication and information for parents, whether screening outcomes are better in areas of higher prevalence, and laboratory organizational issues (115).

Implementing New Genetics Services in the United Kingdom

The last 5 years has seen an increasing emphasis on the development of national policies and approaches in relation to genetics and health care in the United Kingdom. The most recent body to be established is the Genetics Commissioning Advisory Group, which is intended to provide advice and support for commissioners on the appropriateness and implications of new potential service developments (116). This adds to, and complements, the work of other bodies such as the Human Genetics Commission, the Genetics and Insurance Committee, and the Gene Therapy Advisory Committee (117).

It is apparent that a centrally driven, top-down approach is being taken to the development and implementation of new genetics services in the United Kingdom (118). This may have begun as a reaction to incremental, bottom-up service developments across the country, but it has now gathered momentum at the national political level. The main visible effects are the rationalization of clinical and laboratory genetics services across the NHS, with concentration of expertise in relatively few centers, specialization of DNA diagnostic laboratories to promote quality assurance, and co-operatives or consortia working to maximize economies of scale while maintaining universal availability of diagnostic genetic technologies (119).

New funding has resulted in the establishment of two U.K. reference laboratories whose role includes offering national support for technology assessment in the face of an ever-changing technical scene, providing quality assurance reagents and schemes, offering services for very rare conditions, and serving as a horizon scanning service (118). In addition, six "Knowledge Parks" have been funded to provide a focus for research and interaction with industry as well as a forum for public education (120).

Conclusions

The U.K. NHS faces considerable challenges in the effective and efficient implementation of genetic tests and interventions. Issues which are already of concern and will need to be addressed include:

- The implications for resource use if further organizational expansion or restructuring is required to meet future needs,
- The ability of health professional groups outside specialist clinics to provide appropriate and effective genetic advice and counseling,
- The effect on demand and outcomes of unrealistic expectations on the part of (potential and actual) service users.

The issues cut across different levels (Table 17.1) and require both a coordinated approach from different sectors of the health system, and effective national leadership and oversight mechanisms.

Human genome epidemiologic evidence encompasses a spectrum, from basic science to the application of knowledge to produce population benefit (121). The discussion of genetics services in the United Kingdom illustrates the point that, to ensure successful implementation of evidence-based interventions, an understanding is required of the interconnecting web of epidemiologic, clinical, cultural, ethical, professional, organisational, and political issues. In genetics, as in any other health emerging technology, the benefits achieved in practice are dependent on effective implementation strategies within a broader health-care and social system. It has recently been argued that evidence-based health care has the potential to make a substantial contribution to population health, just as basic public health measures did in the last century (122). If this prediction is correct, there is a compelling case for the development of an firm foundation of evidence to ensure the effective implementation of genetics knowledge in health care.

References

1. Department of Health. The NHS Plan. London: Department of Health, 2000 (Cmnd 4818-I).
2. Scottish Executive Health Department. Our national health—a plan for action, a plan for change. Edinburgh: Scottish Executive, 2000.

3. Secretaries of State for Social Services, Wales, Northern Ireland and Scotland. Working for Patients. London: Her Majesty's Stationery Office, 1989.
4. Department of Health. Improvement, expansion and reform: the next 3 years. Priorities and planning framework 2003–2006. London: Department of Health, 2002.
5. Scottish Executive Health Department. NHS Scotland: guidance on regional planning for health care services. HDL(2002)10.
6. Coventry PA, Pickstone JV. From what and why did genetics emerge as a medical specialism in the 1970s in the UK? A case-history of research, policy and services in the Manchester region of the NHS. Soc Sci Med 1999;49:1227–1238.
7. Brock DJ. A consortium approach to molecular genetic services: Scottish Molecular Genetics Consortium. J Med Genet 1990;27:8–13.
8. Kinmonth AL, Reinhard J, Bobrow M, Pauker S. The new genetics: implications for clinical services in Britain and the United States. BMJ 1998;316:767–770.
9. Donnai D, Elles R. Integrated regional genetics services: current and future provision. BMJ 2001;322:1048–1052.
10. Genetic Interest Group. The present organisation of genetic services in the United Kingdom. London: Genetic Interest Group, 1995.
11. Skirton H, Barnes C, Curtis G, Walford-Moore J. The role and practice of the genetic nurse: report of the AGNC Working Party. J Med Genet 1997;34:141–147.
12. Skirton H, Barnes C, Guilbert P, et al. Recommendations for education and training of genetic nurses and counsellors in the United Kingdom. J Med Genet 1998;35:410–412.
13. Skirton H, Patch C. Genetics for healthcare professionals—a life stage approach. London: BIOS Science Publishing Ltd, 2002.
14. Working Group for the Chief Medical Officer. Genetics and cancer services. Report of a working group for the Chief Medical Officer, Department of Health. London: Department of Health, 1998.
15. Emery J, Watson E, Rose P, et al. A systematic review of the literature exploring the role of primary care in genetic services. Fam Pract 1999;16:426–445.
16. Starfield B, Hotzman NA, Roland MO, et al. Primary care and genetic services: health care in evolution. Eur J Pub Health 2002;12:51–56.
17. Eccles DM, Evans DGR, Mackay J, on behalf of the UK Cancer Family Study Group. Guidelines for a genetic risk based approach to advising women with a family history of breast cancer. J Med Genet 2000;37:203–209.
18. Fry A, Campbell H, Gudmundsdottir H, et al. GPs' views on their role in cancer genetics services and current practice. Fam Pract 1999;16:468–474.
19. Emery J, Walton R, Murphy M, et al. Computer support for interpreting family histories of breast and ovarian cancer in primary care: comparative study with simulated cases. BMJ 2000;321:28–32.
20. Bankhead CR, Emery J, Qureshi N, Campbell H, Austoker J, Watson E. New developments in genetics—knowledge, attitudes and information needs of practice nurses. Fam Pract 2001;18:475–486.
21. Summerton N, Garrood PVA. The family history in family practice: a questionnaire study. Fam Pract 1997;14:285–288.
22. Elwyn G, Gray J, Iredale R. General practitioners in south Wales are unconvinced of their role in genetics services. BMJ 2000;321:240.
23. Watson E, Shickle D, Qureshi N, et al. The "new genetics" and primary care: GPs' views on their role and their educational needs. Fam Pract 1999;16:420–425.
24. Kumar S, Gantley M. Tensions between policy makers and general practitioners in implementing new genetics: grounded theory interview study. BMJ 1999;319:1410–1413.
25. General Medical Services Committee (BMA). Defining core services in general practice—reclaiming professional control. London: British Medical Association, 1996.

26. Helliwell CD, Carney TA. General practitioners' workload in primary care led NHS: workload for chronic disease management has increased substantially. BMJ 1997; 315:546.

27. Higson N, Cembrowicz S, Baum M. A family history of breast Ca. The Practitioner 1997;241:59–64.

28. Tudor Hart J. General practitioners' workload in primary care led NHS: policies of comprehensive anticipatory care require extra doctors and staff. BMJ 1997;315:546.

29. Sidford I. General practitioners' workload in primary care led NHS: practice's consultation rates have increased by three quarters in past 25 years. BMJ 1997;315:546.

30. Ferlie EB, Shortell SM. Improving the quality of health care in the United Kingdom and the United States: a framework for change. The Millbank Quarterly 2001;79: 281–315.

31. NHS Executive. Faster access to modern treatment: how NICE appraisal will work. Leeds: NHSE, 1999

32. Evans D, Haines A. Implementing evidence based changes in healthcare. Abingdon, UK: Radcliffe Medical Press, 2000.

33. Klein R. From evidence based medicine to evidence based policy? J Health Serv Res Policy 2000;5:65–66.

34. Cochrane AL. effectiveness and efficiency: random reflections on the health service. London: BMJ Publishing Group, 1989.

35. Royal Commission on the National Health Service. Report (Chairman Sir AW Merrison). London: HMSO, 1979. (Cmnd 7615).

36. Sassi F, Archard L, Le Grand J. Equity and the economic evaluation of health care. Health Technol Assess 2001;5(3).

37. Wagstaff A. QALYs and the equity-efficiency trade-off. J Health Econ 1991;10:21–41.

38. Grol R. Beliefs and evidence in changing clinical practice. BMJ 1997;315:418–421.

39. Agency for Health Research and Quality. Translating research into practice (TRIP)-II. Washington, DC: Agency for Health Research and Quality, 2001.

40. Griffiths R. NHS management enquiry: report. London: Department of Health, 1983.

41. Ferlie EB, Fitzgerald L, Wood M. Achieving change in clinical practice. J Health Serv Res Policy 2000;2:96–102.

42. NHS Executive. Clinical effectiveness resource pack. Leeds: Department of Health, 1997.

43. Calman K. Developing screening in the NHS. J Med Screen 1994;1:101–105.

44. NHS Executive. Research and development: towards an evidence-based health service. Leeds: Department of Health, 1996.

45. NHS Executive. Faster access to modern treatment: how NICE appraisal will work. Leeds: NHSE, 1999.

46. Wonderling D, Hopwood P, Cull A, et al. A descriptive study of UK cancer genetics services: an emerging clinical response to the new genetics. Br J Cancer 2001;85: 166–170.

47. Working Group for the Chief Medical Officer. Genetics and cancer services. Report of a working group for the Chief Medical Officer, Department of Health. London: Department of Health, 1998.

48. Priority Areas Cancer Team. Cancer genetics in Scotland. Edinburgh: The Scottish Office Department of Health, 1998.

49. Morrison PJ, Nevin NC. Cancer genetics services in Northern Ireland. Disease Markers 1999;15:37–40.

50. Wilfond BS, Thomson EJ. Models of public health genetic policy development. In: Khoury MJ, Burke W, Thomson EJ, editors. Genetics and public health in the 21st century. New York: Oxford University Press Inc, 2000:61–81.

51. Moller P, Evans G, Haites N, et al. Guidelines for follow-up of women at high risk for inherited breast cancer: consensus statement from the Biomed 2 Demonstration Programme on Inherited Breast Cancer. Disease Markers 1999;15:207–211.
52. Hartmann LC, Schaid DJ, Woods JE, et al. Efficacy of bilateral prophylactic mastectomy in women with a family history of breast cancer. N Engl J Med 1999;340:77–84.
53. Schrag D, Kuntz KM, Garber JE, Weeks JC. Life expectancy gains from cancer prevention strategies for women with breast cancer and BRCA1 or BRCA2 mutations. JAMA 2000;283:617–624.
54. Meijers-Heijboer H, van Geel B, van Putten WLJ, et al. Breast cancer after prophylactic mastectomy in women with a BRCA1 or BRCA2 mutation. N Engl J Med 2001;345:159–164.
55. Kauff ND, Satagopan JM, Robson ME, et al. Risk-reducing salpingo-oophorectomy in women with a BRCA1 or BRCA2 mutation. N Engl J Med 2002;346;1609–1615.
56. Rebbeck TR, Lynch HT, Neuhausen SL, et al. Prophylactic oopherectomy in carriers of BRCA1 or BRCA2 mutations. N Engl J Med 2002;346:1616–1622.
57. Moller P, Borg A, Evans DG, et al. Survival in prospectively ascertained familial breast cancer: analysis of a series stratified by tumour characteristics, BRCA mutations and oophorectomy. Int J Cancer 2002;101:555–559.
58. Ashley-Koch A, Yang Q, Olney RS. Sickle hemoglobin (Hb S) allele and sickle cell disease: A HuGE review. Am J Epidemiol 2000;151:839–845.
59. Mann JR. Sickle cell haemoglobinopathies in England. Arch Dis Child 1981;56: 676–683.
60. Murtaza LN, Stroud CE, Davis LR, et al. Admissions to hospital of children with sickle-cell anaemia: a study in south London. BMJ 1981;282:1048–1051.
61. Brozovic M, Aniowu EN. Sickle cell disease in Britain. J Clin Pathol 1984;37:1321–1326.
62. Brozovic M, Davies SC, Brownell AI. Acute admissions of patients with sickle cell disease who live in Britain. BMJ 1987;294:1206–1208.
63. Gray A, Aniowu EN, Davies SC, et al. Patterns of mortality in sickle cell disease in the United Kingdom. J Clin Pathol 1991;44:459–463.
64. Weatherall D, Clegg JB. The thalassaemia syndromes. Oxford: Blackwell Scientific, 1981:148–319.
65. Cao A, Galanello R, Rosatelli MC, et al. Clinical experience of management of thalassemia: the Sardinian experience. Semin Hematol 1996;33:66–75.
66. Zurlo MG, De Stefano P, Borgna Pignatti C, et al. Survival and causes of death in thalassaemia major. Lancet 1989;ii:27–30.
67. Modell B, Letsky EA, Flynn DM, et al. Survival and desferrioxamine in thalassaemia major. BMJ 1982;284:1081–1084.
68. Olivieri NF, Nathan DG, MacMillan JH, et al. Survival in medically treated patients with homozygous beta-thalassemia. N Engl J Med 1994;331:574–578.
69. Brittenham GM, Griffith PM, Nienhuis AW, et al. Efficacy of deferoxamine in preventing complications of iron overload in patients with thalassemia major. N Engl J Med 1994;331:567–573.
70. Davies SC, Wonke B. The management of haemoglobinopathies. Ballière's Clin Haematol 1991;4:31–89.
71. Rund D, Rachmilewitz E. Thalassemia major 1995: older patients, new therapies. Blood Rev 1995;9:25–32.
72. Hickman M, Modell B, Greengross P, et al. Mapping the prevalence of sickle cell and beta thalassaemia in England: estimating and validating ethnic-specific rates. Br J Haematol 1999;104:860–867.
73. Davies SC, Cronin E, Gill M, et al. Screening for sickle cell disease and thalassaemia: a systematic review with supplementary research. Health Technol Assess 2000;4(3).

74. Zeuner D, Ades AE, Karnon J, et al. Antenatal and neonatal haemoglobinopathy screening in the UK: review and economic analysis. Health Technol Assess 1999;3(11).

75. Karnon J, Zeuner D, Ades AE, et al. The effects of neonatal screening for sickle cell disorders on lifetime treatment costs and early deaths avoided: a modelling approach. J Public Health Med 2000;22:500–511.

76. Gaston MH, Verter JI, Woods G, et al. Prophylaxis with oral penicillin in children with sickle cell anemia. A randomized trial. New Engl J Med 1986;314:1593–1599.

77. British Society for Haematology. Guidelines for haemoglobinopathy screening. Clin Lab Haematol 1988;10:87–94.

78. Department of Health. Standing Medical Advisory Committee Working Party report on sickle cell, thalassaemia and other haemoglobinopathies. London: Her Majesty's Stationery Office, 1993.

79. Modell B, Anionwu EN. Guidelines for screening for haemoglobin disorders: service specifications for low- and high-prevalence district health authorities. In: Ethnicity and health: reviews of literature and guidance for purchasers in the areas of cardiovascular disease, mental health and haemoglobinopathies. York: NHS Centre for Reviews and Dissemination, University of York, 1996: 127–134.

80. Chief Medical Officer. Chief Medical Officer's update 2. Screening policy in the NHS. London: Department of Health, 1994.

81. Wilson JMG, Jungner G. Principles and practice of screening for disease. Geneva: World Health Organization, 1968.

82. National Screening Committee. First report of the National Screening Committee. Leeds, Department of Health, 1998.

83. Department of Health. The NHS Plan. London: Department of Health, 2000 (Cmnd 4818-I).

84. Pasvol G, Wilson R. Red cells and malaria. Br Med Bull 1982;38:133–140.

85. Flint J, Harding R, Boyce A, et al. The population genetics of the haemoglobinopathies, in The Haemoglobinopathies, Higgs D, Weatherall D, eds. London: Ballière Tindall 1993:215–262.

86. Angastiniotis M, Modell B. Global epidemiology of hemoglobin disorders. Ann NY Acad Sci 1998;850:251–269.

87. Modell B, Petrou M, Layton M, et al. Status report: audit of prenatal diagnosis for haemoglobin disorders in the United Kingdom: the first 20 years. BMJ 1997;315:779–784.

88. Modell B, Harris R, Lane B, et al. Informed choice in genetic screening for thalassaemia during pregnancy: audit from a national confidential enquiry. BMJ 2000;320:337–341.

89. Anionwu EN. Sickle cell and thalassaemia: community experiences and official response. In: Ahmad WIU, editor. Race and health in contemporary Britain. Buckingham: Open University Press, 1993:76–95.

90. Streetly A, Grant C, Pollitt RJ, et al. Survey of the scope of neonatal screening in the United Kingdom. BMJ 1995;311:726.

91. Streetly A, Maxwell K, Mejia A. Sickle cell disorders in Greater London: a needs assessment of screening and care services. Fair shares for London report. London: United Medical and Dental Schools, Department of Public Health Medicine, 1997.

92. Atkin K, Amhad WIR, Anionwu ER. Screening and counselling for sickle cell disorders and thalassaemia; the experiences of parents and health professionals. Soc Sci Med 1998;47:1639–651.

93. Bain BJ, Chapman C. A survey of current United Kingdom practice for antenatal screening for inherited disorders of globin chain synthesis. UK Forum for Haemoglobin Disorders. J Clin Pathol 1998;51: 382–389.

94. Angastiniotis M, Hadjiminas MG. Prevention of thalassaemia in Cyprus. Lancet 1981;i 369–370.

95. Cao A, Rosatelli MC. Screening and prenatal diagnosis of the haemoglobinopathies. Ballière's Clinical Haematology, 1993;6:263–286.

96. Petrou M, Brugiatelli M, Ward RHT, et al. Factors affecting the uptake of prenatal diagnosis for sickle cell disease. J Med Genet 1992;29:820–823.

97. Marteau T, Anionwu E. Evaluating carrier testing: objectives and outcomes, in The Troubled Helix: Social and Psychological Implications of the New Human Genetics. Marteau T, Richards M, eds. Cambridge: Cambridge University Press 1996:123–139.

98. Petrou M, Brugiatelli M, Old J, et al. Alpha thalassaemia hydrops fetalis in the UK: the importance of screening pregnant women of Chinese, other South East Asian, and Mediterranean extraction for alpha thalassaemia trait. Br J Obstet Gynaecol 1992;99: 985–989.

99. Wang X, Seaman C, Paik M, Chen T, Bank A, Piomelli S. Experience with 500 prenatal diagnoses of sickle cell diseases: the effect of gestational age on affected pregnancy outcome. Prenat Diagn 1994;14:851–857.

100. Modell B, Bulyzhenkov V. Distribution and control of some genetic disorders. World Health Stat Q 1988;41:209–218.

101. World Health Organization, Hereditary Diseases Program. Guidelines for the control of haemoglobin disorders. (WHO/HDP/G/94.1) Geneva:WHO, 1994.

102. NHS Haemoglobinopathy Screening Programme. Policy decision for implementing neonatal screening for sickle cell disease. NHS Haemoglobinopathy Screening Programme, 2002. Accessed October 2002, from www-phm.umds.ac.uk/haemscreening/whatsnew.htm

103. NHS London Regional Office Research & Development Directorate. Research to support the implementation of the NHS Plan relating to the rollout of haemoglobinopathy and cystic fibrosis screening. (www.doh.gov.uk/research/london/haemocf.doc, accessed April 2002).

104. Sedgwick J, Streetly A. A survey of haemoglobinopathy screening policy and practice in England. London: NHS Haemoglobinopathy Screening Programme, 2001. Accessed October 2002, from http://www-phm.umds.ac.uk/haemscreening/publications.htm

105. Shickle D, May A. Knowledge and perceptions of haemoglobinopathy carrier screening among general practitioners in Cardiff. J Med Genet 1989;26:109–112.

106. Dyson SM, Fielder AV, Kirkham MJ. Midwives' and senior student midwives' knowledge of haemoglobinopathies in England. Midwifery 1996;12:23–30.

107. Modell M, Wonke B, Anionwu E, et al. A multidisciplinary approach for improving services in primary care: randomised controlled trial of screening for haemoglobin disorders. BMJ 1998;317:788–791.

108. NHS Haemoglobinopathy Screening Programme. Antenatal screening programme policy (interim policy statement). NHS Haemoglobinopathy Screening Programme, 2002 Accessed October 2002, from http//www-phm.umds.ac.uk/haemscreening/whatsnew.htm

109. Cruickshank JK, Beevers DG, eds. Ethnic factors in health and disease. London: Butterworth, 1989.

110. Sheldon TA, Parker H. Race and ethnicity in health research. J Public Health Med 1992;14:104–110.

111. Senior PA, Bhopal R. Ethnicity as a variable in epidemiological research. BMJ 1994;309:327–330.

112. McKenzie KJ, Crowcroft NS. Race, ethnicity, culture and science. BMJ 1994;309: 286–287.

113. Anionwu EN. Ethnic origin of sickle and thalassaemia counsellors: does it matter? in Research in Cultural Differences in Health, Kelleher D, Hillier D, (eds.). London: Routledge, 1996.

114. Laird L, Dezateux C, Anionwu EN. Neonatal screening for sickle cell disorders: what about the carrier infants? BMJ 1996;313:4107–4111.

115. NHS Haemoglobinopathy Screening Programme. Technical workshop with a focus on the implementation of antenatal haemoglobinopathy screening. London: NHS Haemoglobinopathy Screening Programme, 2001 Accessed October 2002, from www-phm.umds.ac.uk/haemscreening/.

116. Milburn A. Speech at the Institute of Human Genetics, International Centre for Life, Newcastle-upon-Tyne, 19 April 2001 Accessed October 2002, from http//www.doh.gov.uk/speeches/apr2001milburngenetics.htm

117. Accessed October 2002, from http://www.doh.gov.uk/genetics.

118. Milburn A. Genetics and health—a decade of opportunity (speech, 1/16/2002). Accessed October 2002, from http//www.doh.gov.uk/speeches/jan2002milburngenetics.htm

119. Expert Working Group. Laboratory services for genetics: report of an expert working group to the NHS Executive and the Human Genetics Commission. Leeds: NHS Executive, 2000.

120. Department of Health. The Genetics Knowledge Parks Network—Overview. London: Department of Health, 2002.

121. Khoury MJ, Burke W, Thomson EJ. Genetics and public health: a framework for the integration of human genetics into public health practice, in Genetics and Public Health in the 21st Century: Using Genetic Information to Improve Health and Prevent Disease, Khoury MJ, Burke W, and Thomson EJ (eds.). New York: Oxford University Press, 2000:3–23.

122. Watt G. The inverse care law today. Lancet 2002;360:252–255.

IV

CASE STUDIES: USING HUMAN GENOME EPIDEMIOLOGY INFORMATION TO IMPROVE HEALTH

18

Paraoxonase polymorphisms and susceptibility to organophosphate pesticides

Kathryn Battuello, Clement Furlong, Richard Fenske, Melissa A. Austin, and Wylie Burke

Many workplaces involve exposure to industrial chemicals and other potentially harmful agents. Government safety standards seek to protect workers by defining levels of acceptable exposure and instituting workplace safety measures such as protective clothing. However, standards based on the average worker may not fully account for genetic variation in response to environmental exposures. For most toxic agents, some workers will have inherited susceptibilities that are many-fold higher than average, because of variation in absorption, metabolism, or excretion. As the genetic variants accounting for these responses are identified, susceptible workers could be identified through genetic and functional genomic tests, and additional protective measures could be taken.

In this chapter we consider the example of polymorphisms in the *PON1* gene that result in an increased susceptibility to organophosphate pesticide (OP) toxicity. *PON1* polymorphisms have potential implications for workers whose jobs involve repeated or prolonged pesticide exposure, such as pesticide applicators in agricultural and other settings (1–4). We review current knowledge about this gene–environment interaction and consider the ethical, legal, and policy implications of genetic screening as a means to reduce the adverse health effects of workplace pesticide exposure.

The Clinical Problem

OPs inhibit the enzyme cholinesterase, leading to a variety of adverse health consequences. Acute toxic symptoms include shortness of breath, confusion, tremor, impaired coordination (ataxia), nausea, diarrhea, muscle weakness and cramping, headache, excessive salivation, increased sweating, and bradycardia (5). Death can occur from respiratory failure caused by paralysis of respiratory muscles, increased bronchial secretions, and depression of the respiratory center in the brainstem (4,5). Persons with respiratory sensitivities such as asthma may be particularly susceptible to the respiratory symptoms resulting from acute exposure (4). Systemic symptoms result from inhaled, ingested, or transdermal exposure, and local irritation of mucus membranes and conjunctival surfaces can also occur (4). Several studies indicate that chronic neurobehavioral effects can result from acute OP intoxications (6,7,9). Whether such effects are seen following chronic exposures remains a subject of debate (7–11).

Although OP toxicity is considered an important public health problem (12,13), robust epidemiologic data on the nature and extent of pesticide poisonings are limited, making it difficult to quantify the incidence of OP-related illness (14). On a yearly basis, the World Health Organization estimates that there are approximately 3 million cases of pesticide poisonings overall, resulting in approximately 220,000 deaths (13,15,16). OP toxicity in humans is described in a number of case reports; there are also data from state mandated pesticide incident reporting systems (5,17–20).

In several studies, cholinesterase levels have been monitored as a measure of unsafe exposures to pesticides, because suppression of serum cholinesterase is one of the earliest measurable effects of OP toxicity (20). A series of these studies relied upon data from the state-mandated cholinesterase biomonitoring program in California that applies to agricultural pesticide workers. One of these studies found that approximately 20% of pesticide applicators (presumed to be those with highest exposure) had reductions in cholinesterase levels indicating pesticide overexposure, 5% had levels below the state threshold for removal from the workplace, and 1.5% exhibited clinical symptoms of pesticide exposure (21). A second small, retrospective, cohort study found that 24% of workers were temporarily removed from the workplace because of low plasma cholinesterase levels, and 5% exhibited mild symptoms of toxicity (3). The generalizability of these data is uncertain because enzyme suppression can be highly dependent on the duration and intensity of pesticide exposure, which is insufficiently documented in these studies to extrapolate dose response.

A clinic-based study comparing farmworkers and nonfarmworkers in North Carolina also suggests that cholinesterase suppression and attendant health affects are associated with agricultural OP exposure (22). Mean cholinesterase levels were significantly lower in farmworkers who reported OP exposure than in nonfarm workers; 12% of farmworkers had very low levels, compared with 0% of nonfarm-workers. Similarly, the mean serum cholinesterase among a group of Australian termiticide applicators was 52% of control levels (23); and urinary metabolites provided evidence for significant exposure to the OP chlorpyrifos among a group of termiticide applicators in North Carolina (24).

The Environmental Exposure

In considering genetic screening related to OP exposure, the OP chlorpyrifos is an important agent to consider because it is widely used and because a substantial body of research in genetics, ecogenetics, and environmental health suggests that poly-morphisms in the *PON1* gene mediate individual susceptibility to chlorpyrifos tox-icity. Chlorpyrifos is a broad-spectrum chlorinated OP used on a variety of agri-cultural crops including wheat and several types of fruits, nuts, and vegetables. Until recently chlorpyrifos was one of the major pesticides used in residential, commer-cial, and public settings for pest control. Approximately 21 million to 24 million pounds are used annually in the United States, of which approximately 13 million pounds are applied in agricultural settings (25).

Chlorpyrifos is one of several OPs that are manufactured as organophospho-rothioate compounds. Other organophosphorothioate compounds commonly used in pesticides include parathion and diazinon. These "parent" compounds are in-herently weak anticholinesterase agents. However, their oxon analogs, which are thought be produced both endogenously during phase I metabolism and exoge-nously when residue of the parent compound undergoes an oxidation reaction dic-tated by climatic factors (25), are significantly more toxic than the parent com-pound, posing a greater risk of neurotoxicity (26). For example, in one published study the oxon analog of chlorpyrifos exhibited a thousand-fold greater rate of cholinesterase inhibition than the parent compound (27). Animal studies using transgenic mice suggest that genetic variation in the *PON1* gene primarily influ-ences individual susceptibility to the exogenous oxon of this subset of OPs (28). This observation could have important implications for agricultural workers whose work in the fields provides routine exposure to OP residues on plant foliage. A survey of published values for oxon percentages in foliar residues suggests oxon content ranges from less than 1% to more than 90% of all residue (29). Levels of chlorpyrifos oxon residues in spray drift were monitored in a California study with similar variation observed (30).

The Genetic Mediator

The *PON1* gene and its regulatory elements determine the activity and expression of PON1, an enzyme involved in the metabolism of certain OPs, including chlor-pyrifos. The gene coding for PON1 is located on the long arm of chromosome 7 (7q21-7q22), near the cystic fibrosis gene (31,32). This gene (designated *PON1*) spans approximately 26 kb, and its coding sequence comprises nine exons, coding for 355 amino acids (33). *PON1* is a member of a multigene family that includes *PON2* and *PON3* (34). The functions of *PON2* and *PON3* are still under study, but research suggests neither gene mediates OP metabolism.

The PON1 enzyme—historically referred to as paraoxonase or arylesterase—is predominantly synthesized in the liver and circulates in the serum as a component of high-density lipoprotein (HDL) particles (35). It is involved in Phase II metabo-lism of the oxon metabolites of several OPs, as well as the nerve agents soman and sarin (36). The physiologic function of PON1 is still under investigation; recent studies suggest it may play an important role in metabolism of oxidized lipids (37,38).

Variation in PON1 metabolic activity occurs by two mechanisms. First, a well-characterized polymorphism at position 192 of the gene sequence controls the catalytic efficiency of PON1 hydrolysis (31,32,35,36). Second, promoter polymorphisms appear to affect enzyme expression. *PON1* expression is constitutive but demonstrates at least a 15-fold interindividual variation in enzyme level that is presumed to result in part from promoter polymorphisms (36,39–41). Efforts to characterize these polymorphisms and their influence on enzyme expression are ongoing and suggest that a polymorphism in the *PON1* regulatory region at position −108 (−108C/T) strongly influences PON1 levels, with the −108C allele generating on average twice the level of plasma PON1 as the −108T allele (42–45). The combined effect of the position 192 genotype on metabolism of a given OP and the level of enzyme expression is referred to as "PON1 status (46)." People with low PON1 status are hypothesized to have increased susceptibility to toxicity of specific OP compounds. A newly identified variant at codon 194 inactivates the *PON1* gene bearing these polymorphisms (Jarvik et al., unpublished data).

Age appears to be another important determinant of PON1 activity and susceptibility to OP toxicity, with markedly increased susceptibility in infants (26,47,48). While we focus on susceptibility among pesticide applicators because they represent a group with the high potential exposure, their children also may suffer indirectly from exposure to clothing carrying pesticide residues, or from airborne exposures in residential settings close to farmland where pesticides are applied (49). There is also concern for the fetuses of mothers with low PON1 status when the mother experiences workplace exposure or is exposed to clothing carrying pesticide residues.

Association between *PON1* Status and Clinical Outcome

Data demonstrating an association between variations in the *PON1* gene and OP toxicity are primarily derived from animal studies demonstrating that variation in OP toxicity, and chlorpyrifos toxicity in particular, is mediated by a gene–environment interaction.

First, animal studies suggest that individuals rely on at least three important metabolic pathways to detoxify the oxon analogs of chlorpyrifos and related OPs: (*1*) the oxons can be hydrolyzed by PON1 enzymes in the liver and the blood (26,50,51); (*2*) the oxons can bind irreversibly to carboxylesterase or butyrylcholinesterase enzymes (52); (*3*) the multigene family of enzymes collectively referred to as the cytochromes P450 (CYPP450s) are involved in bioactivating chlorpyrifos and detoxifying both the parent organophosphorothioate compounds and the oxons (52,53). The importance of the PON1-mediated hydrolysis reaction in detoxifying OPs appears to depend in part upon the OP substrate and the level of the oxon present in the exposure. (*28*).

Second, animal studies and limited studies involving humans indicate individual variation in dose-response to OPs (28,54–57). Data from two studies of humans conducted by a leading chlorpyrifos manufacturer suggest that humans may be more susceptible than other mammals to chlorpyrifos and exhibit variable responses to a given dose, measured either by cholinesterase inhibition or clinical signs (25). It should be noted that studies carried out by the manufacturers generally use 99% pure parent compound (58), whereas occupational exposure can contain significant percentages of the more toxic oxon residue under some circumstances (29).

Third, animal studies support the hypothesis that PON1 status plays an important role in metabolism of OPs, including chlorpyrifos. Studies measuring the protective effect of pretreatment with injected purified rabbit PON1 in rats and mice demonstrated that increased serum PON1 activity was protective (46,55,56,59,60). For example, intravenous injection of rabbit PON1 increased serum PON1 activity in both rats and mice (61), providing increased protection against toxicity from chlorpyrifos and chlorpyrifos oxon, with the greatest protection afforded to brain and diaphragm tissues. Achieving the most pronounced and consistent protective effect in brain and diaphragm tissue is important because OP induced neurotoxicity occurs primarily because of accumulation of acetylcholine in the brain and diaphragm (59).

Studies using *PON1* knockout mice and transgenic mice in which mouse *PON1* has been replaced with either $PON1_{R192}$ or $PON1_{Q192}$ have provided useful insights into the role of PON1 status in determining sensitivity or resistance to chlorpyrifos oxon and diazoxon exposures. Two published studies using *PON1* knockout mice suggest that human *PON1* status may influence individual susceptibility to chlorpyrifos oxon toxicity (28,54). In the first study, *PON1* knockout mice ($PON1^{-/-}$) had dramatically higher sensitivity to chlorpyrifos oxon than did wild-type mice ($PON1^{+/+}$), at each of three exposure levels (1.5 mg, 3 mg, and 6 mg) (54). The two highest doses produced clinical symptoms in the knockout mice within 1 to 2 hours of exposure and death within 2 to 4 hours. The wild-type mice remained asymptomatic with only mild suppression of acetylcholinesterase. The lowest dose inhibited cholinesterase activity in the knockout mice but had no effect in the wild-type mice.

In the second study (28), *PON1* knockout mice exhibited a dramatic increase in sensitivity to diazoxon (28) and chlorpyrifos oxon (54) compared with wild-type controls. Hemizygotes (a hemizygote has only one copy of a gene) showed intermediate sensitivity to diazoxon (28). Similarly, doses of diazoxon that did not inhibit brain cholinesterase in wild-type controls were lethal to the knockout mice. Furthermore, Li et al (28) demonstrated that injecting either human $PON1_{192}$ isoform into wild-type mice restored resistance to chlorpryifos oxon and diazoxon. The degree of resistance depended on the catalytic efficiency (Vm/Km) of the specific $PON1_{192}$ isoform for the specific OP substrate. For example, the $PON1_{R192}$ isoform provided greater protection against chlorpyrifos oxon. Two of the mice injected with $PON1_{Q192}$ developed clinical signs of OP toxicity after exposure, whereas no signs were observed in the $PON1_{R192}$-injected mice. In this study, *PON1* knockout mice had essentially no hydrolytic activity in either the plasma or the liver, and hemizygous mice exhibited 40% of the activity of the wild-type mice. When transgenic mice (*human PON1*$_{R192}$-Tg and *human PON1*$_{Q192}$-Tg on the *PON1*$^{-/-}$ background) were exposed to chlorpyrifos oxon, mice with the *PON1*$_{Q192}$ genotype were nearly as sensitive to chlopyrifos oxon as knockout mice lacking the *PON1* gene (62). This finding is important for public health because the prevalence of *PON1*$_{Q192}$ homozygotes is high: about 50% among northern Europeans (63), about 36% among Hispanic populations (36), and about 25% among Asians (63).

In summary, the data from these animal studies indicate that both *PON1* genotype and PON1 level are important in determining sensitivity to chlorpyrifos oxon, whereas PON1 level is the main determinant of sensitivity to diazoxon (26).

Potential Public Health Contribution of Screening for PON1 Status

Occupational illness resulting from workplace chemicals and other toxic substances is an important public health problem, and genetic testing has the potential to mitigate risk for those workers who are predisposed to react adversely to the work environment. Because chlorpyrifos is a widely used agricultural pesticide, testing for PON1 status could potentially reduce pesticide toxicity among agricultural workers by identifying pesticide applicators with genetically increased susceptibility and reassigning them to other work. However, historical experience with the use of genetic susceptibility testing in the workplace demonstrates that inappropriate or premature application can, in effect, discriminate, stigmatize and unnecessarily exclude workers (64–67).

Legal, ethical, and policy considerations suggest that a workplace screening program for PON1 status could be timely and appropriate if (*1*) a scientifically valid test with demonstrated analytic and clinical validity exists to identify workers for whom appropriate preventive measures can be taken (68–73) and (2) a test administration process exists that respects worker autonomy and provides informed consent, appropriate counseling, and protection of confidentiality (67,72–79).

Analytic and Clinical Validity of PON1 Status Testing
The first issue for consideration is whether current testing methods provide an accurate and relevant method for identifying workers with genetically increased susceptibility to chlorpyrifos and other OPs. Richter and Furlong (39) have developed a high-throughput, two-dimensional enzyme analysis that takes into account both catalytic efficiency and level of enzyme expression providing a measure of PON1 status (activity level and an accurate *PON1*$_{192}$ functional genotype) (39). With respect to determining *PON1*$_{R/Q192}$ functional genotype and activity levels, this testing method demonstrates high analytic validity in the academic research setting where sensitivity, specificity, and predictive value approach 100%. When study samples underwent follow-up genotyping with the use of polymerase chain reaction (PCR), 316 of 317 samples (99.7%) showed agreement between the PCR genotype and the genotype inferred from the enzyme assay (43). Studies of the PON1 assay

have demonstrated the superiority of this genomic functional (or phenotypic) assay over DNA-based testing, because the assay measures the combined effect of PON1 levels, $PON1_{192}$ genotype, and other relevant deleterious alleles. This assay, however, has not yet been tested outside a research setting and is not FDA-approved as a commercial test.

More importantly, the clinical validity of testing for PON1 status—that is, the degree to which low PON1 status predicts adverse events among workers exposed to pesticides—is not well characterized. Epidemiologic studies to determine the association between PON1 status and adverse outcomes among workers exposed to pesticides are urgently needed to verify and extend observations made in animal models. Pending definitive data among exposed workers, rough estimates of clinical validity can be generated by extrapolating from animal data, *PON1* genotype frequencies in human populations and chlorpyrifos toxicity incidence data from the California biomonitoring program. These estimates underscore the need for field studies to assess the clinical implications of low PON1 status in well-defined conditions of pesticide exposure.

Estimates (Table 18.1) are derived from the following assumptions. First, the frequency of low PON1 status ranges from 12.5% to 50%. This range is based on the prevalence of the $PON1_{Q192}$ homozygous genotype (25% to 50% in different ethnic populations) and the projection that 50% to 100% of $PON1_{Q192}$ homozygotes have sufficiently reduced PON1 activity to be at increased risk from chlorpyrifos exposure (35,39). Second, a rate of adverse events among exposed workers is 1.5% for clinical events (acute symptomatic episodes), 5% for severe cholinesterase suppression, and approximately 20% for moderate cholinesterase suppression. These adverse event rate estimates are based on data from the California cholinesterase biomonitoring program (2,21), which encompasses events related to all OP and carbamate exposures, and thus provide only a crude estimate of adverse events related specifically to chlorpyrifos exposure.

The relative risk (RR) for adverse events after exposure to a given OP among people with low PON1 status compared to those with higher *PON1* status is not known. Thus, for our estimates, we calculate the clinical validity of *PON1* status testing across a range of RR values (RRs of 3, 10, 40, and 80), using the assumptions noted above. The RRs of 3 and 10 represent the recommended minimum thresholds for workplace genetic testing suggested by Omenn (69). The RRs of 40 and 80 reflect the maximum differences in susceptibility seen in animals with low *PON1* status and normal *PON1* status in laboratory studies of dose–response (28,54,59).

The predictive value of low PON1 status is summarized in Table 18.1 for prevalences of 12.5% and 50%. As expected, the sensitivity of low PON1 status rises with the RR at both prevalence values. At RR of 3, the sensitivity and positive predictive value of the test are low, even for risk of moderate cholinesterase suppression. At RR of 80, sensitivity for moderate cholinesterase suppression is 56% and positive predictive value is 90% when low PON1 status has a prevalence of 12.5%. At the higher prevalence of 50%, the sensitivity is 98% and the positive predictive

Table 18.1 Clinical Validity of *PON1* Testing at Different Relative Risk (RR) Levels

| | | | CLINICAL VALIDITY | |
| | | | Severe Cholinesterase | Moderate Cholinesterase |
Prevalence of Low PON1	*Test Properties*	*Clinical Events (%)*	*Suppression (%)*	*Suppression (%)*
12.5%	RR = 3			
	PPV	3.6	11.4	39.2
	NPV	98.8	95.9	82.7
	Sensitivity	30.0	28.4	24.5
	Specificity	87.8	88.3	90.5
	RR = 10			
	PPV	6.9	21.4	62.0
	NPV	99.3	97.3	86.0
	Sensitivity	57.3	53.6	38.8
	Specificity	88.2	89.7	94.1
	RR = 40			
	PPV	10.1	31.9	83.2
	NPV	99.7	98.8	89.0
	Sensitivity	83.8	79.8	52.0
	Specificity	88.6	91.0	97.4
50%	RR = 80			
	PPV	10.9	35.3	89.9
	NPV	99.8	99.3	90.0
	Sensitivity	91.1	88.2	56.2
	Specificity	88.7	91.5	98.4
	RR = 3			
	PPV	2.2	7.5	28.4
	NPV	99.2	97.5	88.4
	Sensitivity	74.9	74.6	71.0
	Specificity	50.4	51.3	55.3
	RR = 10			
	PPV	2.7	9.0	34.9
	NPV	99.7	99.0	94.9
	Sensitivity	90.7	90.2	87.3
	Specificity	50.6	52.1	59.3
	RR = 40			
	PPV	2.9	9.7	38.5
	NPV	99.9	99.7	98.5
	Sensitivity	97.5	97.3	96.2
	Specificity	50.7	52.5	61.5
	RR = 80			
	PPV	3.0	9.9	39.2
	NPV	100.0	99.9	99.2
	Sensitivity	98.7	98.6	98.0
	Specificity	50.7	52.6	62.0

value is 39%. As expected, the sensitivity is consistently lower and specificity higher when low PON1 status is less prevalent.

In Table 18.2, the outcome of testing in 1000 exposed workers is estimated for each prevalence value and RR level. At low RR and low prevalence, the test fails to detect most of the workers who experience adverse outcomes and labels many

Table 18.2 Estimated Outcome of Genetic Testing for PON1 Status in 1000 Workers at Different Levels of Relative Risk (RR)

Prevalence of Low PON1		Clinical Events	Severe Cholinesterase Suppression	Moderate Cholinesterase Suppression	No Clinical Events or Cholinesterase Suppression
12.5%	RR = 3				
	Low PON1	5	9	35	76
	Normal PON1	11	25	115	724
	RR = 10				
	Low PON1	9	18	51	48
	Normal PON1	6	17	100	753
	RR = 40				
	Low PON1	13	27	64	21
	Normal PON1	2	8	86	779
	RR = 80				
	Low PON1	14	30	68	12
	Normal PON1	1	5	82	787
50%	RR = 3				
	Low PON1	11	26	105	358
	Normal PON1	4	9	45	442
	RR = 10				
	Low PON1	14	31	130	325
	Normal PON1	1	4	20	475
	RR = 40				
	Low PON1	15	34	143	308
	Normal PON1	<1	<1	7	492
	RR = 80				
	Low PON1	15	34	147	304
	Normal PON1	<1	<1	39	496

who will not experience any clinical effect of pesticide exposure. Testing is more accurate at RRs of 40 and 80, but most of those with a positive test will experience only moderate cholinesterase suppression and some will experience no clinical effects. At a prevalence of 50%, most workers experiencing adverse events will have low PON1 status, even at low RR, but the majority of workers with low PON1 status will suffer no adverse events (Table 18.2).

This exercise underscores the need for additional epidemiologic data, as well as informed policy debate about appropriate thresholds for testing. The outcome of testing for PON1 status among workers exposed to OP pesticides cannot be accurately predicted until additional studies are done to determine the RR for adverse outcomes among exposed workers with low PON1 status. In particular, the RR of 40 to 80 predicted by animal studies needs to be either confirmed or disproved for human exposures to OPs. Studies are needed that combine PON1 status assessment with documentation of pesticide exposure, with the use of both field measurements and measurement of urinary metabolites in exposed workers. These studies should incorporate prospective study designs to assess short-term and long-term cholinesterase suppression and clinical outcomes.

If the risk associated with low PON1 status is confirmed to be high among exposed workers, the threshold for test use needs to be debated. At what level of clinical validity is the test acceptable? Specifically, what proportion of false-positive and false-negative test results is acceptable in a workplace genetic testing program, recognizing that some false-positive and false-negative test results will occur even at high RRs?

Feasibility of a Workplace Genetic Testing Program

The second issue for consideration is whether a genetic screening program that meets accepted ethical, legal, and social criteria is feasible. Several experts and consensus groups have considered genetic screening in the context of concerns for worker autonomy, informed consent, and confidentiality (68–73). They suggest that any test administration process must, at a minimum, be voluntary and include meaningful informed consent, genetic counseling, job retraining or relocation opportunities for workers who test positive, and established confidentiality procedures.

These criteria, however, compete with the business, liability, and worker safety concerns that make genetic screening programs attractive to employers. Employers are thought to view genetic screening as a cost-effective mechanism for reducing worker injury and illness because removal of susceptible workers may reduce occupational disease without extensive modifications to the workplace or to production practices (80,81). If responsible for funding a genetic screening program, yet unable to be assured of using test results to remove high-risk workers from areas of exposure or avoid liability for workers' compensation payments, employers might be unwilling to implement genetic screening programs. Some of these issues have been explored with respect to chronic beryllium testing (82). They indicate that a genetic testing program faces many challenges and underscore the need for careful studies to measure the risks and benefits of a testing program before it is undertaken.

Discussion

The assay for *PON1* status represents an important potential tool in addressing the public health problem of OP toxicity. However, the rough estimates of test properties shown in Tables 18.1 and 18.2 demonstrate the need for more information about the predictive value of the test in exposed workers. A test with limited predictive value would result in misclassification of many workers and could unnecessarily increase worker concerns or perceptions regarding risk. Epidemiologic data confirming the clinical validity of the test for PON1 status are needed before screening to identify workers at high risk is considered. Studies assessing the association between PON1 status and OP toxicity in populations of exposed workers are urgently needed to determine the strength of the association, the degree of risk for pesticide toxicity, and the clinical validity of workplace screening under actual exposure conditions.

Characterization of the association between PON1 status and OP toxicity should ideally also take into account variation in other metabolic pathways. The body's disposition of hazardous substances involves multiple metabolizing enzymes, all of which are under some degree of genetic control and many of which are polymorphic. As a result, complex and varied interactions among several genes undoubtedly influence variations in dose response and risk for toxicity (83), posing a significant challenge to efforts to pinpoint "the" highly susceptible genotype. The available animal studies do, however, suggest a high level of sensitivity among $PON1_{Q192}$ homozygotes and among other individuals with other $PON1_{192}$ genotypes ($_{Q/R;\ R/R}$) who have low enzyme levels.

Until further epidemiologic data are available to assess the public health implications of testing for PON1 status, it will be difficult to justify the expense that attends an occupational genetic screening program. These expenses, which are borne by workers, employers, and society, include worker education, testing, genetic counseling, workers' compensation, unemployment benefits, delays in production, and lost opportunities to focus on overall reduction of workplace hazards that will attend an occupational genetic screening program. The expense government programs might incur as a result of job loss needs to be taken into account as well, particularly in the case of agricultural pesticide workers for whom comparable alternative jobs may not exist.

Employers are likely to prefer mandatory screening programs that provide them the prerogative to exclude high-risk and potentially costly workers and offer an alternative option to expensive programs designed to reduce overall exposure for all workers. However, these programs are likely to be acceptable only when a genetic test is highly predictive for occupational illness, with persuasive arguments that exposure-related illness could not be averted in the absence of genetic testing. Even when these conditions occur, ethical concerns for worker autonomy would argue against mandatory testing and routine disclosure of results to employers (70,72–79). Moreover, given the multifactorial complexities of most gene–environment interactions such highly predictive tests are likely to be rare. Animal studies of PON1 and OP pesticide exposure suggest that testing for PON1 status might meet this standard, but data are currently insufficient to make this conclusion. The acceptability of a testing program is likely to depend to a great degree on context, however. Thus, in a work environment where reassignment to a safer work site is readily available without loss of income, a testing program might be acceptable even if the predictive value of testing were limited.

We recognize that large-scale studies to address the epidemiology of pesticide exposure, association between PON1 status and health outcomes in exposed workers, and cost and feasibility of the different methods to monitor and reduce pesticide exposure would be costly. Analytic methods to identify the most critical research questions could help to focus research efforts. Tools such as risk assessment and decision analysis, developed to assist policymakers, can provide a model for assessing the effect of variability of key parameters on predicted costs and benefits (82). For example, studies assessing the practical value of gene-based interventions to minimize adverse health outcomes from occupational exposure to beryllium and benzene provide information about the effect of differences in sensitivity, specificity, and predictive value of the tests; strength of the genotype–disease association; participation rate in screening or biomonitoring programs; and cost factors (82,84). These studies help to pinpoint the variables that have the greatest impact on testing outcomes. In the case of PON1 status, these variables include definitive measurement of the prevalence of low PON1 status and of the association between PON1 status and health outcomes in exposed workers. However, there are additional cost/benefit variables impacting the practical value of workplace screening programs for PON1 status that could be identified and assessed by using these tools.

Decision analysis traditionally emphasizes measurable health outcomes and the economic costs of testing programs. Our analysis indicates the importance of additional ethical, legal, and social factors, some of which must be assessed qualitatively. Examples include perceptions about the meaning of positive genetic test results and social outcomes rarely studied in clinical trials, such as the impact of test results on employability and insurability. If we accept the primacy of public health and safety over considerations of cost and feasibility as a moral imperative (85,86), these factors, which contribute to the risks associated with workplace genetic testing, must also be incorporated into future studies. With appropriate estimates, outcomes related to these qualitative and social factors can be incorporated into a decision model.

Conclusions

Oganophosphate pesticide toxicity is an important public health problem, and extensive scientific research suggests that genetic variation influences individual susceptibility to chlorpyrifos and other OPs. A measurement of susceptibility is available, in the form of the assay for PON1 status, and animal studies strongly support an association between low PON1 status and vulnerability to OP pesticide exposure. However, there are currently no data relating PON1 status to clinical outcomes in workers exposed to pesticides. As a result, it is difficult to estimate the predictive value of the test for PON1 status; with limited predictive value, the test might result in misclassification of many workers. In the absence of robust data confirming the clinical validity of the available test for PON1 status, consideration of genetic screening to identify workers at high risk would be premature. Epidemiologic studies assessing the association between PON1 status and OP toxicity in populations of exposed workers are urgently needed, to determine the strength of the association, the degree of risk for different clinical outcomes of pesticide toxicity, and the clinical validity of workplace screening under actual exposure conditions.

Acknowledgments
This work was supported by the UW NIEHS sponsored Center for Ecogenetics and Environmental Health, Grant No. NIEHS P30ES07033. Additional support was provided through the following grants: ES09883, ES07033, ES09601/EPA-R82886, and ES11387.

References

1. Coye MJ. The health effects of agricultural production: I. The health of agricultural workers. J Public Health Policy 1985;6:349–370.
2. Ames RG, Brown SK, Mengle DC, et al. Cholinesterase activity depression among California agricultural pesticide applicators. Am J Ind Med 1989;15:143–150.
3. Fillmore CM, Lessenger JE. A cholinesterase testing program for pesticide applicators. J Occup Med 1993;35:61–70.
4. PIRT (Pesticide Incident Reporting and Tracking Review Panel). 1999 Annual Report. Olympia, WA: Washington State Department of Health, 2000.
5. Reigart R, Roberts J. Recognition and Management of Pesticide Poisonings. Washington D.C.: Office of Pesticide Programs, U.S. Environmental Protection Agency. Available on line at http://www.epa.gov/oppfod01/safety/healthcare/handbook/handbook.htm, 1999.
6. Fiedler N, Kipen H, Kelly-McNeil K, et al. Long-term use of organophosphates and neuropsychological performance. Am J Ind Med 1997;32:487–496.
7. Daniell W, Barnhart S, Demers P, et al. Neuropsychological performance among agricultural pesticide applicators. Environ Res 1992;59:217–228.
8. Rosenstock L, Keifer M, Daniell WE, et al. Chronic central nervous system effects of acute organophosphate pesticide intoxication. The Pesticide Health Effects Study Group. Lancet 1991;338:223–227.
9. Keifer MC, Mahurin RK. Chronic neurologic effects of pesticide overexposure. Occup Med 1997;12:291–304.
10. Savage EP, Keefe TJ, Mounce LM, Heaton RK, Lewis JA, Burcar PJ. Chronic neurological sequelae of acute organophosphate pesticide poisoning. Arch Environ Health 1988;43:38–45.
11. Kilburn KH. Evidence for chronic neurobehavioral impaairment from chlorpyrifos an organophosphate insecticide (Dursban) used indoors. Environmental Epidemiology and Toxicology 1999:1–10.
12. Wilson BW, Sanborn JR, O'Malley MA, et al. Monitoring the pesticide-exposed worker. Occup Med 1997;12:347–363.
13. World Health Organization (WHO). Public health impact of pesticides used in agriculture. Geneva: World Health Organizationn (WHO), 1990.
14. Blondell J. Epidemiology of pesticide poisonings in the United States, with special reference to occupational cases. Occup Med 1997;12:209–220.
15. World Health Organization (WHO). Geneva: WHO/VBC, 1986.
16. World Health Organization (WHO). Public health imipact of pesticides used in agriculture. Geneva: World Health Organization, 1990.
17. Brown SK, Ames RG, Mengle DC. Occupational illnesses from cholinesterase-inhibiting pesticides among agricultural applicators in California, 1982–1985. Arch Environ Health 1989;44:34–39.
18. Keim SA, Alavanja MC. Pesticide use by persons who reported a high pesticide exposure event in the agricultural health study. Environ Res 2001;85:256–259.
19. Lessenger JE, Estock MD, Younglove T. An analysis of 190 cases of suspected pesticide illness. J Am Board Fam Pract 1995;8:278–282.
20. Lessenger JE, Reese BE. Rational use of cholinesterase activity testing in pesticide poisoning. J Am Board Fam Pract 1999;12:307–314.
21. Ames RG, Brown SK, Mengle DC, Kahn E, Stratton JW, Jackson RJ. Protecting agricultural applicators from over-exposure to cholinesterase-inhibiting pesticides: perspectives from the California programme. J Soc Occup Med 1989;39:85–92.
22. Ciesielski S, Loomis DP, Mims SR, Auer A. Pesticide exposures, cholinesterase depres-

sion, and symptoms among North Carolina migrant farmworkers. Am J Public Health 1994;84:446–451.

23. Dyer SM, Cattani M, Pisaniello DL, Williams FM, Edwards JW. Peripheral cholinesterase inhibition by occupational chlorpyrifos exposure in Australian termiticide applicators. Toxicology 2001;169:177–185.

24. Hines CJ, Deddens JA. Determinants of chlorpyrifos exposures and urinary 3,5,6-trichloro-2-pyridinol levels among termiticide applicators. Ann Occup Hyg 2001;45: 309–321.

25. EPA (Environmental Protection Agency). U.S. Environmental Protection Agency Office of Pesticide Programs Website. www.epa.gov/pesticides/ops/chlorpyrifos/summary.htm, 2000.

26. Atterberry TT, Burnett WT, Chambers JE. Age-related differences in parathion and chlorpyrifos toxicity in male rats: target and nontarget esterase sensitivity and cytochrome P450-mediated metabolism. Toxicol Appl Pharmacol 1997;147:411–418.

27. Huff RA, Corcoran JJ, Anderson JK, et al. Chlorpyrifos oxon binds directly to muscarinic receptors and inhibits cAMP accumulation in rat striatum. J Pharmacol Exp Ther 1994;269:329–335.

28. Li WF, Costa LG, Richter RJ, et al. Catalytic efficiency determines the in-vivo efficacy of PON1 for detoxifying organophosphorus compounds. Pharmacogenetics 2000;10: 767–779.

29. Yuknavage KL, Fenske RA, Kalman DA, Keifer MC, Furlong CE. Simulated dermal contamination with capillary samples and field cholinesterase biomonitoring. J Toxicol Environ Health 1997;51:35–55.

30. California Air Resources Board (CARB). Report for the application and ambient air monitoring of chlorpyrifos (and the oxon analog) in Tulare County during spring/summer, 1998. Project Nos. C96-040. Sacramento: Air Resournces Board, California Environmental Protection Agency, 1998.

31. Humbert R, Adler DA, Disteche CM, et al. The molecular basis of the human serum paraoxonase activity polymorphism. Nat Genet 1993;3:73–76.

32. Hassett C, Richter RJ, Humbert R, et al. Characterization of cDNA clones encoding rabbit and human serum paraoxonase: the mature protein retains its signal sequence. Biochemistry 1991;30:10141–10149.

33. Clendenning JB, Humbert R, Green ED, et al. Structural organization of the human PON1 gene. Genomics 1996;35:586–589.

34. Primo-Parmo SL, Sorenson RC, Teiber J, et al. The human serum paraoxonase/arylesterase gene (PON1) is one member of a multigene family. Genomics 1996;33:498–507.

35. Furlong CE, Li WF, Richter RJ, et al. Genetic and temporal determinants of pesticide sensitivity: role of paraoxonase (PON1). Neurotoxicology 2000;21:91–100.

36. Davies HG, Richter RJ, Keifer M, Broomfield CA, Sowalla J, Furlong CE. The effect of the human serum paraoxonase polymorphism is reversed with diazoxon, soman and sarin. Nat Genet 1996;14:334–336.

37. Shih DM, Reddy S, Lusis AJ. CHO and atherosclerosis: human epidemiological studies and transgenic mouse models. In: Costa LG FC, ed. Paraoxonase in Health and Disease: Basic and Clinical Aspects. Boston: Kluwer Academic Publishers, 2002:93–123.

38. Navab M, Hama SY, Wagner AC (2000). Protective action of HDL-associated PON1 against LDL oxidation, in Paraoxonase in Health and Disease: Basic and Clinical Aspects, Costa LG FC, ed. Boston: Kluwer Academic Publishers, pp. 125–136.

39. Richter RJ, Furlong CE. Determination of paraoxonase (PON1) status requires more than genotyping. Pharmacogenetics 1999;9:745–753.

40. Zech R, Zurcher K. Organophosphate splitting serum enzymes in different mammals. Comp Biochem Physiol B 1974;48:427–433.

41. Furlong CE, Richter RJ, Seidel SL, et al. Spectrophotometric assays for the enzymatic hydrolysis of the active metabolites of chlorpyrifos and parathion by plasma paraoxonase/arylesterase. Anal Biochem 1989;180:242–247.

42. Brophy VH, Hastings MD, Clendenning JB, et al. Polymorphisms in the human paraoxonase (PON1) promoter. Pharmacogenetics 2001;11:77–84.

43. Brophy VH, Jarvik GP, Richter RJ, et al. Analysis of paraoxonase (PON1) L55M status requires both genotype and phenotype. Pharmacogenetics 2000;10:453–460.

44. Leviev I, James RW. Promoter polymorphisms of human paraoxonase PON1 gene and serum paraoxonase activities and concentrations. Arterioscler Thromb Vasc Biol 2000; 20:516–521.

45. Suehiro T, Nakamura T, Inoue M, et al. A polymorphism upstream from the human paraoxonase (PON1) gene and its association with PON1 expression. Atherosclerosis 2000;150:295–298.

46. Li WF, Costa LG, Furlong CE. Serum paraoxonase status: a major factor in determining resistance to organophosphates. J Toxicol Environ Health 1993;40:337–346.

47. Augustinsson KB, Barr M. Age variation in plasma arylesterase activity in children. Clinica Chimica Acta 1963;8:568–573.

48. Ecobichon DJ, Stephens DS. Perinatal development of human blood esterases. Clin Pharmacol Ther 1973;14:41–47.

49. Lu C, Fenske RA, Simcox NJ, Kalman D. Pesticide exposure of children in an agricultural community: evidence of household proximity to farmland and take home exposure pathways. Environ Res 2000;84:290–302.

50. Butler EG, Eckerson HW, La Du BN. Paraoxon hydrolysis vs. covalent binding in the elimination of paraoxon in the rabbit. Drug Metab Dispos 1985;13:640–645.

51. Aldridge WN. Serum esterases 1. Two types of esterase (A and B) hydrolysing p-nitrophenyl acetate, proprionate and butyrayate and a method for their determination. Biochem J 1953;53:117–124.

52. Wormhoudt LW, Commandeur JN, Vermeulen NP. Genetic polymorphisms of human N-acetyltransferase, cytochrome P450, glutathione-S-transferase, and epoxide hydrolase enzymes: relevance to xenobiotic metabolism and toxicity. Crit Rev Toxicol 1999;29: 59–124.

53. Eaton DL. Biotransformation enzyme polymorphism and pesticide susceptibility. Neurotoxicology 2000;21:101–111.

54. Shih DM, Gu L, Xia YR, et al. Mice lacking serum paraoxonase are susceptible to organophosphate toxicity and atherosclerosis. Nature 1998;394:284–287.

55. Brealey CJ, Walker CH, Baldwin BC. A-esterase activities in relation to the differential toxicity of pirimiphos-methyl to birds and mammals. Pestic Sci 1980;11:546–554.

56. Main AR. The role of a-esterase in the acute toxicity of paraoxon, TEPP and parathion. Canadian J Biochemistry and Physiology 1956;34:197–216.

57. Costa LG, Richter RJ, Murphy SD, et al. (1987) Species differences in serum paraoxonase correlate with sensitivity to paraoxon toxicity. In Toxicology of Pesticides: Experimental, Clinical and Regulatory Aspects, Costa LG et al, ed. Berlin: Springer-Vrelag, pp. 263–266.

58. Nolan RJ, Rick DL, Freshour. Chlorpyrifos: pharmacokinetics in human volunteers following single oral and dermal doses. Toxicol Appl Pharmacol 1984;73:8–15.

59. Costa LG, McDonald BE, Murphy SD, et al. Serum paraoxonase and its influence on paraoxon and chlorpyrifos-oxon toxicity in rats. Toxicol Appl Pharmacol 1990;103: 66–76.

60. Li WF, Furlong CE, Costa LG. Paraoxonase protects against chlorpyrifos toxicity in mice. Toxicol Lett 1995;76:219–226.
61. Costa LG (2000). Role of PON1 in organophosphate toxicity, in Paraoxonase in Health and Disease: Basic and Clinical Aspects, Furlong CE, ed. Boston: Kluwer Academic Publishers.
62. Furlong CE, Cole T, Li W-F, et al. OP toxicity in mice expressing the human 192R or 192Q isoforms of paraoxonase (PON1). Toxicol Sci 2002;66:S35.
63. Brophy VH, Jarvik GP, Furlong CE (2002). PON1 polymorphisms, in Paraoxonase in Health and Disease: Basic and Clinical Aspects, Furlong CE, ed. Boston: Kluwer Academic Publishers, pp. 51–77.
64. Draper E. Genetic secrets: social issues of medical screening in a genetic age. Hastings Cent Rep 1992;22:S15–8.
65. Schill AL. Genetic information in the workplace. Implications for occupational health surveillance. Aaohn J 2000;48:80–91.
66. King PA (1992). The past as prologue: race, class and gene discrimination, in Gene Mapping, Elias S (ed). pp. 94–98.
67. Rothstein MA. Genetics and the workforce of the next hundred years. Columbia Bus. L. Rev. 2000;2000:371.
68. Newill CA, Khoury MJ, Chase GA. Epidemiological approach to the evaluation of genetic screening in the workplace. J Occup Med 1986;28:1108–1111.
69. Omenn GS. Predictive identification of hypersusceptible individuals. J Occup Med 1982;24:369–374.
70. Vineis P, Schulte PA. Scientific and ethical aspects of genetic screening of workers for cancer risk: the case of the N-acetyltransferase phenotype. J Clin Epidemiol 1995; 48:189–197.
71. Van Damme K, Casteleyn L, Heseltine E, et al. Individual susceptibility and prevention of occupational diseases: scientific and ethical issues. J Occup Environ Med 1995; 37:91–99.
72. U.S. Congress Office of Technology Assessment (OTA). The Role of Genetic Testing in the Prevention of Occupational Disease. Washington D.C.: Office of Technology Assessment, 1983.
73. U.S. Congress Office of Technology Assessment (OTA). Genetic Moniitoring and Screening in the Workplace. Washington D.C.: Government Printing Office, 1990.
74. Lappe M. Ethical issues in testing for differential sensitivity to occupational hazards. J Occup Med 1983;25:797–808.
75. Rothenberg K, Fuller B, Rothstein M, et al. Genetic information and the workplace: legislative approaches and policy changes. Science 1997;275:1755–1757.
76. U.S. Department of Labor (DOL), Department of Health and Human Services, Equal Employment Opportunity Commission, Department of Justice. Genetic Information and the Workplace. http://www.dol.gov/dol/asp/public/programs/history/herman/reports/genetics.htm, 1998.
77. Use of genetic testing by employers. Council on Ethical and Judicial Affairs, American Medical Association. Jama 1991;266:1827–1830.
78. ACOEM (American College of Occupational and Environmental Medicine). Position Statement: Genetic Screening in the Workplace. http://www.acoem.org/paprguid/papers/gensc.htm, 1994.
79. APHA (American Medical Association). Guidelines for genetic testing in industry. Am J Publ Health 1984;74:281–282.
80. Draper E. Risky Business. Genetic Testing and Exclusionary Practices in the Hazardous Workplace. New York: Cambridge University Press, 1991.

81. Draper E. The Screening of America: The Social and Legal Framework of Employer's Use of Genetic Information. Berkeley Journal of Employment and Labor Law 1999;20: 286.

82. Bartell SM, Ponce RA, Takaro TK, Zerbe RO, Omenn GS, Faustman EM. Risk estimation and value-of-information analysis for three proposed genetic screening programs for chronic beryllium disease prevention. Risk Anal 2000;20:87–99.

83. Hirvonen A. Combinations of susceptible genotypes and individual responses to toxicants. Environ Health Perspect 1997;105(suppl 4):755–758.

84. Nicas M, Lomax GP. A cost-benefit analysis of genetic screening for susceptibility to occupational toxicants. J Occup Environ Med 1999;41:535–544.

85. Bates DV. Environmental Health Risks and Public Policy. Seattle: University of Washington Press, 1994.

86. Moreno JD, Bayer R. The limits of the ledger in public health promotion. Hastings Cent Rep 1985;15:37–41.

19

Factor V Leiden, oral contraceptives, and deep vein thrombosis

Jan P. Vandenbroucke, Frits R. Rosendaal, and Rogier M. Bertina

The combination of factor V Leiden mutation (FVL) and the use of oral contraceptives results in a much higher risk for venous thrombosis than would be expected from the effect of each risk factor alone. This finding has led to new etiologic insights in the mechanisms by which oral contraceptives induce venous thrombosis. Biochemically, the increased risk is apparent by effects on activated protein C resistance ("APC resistance") and by several other hemostatic changes. This review describes the epidemiologic and hemostatic background to these findings and indicates the main papers in the literature. We will explain APC resistance, both the genetic form (by factor V Leiden mutation) and the acquired form (by female hormones), the hemostatic and epidemiologic interactions, and the consequences to public health and clinical medicine.

The Discovery

The discovery of the interaction between factor V Leiden and oral contraceptives was originally made by looking at the data of a case-control study on hereditary and other risk factors for venous thrombosis (1). In that study, which included patients of both sexes from age 15 to 70, we found a handful of patients who were homozygous carriers of the mutation; almost all of these homozygotes were young women. This struck us as unusual, since we saw no reason why a group of consecutive patients with venous thrombosis, in whom an hitherto unknown mutation on an autosomal chromosome was found, would show a predominance of homozygotes among the younger women. Because many of these young women used oral contraceptives, we used this fact as the starting point for an analysis of the interaction between oral contraceptives and FVL in the development of venous thrombosis. In the analysis we combined heterozygous and homozygous carriers of the mu-

322

tation, since the heterozygous state is present in a larger number of patients and was recognized as having a strongly elevated risk by itself. We found that use of oral contraceptives increased the risk of venous thrombosis about fourfold and factor V Leiden about sevenfold, but that the joint effect was a more than 30-fold increase (1).

Background

Deep Vein Thrombosis as a Side Effect of Oral Contraceptives

In the 1960s, venous thrombosis was described in multiple case reports and epidemiologic studies as a side effect of combined oral contraceptives (combination of estrogen and progestin). This produced the second great wave of epidemiologic adverse effect research—the first being smoking and lung cancer—and both general epidemiology and pharmaco-epidemiology matured in the debates about the findings. An authoritative review was published by Stadel in 1981 (2). He concluded that use of oral contraceptives increased the risk of deep vein thrombosis about fourfold. This conclusion remained a long-held consensus, although there were some uncertainties. For example, it was believed that "diagnostic bias" might have led to an overestimation of the risk and that the lower-dose pills (lower dose of estrogens from the 1980s onward) might have had much less adverse effects. There was lasting uneasiness as no clear biologic basis for this adverse effect was found.

From 1995 onward, a new wave of epidemiologic and hemostatic studies about oral contraceptives and deep vein thrombosis has been published. There were several new findings: the relative risk for deep vein thrombosis was still about a fourfold increase and some of the newer pill brands with a new generation of progestins (so-called third generation progestins) showed a larger risk—twice that of the older second-generation pills (3,4). Third-generation pills thus yielded overall relative risks that came close to the risks of the earliest very high-dose pills of the 1960s. The newest progestins had apparently changed the balance of the steroids and made the combined pill "more estrogenic" again. In part the estrogenicity was the purpose of pharmacologic design: to increase the level of HDL-cholesterol in women who take the pills in an effort to lower the increased risk for arterial thrombosis (myocardial infarction and stroke) that also exists as a side effect of the pills. Although the intended increase of HDL-cholesterol was achieved with third-generation pills, the new steroid combination was shown to be associated with a larger prothrombotic effect on hemostasis (3). This new information helped to elucidate the mechanisms by which oral contraceptives induce venous thrombosis.

A "Teleological" Description of Hemostasis

For readers not familiar with hemostasis, a brief "teleological" description is useful. Blood needs to coagulate quickly, at the slightest provocation (e.g., after damage to vessel walls) in order to maintain the integrity of the circulation. Otherwise, even minor wounds might be fatal. All animals who depend on a circulation system have some form of hemostasis. On the other hand, blood clots should not grow too large; they should only patch the damaged vessels—if there were no "check," all vessels might be completely obliterated once the blood starts coagulating. There are three enzyme cascades involved in hemostasis: the procoagulant cascade, the anticoagulant cascade, and the fibrinolytic cascade. The procoagulant cascade is activated by signals from the damaged endothelium. One of the main anticoagulant systems that controls coagulation (i.e., "keeps it in check") is the protein C system. It is activated by the binding of thrombin, the end product of the prococoagulant cascade, to normal endothelium and results in the formation of "activated protein C" or APC. The "activated protein C" inactivates and thus downregulates two important procoagulant factors (factor V and factor VIII). Lastly, there is fibrinolysis to dissolve clots that have been formed. The mechanisms that play a role in this review center on the role of the protein C anticoagulant system.

Discovery of APC Resistance and Factor V Leiden as Prothrombotic States

Resistance to "activated protein C" was first described in 1993 by the biochemical observations of Dahlbäck (5,6). Working with several members of a large family with multiple venous thrombosis he found that adding APC to plasma did not result in the expected prolongation of an in vitro clotting time measurement, activated partial thromboplastin time (aPTT). A prolongation of the clotting time was expected because normally APC leads to inactivation of clotting. He called the phenomenon "resistance to activated protein C." The main genetic cause of this resistance was later found to be a mutation in procoagulant factor V (factor V G1691A, also known as factor V Leiden) (7). Activation of factor V into factor Va ("a" stands for activated) is a key reaction in the acceleration of clot formation. Thus, the inactivation of Va by APC is important in the downregulation of coagulation. The factor V Leiden mutation occurs at one of the sites where factor Va is cleaved by APC, which makes this mutant factor V relatively insensitive to inactivation by APC. Thereby, the mutation causes prolonged activity of factor Va, which results in a tendency to thrombosis (a prothrombotic state).

Molecular Aspects of Action of Factor V and Factor V Leiden

Factor V is a glycoprotein with a molecular weight of 330.000 daltons. It is built of 2196 amino acids and circulates in the blood—in a concentration of 20 nM—as the procofactor of coagulation factor Xa. In the early phase of the coagulation process, after a signal has been received—from, for example, a damaged vessel wall—factor V is activated by thrombin via proteolysis of peptide bonds after Arg709, Arg1018, and Arg1545. Factor Va forms a complex with Xa and calcium ions on the surface of activated platelets, which then converts prothrombin into thrombin. The latter reaction proceeds about 1000 times faster in the presence of factor Va than in its absence, which explains why the inactivation of factor Va is an important physiologic process in downregulating coagulation. Inactivation of factor Va takes place by "activated protein C (APC)" that is formed from its zymogen under the influence of the thrombin–thombomodulin complex on normal endothelium (the undamaged parts of the blood vessel). APC cuts the protein chain of factor Va in two places, after Arg506 and after Arg306. The first cut is fast but results only in partial inactivation of Va; the second cut is much slower and results in complete inactivation of Va.

Factor V Leiden is a mutated factor V, to the effect that the triplet coding for Arg506 (CGA) is replaced by CAA, which codes for Gln506. This factor V Q506 is activated in a normal way by thrombin and has normal Xa co-factor activity. However, its inactivation by APC is much slower because the mutation prevents the fast cut of factor Va, so that the inactivation becomes completely dependent upon the much slower second cut. The result is that factor V Leiden, after activation, can express its Xa-cofactor activity for a much longer period. Thus, the same stimulus will result in the formation of more thrombin and more fibrin (larger clots) in a factor V Leiden carrier than in carriers of normal factor V. Factor Va Leiden is thus relatively insensitive to inactivation by APC. Hence, the observed "resistance to APC" in plasma of carriers of factor V Leiden.

Oral Contraceptives and APC Resistance

Early results by several researchers indicated that the aPTT-based test for APC resistance gave different results according to hormonal status; increased APC resistance was found with oral contraceptive use and pregnancy, which made it more difficult to discriminate whether the person would be a carrier of factor V Leiden. APC determinations in oral contraceptives users and pregnant women overlapped with those carrying factor V Leiden (8–11). This led to the concept of "acquired" APC resistance. The "acquired" form of APC resistance was more readily demonstrated by using a newer APC-resistance test based on the measurement of "tissue factor induced thrombin generation," (variously called the endogenous thrombin potential (ETP), or "thrombin generation" test) (12). This test appeared equally sensitive for FVL, but was more sensitive for the effect of female hormones than the aPTT-based test. Women who used oral contraceptives had higher levels of APC resistance in the thrombin generation test than did women who did not use contraceptives. Women who used third-generation oral contraceptives had a degree of APC resistance that was close to that of carriers of FVL when measured with the thrombin generation test, and this was significantly higher than for women on second-generation pills (12). Finally, women who used the pill and were at the same time heterozygous carriers of factor V Leiden, showed a degree of APC resistance close to that of homozygous factor V Leiden carriers, who have the highest risk of venous thrombosis (13). These hemostatic data were completely in line with the epidemiologic findings. A randomized cross-over study confirmed the effect of oral contraceptives on APC resistance in the thrombin generation test, and demonstrated many other effects of oral contraceptives on procoagulant, anticoagulant, and fibrinolytic factors (14). All effects pointed to the induction of a prothrombotic state for oral contraceptives in general, and all were stronger for third-generation oral contraceptives than for second-generation pills (3,14). In the meantime, it was also shown that in the absence of pill use or factor V Leiden, even a mild reduction in the sensitivity for APC is a risk factor for venous thrombosis (15). Similar effects exist during use of hormonal replacement therapy and pregnancy (16,17).

Epidemiologic Interactions Between Female Hormones and Prothrombotic Mutation

Factor V Leiden has an interesting geographic distribution; its prevalence is about 5% in Caucasians, but with local differences that presumably point to founder effects (e.g., highest prevalences are found in the south of Sweden and in some Middle Eastern countries, while the prevalence is much lower in Italy and Spain). Factor V Leiden is virtually absent in African and Oriental populations (18,19). Factor V Leiden results in a four- to ninefold increase in venous thrombosis incidence; among homozygous persons, the increased risk is up to 50-fold (13,20). The mutation is found in about 20% of Caucasian patients with venous thrombosis and in about 50% from families with multiple and early venous thrombosis, heritable thrombophilia (21).

The original findings on the interaction of FVL and oral contraceptives came from a case-control study with consecutive patients in three geographically distinct anticoagulation clinics in the Netherlands (these clinics monitor anticoagulation treatment, e.g., following a venous thrombosis). Further details can be found in the original studies (1,22). The case control findings can be regrouped as in Table 19.1.

One initial difficulty in the analysis was the small number of persons with joint exposure in the control group. However, by using the overall prevalences in the control group, we could check whether the partial contrasts (the effect of FVL and oral contraceptives) held by themselves. Control subjects are individuals who have not

Table 19.1 Distribution of Women with Venous Thrombosis and of Control Subjects by Oral Contraceptive Use and Presence of Factor V Leiden Mutation

	EXPOSURE			
Group	$OC(-)^a$ $FVL (-)^a$	$OC(+)$ $FVL (-)$	$OC(-)$ $FVL (+)$	$OC(+)$ $FVL (+)$
Cases	36	84	10	25
Controls	100	63	4	2
Odds ratio	baseline = 1	3.7	6.9	34.7
95% CI[b]	—	2.3 to 6.1	2.3 to 20.8	11.8 to 101.8
Mantel-Haenszel odds ratio for OC, controlling for FVL			3.8 (2.4 to 6.1)	
Mantel-Haenszel odds ratio for VL, controlling for OC			8.2 (3.5 to 19.1)	

[a]OC, oral Contraceptives; FVL, factor V Leiden.
[b]95% confidence interval (test-based)
Source: Adapted from Vandenbroucke JP et al. Increased risk of venous thrombosis in oral-contraceptive users who are carriers of factor V Leiden mutation. Lancet 1994;344:1453–1457. (1)

yet experienced thrombosis, and in them an association between FVL and oral contraceptives was unlikely. In other words, the prevalence of oral contraceptives use among control subjects was considered equal for those with and without factor V Leiden and vice versa. Use of these marginal distributions of FVL and oral contraceptives did not materially alter our interaction estimates. (See discussion (1)).

Since the population base was geographically defined, we could back-calculate incidences. These illustrate how the joint presence of factor V Leiden and oral contraceptives use gives rise to a much higher incidence of deep vein thrombosis than is expected by summing the separate effects (Table 19.2).

Interaction effects of the much rarer prothrombin mutation with oral contraceptives use were established by Martinelli et al (23). An overview and pooled analysis of studies with information about FVL and the prothrombin mutation confirmed the interaction effects with oral contraceptives use (20). The highest risk was in those who carried both mutations and used oral contraceptives as well. In the pooled analysis, the overall risks were somewhat lower than in several original investigations, presumably due to some heterogeneity in the case selection of different studies (20).

Finally, women who carry FVL not only develop deep vein thrombosis more often when they use oral contraceptives, but they develop it sooner. Their risk of deep vein thrombosis in the first year of use is more than 10 times that in later years (24). Apparently, the presence of the genetic mutation puts them at a disadvantage.

Hormone Replacement Therapy

For hormone replacement therapy (HRT), an interaction also exists between FVL and hormonal substitution in the development of venous thrombosis. This interaction was found in the reanalysis of an earlier case-control study in which additional DNA and hemostatic determinations were performed (Table 19.3) (16).

Although the relative risks for women who use HRT are somewhat smaller than for those who use oral contraceptives, the absolute impact of these risks will be

Table 19.2 Estimated Population Incidence of First Venous Thrombosis in Women Aged 15–49, According to Presence of Factor V Leiden Mutation and Use of Oral Contraceptives

	Patients	Person-years[a]	Incidence Per 10,000 Person-Years
Factor V Leiden negative			
No OC use	36	437,870	0.8
Current OC use	84	275,858	3.0
Factor V Leiden positive			
No OC Use	10	17,515	5.7
Current OC use	25	8,757	28.5

OC, oral contraceptives.
[a]A total of 740 000 person-years (yielding 155 patients) was partioned according to the distribution of the control group: 100/63/4/2.
Source: Vandenbroucke JP et al. Increased risk of venous thrombosis in oral-contraceptive users who are carriers of factor V Leiden mutation. Lancet 1994:344:1453–1457. (1)

Table 19.3 Distribution of Women with Venous Thrombosis and Control Subjects by Hormonal Replacement Therapy and Presence of Factor V Leiden Mutation

Factor V Leiden	Hormonal Replacement Therapy	Patients	Control	Odds Ratio	95% Confidence Interval
−	−	30	116	1	
+	−	8	8	3.9	1.3 to 11.2
−	+	31	37	3.2	1.7 to 6.0
+	+	8	2	15.5	3.1 to 76.7

Source: From Rosendaal FR et al. (16)

higher, since the baseline incidence of venous thrombosis in middle-aged women is higher than in younger women.

Cerebral Venous Thrombosis

Cerebral venous thrombosis is a much rarer condition than venous thrombosis, but it was also described in the past as having a clear clinical associaton with pregnancy and oral contraceptive use. Martinelli et al. first found that FVL and the prothrombin mutation are risk factors for cerebral venous thrombosis, the latter mutation in particular in the presence of oral contraceptive use (25).

The interaction between pill use and prothrombotic mutations (factor V Leiden, and others like the much rarer protein C and antithrombin mutations) in the development of venous thrombosis was confirmed in a case-control study in the Netherlands. A nationwide series of patients with cerebral sinus thrombosis was contrasted with a "virtual control group" that was created by calculating the expected frequency of all combinations of carriership of prothrombotic mutations and pill use in the general population of the Netherlands. It was found again that the joint effect of both exposures was higher than were the individual effects (26).

Hemostatic Interaction of Genetic and Acquired APC Resistance

A recent randomized trial in which two groups of young women were recruited from the general population (women with normal factor V and women with factor V Leiden) showed the existence of acquired APC resistance, induced by oral contraceptives, that adds itself to genetic APC resistance by FVL (27). Both groups of women were first randomized between second- and third-generation combined oral contraceptives, and after a wash-out period, the estrogen component of the pill was stopped, and women continued with progestin only pills (either second or third generation). The baseline level of APC resistance was markedly higher in the group with factor V Leiden. The APC resistance effect of the pill added itself upon these different baselines. The level of APC resistance of a woman with factor V Leiden at baseline was already about as high as that of a woman with normal factor V who used third-generation oral contraceptives, and it increased further by oral contraceptive use. In both the FVL and the normal group, the effect of third-generation oral contraceptives was significantly higher than that of second-generation contraceptives. It is intriguing that an additive effect on APC resistance apparently gives rise to an exponential increase in the risk (as the additive effect on APC resistance multiplies the risk of venous thrombosis).

Venous Thrombosis as a Multicausal Disease

Perhaps the greatest insight into the influence of genetic and environmental factors on the development of venous thrombosis came from the realization that in families with multiple and early venous thrombosis (thrombophilia) several genetic factors often were present. In families with known rare thrombophilic mutations (protein S, antithrombin, and protein C deficiencies), the newly found factor V Leiden was more often present than in the general population (21,28,29). Among families with factor V Leiden, the even newer prothrombin mutation was also more often present (20). Highest risks are found in women who have two mutations and use oral contraceptives (20,23). Looking for additional genetic factors in families with multiple venous thrombosis at young ages has been advocated as a useful research strategy (30). Apparently, several risk factors have to be present together in order for the disease to develop, a condition that is clearest among the youngest patients (31). These risk factors are not only genetic, but also involve classic risk factors such as surgery and trauma that have relative risks that often exceed those of the mutations. Even simple immobilization, such as in persons sitting for a long time in cars, might interact with factor V Leiden, as was observed during Metro strikes in Paris (32). Mental stress, which increases procoagulation (33) might be an additional factor. In a single study of women who used oral contraceptives, the risk for factor V Leiden and long distance travel was found to be multiplicative, apparently over and beyond the effect of the oral contraceptives (34).

Clinical and Public Health Consequences

The question whether screening before prescribing oral contraceptives is beneficial remains controversial in the absence of any data on the effect of such a program. We have argued that more than half a million women would need to be screened for factor V Leiden, and tens of thousands should be denied the use of oral contraceptives in order to prevent a single death. Venous thrombosis remains a relatively rare event, and mortality from venous thrombosis is low in young women (35). Others have taken the position that if all costs of the treatment of venous thrombosis were taken into account, screening might become cost-effective when the cost of a screening test would be lower than about 9 U.S. Dollars (see ref. 3). It is clear that there are not only medical and financial aspects to be considered, but also issues related to quality of life, risk for unwanted pregnancy, and insurance.

A difficult problem is what to do when a woman has a family history of venous thrombosis. Again, the picture is not clear. Even for the patients in such families, it is uncertain whether they benefit from knowing their factor V Leiden status; it as yet uncertain whether there is a higher risk of recurrence, and it is equally unclear whether patients with a prothrombotic mutation should be treated any differently (i.e., should receive anticoagulation for longer periods) (36,37). A basic question is what one should screen for: for a positive family history—irrespective of the presence of a mutation, or for the mutations. A positive family history has poor predictive value for the presence of factor V Leiden mutation (38). However, the most meaningful information for judging the background risk of a person might be the positive family history itself (39). Indeed, even the effect of a prothrombotic mutation differs according to familial background; in patients who belong to multicase families, the first venous thrombosis in gene carriers was on average at age 30—in contrast, in consecutive patients (in whom those from multicase families consitute a small minority) average age at first thrombosis in persons carrying exactly the same prothrombotic mutations was at age 45 (40). This 15-year age difference exemplifies the effect of all other risk factors that cumulate in multicase families, both genetic and environmental. An obvious next question that necessitates careful definition is what constitutes a positive family history (41). In the end, for the asymptomatic woman from a family with multiple thrombosis, there are no "evidence-based" guidelines, and a decision will have to be reached on the basis of clinical and physiologic reasoning.

Conclusions

The discovery of the interaction of FVL and oral contraceptives in the development of venous thrombosis is intellectually exciting. For the first time since the 1960s, we begin to understand why and how oral contraceptives may lead to venous thrombosis, grace to much complementary and joint research activity between epidemiology and hemostasis. APC resistance, either genetic or acquired, plays a key role, but not the only one, since oral contraceptives also cause other prothrombotic changes in the pro- and anticoagulant pathways and even in the fibrinolytic system (3). Acquired APC resistance also offers the beginning of an explanation for the higher incidence of venous thrombosis during pregnancy, which is the first medical condition in which venous thrombosis was clearly described (as the "milk leg" in the 18th century). However, at this time, neither the epidemiology nor the hemostatic findings on this single genetic and environmental factor have immediate consequences for patients or the general public. The continued study of all the gene–gene and gene–environment interactions in venous thrombosis has taught us that venous thrombosis is a multicausal disease and that several factors have to be present for the disease to develop (42).

References

1. Vandenbroucke JP, Koster T, Briet E, et al. Increased risk of venous thrombosis in oral-contraceptive users who are carriers of factor V Leiden mutation. Lancet 1994;344: 1453–1457.

2. Stadel BV. Oral contraceptives and cardiovascular disease. Review. N Engl J Med 1981; 305:612–618.
3. Vandenbroucke JP, Rosing J, Bloemenkamp KWM, et al. Oral contraceptives and the risk of venous thrombosis. Review. N Engl J Med 2001;344:1527–1535.
4. Kemmeren JM, Algra A, Grobbee DE. Third generation oral contraceptives and risk of venous thrombosis: meta-analysis. BMJ 2001;323:131–134.
5. Dahlbäck B, Carlsson M, Svensson PJ. Familial thrombophilia due to a previously unrecognised mechanism characterized by poor anticoalgulant response to activated protein C: prediction of a cofactor to activated protein C. Proc Natl Acad Sci U S A 1993;90: 1004–1008.
6. Dahlbäck B. Thrombophilia: the discovery of activated protein C resistance. Review. Adv Genet 1995;33:135–175.
7. Bertina RM, Koeleman BPC, Koster T, et al. Mutation in blood coagulation factor V associated with resistance to activated protein C. Nature 1994;369:64–67.
8. Østerud B, Robertsen R, Åsvang GB, et al. Resistance to activated protein C is reduced in women using oral contraceptives. Blood Coagul Fibrinolysis 1994;5:853–854.
9. Olivieri O, Friso S, Manzato F, et al. Resistance to activated protein C in healthy women taking oral contraceptives. Br J Haematol 1995;91:465–470.
10. Henkens CMA, Bom VJJ, Seinen AJ, van der Meer J. Sensitivity to activated protein C: influence of oral contraceptives and sex. Thromb Haemost 1995;73:402–404.
11. Lowe GD, Rumley A, Woodward M, et al. Activated protein C resistance and the FV:R506Q mutation in a random population sample–associations with cardiovascular risk factors and coagulation variables. Thromb Haemost 1999;81:918–924.
12. Rosing J, Tans G, Nicolaes GAF, et al. Oral contraceptives and venous thrombosis: different sensitivities to activated protein C in women using second- and third-generation oral contraceptives. Br J Haematol 1997;97:233–238.
13. Rosendaal FR, Koster T, Vandenbroucke JP, et al. High risk of thrombosis in patients homozygous for factor V Leiden (activated protein C resistance). Blood 1995;85:1504–1508.
14. Rosing J, Middeldorp S, Curvers J, et al. Low-dose oral contraceptives and acquired resistance to activated protein C: a randomised cross-over study. Lancet 1999;354:2036–2040.
15. De Visser MCH, Rosendaal FR, Bertina RM. A reduced sensitivity for activated protein C in the absence of factor V Leiden increases the risk of venous thrombosis. Blood 1999;93:1271–1276.
16. Rosendaal FR, Vessey M, Rumley A, et al. Hormonal replacement therapy, prothrombotic mutations and the risk of venous thrombosis. Brit J Haematol, in press, 2002.
17. Hellgren M, Svensson PJ, Dahlbäck B. Resistance to activated protein C as a basis for venous thromboembolism associated with pregnancy and oral contraceptives. Am J Obstet Gynecol 1995;173:210–213.
18. Rees DC, Cox M, Clegg JB. World distribution of factor V Leiden. Lancet 1995;346: 1133–1134.
19. Ridker PM, Miletich JP, Hennekens CH, et al. Ethnic distribution of factor V Leiden in 4047 men and women: implications for venous thromboembolism screening. JAMA 1997;277:1305–1307.
20. Emmerich J, Rosendaal FR, Cattaneo M, et al. Combined effect of factor V Leiden and prothrombin 20210A on the risk of venous thromboembolism. Pooled analysis of 8 case-control studies including 2310 cases and 3204 controls. Thromb Haemost 2001;86:809–816.
21. Bertina RM. Genetic approach to thrombophilia. Thromb Haemost 2001;86:92–103. Review.
22. Koster T, Rosendaal FR, de Ronde H, et al. Venous thrombosis due to poor anticoagulant response to activated protein C: Leiden Thrombophilia Study. Lancet 1993;342: 1503–1506.

23. Martinelli I, Taioli E, Bucciarelli P, et al. Interaction between the G20210A mutation of the prothrombin gene and oral contraceptive use in deep vein thrombosis. Arterioscler Thromb Vasc Biol 1999;19:700–703.

24. Bloemenkamp KWM, Rosendaal FR, Helmerhorst FM, et al. Higher risk of venous thrombosis during early use of oral contraceptives in women with inherited clotting defects. Arch Intern Med 2000;160:49–52.

25. Martinelli I, Sacchi E, Landi G, et al. High risk of cerebral-vein thrombosis in carriers of a prothrombin-gene mutation and in users of oral contraceptives. N Engl J Med 1998;338:1793–1797.

26. de Bruijn SF, Stam J, Koopman MM, et al. Case-control study of risk of cerebral sinus thromobosis in oral contraceptive users and in [correction of who are] carriers of hereditary prothrombotic conditions. The Cerebral Venous Sinus Thrombosis Study Group. BMJ 1998;316:589–592.

27. Kemmeren JM. Vascular effects of second and third generation contraceptives: drug-drug, drug-gene and drug-environment interactions. PhD Thesis. Utrecht, The Netherlands, 2001. pp. 43–58.

28. Van Boven HH, Vandenbroucke JP, Briet E, et al. Gene-gene and gene-environment interactions determine risk of thrombosis in families with inherited antithrombin deficiency. Blood 1999;94:2590–2594.

29. Lensen R, Bertina RM, Vandenbroucke JP, et al. High factor VIII levels contribute to the thrombotic risk in families with factor V Leiden. Br J Haematol 2001;114:380–386.

30. Majerus PW. Human genetics. Bad blood by mutation. Nature 1994;369:14–15.

31. Rosendaal FR. Thrombosis in the young: epidemiology and risk factors. A focus on venous thrombosis. Review. Thromb Haemost 1997;78:1–6.

32. Eschwege V, Robert A. Strikes in French public transport and resistance to activated protein C. Lancet 1996;347:206.

33. Jern C, Eriksson E, Tengborn L, Risberg B, et al. Changes of plasma coagulation and fibrinolysis in response to mental stress. Thromb Haemost 1989;62:767–771.

34. Schambeck CM, Schwender S, Haubitz I, et al. Selective screening for the Factor V Leiden mutation: is it advisable prior to the prescription of oral contraceptives? Thromb Haemost 1997 Dec;78:1480–1483.

35. Vandenbroucke JP, van der Meer FJM, Helmerhorst FM, et al. Factor V Leiden: should we screen oral contraceptive users and pregnant women? BMJ 1996; 313:1127–1130.

36. Middeldorp S, Meinardi JR, Koopman MM, et al. A prospective study of asymptomatic carriers of the factor V Leiden mutation to determine the incidence of venous thromboembolism. Ann Intern Med 2001;135:322–327.

37. Bauer KA. The thrombophilias: well-defined risk factors with uncertain therapeutic implications. Ann Intern Med 2001;135:367–373.

38. Cosmi B, Legnani C, Bernardi F, et al. Value of family history in identifying women at risk of venous thromboembolism during oral contraception: observational study. BMJ 2001;322:1024–1025.

39. Vandenbroucke JP, van der Meer FJ, Helmerhorst FM, et al. Family history and risk of venous thromboembolism with oral contraception. Family history is important tool. BMJ 2001;323:752.

40. Lensen RP, Rosendaal FR, Koster T, et al. Apparent different thrombotic tendency in patients with factor V Leiden and protein C deficiency due to selection of patients. Blood 1996;88:4205–4208.

41. Briet E, van der Meer FJ, Rosendaal FR, et al. The family history and inherited thrombophilia. Br J Haematol 1994;87:348–852.

42. Rosendaal FR. Venous thrombosis: a multicausal disease. Review. Lancet 1999;353:1167–1173.

20

Methylenetetrahydrofolate reductase gene (*MTHFR*), folate, and colorectal neoplasia

Linda Sharp and Julian Little

An estimated 945,000 new cases of colorectal cancer were diagnosed worldwide in 2000, and 492,000 people died from the disease (1). Two-thirds of the incident cases occurred in developed countries, where colorectal cancer is the third most common cancer in men and second most common in women. In developing countries, it is the fifth most common tumor in both sexes. Unlike many cancers, invasive tumors of the large bowel are diagnosed in almost as many women as men. Almost all colorectal tumors are adenocarcinomas, and between 60% and 70% occur in the colon (2).

During 1988–1992 international age-standardized incidence rates of colorectal cancer varied widely (2). In men, incidence rates of less than 10 per 100,000 occurred in Africa, India, Thailand, and Vietnam. The highest rates, around 45 per 100,000 and above, occurred in Australia, the non-Maori population of New Zealand, and the United States. Across Japan the incidence ranged from 30 to 50 per 100,000. Rates in most of western Europe were around 40 per 100,000, and were somewhat lower (20–30) in Spain and parts of northern and eastern Europe. For women, the geographical pattern was similar, but the age-standardized rates were about 60%–75% of those in men.

In 1971, Haenzel and Correa (3) noted that colon cancer incidence was slowly increasing. Since then, moderate increases in incidence have been observed in many populations (4). These include populations that, in earlier decades, had intermediate or high rates of colorectal cancer—such as Denmark, Spain, Australia, the United Kingdom, and the United States—and those with low rates—such as Japan. Although the general pattern is similar, the magnitude of the increase differed between populations, as did the timing. In most, the increase was either more pronounced in men than women or observed only in men. These trends are likely to be only in part an artifact of improvements in the efficiency of cancer registration and increased detection rates resulting from the introduction of newer diagnostic tools. They indicate strong sex-specific cohort effects, most likely associated with changes in exposures to environmental and lifestyle risk factors (e.g., diet) for the disease.

Tumors of the right (proximal) colon have been reported to have become more common over the past 60 years (see, for example, refs. 5 and 6). These observations are difficult to interpret for a number of reasons including selection bias, demographic changes, and increasing use of colonoscopy and flexible sigmoidoscopy (4). Despite this, speculation persists that tumors of the left and right colon have different etiologies and pathogenesis (7).

The extent of disease at diagnosis is a strong predictor of survival. SEER Program data for patients diagnosed with colon cancer in 1992–1997 show 5-year relative survival of 91% for those whose disease was localized at diagnosis, 67% for those presenting with regional spread, and 9% for those with distant metastasis (8). For rectal cancer the corresponding figures were 87%, 57%, and 8%. Of all those for whom the extent of disease was known, 39% had localized disease, 40% regional spread, and 21% distant metastasis. In developed countries, colorectal cancer death rates have declined steadily during the past 20 to 30 years (9). At least in part, this is the result of a decline over time in the proportion of patients presenting with more advanced disease (10,11), which is likely to have been a consequence of increased availability and use of sigmoidoscopy, colonoscopy, and, possibly, fecal occult blood testing.

Groups at High Risk for Colorectal Cancer

Several groups have an increased risk of developing colorectal cancer: those with inflammatory bowel disease or colorectal polyps, individuals in families affected by the autosomal dominant conditions hereditary nonpolyposis colorectal cancer (HNPCC) and familial adenomatous polyposis (FAP), and individuals who have a family history of colorectal neoplasia but are not part of families affected by HNPCC or FAP. The risk of cancer in patients with longstanding ulcerative colitis or Crohn's disease is hard to quantify, but is thought to be similar for patients with the two conditions (12). A meta-analysis of 116 studies estimated that the cumulative probability of cancer in a patient with ulcerative colitis was 2% by 10 years, 8% by 20 years, and 18% by 30 years (13).

A few studies of colorectal adenomatous polyps left in situ show progression from adenoma to cancer (14). These observations, coupled with indirect evidence, support the view that most colorectal carcinomas develop from adenomas. Although these lesions are usually removed when detected, the risk for recurrence 3 years after colonoscopic polypectomy is 30% to 40% (15,16). Hyperplastic polyps also may exhibit malignant potential. These, and serrated adenomas, may be precursors of some right-sided colon cancers (17). Investigation of factors related to both occurrence and recurrence of polyps potentially provides information about the role of exposures in the earlier stages of the adenoma–carcinoma sequence.

Fewer than 10% of incident colorectal cancers are due to HNPCC and FAP (18). Even when these syndromes are excluded, carcinomas and adenomas aggregate in families, and individuals who have a first-degree relative with colorectal cancer have around a twofold increased risk of developing the disease themselves (19,20). This pattern is probably not entirely explained by familial clustering of environmental factors (21). This points to the potential importance of genetic susceptibility factors, and the interaction of these with each other and with environmental factors, in causing the disease.

Environmental and Lifestyle Factors in Colorectal Neoplasia

The classic studies of Japanese migrants to the United States conducted in the 1960s pointed to the overwhelming importance of environmental factors in colorectal cancer etiology (22). Higher levels of physical activity are consistently associated with reduced colon cancer risk, with the most active persons having 50% of the risk of the least active (23). Studies of adenoma risk are compatible with this. Excess weight raises risk of developing colon cancer, with an increase of 15% in risk for an overweight person and 33% for an obese person (24). Evidence consistently shows that tobacco smoking increases risk for adenomas (14). For carcinomas, recent studies also suggested that smoking moderately elevates risk (25), with an induction period of 35 to 40 years. Alcohol intake also appears to be positively associated with risk of adenomas (14); the evidence for colon and rectal cancer is less consistent (26,27). Results of observational studies suggested that the use of aspirin or other nonsteroidal anti-inflammatory drugs may reduce risk for colorectal cancer by as much as 50% (28). In the majority of studies, use of postmenopausal estrogen replacement therapy has been associated with reduced risk of colon cancer, although there is concern that the association may be due to uncontrolled confounding (29).

Folate and Colorectal Neoplasia

Although diet is thought to have a major role in colorectal cancer etiology (27), the specific components involved have been difficult to identify. Observational epidemiologic studies suggested that high intakes of vegetables were associated with a decreased colorectal cancer risk (27), but recent evidence suggests that the relation is complex (30–32). Vegetables, particularly green leafy vegetables, are a major source of folate. This, supported by proposed mechanisms, has prompted the investigation of the role of folate—and, its synthetic form, folic acid—in colorectal neoplasia.

The first mechanism by which folate deficiency might increase disease risk is through methylation (Fig. 20.1). Folic acid, in the form of 5-methyltetrahydrofolate, is important for the production of S-adenosylmethionine (SAM), the primary methyl donor for DNA methylation. Decreased DNA methylation (hypomethylation) ap-

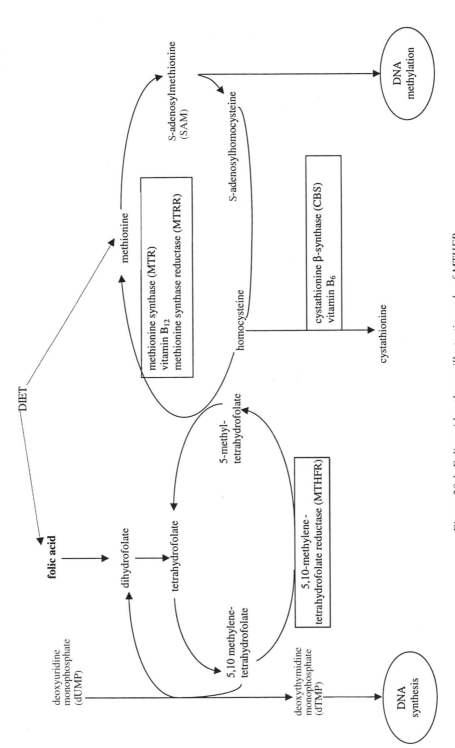

Figure 20.1 Folic acid pathway, illustrating role of MTHFR

pears to be an early step in colorectal carcinogenesis (33,34). Folate deficiency may deplete cellular SAM levels and cause DNA hypomethylation and inappropriate activation of proto-oncogenes (35). The second possible mechanism relates to the role of folate in DNA synthesis. Folate deficiency may induce uracil misincorporation in the conversion of deoxyuridine monophosphate (dUMP) to thymidine monophosphate (TMP), leading to catastrophic DNA repair, DNA strand breakage, and chromosome damage (36). Human data in support of these mechanisms are still limited (35,36). Furthermore, how far these mechanisms apply in conditions of mild folate depletion (such as relatively low dietary intake) rather than overt systemic folate deficiency is not clear.

The majority of observational studies of folate—either measured in the blood or assessed from reported intake—and colon cancer are compatible with an inverse association between folate level and risk. Two of three prospective studies found an increased risk for colorectal cancer among people with reduced levels of serum or plasma folate (37–39), a short-term marker of folate intake. Folate levels in red cells indicate folate status, determined by intake, absorption, and metabolism, over a 3- to 4-month period. Two studies reported an inverse association between red cell folate and adenoma risk (40,41).

Inverse associations were observed between reported dietary folate intake and colon cancer risk in all seven published prospective studies (37,39,42–46). Of the nine case-control studies, two reported substantially reduced relative risks for colon cancer in the highest intake group; four others found reduced risk in subgroups; two were compatible with modestly reduced risk for colon and rectal tumors combined; and one was null (47–55). No consistent association between cancer of the rectum and folate intake has been observed (37,43,45,48,50,51). The inconsistencies in studies may have in part resulted from the difficulties in assessing dietary folate intake, leading to misclassification (56). Use of dietary supplements containing folic acid—particularly supplements containing higher levels of folic acid (57) or long-term use (43)—has been associated with reduced risk for colon cancer.

For colorectal adenomas, the studies were remarkably consistent, with all five case-control studies and two cohort studies showing decreased risk associated with higher folate intake (40,51,58–62). However, in an observational analysis of data from a trial of adenoma recurrence, dietary folate intake at recruitment was not associated with risk of recurrence (63).

In a recent analysis from the Nurses Health Study cohort, a higher level of total folate intake had only a minimal protective effect on colon cancer risk among women without family histories of colorectal cancer in first-degree relatives but was associated with a substantial reduction in risk among women with family histories of the disease (among women with family histories: relative risk [RR] for folate intake of >400 μg/day vs. ≤200 μg/day = 0.48, 95% confidence interval [CI], 0.28–0.93) (64). Although it requires confirmation, this finding supported gene–environment interaction in the etiology of colorectal lesions. It also suggested a group on whom chemoprevention strategies could focus.

Evidence from intervention studies is limited. In the single randomized controlled trial, of 60 persons from whom polyps had been removed, supplementation with 1 mg/day of folate resulted in a reduced recurrence rate (65). Three small studies have been conducted of adenoma patients, taking as their endpoints possible intermediate biomarkers. In two, supplementation of 5 mg/day and 10 mg/day of folic acid was reported to reverse genomic DNA hypomethylation (66,67). However, in a recent small (n = 20) randomized controlled trial, Kim et al. (68), showed that, although 5 mg/day folic acid increased genomic DNA methylation and reduced DNA strand breaks compared with placebo at 6 months, increased methylation and decreased stand breaks were observed in both arms at 12 months.

Mandatory fortification of enriched cereal grain products with folic acid was introduced in the United States at the start of 1998 (fortification was optional during March 1996–December 1997) (69). It will be interesting to monitor the rates of colorectal lesions in this population before and after fortification.

Other Dietary Factors in the Folate Pathway

Other nutrients, including vitamins B_6 and B_{12}, and methionine, are involved in the folate pathway (Figure 20.1). Methionine is required for production of S-adenosylmethionine, vitamin B_{12} is a cofactor in the methylation of homocysteine to methionine, and vitamin B_6 is required to convert homocysteine to cystathionine. Hypothetically, deficiency, depletion, or low levels of intake, of any of these might affect colorectal cancer risk through the same mechanisms as proposed for folate (70) or might act together with diminished folate levels and genetic variations to affect risk. So far, epidemiologic findings on the relations between methionine (42,45,46,50,53,61), vitamin B_6 (48,52,53,55) and vitamin B_{12} (48,53,55,58) and colorectal cancer have been limited and inconsistent.

Alcohol adversely affects the metabolism of folate (71), affecting intestinal absorption and renal excretion. This has prompted interest in whether a composite dietary profile of lower folate and higher alcohol intake, together with low intakes of methionine and vitamins B_6 and B_{12} (a "low methyl" diet), is associated with colorectal neoplasia. A number of studies have investigated this, but their results are difficult to interpret because of the different combinations of factors investigated, different methods of analysis, and small numbers in the strata. Nonetheless, four studies appeared consistent with the hypothesis that persons with a combination of high alcohol, low folate, and low methionine (or protein, as a marker for methionine) intake had higher risk for colon cancer than do those with low alcohol and high folate and methionine intake, at least in some subgroups (defined, for example, by age, sex or tumor location) (37,42,44,53).

The *MTHFR* Gene

The enzyme 5,10-methylenetetrahydrofolate reductase (MTHFR) plays a central role in folate metabolism (Fig. 20.1), operating at a branch point determining the direction of the folate pool, toward remethylation or DNA synthesis. The enzyme irreversibly converts 5,10 methylenetetrahydrofolate to 5-methyltetrahydrofolate, the primary circulating form of folate. A defect in MTHFR activity could influence DNA methylation and/or synthesis.

The *MTHFR* gene is located at the end of the short arm of chromosome 1 (1p36.3). Two common polymorphisms have been reported: (*1*) a C to T substitution at nucleotide 677, leading to an alanine to valine conversion in the protein (72); and (*2*) an A to C transition in exon 7, leading to an alanine to glutamate protein change (73, 74). The two polymorphisms are close together on the gene—at a physical distance of 2.1 kb. Other polymorphisms also have been reported but are silent (e.g., T1059C [75]; T1317C [76]) and have not been investigated in colorectal neoplasia.

Both the C677T and A1298C polymorphisms have functional consequences. For C677T, compared with homozygous wild-type, heterozygotes have 65% of enzyme activity levels in vitro and homozygous variants 30% (76). For the A1298C polymorphism, MTHFR enzyme activity in lymphocytes is lower in homozygotes and, to a lesser extent, in heterozygotes, than in the wild type (73). Nondiseased persons with the 677 TT genotype have lower red cell folate, plasma folate, and vitamin B_{12} and higher homocysteine levels than do persons with other genotypes (38,77,78), although two studies have suggested that the association with homocysteine holds only when folate status is low (79,80). The studies of A1298C genotype and plasma folate and homocysteine levels have not been consistent (73,81–84), possibly because of methodologic reasons (e.g., a non population-based study series, small sample size), or a relation may exist that depends on folate status.

Prevalence of Polymorphisms in *MTHFR*

In a systematic review, Botto and Yang (85) summarized the considerable ethnic and geographic variation in the frequency of the 677 alleles. However, few studies were population-based; many relied on convenience samples and were likely to be affected by selection bias. The highest frequency of the TT genotype, in excess of 20%, was observed in U.S. Hispanics, Colombians, and Amerindians in Brazil. The prevalence in white populations in Europe, North America, and Australia ranged from 8% to almost 20%. Within Europe the variant appeared to increase in frequency from north to south; homozygosity was around 10% in The Netherlands, rising to almost 20% in Italy. In Japan the frequency of TT homozygotes was 12%. In black populations, the polymorphism was rare, occurring in 2% or fewer of U.S. blacks, sub-Saharan Africans, or blacks in South America. However, black populations have been relatively little studied.

The studies of A1298C genotype frequencies have been reviewed by Robein and Ulrich (86). These, too, were potentially limited because few were population-based. The most well-studied groups were white populations in North America and Europe; data from other ethnic groups and geographic areas were limited or lacking. In the North American series, which included mainly white persons, the frequency of homozygous variant persons ranged from 7% to 12%. In the two small studies that comprised mainly Hispanic persons, a frequency of 4% to 5% was observed (87,88). In Europe, the prevalence of homozygotes for the C allele ranged from 4% to 12%. In the two studies in China and one in Japan, the frequency was lower than in other populations, ranging from 1% to 4%.

Studies of *MTHFR* and Colorectal Neoplasia

Gene–Disease Associations

As the C677T variant allele is associated with a reduced enzyme activity in vitro and low red cell and plasma folate, and various measures of high folate status are associated with reduced risk for colorectal neoplasia (see above), it might be expected that the variant would be associated with an increased risk for colorectal cancer. In fact, the opposite was found, i.e., homozygosity for the variant was associated with decreased cancer risk. A similar relation has been found in the studies of A1298C and colorectal cancer risk, while studies of C677T and adenoma risk have been inconsistent.

C677T and Cancer. Eight studies have been published of *MTHFR* C677T genotype and colorectal cancer, four in the United States and one each in Scotland, Australia, Mexico and Korea (Table 20.1). Six are consistent with a lower risk in homozygous variant persons than in homozygous wild types (38,55,89–92). Relative risks ranged from 0.45 to 0.9, but mostly did not reach statistical significance. The strongest relations were observed in the two earliest studies, both in predominantly white populations in the United States. A significant trend of decreasing risk with increasing number of T alleles has been observed (55). One study was null overall (93) but found an association with genotype in a subgroup of cases (see below). The final study, in Mexico, where the population frequency of the T allele is relatively high, reported a nonsignificantly increased risk in carriers of the T allele (94), but this study was based on small numbers of persons and may have been biased in subject selection.

One study observed that the inverse association with the homozygous variant genotype seemed to be stronger in older (60–84 years) than in younger (40–59 years) persons, but this was not statistically significant (38). The same study also reported that the inverse association held for tumors in both the colon and rectum. In terms of location within the colon, the results of Slattery et al. (90) suggested that the TT genotype was associated with reduced risk in persons with proximal tumors (odds ratio (OR) TT vs. CC = 0.8, 0.6–1.1), but not those with distal (OR = 1.0, 0.7–1.3) tumors. In the only study to report results by ethnic group, in Hawaii, Le Marchand et al. (55) found that although a modest reduced risk was associated with the TT genotype in persons of Japanese origin and Caucasians, it was not observed in Hawaiian persons. However, only nine Hawaiians were homozygous for the variant.

Table 20.1 Studies of *MTHFR* C677T Genotype and Colorectal Carcinoma, with Relative Risks and 95% Confidence Intervals

Area of Study	CASES		COMPARISON GROUP				Comparison	Relative Risk (95% CI)	Adjustment Factors	References
	Type	No.	Type	No.	TT (%) (95% CI)					
Australia	Patients undergoing surgery for colorectal cancer at a hospital in Western Australia during 1985–1998; Duke's stage B or C; 46% male; 48% aged <70 years	501	"Healthy" persons from Western Australia; 20–92 years, 81% aged <70 years	1207	11.0 (8.4–14.0)		TT vs. CC / CT vs. CC	1.03 (0.71–1.49)[a] / 0.75 (0.60–0.95)[a]	—	93
Korea	Patients undergoing operation for colorectal cancer at two centers; 51% male	200	"Healthy" unrelated adults without colorectal cancer; source not stated	460	16.1 (12.8–19.8)		TT vs. CC / CT vs. CC	0.81 (0.46–1.42)[a] / 0.94 (0.64–1.39)[a]	—	91
Mexico	Patients with colorectal cancer	74	Normal samples	110	21.8 (14.5–30.7)		TT vs. CC / CT vs. CC	1.61 (0.62–4.19)[a] / 1.83 (0.84–4.11)[a]	—	94
United Kingdom, Scotland	Residents of Grampian who had a first primary, histologically confirmed, colorectal cancer diagnosed in 1998–2000; 57% male; median age = 70 years	251	Persons randomly selected from lists of all those registered with general practitioners in Grampian; frequency matched to cases on age and sex; 51% male; median age = 62 years	394	11.9 (8.9–15.5)		TT vs. CC / CT vs. CC	0.93 (0.66–1.32) / 0.72 (0.41–1.28)	Age, sex	92,96

(continued)

Table 20.1 Studies of *MTHFR* C677T Genotype and Colorectal Carcinoma, with Relative Risks and 95% Confidence Intervals (*Continued*)

Area of Study	Cases		Comparison Group		TT (%) (95% CI)	Comparison	Relative Risk (95% CI)	Adjustment Factors	References
	Type	No.	Type	No.					
United States	Men enrolled in Health Professionals Follow-up study in 1986 and who gave a blood sample in 1993–1994; self-reported colorectal cancer, confirmed from medical records, and diagnosed in 1986–1994; subjects were aged 40–75 at enrolment in 1986; cohort is predominantly white	144	Male controls selected from same cohort, from among those who gave blood sample in 1993–1994, but who did not report a diagnosis of colorectal cancer	627	13.4 (10.8–16.3)	TT vs. CT/CC	0.57 (0.30–1.06)	Age, family history, and intake of folate, methionine, and alcohol	89
United States	Male physicians, participating in Physicians Health Study trial (exclusion criteria included history of myocardial infarction, stroke, or ischemic heart disease, cancer, current renal or liver disease, peptic ulcer or gout) who gave a blood sample at baseline in 1982; and who reported colorectal cancer in 1982–1985, which was confirmed in medical records; mean age 60 (±9) years	202	Male controls selected from same cohort; matched to cases on age and smoking status; alive and free of colorectal cancer when matched case was diagnosed; mean age 57 (±8) years	326	15.0 (11.3–19.4)	TT vs. CC	0.45 (0.24–0.86)	Age, smoking status, alcohol intake, multivitamin use, exercise, body mass index, aspirin use	38
						CT vs. CC	0.98 (0.67–1.45)		

Location	Cases	No. cases	Controls	No. controls	%	Comparison	OR (95% CI)	Adjustment factors	Ref.
United States, Utah, California, and Minnesota	Participants in Kaiser Permanente Medical Care Program (KPMCP) in Northern California, and residents of eight counties of Utah and Twin Cities area of Minnesota; diagnosed with first primary colon cancer in 1991–1994; age 30–74 years at diagnosis; 56%male; ethnic group of entire study population 4.2% black, 4.4% Hispanic, 91.4% white	1467	Controls (a) randomly selected from KPMCP lists, and (b) identified by random-digit dialling and lists with drivers license or state identification in Minnesota and Utah (under 65) and (c) randomly selected from Medical Care Financing lists in Utah (65+)	1821	11.4 (9.9–12.9)	TT vs. CC; CT vs. CC	0.9 (0.7–1.1); 1.0 (0.9–1.2)	Age, body mass index, long-term vigorous physical activity, energy intake, dietary fibre, usual number of ciagrettes	90
United States, Hawaii	Persons with primary adeno-carcinoma of the colon or rectum diagnosed 1994–1998; identified through tumour registry; at least 75% Japanese or Caucasian or any percentage Hawaiian ancestry; 61% male; median age 66; 59% Japanese, 27% Caucasian, 14% Hawaiian	548	Controls selected from (a) participants in ongoing health survey among 2% random sample of state households and (b) for those over 65, Health Care Financing Administration rolls; 61% Japanese, 26% Caucasian, 13% Hawaiian	656	15.9 (13.1–18.9)	TT vs. CC; CT vs. CC	0.7 (0.5–1.0); 0.8 (0.6–1.1)	Age, sex, ethnicity, smoking, physical activity, aspirin use, body mass index, schooling, intakes of non-starch polysaccharides and calcium	55

[a] Unmatched OR, computed by us from data in paper.

From 13% to 15% of colorectal cancers show evidence of genetic instability at microsatellite loci. Methylation may be involved (95). Shannon et al. (93) stratified their case group into those showing microsatellite instability (MSI+) and those not (MSI−). The TT genotype was associated with significantly raised risk in the MSI+ group (unadjusted OR TT vs. CC = 2.6, 1.08–5.82)[1] while in the MSI-group the OR for TT vs CC was 0.8 (0.52–1.22). The MSI+ tumors were exclusively in the proximal colon, and the patients tended to be older. Further investigation is needed to unravel the independent and joint effects of MSI, age, and tumor location.

A1298C and Cancer. Only three studies, two in the United States (55,84) and one in Scotland (96), have investigated the role A1298C in colorectal cancer. All three found the risk was modestly lower in people with the CC genotype than in those with the AA genotype; relative risks ranged from 0.7–0.8 but did not reach statistical significance. Chen et al. (84) reported that the result did not appear to be due to confounding by the C677T polymorphism. Le Marchand et al. (55) investigated joint effects of the C677T and A1298C polymorphisms and found that, compared with people who were 677CC/1298AA, those who carried the 677 T allele and the 1298 C allele had the lowest risk (RR = 0.6, 0.4–0.9).

C677T and Polyps. Six studies have reported on C677T and adenomatous polyps, three in the United States and one each in Japan, Norway and Mexico (Table 20.2). The studies were inconsistent, with none finding a statistically significant association between genotype and risk overall (61,94,97–100). Ulrich et al. (97) observed a modestly reduced risk for the TT genotype compared with CC among people with two or more polyps but not among people with one polyp only. Results were inconsistent on whether genotype was related to risk for large or small adenomas (99,100). These observations raise the possibility that *MTHFR* genotype may be relevant only in the later stages of the adenoma–carcinoma process, for example, in determining those people with adenoma who will develop carcinoma. In two studies of hyperplastic polyps, no association existed between genotype and overall risk (100,101).

C677T and Inflammatory Bowel Disease. Intriguingly, in a study of *MTHFR* and inflammatory bowel disease, the TT genotype occurred twice as often among 174 people with established disease than among 89 hospital staff controls (102). The unadjusted odds ratio for homozygous variant versus homozygous wild-type was 2.5 (0.96–7.30).[1]

Methodologic Comments. Five of the cancer studies and four of the adenoma studies included fewer than 300 cases, thus limiting statistical power particularly for subgroup analyses (and investigation of interactions—see below). In some studies, bias is possible because it is not clear whether the controls came from the pop-

Table 20.2 Studies of *MTHFR* C677T Genotype and Colorectal Polyps, by Type of Polyp, with Relative Risks and 95% Confidence Intervals

Area of Study	CASES		COMPARISON GROUP				Comparison	Relative Risk (95% CI)	Adjustment Factors	References
	No.	Type	No.	Type	TT (%) (95% CI)					
STUDIES OF ADENOMATOUS POLYPS										
United States	257	Women enrolled in Nurses Health Study in 1976 and who provided a blood sample in 1989–1990; with first incident proximal or distal colorectal adenoma diagnosed in time from blood specimen to June 1994; approximately 95% of NHS cohort is white	713	(a) Cohort members in whom colorectal adenoma had not been diagnosed and who were born in same year as matched case and had had a sigmoid-scopy since blood sample taken (n = 257); plus (b) female cohort members who had served as controls for a breast cancer study; of whom 71% had not had a sigmoidoscopy and 29% had had a sigmoid-oscopy and did not have an adenoma	9.3 (7.2–11.6)	TT vs. CT/CC	1.35 (0.84–21.7)	Age, family history, smoking status, body mass index and intakes of folate, methionine, alcohol, fibre and saturated fat	61
United States, California	471	Diagnosed for first time with one or more histologically confirmed adenomas; 65% male; 55% white, 17% black, 17% hispanic, 11 Asian; mean age = 67 years	510	Subjects undergoing screening sigmoidoscopy at two medical centres during 1991–1993; aged 50–74; with no evidence of prior bowel disease and no previous bowel surgery. Without any adenoma at sigmoidoscopy, and no history of adenomas; matched to cases on sex, sex, date of sigmoidoscopy, and clinic	9.6 (7.2–12.5)	TT vs. CC CT vs. CC	1.11 (0.71–1.71) 0.85 (0.65–1.13)	Age, race, sex, clinic, date of signoidoscopy	98

(continued)

Table 20.2 Studies of *MTHFR* C677T Genotype and Colorectal Polyps, by Type of Polyp, with Relative Risks and 95% Confidence Intervals (*continued*)

Area of Study	Cases			Comparison Group			Comparison	Relative Risk (95% CI)	Adjustment Factors	References
	Type	No.	TT (%) (95% CI)	Type	No.	TT (%) (95% CI)				
United States, Minneapolis	Subjects recruited from private gastroenterology practice undertaking colonoscopies in 10 hospitals; having undergone colonoscopy 1991–94; English speaking, without known genetic syndromes predisposing to colorectal cancer; no history of cancer or inflammatory bowel disease; aged 30–74 years; participation rate 68%									97
	With first diagnosis of colon or rectal adenomas; 62% male; mean age 58.1 ($\pm$9.7) years; 98% white	527		Free of all polyps at colonoscopy; 38% male; mean age 52.8 ($\pm$10.9) years; 97% white	645	11.0 (8.7–13.7)	TT vs. CC CT vs. CC	0.8 (0.5–1.3) 0.9 (0.7–1.2)	Age, sex, body mass index, use of hormone replacement therapy, percent of calories from fat, dietary fiber, folate, vitamin B_{12}, vitamin B_6, methionine, alcohol	
Japan	Male self-defense officials undergoing preretirement health examination at two hospitals; had partial or total colonoscopy and provided blood sample; aged 47–55 years; no prior history of colectomy, polypectomy, malignant neoplasia									99
	Histologically confirmed colorectal adenoma without in situ or invasive carcinoma	205		Normal total colonoscopy	220	11.8 (7.9–16.8)	TT vs. CC CT vs. CC	0.87 (0.56–1.34) 1.17 (0.61–2.23)	Hospital, employment, rank, smoking, alcohol use	

Country		n		Comparison	OR (95% CI)	Adjustment for potential confounding
Norway	Participants in Telemark I study; born 1924–33, selected from population register in 1983 and randomised to endoscopy or control group; 799 participated; in 1996 offered colonoscopy and removal of polyps; results available for 443 (229 male, 214 female); median age = 67					Age, sex, red blood cell folate, use of non-steroidal anti-inflammatory drugs, flexible sigmoidoscopy in 1983, body mass index, current smoking
	With "high-risk" colorectal adenomas (≥10mm or severe dysplasia or villous components)	47				
	Without polyps (n = 116) or with hyperplastic polyps or "low risk" adenomas (n = 278)	394	7.1 (4.8–10.1)	TT vs. CC	2.41 (0.82–7.06)	
				CT vs. CC	1.51 (0.76–2.99)	
Mexico	Patients with colorectal adenomas	32				—
	Normal samples	110	21.8 (14.5–30.7)	TT vs. CC	1.65 (0.41–6.73)[a]	
				CT vs. CC	0.98 (0.28–3.67)[a]	

STUDIES OF HYPERPLASTIC POLYPS

Country		n		Comparison	OR (95% CI)	Adjustment for potential confounding
Norway	Participants in Telemark I study; born 1924-1933, selected from population register in 1983 and randomised to endoscopy or control group; 799 participated; in 1996 offered colonoscopy and removal of polyps; results available for 443 (229 male, 214 female); median age = 67					—
	With "high-risk" hyperplastic polyps (n ≥ 3)	91				
	Without polyps (n = 116) or with adenomas or "low risk" hyperplastic polyps (n = 233)	349	7.1 (4.8–10.1)	TT/CT vs. CC	1.43 (0.87–2.33)[a]	

(continued)

Table 20.2 Studies of *MTHFR* C677T Genotype and Colorectal Polyps, by Type of Polyp, with Relative Risks and 95% Confidence Intervals *(continued)*

Area of Study	CASES			COMPARISON GROUP			Comparison	Relative Risk (95% CI)	Adjustment Factors	References
	Type	No.	TT (%) (95% CI)	Type	No.	TT (%) (95% CI)				
United States, Minneapolis	Diagnosis of colon or rectal hyperplastic polyps; 97% white; 57% male; mean age 53.7 years	200		Subjects recruited from private gastroenterology practice undertaking colonoscopies in 10 hospitals; having undergone colonoscopy 1991–1994; English speaking, without known genetic syndromespredisposing to colorectal cancer; no history of cancer or inflammatory bowel disease, aged 30–74 years Free of all polyps at colonoscopy; 97% white, 38% male; mean age 52.8 (±10.9) years	645	11.0 (8.7–13.7)	TT vs. CC CT vs. CC	0.9 (0.5–1.6) 0.8 (0.6–1.2)	Age, sex, body mass index, use of hormone replacement therapy, smoking, percent of calories from fat, dietary fiber, folate, vitamins B_{12} and B_6, methionine, alcohol	101

ulation that gave rise to the cases. Few of the studies provide information about participation rates, making assessment of biases and generalizibility difficult. A proportion of the controls in the cancer studies may have been harboring undiagnosed polyps. Depending on the relation between adenomas and *MTHFR*, this could have introduced random error or bias. In the nonprospective studies, the case series were limited to those who were still alive to provide a DNA sample. If *MTHFR* genotype is associated with survival, this could have biased the results. Finally, the possibility cannot be discounted that the findings do not reflect an association between *MTHFR* and colorectal cancer but an association with another gene in linkage disequilibrium with *MTHFR*.

Gene–Environment Interactions

The original hypothesis about interaction between *MTHFR* genotype and environment specified that individuals who inherited the C677T variant genotype would "respond differently" to the methyl content of their diet than persons with other genotypes (89). Therefore investigators have explored interactions between C677T genotype and folate, methionine, vitamins B_6 and B_{12}, and alcohol, looking at these factors individually and in combination. No data have as yet been published on the joint effects of A1298C and folate, or related dietary factors, and risk for colorectal neoplasia.

In four studies, the lowest risk for colorectal cancer was among persons with the TT genotype who had higher levels of folate and/or other indicators of the "high methyl diet." Two of these studies further suggested that the "protection" afforded by the TT genotype was lost when folate or methyl levels were low. These observations suggest that a gene–environment interaction exists. In the first published study on C677T and colorectal cancers, Chen et al. (89) reported that the inverse association with the TT genotype was strongest among people in the highest tertiles of folate and methionine intake. The results of Ma et al. (38), who examined plasma folate, and Le Marchand et al. (55), who analyzed both food and total folate intake, were consistent with this. Slattery et al. (90) categorized subjects into low-, intermediate-, and high-methyl diets based on combination of folate, methionine, and alcohol intake. The lowest odds ratio was observed for persons with the TT genotype who consumed a high-methyl diet (OR for high-methyl diet and TT genotype vs. low-methyl diet and CC = 0.4, 0.1–0.9).

With regard to those with low folate levels, Ma et al. (38) reported that that colorectal cancer risk was increased in those who were folate deficient (plasma folate <3.0 ng/mL) irrespective of genotype. Consistent with this, Slattery et al. (90) reported that the odds ratios for persons consuming a low methyl diet did not vary by genotype. These observations could be relevant to disease prevention

Little work on *MTHFR*–diet interaction and adenoma risk has been done. In the two studies so far, the stratum of highest risk comprised TT persons who had the lowest red cell or plasma folate levels (98) or lowest intakes of folate, methionine, vitamin B_6, or vitamin B_{12} (97), but the formal tests for gene–nutrient interaction were not statistically significant.

Ma et al. (38) and Chen et al. (89) found a significant interaction between alcohol intake and genotype for colorectal cancers. High alcohol intake abolished the negative association with the TT genotype to the extent that TT persons drinking the largest amounts of alcohol were at greatest risk (greater even than those without the T allele who were in the highest alcohol group). Two studies of adenomas found a similar pattern (97,98).

Gene–Gene Interactions

The metabolism of any exposure most likely depends on the balance between the relative activities of all the enzymes active within the metabolic pathway (103). Other polymorphic genes in the folate pathway—methionine synthase (*MTR*), methionine synthase reductase (*MTRR*), cystathionine β-synthase (*CBS*) (Fig. 20.1)—have been investigated only to a limited extent in colorectal neoplasia, and only one study has looked at the combined effects of these and *MTHFR* (55). In addition, little work has been done on the interaction of these polymorphisms with folate or other dietary factors. Moreover, the population frequencies and functional impact of the polymorphisms are not well described.

The studies that have been done are summarized in Table 20.3. An A-G polymorphism at position 2756 in the protein-binding region of the *MTR* gene has been investigated in three studies (55,61,78). Two studies, one of cancer and one of adenomas, found a slightly reduced risk among persons homozygous for the variant genotype (61,78). The third study, of cancer, found no effect of the *MTR* genotype alone, but did observe a significant interaction between *MTHFR* C677T and *MTR* genotypes (55). The *MTHFR* T allele was most protective among persons with the *MTR* G allele (OR CT/TT and AG/GG vs. CC and AA = 0.7, 0.5–1.0; *p* interaction = 0.05).

A 68 bp insertion in the exon 8 coding region of the *CBS* gene has been investigated in two studies. Persons heterozygous for the insertion were twice as frequent among controls as among cancer cases in one study (93). In the other, there was a suggestion that the variant was associated with reduced cancer risk and might interact with *MTHFR* C677T genotype (55). In a single study the A66G polymorphism in the *MTRR* gene was not associated with cancer risk and did not appear to interact with *MTHFR* C677T genotype (55).

Conclusions

The observed association of the *MTHFR* homozygous variant genotype with reduced risk of carcinoma was the opposite of might have been expected a priori. This has led investigators to reconsider the folate metabolism pathway, putting a greater emphasis on the functions of folate and MTHFR in DNA synthesis. The epidemiologic evidence so far is compatible with interactions between *MTHFR* genotype and folate, alcohol, and/or related nutrients in relation to risk of colorectal neoplasia. The few studies of other folate pathway genes suggest the possibility of gene–gene interactions, although further studies are needed to confirm these initial findings. Altogether this suggests that the roles of *MTHFR*, folate and related dietary factors and genes in colorectal neoplasia are complex.

Table 20.3 Summary of Studies of Other Folate Pathway Genes and Colorectal Neoplasia

| | | | GENE–DISEASE ASSOCIATIONS | | |
Gene	Polymorphism	Study Area, Design Case Subjects[a]	Comparison	Relative Risk (95% CI)	Gene–Gene Interactions	References
MTR (methionine synthase)	A2756G	United States, nested case-control, adenoma	GG vs. AA	0.66 (0.26–1.70)	No interaction with *MTHFR* C677T	61
		United States, case-control, carcinoma	GG vs. AA	1.1 (0.6–2.2)	Significant interaction with *MTHFR* C677T	55
		United States, nested case-control, carcinoma	GG vs. AA	0.59 (0.27–1.27)	—[c]	78
MTRR (methionine synthase reductase)	A66G	United States, case-control, carcinoma	GG vs. AA	1.4 (0.9–2.0)	No interaction with *MTHFR* C677T	55
CBS (cystathionine β-synthase)	68 bp insertion	United States, case-control, carcinoma	Weak inverse association with presence of insertion		Weak suggestion of interaction with *MTHFR* C677T	55
		Australia, case-control, carcinoma[b]	Frequency of heterozygotes in controls (10%) vs. cases (5%)		—[c]	93

[a]See Tables 20.1 and 20.2 for further details.
[b]Only 155 of original control series was included in this analysis.
[c]None investigated.

A major limitation of the studies conducted so far is that most have been in mainly white populations. Population-based studies are needed in populations and ethnic groups that have been little researched and where the genotype frequencies or dietary patterns differ from those among United States whites. Since folate is consistently associated with cancer of the colon, but not of the rectum, results for these should be reported separately. Investigation of different subgroups of tumors—for example, those showing microsatellite instability or loss of heterozygosity—may be valuable. Clearly studies should be large enough to have adequate statistical power for the investigation of subgroups or interactions. Pooled analyses of published studies might help to achieve this.

No studies have been conducted of *MTHFR* polymorphisms and risk of recurrence of colorectal adenomas, and little work is available on adenomas with advanced pathologic features; such data might help to clarify the apparently different findings with regard to *MTHFR* and risk for adenomas and carcinomas. In addition, further investigations of the relation between *MTHFR* and inflammatory bowel disease, particularly in relation to the risk of developing colorectal cancer, may be informative.

Other genetic polymorphisms may influence the metabolism of folate and related nutrients (104). There is no published research on these and colorectal neoplasia. In addition, several of the genes involved in the metabolism of alcohol are polymorphic and these may act with *MTHFR* and dietary factors to affect disease risk. These factors should also be investigated.

Characteristics of Available Tests for Polymorphisms in *MTHFR*

Detecting the *MTHFR* polymorphisms is relatively straightforward. Amplification of DNA using polymerase chain reaction (PCR) is followed by restriction fragment-length polymorphism analysis (RFLP), with *Hinf*I for C677T and *Mbo*II for A1298C, and gel electrophoresis to separate the fragments (72,105). Use of *Mbo*II may also detect the silent T1317C polymorphism on the same exon as A1298C (73), thus resulting in genotype misclassification, the extent of which will depend on the population frequency of the T1317C polymorphism (which is not well described). A refinement to overcome this problem has been proposed, employing a different artificially created restriction site (73). No other data are available on the sensitivity, specificity, and predictive value of these tests for classifying the underlying genotype (85).

Potential Contribution of Information on *MTHFR*, Folate, and Related Dietary Factors to Improving Health Outcomes

The *MTHFR*–folate relation could be relevant to improved health outcomes in two main areas: (*1*) with regard to the primary prevention of colorectal cancer and (*2*) with regard to tertiary prevention in persons in whom an invasive colorectal tumor has been diagnosed.

MTHFR, Folate, and Primary Prevention

Because folate levels are potentially modifiable, the obvious first option to consider with regard to prevention of colorectal neoplasia is increasing folate levels. This could be formulated as either a whole population approach or a strategy targeted toward persons at higher risk for colorectal neoplasia. Three high-risk groups might be considered: (*1*) persons from whom an adenoma has been removed—the aim would be to prevent recurrence; (*2*) persons with inflammatory bowel disease—the aim would be to prevent development of a colorectal tumor; and (*3*) persons with a family history of colorectal neoplasia—the aim would be to prevent adenoma or carcinoma. The aim of a whole population strategy would be either (*1*) prevention of the first formation of an adenoma or, in recognition of the fact that not all colorectal cancers develop from adenomas, (*2*) prevention of the occurrence of de novo colorectal tumors. Where *MTHFR* may be relevant is with regard to whether a folate intervention should be aimed at everyone in (any of) the above groups or whether it would be targeted at, or tailored to, those with particular genotypes. For example, one might "screen" patients who have had an adenoma removed for *MTHFR* genotype and either (*1*) make the folate intervention only in those with a particular genotype or (*2*) advocate a different level of folate according to genotype.

General Issues Relevant to Prevention Strategies. Several issues are relevant to any prevention strategy based on increasing folate levels.

Options for Increasing Folate Intake. Three options exist for increasing folate intake in a population: (*1*) encourage increased intake of folate-rich foods; (*2*) fortify common foodstuffs with folic acid; or (*3*) promote use of supplements that contain folic acid. The second two are attractive because changes in population dietary patterns are difficult to achieve, and, compared with natural folates, synthetic folic acid is extremely stable and totally bioavailable. The main disadvantage of supplements as a preventive agent is that usually only a minority of people will take these on a regular basis (106), although this may well vary according to the group who are advised to take the supplements. For example, one might expect that persons with a clinical condition such as adenoma or ulcerative colitis would be more likely to take supplements if advised to do so than would members of the general population. Folate fortification of staple foods appears an attractive option. The main disadvantage here is that to deliver the target dose, those persons on higher intakes of the staple food might be exposed to 10 times that amount (106), which could well have adverse consequences for particular subgroups of the population.

The results of the studies of *MTHFR*, particularly the initially surprising observation of a reduced risk with the TT genotype, have suggested the possibility that folic acid per se may not be the best chemopreventive agent. Other forms of folate might be more effective (107). Work in this area is at an early stage, but preliminary findings in cell lines imply that different coenzymatic forms of folate might affect cellular proliferation in different ways (108). Because this area is under development, the reminder of this discussion will pertain to a prevention strategy based on folic acid. However, the general issues would be relevant to supplementation or fortification with other forms of folate.

Effects of Increasing Folate Intake. Several general questions are relevant: would raising levels of folate intake in the population (irrespective of how this is done) result in increased circulating folate concentrations and increased folate levels in colonic mucosa, and if so, by how much, and would this hold irrespective of individual *MTHFR* genotype? Increased intakes of folate, or folic acid, do increase systemic blood folate concentrations, with natural folates having a smaller effect on serum folate than folic acid (109). In an analysis of 13 folic acid supplement trials, Wald et al. (110) estimated that for every 0.1 mg/day increase in folic acid intake, serum folate concentration rose by 0.94 ng/mL in women aged 20 to 35 years and by 2.5 ng/mL in people aged 40 to 65 years. The fortification of grains in the United States has reduced the percentage of the population with low plasma folate levels (<3 μg/L) from 22.0% to 1.7% (111). The incremental impact of supplementation in addition to fortification is currently unclear.

In humans, folate depletion in the colorectal mucosa may predispose the mucosa to malignant transformation (112). Limited evidence suggests that supplemental folic acid intake can significantly increase colonic mucosal concentrations. In nine persons from whom adenomas had been removed, supplementation by 5 mg/day of folic acid for 1 year increased colonic mucosal concentrations twofold to fourfold above baseline values (68).

A few studies have reported on the effects of folic acid supplementation by C677T *MTHFR* genotype (113–118). Some tentatively suggest that folate supplementation can result in a greater increase in plasma and, possibly, red cell folate concentrations in persons with the TT genotype than other genotypes. These observations should be viewed with caution, however, because the studies were small, with some methodologic flaws and limited generalisibility. A1298C genotype, combined 677/1298 genotype, baseline folate status, and use of multivitamins during the supplementation period also might be relevant. Further research is needed in this area.

Potential Adverse Effects. Folate fortification or supplementation may, potentially, have adverse effects. In people with vitamin B_{12} deficiency, high levels of folate intake can reverse the anemia associated with the deficiency but may precipitate neurologic complications. No data exist on how commonly this might occur in a supplemented population (119). To prevent vitamin B_{12} deficiencies, a staple food could be fortified with high levels of B_{12} (around 1 mg/day), but the effects of this in combination with folic acid supplementation are not known (120). Folate supplementation interferes with zinc absorption (121). Although zinc's role in colorectal neoplasia is not clear, it is necessary for DNA and RNA synthesis, and compromised zinc metabolism may have some adverse consequences. Finally, high levels of folate interfere with metabolism of antiepileptic drugs, and this clinical group must be of concern in any colorectal neoplasia prevention strategy.

The Need for Randomized Controlled Trials. Although increasing folate intake through diet, fortification, or supplementation clearly results in increased systemic folate concentrations, little direct evidence exists on whether this influences the rates of formation of adenomas or carcinomas in any population or subgroup. (There is only one small trial of adenoma recurrence by Paspatis et al. [65]; see above.) Hence, whether this strategy would improve health outcomes cannot be concluded. Whether incorporating *MTHFR* genotyping would improve efficacy, or effectiveness, of a folate intervention also is unclear. The epidemiologic data reviewed above with regard to cancer suggested that the effect of a low folate–methionine diet overrides the effect of genotype, but this is based on a limited number of observational studies. Moreover, two studies of adenomas suggested the opposite. A further concern is the cost–benefit ratio of genotyping in this context. Vineis et al. (122) highlighted concerns about genetic screening, particularly in the situation of a low-penetrance gene. They argued that the number needed to screen to prevent one case of cancer often will be large. The number depends on the cumulative risk of the disease in persons with the different genotypes, the genotype frequencies, the relative risk associated with genotype, and the risk reduction likely to be achieved with an intervention (hypothetically, chemoprevention). This implies that few people screened will benefit, and a large number of screening results will be false-positive, which has consequences both for health services (in terms of costs and resources) and for the individuals involved (in terms of psychosocial impact, unnecessary treatment, and possibly costs). This situation is further complicated by the fact that several (many) polymorphisms may affect risk of a single disease, and that one particular polymorphism may be associated with increased risk of one disease and reduced risk of another.

Large randomized controlled trials of folate supplementation for the prevention of colorectal neoplasia, including stratification by *MTHFR* genotype, are needed as a first step. The results of four moderately sized trials of folic acid supplements and adenoma recurrence, three in the United States and one in Europe are awaited. Others are necessary, particularly in populations with different gene frequencies and baseline levels of folate intake. Because current evidence does not indicate what level of folate supplementation might be required—and this might differ by population—a range of levels should be considered (and, in fact, the trials currently under way are investigating levels in the range 0.5–5 mg/day). The addition of B_{12} and related nutrients also should be considered. Forms of folate other than folic acid also are appropriate to consider. Such trials might consider other high-risk groups such as persons with inflammatory bowel disease or family histories of colorectal neoplasia. In view of the evidence that *MTHFR* genotype may be relevant in development of cancer, but not adenoma, a long-term future objective might be to undertake whole population trials of the prevention of adenomas and carcinomas. Both effectiveness and cost-effectiveness would need to be assessed, with a particular focus on the value of testing for *MTHFR*. A further consideration should be the psychosocial impact and acceptability of genotyping in the populations studied.

In some populations an issue for consideration would be the relation of such a chemopreventive strategy to existing programs or services for early detection of colorectal lesions (e.g., fecal occult blood or flexible sigmoidoscopy based screening services). Chemoprevention might operate in a variety of ways, as either an alternative or an adjunct to existing services. Relative effectiveness and cost-effectiveness in either scenario need to be assessed.

Additional Health Benefits. Additional health benefits could accrue from a population policy resulting in increased folate levels because low folate is involved in the etiology of neural tube defects and perhaps other congenital anomalies (85), cardiovascular disease (123), Alzheimer's disease (124), and possibly also other forms of cancer (119,125).

MTHFR, Alcohol, and Primary Prevention

The limited epidemiologic data suggest that, with regard to risk of developing an adenoma or carcinoma, consumption of excess amounts of alcohol may be particularly inadvisable for persons with the TT genotype. Therefore, an alternative public health approach might be to consider screening for genotype as part of a health promotion programme aimed at restricting alcohol intake, such as those in place in the United Kingdom and United States. Whether this would improve effectiveness and cost-effectiveness is not known, and it would be premature to suggest that this approach be adopted without considerable further evidence.

Folate, MTHFR, and Prognosis of Patients with Colorectal Cancer

The final group in whom knowledge of *MTHFR* could potentially be used to improve health outcomes is persons diagnosed with colorectal cancer. The drug 5-fluorouracil (5-FU), commonly used in colorectal cancer chemotherapy, is a thymidylate synthase inhibitor and can cause severe folate depletion. For that reason, patients undergoing chemotherapy may be given folic acid supplements. Toxic effects of treatment are common. Knowledge of patient *MTHFR* genotype potentially could be used to tailor chemotherapy regimes to (*1*) avoid folate depletion and minimize toxicity and side effects, thus improving quality of life and (2) increase the effectiveness of treatment and ultimately lengthen survival. Evidence in this area is extremely limited. One study attempted to address whether the effectiveness of treatment with 5-FU and leucovorin (folinic acid) was lower in patients with the TT genotype than in patients with other genotypes (126). In 51 patients with stage III colon cancer, presence of the T allele appeared to have little effect on probability of death or length of survival in those who had died, except in 12 patients with rectosigmoid colon cancer; the analysis was not, however, adjusted for other prognostic indicators. Shannon et al. (93) reported that the TT genotype was associated with improved survival in 365 nonadjuvant treated patients (hazard ratio = 0.77, 0.6–0.99), but this did not persist after adjustment for stage. Park et al. (91) found a significantly higher average number of cancer-positive lymph nodes in patients with the TT than the CC genotype; the authors suggested that genotype might contribute to tumor spread but how many nodes were tested in the patients was not clear. The *MTHFR* genotype modifies responses of bone marrow transplantation patients to methotrexate, another antifolate chemotherapy agent; persons with the TT genotype appeared to be at higher risk for toxicity (127). In a very small study of breast cancer patients being treated with a regimen of fluorouracil, methotrexate, and cyclophosphamide, five of six patients developing severe acute toxicity had the TT genotype (128). Overall, this would appear to be a promising area for further research.

Comments

The folate–*MTHFR*–colorectal neoplasia relation provides an excellent example of the challenges facing genetic epidemiology. Several genes in addition to *MTHFR* and several dietary factors in addition to folate are potentially relevant. All of these might act together to determine risk. Thus, we may be dealing with not just gene–environment interaction, but rather with interactions between multiple genes and multiple environmental factors. The demonstrated dilution in the observed effects of a gene and an environmental factor when an underlying interaction between the factors has not been taken into account (129) also may hold for interactions between multiple genes and exposures in a pathway. Therefore, the published results may be biased or simply misleading. This problem is further compounded by a lack of a clear functional understanding of the genes and a lack of knowledge about the patterns of interaction that might be expected. Together these issues produce a level of complexity far beyond that which has been investigated so far and one that is beyond current theoretical and methodologic frameworks for investigation. Methods are needed for specifying hypotheses around gene–environment pathways, for clarifying functional effects, and for statistical analysis if we are to advance in our understanding of the etiology of complex diseases and translate this understanding into prevention strategies that will improve health outcomes.

Notes

[1]Calculated by us from data contained in the reference.

References

1. Ferlay J, Bray F, Pisani P, et al. Globocan 2000: Cancer Incidence, Mortality and Prevalence Worldwide. International Agency for Research on Cancer, World Health Organization. Lyon, France: IARC Press, 2001.

2. Parkin DM, Whelan SL, Ferlay J, et al. Cancer Incidence in Five Continents. Volume VII. Lyon, France: IARC Press, 1997.
3. Haenszel W, Correa P. Cancer of the colon and rectum and adenomatous polyps: a review of epidemiologic findings. Cancer 1971;28:14–24.
4. Sharp L. Current trends in colorectal cancer: what they tell us and what we still do not know. Clin Oncol 2001;13:444–447.
5. Devesa SS, Chow WH. Variation in colorectal cancer incidence in the United States by subsite of origin. Cancer 1993;71:3819–3826.
6. dos Santos Silva I, Swerdlow AJ. Sex differences in time trends of colorectal cancer in England and Wales: the possible effect of female hormonal factors. Br J Cancer 1996;73:692–697.
7. Iacopetta B. Are there two sides to colorectal cancer? Int J Cancer 2002;101:403–408.
8. Reis LAG, Wingo PA, Miller DS, et al. The annual report to the nation on the status of cancer, 1973–19977, with a special section on colorectal cancer. Cancer 2000;88:2398–2424.
9. Coleman MP, Estève J, Damiecki P, et al. Trends in Cancer Incidence and Mortality. Lyon, France: IARC Press, 1993.
10. Robinson MHE, Thomas WM, Hardcastle JD, Chamberlain J, Mangham CM. Change towards earlier stage at presentation of colorectal cancer. Br J Surg 2002;80:1610–1612.
11. Chu KC, Tarone RE, Chow W-H, Hankey BF, Reis LAG. Temporal patterns in colorectal cancer incidence, survival and mortality from 1950 through 1990. J Nat Cancer Inst 2002;86:997–1006.
12. Gillen CD, Walmsley RS, Prior P, et al. Ulcerative colitis and Crohn's disease: a comparison of the colorectal cancer risk in extensive colitis. Gut 1994;35:1590–1592.
13. Eaden JA, Abrams KR, Mayberry JF. The risk of colorectal cancer in ulcerative colitis: a meta-analysis. Gut 2001;48:526–535.
14. Cotton S, Sharp L, Little J. The adenoma-carcinoma sequence and prospects for the prevention of colorectal neoplasia. Crit Rev Oncogen 1996;7:293–342.
15. Winawer SJ, O'Brien MJ, Waye JD, et al. Risk and surveillance of individuals with colorectal polyps. Bulletin of the World Health Organisation 1990;68:789–795.
16. Peipins LA, Sandler RS. Epidemiology of colorectal adenomas. Epidemiol Rev 1994;16:273–297.
17. Hawkins NJ, Ward RL. Sporadic colorectal cancers with microsatellite instability and their possible origin in hyperplastic polpys and serrated adenomas. J Natl Cancer Inst 2001;93:1307–1313.
18. Mecklin JP, Ponz de Leon M. Epidemiology of HNPCC. Anticancer Res 1994;14:1625–1629.
19. Fuchs CS, Giovannucci EL, Colditz GA, Hunter DJ, Speizer FE, Willett WC. A prospective study of family history and the risk of colorectal cancer. N Engl J Med 1994;331:1669–1674.
20. Winawer SJ, Zauber AG, Gerdes H, et al. Risk of colorectal cancer in the families of patients with adenomatous polyps. N Engl J Med 1996;334:82–87.
21. Khoury MJ, Beaty TH, Liang KY. Can familial aggregation of disease be explained by familial aggregation of environmental risk factors? Am J Epidemiol 1988;127:674–683.
22. Haenszel W, Kurihara M. Studies of Japanese migrants: mortality from cancer and other diseases among Japanese in the United States. J Natl Cancer Inst 1968;40:43–68.
23. IARC Working Group. IARC Handbooks of Cancer Prevention, Volume 6: The Role of Weight Control and Physical Activity in Cancer Prevention. Lyon, France: IARC Press, 2002.
24. Bergström A, Pisani V, Tenet V, Wolk A, Adami HO. Overweight as an avoidable cause of cancer in Europe. Int J Cancer 2001;91:421–430.

25. Giovannucci E. An updated review of the epidemiological evidence that cigarette smoking increases risk of colorectal cancer. Cancer Epidemiol Biomarkers Prev 2001; 10:725–731.

26. Corrao G, Bagnardi V, Zambon A, Arico S. Exploring the dose-response relationship between alcohol consumption and the risk of several alcohol-related conditions: a meta-analysis. Addiction 1999;94:1551–1573.

27. World Cancer Research Fund in Association with American Institute for Cancer Research. Food, Nutrition and the Prevention of Cancer: A Global Perspective. Menasha, WI: American Institute for Cancer Research, 1997.

28. IARC Working Group. IARC Handbook of Cancer Prevention, Volume 1: Non-Steroidal Anti-Inflammatory Drugs. Lyon, France: IARC Press, 1997.

29. Beral V, Banks E, Reeves G, et al. Use of HRT and the subsequent risk of cancer. J Epidemiol Biostatist 1999;4:191–215.

30. Michels KB, Giovannucci E, Joshipura KJ, et al. Prospective study of fruit and vegetable consumption and incidence of colon and rectal cancers. J Natl Cancer Inst 2000; 92:1740–1752.

31. Terry P, Giovannucci E, Michels KB, et al. Fruit, vegetables, dietary fiber and risk of colorectal cancer. J Nat Cancer Inst 2001;93:525–533.

32. Flood A, Velie EM, Chaterjee N, et al. Fruit and vegetable intakes and the risk of colorectal cancer in the Breast Cancer Detection Demonstration Project follow-up cohort. Am J Clin Nutr 2002;75:936–943.

33. Feinberg AP, Vogelstein B. Hypomethylation distinguishes genes of some human cancers from their normal counterparts. Nature 1983;301:89–91.

34. Goelz SE, Vogelstein B, Hamilton SR, Feinberg AP. Hypomethylation of DNA from benign and malignant human colon neoplasms. Science 1985;228:187–190.

35. Duthie SJ. Folic acid instability and cancer: mechanisms of DNA instability. Br Med Bull 1999;55:578–592.

36. Ames BN. DNA damage from micronutrient deficiencies is likely to be a major cause of cancer. Mutat Res 2001;475:7–20.

37. Glynn SA, Albanes D, Pietinen P, et al. Colorectal cancer and folate status: a nested case-control study among male smokers. Cancer Epidemiol Biomarkers Prev 1996;5:487–494.

38. Ma J, Stampfer MJ, Giovannucci E, et al. Methylenetetrahydrofolate reductase polymorphism, dietary interactions and risk of colorectal cancer. Cancer Res 1997;57:1098–1102.

39. Kato I, Dnistrian AM, Schwartz M, et al. Serum folate, homocysteine and colorectal cancer risk in women: a nested case-control study. Br J Cancer 1999;79:1917–1921.

40. Bird CL, Swendseid ME, Witte JS, et al. Red cell and plasma folate, folate consumption, and the risk of colorectal adenomatous polyps. Cancer Epidemiol Biomarkers Prev 1995;4:709–714.

41. Paspatis GA, Kalafatis E, Oros L, et al. Folate status and adenomatous colonic polyps. A colonoscopically controlled study. Dis Col Rect 1995;38:64–68.

42. Giovannucci E, Rimm EB, Ascherio A, et al. Alcohol, low-methionine-low-folate diets, and risk of colon cancer in men. J Natl Cancer Inst 1995;87:265–273.

43. Giovannucci E, Stampfer MJ, Colditz GA, et al. Multivitamin use, folate, and colon cancer in women in the Nurses' Health Study. Ann Intern Med 1998;129:517–524.

44. Su LJ, Arab L. Nutritional status of folate and colon cancer risk: evidence from NHANES I epidemiologic follow-up study. Ann Epidemiol 2001;11:65–72.

45. Terry P, Jain M, Miller AB, et al. Dietary intake of folic acid and colorectal cancer risk in a cohort of women. Int J Cancer 2002;97:864–867.

46. Flood A, Caprario L, Chaterjee N, et al. Folate, methionine, alcohol, and colorectal cancer in a prospective study of women in the United States. Cancer Causes Control 2002;13:551–561.

47. Freudenheim JL, Graham S, Marshall JR, et al. Folate intake and carcinogenesis of the colon and rectum. Int J Epidemiol 1991;20:368–374.

48. Benito E, Stiggelbout A, Bosch FX, et al. Nutritional factors in colorectal cancer risk: a case-control study in Majorca. Int J Cancer 1991;49:161–167.

49. Meyer F, White E. Alcohol and nutrients in relation to colon cancer in middle-aged adults. Am J Epidemiol 1993;138:225–236.

50. Ferraroni M, La Vecchia C, D'Avanzo B, Negri E, Franceschi S, Decarli A. Selected micronutrient intake and the risk of colorectal cancer. Br J Cancer 1994;70:1150–1155.

51. Boutron-Ruault MC, Senesse P, Faivre J, et al. Folate and alcohol intakes: related or independent roles in the adenoma-carcinoma sequence? Nutr Cancer 1996;26:337–346.

52. La Vecchia C, Braga C, Negri E, et al. Intake of selected micronutrients and risk of colorectal cancer. Int J Cancer 1997;73:525–530.

53. Slattery ML, Schaffer D, Edwards SL, et al. Are dietary factors involved in DNA methylation associated with colon cancer? Nutr Cancer 1997;28:52–62.

54. Levi F, Pasche C, Lucchini F, et al. Selected micronutrients and colorectal cancer: a case-control study from the Canton of Vaud, Switzerland. Eur J Cancer 2000;36:2115–2119.

55. Le Marchand L, Donlon T, Hankin JH, et al. B-vitamin intake, metabolic genes, and colorectal cancer risk (United States). Cancer Causes Control 2002;13:239–248.

56. Little J, Sharp L. Colorectal neoplasia and genetic polymorphisms associated with metabolism. Eur J Cancer Prev 2002;11:105–110.

57. White E, Shannon JS, Patterson RE. Relationship between vitamin and calcium supplement use and colon cancer. Cancer Epidemiol Biomarkers Prev 1997;6:769–774.

58. Benito E, Cabeza E, Moreno V, et al. Diet and colorectal adenomas: a case-control study in Majorca. Int J Cancer 1993;55:213–219.

59. Giovannucci E, Stampfer MJ, Colditz GA, et al. Folate, methionine and alcohol intake and risk of colorectal adenoma. J Natl Cancer Inst 1993;85:875–884.

60. Tseng M, Murray SC, Kupper LL, et al. Micronutrients and the risk of colorectal adenomas. Am J Epidemiol 1996;144:1005–1014.

61. Chen J, Giovannucci E, Hankinson SE, et al. A prospective study of methylenetetrahydrofolate reductase and methionine synthase gene polymorphisms, and risk of colorectal adenoma. Carcinogenesis 1998;19:2129–2132.

62. Breuer-Katschinski B, Nemes K, Marr A, et al. Colorectal adenomas and diet: a case-control study. Dig Dis Sci 2001;46:86–95.

63. Baron JA, Sandler RS, Haile RW, et al. Folate intake, alcohol consumption, cigarette smoking, and risk of colorectal adenomas. J Natl Cancer Inst 1998;90:57–62.

64. Fuchs CS, Willett WC, Colditz GA, et al. The influence of folate and multivitamin use on the familial risk of colon cancer in women. Cancer Epidemiol Biomarkers Prev 2002;11:227–234.

65. Paspatis G, Xourgias B, Mylonakou E, et al. A prospective clinical trial to determine the influence of folate supplementation on the formation of recurrent colonic adenomas. Gastroenterology 1994;106:A425.

66. Cravo M, Fidalgo P, Pereira AD, et al. DNA methylation as an intermediate biomarker in colorectal cancer: modulation by folic acid supplementation. Eur J Cancer Prev 1994;3:473–479.

67. Cravo ML, Pinto AG, Chaves P, et al. Effect of folate supplementation on DNA methylation of rectal mucosa in patients with colonic adenomas: correlation with nutrient intake. Clin Nutr 1998;17:45–49.

68. Kim YI, Baik HW, Fawaz K, et al. Effects of folate supplementation on two provisional molecular markers of colon cancer: a prospective, randomized trial. Am J Gastroenter 2001;96:184–195.

69. Erickson JD. Folic acid and prevention of spina bifida and anencephaly. 10 years after the U.S. Public Health Service recommendation. Introduction. MMWR 2002;51 (RR–13):1–3.
70. Fenech M. The role of folic acid and vitamin B_{12} in genomic stability of human cells. Mutat Res 2001;475:57–67.
71. Herbert V. Recommended dietary intakes (RDI) of folate in humans. Am J Clin Nutr 1987;45:661–670.
72. Frosst P, Blom HJ, Milos R, et al. A candidate genetic risk factor for vascular disease: a common mutation in methylenetetrahydrofolate reductase. Nat Genet 1995;10: 111–113.
73. Weisberg I, Tran P, Christensen B, et al. A second genetic polymorphism in methyl-enetetrahydrofolate reductase (MTHFR) associated with decreased enzyme activity. Mol Genet Metab 1998;64:169–172.
74. van der Put NMJ, Gabreëls F, Stevens EMB, et al. A second common mutation in the methylenetetrahydrofolate reductase gene: an additional risk factor for neural-tube de-fects? Am J Hum Genet 1998;62:1044–1051.
75. Trembath D, Sherbondy AL, Vandyke DC, et al. Analysis of select folate pathway genes, *PAX3*, and human *T* in a midwestern neural tube defect population. Teratology 1999;59:331–341.
76. Rozen R. Genetic predisposition to hyperhomocysteinemia: deficiency of methylenete-trahydrofolate reductase (MTHFR). Thromb Haemost 1997;78:523–526.
77. Molloy AM, Daly S, Mills JL, et al. Thermolabile variant of 5,10-methylenetetrahy-drofolate reductase associated with low red-cell folates: implications for folate intake recommendations. Lancet 1997;349:1591–1593.
78. Ma J, Stampfer MJ, Christensen B, et al. A polymorphism of the methionine synthase gene: assocation with plasma folate, vitamin B_{12}, homocyst(e)ine, and colorectal can-cer risk. Cancer Epidemiol Biomarkers Prev 1999;8:825–829.
79. Jacques PF, Bostom AG, Williams RR, et al. Relation between folate status, a common mutation in methylenetetrahydrofolate reductase, and plasma homocysteine concentra-tions. Circulation 1996;93:7–9.
80. Girelli D, Friso S, Trabetti E, et al. Methylenetetrahydrofolate reductase C677T muta-tion, plasma homocysteine, and folate in subjects from northern Italy with or without angiographically documented severe coronary atherosclerotic disease: evidence for an important genetic-environmental interaction. Blood 1998;91:4158–4163.
81. Lievers KJA, Boers GHJ, Verhoef P, et al. A second common variant in the methyl-enetetrahydrofolate reductase (MTHFR) gene and its relationship to MTHFR enzyme activity, homocysteine,and cardiovascular disease risk. J Mol Med 2001;79:522–528.
82. Friedman G, Goldschmidt N, Friedlander Y, et al. A common mutation A1298C in hu-man methylenetetrahydrofolate reductase gene: association with plasma total homocys-teine and folate concentrations. J Nutr 1999;129:1656–1661.
83. Chango A, Boisson F, Barbé F, et al. The effect of 677C-T and 1298A-C mutations on plasma homocysteine and 5,10-methylenetetrahydrofolate reductase activity in healthy subjects. Br J Nutr 2000;83:593–596.
84. Chen J, Ma J, Stampfer MJ, Palomeque C, et al. Linkage disequilibrium between the 677C>T and 1298A>C polymorphisms in human methylenetetrahydrofolate reductase gene and their contributions to risk of colorectal cancer. Pharmacogenetics 2002;12: 339–342.
85. Botto LD, Yang Q. 5,10 methylenetetrahydrofolate reductase gene variants and con-genital anomalies: a HuGE review. Am J Epidemiol 2000;151:862–877.
86. Robien K, Ulrich CM. 5,10-methylenetetrahydrofolate reductase polymorphisms and leukemia risk: a HuGE mini-review. Am J Epidemiol 2003;157(7):571–582.

87. Barber R, Shalat S, Hendricks K, et al. Investigation of folate pathway gene polymorphisms and the incidence of neural tube defects in a Texas Hispanic population. Mol Genet Metabol 2000;70:45–52.

88. Volcik KA, Blanton SH, Tyerman GH, et al. Methylenetetrahydrofolate reductase and spina bifida: evaluation of level of defect and maternal genotypic risk in Hispanics. Am J Med Genet 2000;95:21–27.

89. Chen J, Giovannucci E, Kelsy K, et al. A methylenetetrahydrofolate reductase polymorphism and the risk of colorectal cancer. Cancer Res 1996;56:4862–4864.

90. Slattery ML, Potter JD, Samowitz W, Schaffer D, Leppert M. Methylenetetrahydrofolate reductase, diet, and risk of colon cancer. Cancer Epidemiol Biomarkers Prev 1999;8:513–518.

91. Park KS, Mok JW, Kim JC. The 677>T mutation in 5,10-Methylenetetrahydrofolate reductase and colorectal cancer risk. Genetic Testing 1999;3:233–236.

92. Sharp L, Little J, Brockton N, et al. Genetic polymorphisms in folate metabolism, dietary folate intake and colorectal cancer: a population-based case-control study. J Epidemiol Community Health 2001;55:A27.

93. Shannon B, Gnanasampanthan S, Beilby J, Iacopetta B. A polymorphism in the methylenetetrahydrofolate reductase gene predisposes to colorectal cancers with microsatellite instability. Gut 2002;50:520–524.

94. Delgado-Enciso I, Martinez-Garza SG, Rojas-Martinez A, et al. 677T mutation of the MTHFR gene in adenomas and colorectal cancer in a population sample from the Northeastern Mexico. Rev Gastroenterol Mex 2001;66:32–37.

95. Kane MF, Loda M, Gaida GM, et al. Methylation of the hMLH1 promoter correlates with lack of expression of hMLH1 in sporadic colon tumors and mismatch repair-defective human tumor cell lines. Cancer Res 1997;57:808–811.

96. Sharp L, Little J, Brockton N, et al. Dietary intake of folate and related micronutrients, genetic polymorphisms in *MTHFR* and colorectal cancer: a population-based case-control study in Scotland. J Nutrit 2002;132(115):35425.

97. Ulrich CM, Kampman E, Bigler J, et al. Colorectal adenomas and the C677T *MTHFR* polymorphism: evidence for gene-environment interaction? Cancer Epidemiol Biomarkers Prev 1999;8:659–668.

98. Levine AJ, Siegmund KD, Ervin CM, et al. The methylenetetrahydrofolate reductase 677C → T polymorphism and distal colorectal adenoma risk. Cancer Epidemiol Biomarkers Prev 2000;9:657–663.

99. Marugame T, Tsuji E, Inoue H, et al. Methylenetetrahydrofolate reductase polymorphism and risk of colorectal adenomas. Cancer Lett 2000;151:181–186.

100. Ulvik A, Evensen ET, Lien EA, et al. Smoking, folate and methylenetetrahydrofolate reductase status as interactive determinants of adenomatous and hyperplastic polyps of colorectum. Am J Med Genet 2001;101:246–254.

101. Ulrich CM, Kampman E, Bigler J, et al. Lack of association between the C677T *MTHFR* polymorphism and colorectal hyperplastic polyps. Cancer Epidemiol Biomarkers Prev 2000;9:427–434.

102. Mahmud N, Molloy A, McPartlin J, et al. Increased prevalence of methylenetetrahydrofolate reductase C677T variant in patients with inflammatory bowel disease, and its clinical implications. Gut 1999;45:389–394.

103. Wolf CR, Smith G. Cytochrome p450 CYP2D6. In: Vineis P, Malats N, Lang M, et al., editors. Metabolic Polymorphisms and Susceptibility to Cancer. Lyon, France: IARC Press, 1999: 209–229.

104. Johnson WG. DNA polymorphism-diet-cofactor-development hypothesis and the gene-teratogen model for schizophrenia and other developmental disorders. Am J Med Genet 1999;88:311–323.

105. van der Put NMJ, Blom HJ. Reply to Donnelly. Am J Hum Genet 2000;66:744–745.
106. Scott J (2000). Folate (folic acid) and vitamin B_{12}, in Human Nutrition and Dietetics. Garrow JS, James WPT, Ralph A, eds. Churchill Livingstone, 271–280.
107. Choi SW, Friso S. Is it worthwhile to try different coenzymatic forms of folate in future chemoprevention trials? Nutr 2001;17:738–739.
108. Akoglu B, Faust D, Milovic V, Stein J. Folate and chemoprevention of colorectal cancer: is 5-methyltetrahydrofolate an active antiproliferative agent in folate-treated coloncancer cells? Nutr 2001;17:652–653.
109. Cuskelly GJ, McNulty H, Scott JM. Effect of increasing dietary folate on red-cell folate: implications for prevention of neural tube defects. Lancet 1996;347:657–659.
110. Wald NJ, Law MR, Morris JK, et al. Quantifying the effect of folic acid. Lancet 2001;358:2069–2073.
111. Jacques PF, Selhub J, Bostom AG, Wilson PWF, Rosenberg IH. The effect of folic acid fortification on plasma folate and total homocysteine concentrations. N Engl J Med 1999;340:1449–1454.
112. Kim YI, Fawaz K, Knox T, et al. Colonic mucosal concentrations of folate correlate well with blood measurements of folate status in persons with colorectal polyps. Am J Clin Nutr 1998;68:866–872.
113. Nelen WLDM, Blom HJ, Thomas CMG, Steegers EAP, Boers GHJ, Eskes TKAB. Methylenetetrahydrofolate reductase polymorphism affects the change in homocysteine and folate concentrations resulting from low dose folic acid supplementation in women with unexplained recurrent miscarriages. J Nutrit 1998;128:1336–1341.
114. Woodside JV, Yarnell JW, McMaster D, et al. Effect of B-group vitamins and antioxidant vitamins on hyperhomocysteinemia: a double-blind, randomised, factorial-design, controlled trial. Am J Clin Nutr 1998;67:858–866.
115. Fohr IP, Prinz-Langenohl R, Brönstrup A, et al. 5,10-methylenetetrahydrofolate reductase genotype determines the plasma homocysteine-lowering effect of supplementation with 5-methyltetrahydrofolate or folic acid in healthy young women. Am J Clin Nutr 2002;75:275–282.
116. Hauser AC, Hagen W, Rehak PH, et al. Efficacy of folinic versus folic acid for the correction of hyperhomocysteinemia in hemodialysis patients. Am J Kidney Dis 2001;37:758–765.
117. Pullin CH, Ashfield-Watt PAL, Burr ML, et al. Optimization of dietary folate or low-dose folic acid supplements lower homocysteine but do not enhance endothelial function in healthy adults, irrespective of the methylenetetrahydrofolate reductase (C677T) genotype. J Am Coll Cardiol 2001;38:1799–1805.
118. Silaste ML, Rantala M, Sampi M, Alfthan G, Aro A, Kesaniemi YA. Polymorphisms of key enzymes in homocysteine metabolism affect diet responsivenss of plasma homocysteine in healthy women. J Nutrit 2001;131:2643–2647.
119. Little J. Is folic acid pluripotent? A review of the associations with congenital anomalies, cancer and other diseases. In: Ioannides C, Lewis DFV, editors. Drugs, Diet and Disease. Volume 1. Mechanistic Approaches to Cancer. New York: Ellis Horwood, 1995: 259–308.
120. Eichholzer M, Weil C, Stadiehelin HB, Moser U, Ludiethy J. Folic acid to prevent spina bifida. Should wheat be fortified in combination with vitamin B_6 and vitamin B_{12}? Schweizerische Zeitschrift fur Ganzheitsmedizin 2002;14:64–74.
121. Simmer K, James C, Thompson RPH. Are iron-folate supplements harmful? Am J Clin Nutr 45, 122–125. 1987.
122. Vineis P, Schulte P, McMichael AJ. Misconceptions about the use of genetic tests in populations. Lancet 2001;357:709–712.
123. Verhaar MC, Stroes E, Rabelink TJ. Folates and cardiovascular disease. Arterioscler Thromb Vasc Biol 2002;22:6–13.

124. Molloy AM, Scott JM. Folates and prevention of disease. Public Health Nutr 2001;4: 601–609.
125. Sharp L, Little J, Schofield AC, et al. Folate and breast cancer: the role of polymorphisms in methylenetetrahydrofolate reductase (MTHFR). Cancer Lett 2002;181:65–71.
126. Wisotzkey JD, Toman J, Bell T, Monk JS, Jones D. MTHFR (C677T) polymorphisms and stage III colon cancer: response to therapy. Molecular Diagnosis 1999;4:95–99.
127. Ulrich CM, Yasui Y, Storb R, et al. Pharmacogenetics of methotrexate: toxicity among marrow transplantation patients varies with the methylenetetrahydrofolate reductase C677T polymorphism. Blood 2001;98:231–234.
128. Toffoli G, Veronesi A, Boiocchi M, et al. MTHFR gene polymorphism and severe toxicity during adjuvant treatment of early breast cancer with cyclophosphamide, methotrexate, and fluorouracil (CMF). Ann Oncol 2000;11:373–374.
129. Khoury MJ, Stewart W, Beaty TH. The effect of genetic susceptibility on causal inference in epidemiologic studies. Am J Epidemiol 1987;126:561–567.

21

Apolipoprotein E and Alzheimer disease

Richard Mayeux

Alzheimer disease is a degenerative brain disease affecting an estimated 15 million persons worldwide, making it the most frequent cause of dementia in adults. Over the last 2 decades variant alleles in four genes have been implicated in the cause of this disease. Each of the genes identified is involved in the production or processing of amyloid, a beta pleated sheet peptide.

The earliest manifestation is an insidious impairment of memory. As the disease progresses, other intellectual skills become impaired, and erratic behavior, delusions, and a loss of control over body functions occur. Language deteriorates gradually. For example, word-finding difficulty is prominent in early stages and impairment of verbal and written comprehension and expression occurs in later stages of illness. Disease progression also affects spatial, analytic, and synthetic abilities, and judgment, sometimes accompanied by a loss of insight. Delusions and hallucinations are late manifestations and can include irritability, agitation, verbal or physical aggression, wandering, and a loss of inhibition of emotions. Self-care becomes difficult in the final stages of the disease.

Criteria for the clinical diagnosis of Alzheimer disease were established in 1984 (1). Patients with no associated illnesses are termed *probable* Alzheimer disease; *possible* Alzheimer disease refers to patients meeting these criteria, with other illnesses that could have caused central nervous system dysfunction, such as hypothyroidism or cerebrovascular disease. Alzheimer disease is a typical degenerative disease of the brain and nervous system. Although it presents with a clinically recognizable syndrome, its disease-defining features are pathologic. In fact, the term *definite* Alzheimer disease is reserved for instances in which the disease is confirmed at the postmortem examination.

Microscopic neuropathological examination of the brain reveals deposits of extracellular β-amyloid protein in diffuse plaques and in plaques containing elements of degenerating neurons, termed neuritic plaques. Intracellular changes include deposits of abnormally hyperphosphorylated τ protein, a microtuble assembly protein, in the form of neurofibrillary tangles. Loss of both neurons and synapses is also widespread. Additional features include other neuropil pathology (e.g., neuropil threads), cellular pathology (e.g., granulovacuolar degeneration in the hippocampus), and regional cell losses (particularly in the hippocampus). These degenerative changes occur first in a small region of the hippocampus before involving other brain structures (2). The fundamental pathogenic mechanisms responsible for the development of these changes are unknown.

The degenerative changes in the basal forebrain profoundly reduce its content of acetylcholine and the activities of cholineacetyltransferase and acetylcholinesterase. Although other neurotransmitters can be involved, the loss of acetylcholine occurs early and correlates with the memory impairment. A variety of pharmacological interventions are available to improve the symptoms of the disorder. These medications include cholinesterase inhibitors, which increase central "cholinergic tone" and ameliorate secondary consequences of the disease (3). However, no therapies are proven to affect the course of this disorder.

This review will include a briefly describe the disease and its epidemiology and discusses the genes involved in Alzheimer disease, with the major focus on the role of apolipoprotein-E.

The Frequency of Alzheimer Disease

The prevalence of Alzheimer disease before age 65 years is less than 1%, but it increases so dramatically afterward that by age 85 and older almost 30% of people may have this disease. One of the best studied populations has been in East Boston where Evans et al. (4,5) found that the prevalence of Alzheimer disease increased 15-fold from 3% among people between the ages of 65 to 74 years to 47% among people aged 85 and older. These authors included of the full spectrum of disease from mild to severe forms. Compared with data from Europe and Asian as well as some areas of the United States (6–14), the prevalence of Alzheimer disease in East Boston appeared to be much higher, probably because of the inclusion of mild cases. However, similar rates have been found in the United States among African-American and Hispanic populations (14,15). Lower rates have been observed for Africans in their homeland (16).

Survival with Alzheimer disease varies from 2 to 20 years. Two population-based studies found that the median survival is 3 to 4 years (17,18). Alzheimer disease increases the risk for mortality by twofold (17,19–22), particularly among men.

The incidence rate for Alzheimer disease also increases with advancing age (6,13,15,23–31). In some Asian and African populations, the incidence rates are lower than estimates from more developed countries (13,32), but in at least two studies, people from African-American and Hispanic ethnic groups appear to have higher rates of disease relative to white non-Hispanics (15,30). There are two factors that contribute significantly to the difficulty in establishing accurate estimates of the incidence of Alzheimer disease: (*1*) determining the exact age at onset and (*2*) defining a disease-free population. Despite these difficulties, the average incidence rate increases from approximately 0.5% per year among people aged 65 to 70 years to approximately 6% to 8% for people over aged 85 years.

Environmental and Medical Risk Factors

No specific environmental toxins are known to be associated with Alzheimer disease. However, several medical disorders have been associated with this disease. For example, Alzheimer disease occurs more frequently among people with a history of a prior depressive illness or a traumatic head injury than among those without these disorders (33–39). Cardiovascular disease and dementia are frequent coincident disorders among the elderly. Heart disease and its antecedents, specifically hypertension, ischemic heart disease, hypercholesterolemia, and stroke, may predispose to Alzheimer disease (40–44). Cigarette smoking, once purported to be protective, increases the risk of developing Alzheimer disease, particularly among people without an APOE-ε4 allele (45–47). Socioeconomic factors may also contribute to disease risk because illiteracy and the lack of formal education, and even fewer years of formal education, have been associated with Alzheimer disease (48,49).

Several therapies for common health problems have been associated with a decreased risk for Alzheimer disease. The use of estrogen by postmenopausal women may result in a 50% reduction in occurrence for Alzheimer disease (50–53). Antiinflammatory agents also decrease the risk of Alzheimer disease (54–56). Although not confirmed, wine in moderate amounts each day can reduce the risk of Alzheimer disease (57). Time spent engaged in physical and mental activities during late life has been associated with a lower risk of Alzheimer disease (58–60). Risk was lowest for people with complex activity patterns that included frequent intellectual, passive, and physical activities. In the Canadian Study of Health and Aging, the strongest effects were related to physical activities such as vigorous exercise (59).

Genetic Epidemiology of Alzheimer Disease

Large multigenerational families with Alzheimer disease have been observed for decades, but in the majority of families the inheritance of Alzheimer disease does not fit a Mendelian pattern. First-degree relatives of patients with Alzheimer disease, particularly siblings, have twice the expected lifetime risk of developing the disease (61–63). Alzheimer disease is more frequent among monozygotic than dizygotic twins (64).

Mutations in three genes, the amyloid precursor protein gene on chromosome 21, the presenilin 1 (*PS1*) on chromosome 14, and the presenilin 2 (*PS2*) on chromosome 1, are usually found in families with an autosomal dominant pattern of disease inheritance beginning as early as the third decade of life (65) (Table 21.1). *PS1*

Table 21.1 Genes and Other Chromosomal Locations with Suggestive Variation in Genes Associated with Alzheimer Disease

Chromosome	Gene	Age at Onset	Pattern	Variants/Mutations
ch21q21.3	APP	30 to 60 years	AD	16 (exons 16,17)
ch14q24.13	PS1	30 to 50 years	AD and familial	129 (exons 4–12)+
ch1q31.42	PS2	50 to 70 years	AD	9 (exons 4,5,7,12)
ch19q13.2	APOE	50 to 80+ years	Familial and sporadic	3 Isoforms
ch12p13[a]	?	>65 years	Familial	Unknown
ch10q[a]	?	>65 years	Familial	Unknown
ch9p[a]	?	>65 years	Familial	Unknown

APP, gene encoding the amyloid precursor protein; PS1 and PS2, presenilin 1 and presenilin 2 genes; APOE, apolipoprotein E gene; AD, Mendelian autosomal dominant pattern of inheritance.
[a]Chromosomal location identified by linkage.

mutations are the most frequent cause of familial early-onset Alzheimer disease because there are over 100 known mutations in this gene (66). In a study of 414 patients suspected on having early-onset familial Alzheimer disease and referred for genetic testing, 11% had mutations in *PS1*. Studies of these families show enhanced generation or aggregation of amyloid β peptide in brain in the form of neuritic plaques suggesting a pathogenic role.

One of the most important observations has been the identification of the relation between the $\epsilon 4$ polymorphism or variant allele of the apolipoprotein E (*APOE*) gene on chromosome 19 and both sporadic and familial disease with onset usually after age 65 years (Table 21.1). The frequency of the *APOE-ε4* allele can be as high as 40% among patients with Alzheimer disease compared with 15% to 20% among unaffected individuals similar in age. A single *APOE-ε4* allele can increase the risk of Alzheimer disease by twofold, while the homozygous configuration can be associated with a fivefold increase. The population-attributable risk associated with *APOE-ε4* may be as high as 20%, making it the single most important risk factor for the disease in elderly people. Each *APOE-ε4* allele lowers the age-at-onset by nearly 5 years (67). *APOE-ε4* may also influence the age at onset in families with mutations in the amyloid precursor protein gene (68) and in adults with Down syndrome who develop dementia as they age (69).

Genetic linkage studies show at least three additional putative loci with association to Alzheimer disease (Table 21.1). Pericak-Vance and colleagues (70) identified a locus on chromosome 12p conferring susceptibility to Alzheimer disease. Subsequent confirmation been limited because of locus heterogeneity related to *APOE-ε4* and to clinical heterogeneity as a result of the identification of Lewy bodies, small intracellular inclusion found in the brains of patients (70–74). A locus on chromosome 10q has also been associated with Alzheimer disease and with a putative biomarker of altered amyloid β in plasma of family members (75,76). Other locations on this chromosome have also been identified, but not confirmed (77). Lastly, Pericak-Vance and associates identified a locus on chromosome 9p with linkage to Alzheimer disease restricted to a series of families in whom the diagnosis was confirmed by postmortem examination (72).

Several candidate genes have been associated with Alzheimer disease, but have not been confirmed. Researchers have focused on the gene of proteins intimately involved in the disease pathogenesis, sequenced these genes in search of polymorphisms, and compared the frequency of variants in affected and unaffected people. Given the large number of variant alleles that could exist in the human genome, it is not surprising that investigators have difficulty sorting through them to identify the exceptional one that increases susceptibility to a disease. A statistically significant association between a polymorphism and a disease can occur for one of several reasons: the polymorphism has a true effect on disease risk, the polymorphism is in linkage disequilibrium with the "true" disease-causing polymorphism, the frequency of the polymorphism differs in subgroups of a heterogeneous population that are unevenly sampled in the cases and controls (referred to as population stratification or confounding), or the association occurs simply by chance. Nonetheless, additional variants in other genes will certainly be identified that associate with Alzheimer disease (78).

The relation of the $\epsilon4$ polymorphism in *APOE* to Alzheimer disease remains a unique finding. Association of a genetic variation that occurs in up to 25% of the population with a common illness is unusual. The confirmation of this association has been virtually worldwide. It is less robust among some ethnic groups (79–87). *APOE* genotyping was considered as a possible adjunct in the diagnosis of Alzheimer disease because of the strong association with the *APOE-$\epsilon4$* allele. In a large collaborative study (88) the *APOE* genotype provided a slight benefit in the overall specificity for those patients first meeting NINCDS-ADRDA clinical criteria for Alzheimer disease who later came to autopsy. The *APOE* genotype alone had limited sensitivity or specificity, but when used in combination with clinical criteria the *APOE* genotype did improve specificity.

Consistent with other genes involved in Alzheimer disease, *APOE* may also act through a complex and poorly understood relationship with amyloid β deposition. The ApoE protein is an obligatory participant in amyloid β accumulation, and postmortem data indicate that ApoE isoforms exert at least some of their effects by controlling amyloid β accumulation or the clearance of amyloid β peptides (89). *APOE-$\epsilon4$* is associated with greater amyloid β plaque density than other *APOE* alleles among patients with Alzheimer disease (90,91). *APOE*-deficient mice expressing the APP_{717} mutation that causes an early-onset, autosomal dominant form of Alzheimer disease deposit fewer amyloid β plaques (92) and show less memory impairment than wild-type mice with or with the APP_{717} mutation (93). A direct role for APOE, independent of an interaction with amyloid β, involving both biochemical and neuronal integrity has been suggested in animal models with impaired memory (93,94). Compared with intact mice, *APOE*-deficient mice have decreased synaptic density in cholinergic, noradrenergic, and serotinergic projections to relevant brain regions (95) and perform worse in several types of memory tasks (93,96–98). Therefore, *APOE-$\epsilon4$* has a direct effect on memory in the absence of Alzheimer disease (99).

Apolipoprotein E

In addition to Alzheimer disease, *APOE* has been extensively investigated because of its role in lipid metabolism and ischemic cardiovascular disease (44,100–109). Mortality from ischemic heart disease is related to the presence of the *APOE-ε4* allele (110). Variation at the *APOE* locus has also been related to cerebral hemorrhage and insulin levels (109,111,112).

There are three common alleles of *APOE*: ε2, ε3, and ε4. *APOE-ε3* is by far the most common allele occurring in 60% to 80% of humans; *APOE-ε4* is considered to be the ancestral allele. The frequency of *APOE-ε4* varies worldwide from 40.7% among Pygmies (113) to 8.5% among people from Morocco (Table 21.2). Among European populations, the highest frequency of *APOE-ε4* occurs among Lapps, Swedes, and Finns; the lowest frequency occurs in Greeks and Italians. Sudanese, Nigerians and African Americans have an intermediate *APOE-ε4* frequency of 22% to 29%. The lowest frequency occurs among Asians (114).

APOE is the most upstream member of a large cluster of apolipoprotein genes that are co-regulated, at least in peripheral tissues such as liver and kidney, by the interaction of the promoters of the individual genes with a set of shared enhancer elements (115–117). Evidence suggests that sequence variants in the promoter and enhancer regions of *APOE* or *APOCI*, near *APOE*, associate with Alzheimer dis-

Table 21.2 Selected Distribution of APOE Allele Frequencies in Some Human Populations

Population	APOE-ε2	APOE-ε3	APOE-ε4
AFRICANS			
Pygmies	0.057	0.536	0.407
Nigerians	0.027	0.677	0.296
Moroccans	0.065	0.085	0.085
EUROPEANS			
Lapps	0.050	0.640	0.310
French	0.108	0.771	0.121
Sardinians	0.050	0.898	0.052
Spaniards	0.052	0.856	0.091
Germans	0.077	0.778	0.145
ASIANS			
Malay Aborigines	0.140	0.620	0.240
Chinese	0.105	0.824	0.071
Koreans	0.020	0.870	0.110
NATIVE AMERICANS			
Cayapas	0	0.720	0.280
Amerindians	0	0.816	0.184
Mayans	0	0.911	0.089
OCEANIANS			
Papuans	0.145	0.486	0.368
Polynesians	0.110	0.630	0.260
Aboriginal Australian	0	0.740	0.260

Source: Modified from Table 1 in Corbo and Scacchi (114).

ease (118–122). Lambert et al. (121) suggested that the Th1/E47 polymorphism in the *APOE* regulatory region is not only associated with Alzheimer disease, but it increases *APOE-ε4* gene expression. However, this work needs independent confirmation.

Some investigators have suggested that the major phenotypic effect of *APOE* may be to lower the age at onset of Alzheimer disease (67,123). The presence of an *APOE-ε4* allele results in a much earlier age at onset, even among families that have mutations in the amyloid precursor protein (124,125). In Down syndrome the *APOE-ε4* allele also lowers the age of onset of dementia (69). Alternatively, *APOE* may have more immediate effects on the nervous system. For example, compared to people with other *APOE* genotypes, those with an *APOE-ε4* allele appear to develop hippocampal atrophy (126–128) and are more likely to develop age-related cognitive impairment (99,129,130). Memory decline in the elderly population may well be the direct effect of *APOE-ε4* on hippocampal-based memory systems, rather than incipient Alzheimer disease. The ε4 variant of *APOE* causes a decrease in synapse per neuron ratio (131), developmental defects within the dentate gyrus (132), and increased vulnerability to exogenous neurotoxins(98). Any one of these, or other mechanisms as yet unidentified, may explain the decrease in memory over time among humans with the *APOE-ε4* allele.

Although a large number of studies have examined the potential for gene–environment interaction in Alzheimer disease, little evidence esists of an important environmental risk factor that interacts with *APOE*. Despite the important role of *APOE* on lipid metabolism, no relation between lipid levels, *APOE*, and Alzheimer disease exists (133). Presence of *APOE-ε4* seems to increase the risk for dementia and Alzheimer disease independently of its effect on dyslipidemia and atherogenesis (134). Smoking also increases the risk for Alzheimer disease, but only among people without the *APOE-ε4* polymorphism (46,47). Compared with healthy elderly without an *APOE-ε4* allele or a history of traumatic head injury, patients with Alzheimer disease were 10 times more likely to have both. In a case-control study, head injury alone was not associated with increased risk, and those with an *APOE-ε4* allele had only a twofold increased risk. However, subsequent prospective studies failed to confirm this result, finding either no joint effects of *APOE-ε4* and head injury on Alzheimer disease risk or that the effects are independent and additive (34–36,38).

Diagnostic Tests for Alzheimer Disease

The clinical diagnosis of Alzheimer disease has relied on a set of guidelines published in 1984, by a joint working group the National Institute for Neurological and Communicative Disorders and the Alzheimer Disease and Related Disorders Association (NINCDS-ADRDA) (1). The reliability and consistency of these clinical criteria are quite high (88,135–138). The diagnostic categories of probable and possible Alzheimer disease have been shown to provide a high sensitivity with moderate specificity using autopsy confirmation as the "gold standard" (139–141). Studies evaluating the usefulness of these criteria have suggested that they are accurate and cost-effective (142). Routine blood tests that screen for common metabolic disorders are helpful in excluding other diagnoses, but not in diagnosing Alzheimer disease (136–138,143). Brain imaging that includes magnetic resonance imaging, computed tomography, and functional brain imaging is also useful, and their importance in the early diagnosis has been increasing (136,143,144). Diagnostic tests for Alzheimer disease have also included the recognition of odors (145,146) and measurement of amyloid β peptide and τ protein in cerebrospinal fluid (147,148). Although these tests definitely represent an advance over prior methods, none have yet been demonstrated to improve accuracy over the NINCDS-ADRDA clinical criteria.

Genetic tests are generally not recommended for use in the diagnosis of Alzheimer disease (88,149–152). Several reviews of Alzheimer disease concerning the accuracy, benefits, and risks of genetic testing in patients with the disease or in their asymptomatic family members have been published. These consensus groups agree that limited testing for mutations in the genes associated with early-onset familial Alzheimer disease, amyloid precursor protein, and PS1 and PS2 may be acceptable but only with appropriate counseling (153–157). In contrast, agreement has been almost universal that *APOE* testing should not be recommended because the test does not provide sufficient sensitivity or specificity for the diagnosis (88). The Stanford program on genetic testing had broader and more comprehensive recommendations that took into account the availability of procedures to promote good surrogate decision making for incompetent patients and to safeguard confidentiality. Also they considered it imperative to have access to sophisticated genetic counselors who could communicate complex risk information and effectively convey the social costs and psychological burdens of testing. For example, they discussed the dangers of unintentional disclosure of predictive genetic information to family members. They also considered the need to protect family members from inappropriate advertising and marketing of genetic tests. Finally, they recognized the need for public education about the meaning and usefulness of predictive and diagnostic tests for Alzheimer disease (151).

Contribution of Apolipoprotein E to Alzheimer Disease

Despite the potential hazards described above, few doubt the advantages of having a highly consistent and reliable indicator of disease risk such, as the *APOE-ϵ4* allele in Alzheimer disease. *APOE* genotyping allows for the identification of people destined to become affected or who are in the "preclinical" stages of the illness; reduces disease heterogeneity in clinical trials or epidemiologic studies; allows for a better understanding of natural history of disease with that genotype encompassing the phases of induction, latency, and detection; and also provides target for a clinical trial. Each of the advantages would strongly advocate the use of *APOE* genotyping in all clinical studies. The improvement in validity and precision that is gained would far outweigh the difficulties in obtaining such genotypes from patients or participants in a study. Not only would *APOE* genotyping easily fit into clinical trials, but it would also provide the basis for conducting a therapeutic trial in patients before overt manifestations occur and could provide information about the variability in the response to drugs.

Potential Contribution of *APOE* to the Management of Alzheimer Disease

Population screening for *APOE-ε4* would be impractical because at least half of patients with Alzheimer disease do not have this allele or might not develop the disease. Predictive testing based on *APOE-ε4* might be used to detect presymptomatic susceptibility for the purpose of advanced directives or reproductive planning. However, without a clear-cut therapeutic option, such early detection at this point does not seem beneficial. Moreover, no rationale yet exists for primary or secondary prevention based on the presence or absence of *APOE-ε4*. Drugs or other therapeutic strategies that delay the onset of Alzheimer disease are being investigated. If one or more of these agents prove to be effective, then *APOE*-based risk prediction might be more acceptable. Not unexpectedly, family members of patients with Alzheimer disease have been concerned about their status with regard to *APOE-ε4*, but an unambiguous response to their questions about prognostic implications is difficult to present. The *APOE* genotype may be undeniable as a genetic risk factor for Alzheimer disease, but it does not provide sufficient information to be an adequate predictive genetic test.

Consequences of *APOE* Testing

Stored samples enabled investigators to rapidly study the relation between *APOE-ε4* and Alzheimer disease. Sharing of medical information in research is often stipulated with standard informed consent procedures, particularly for stored samples that may be used at a later date. For people unable to make such decisions, as is often the case in Alzheimer disease, surrogate consent has been obtained for previously acquired information. However, although guidelines for the use of this type of genetic data in research have been developed, they are only now being discussed with regard to clinical practice.

A commercial test is available for Alzheimer disease risk based on the *APOE*. Knowledge that one might develop a chronic disease in late life would be advantageous if treatments or preventive measures were available. However, serious ethical questions arise from the prospect that the genetic information contained in an person's *APOE* genotype may be entered into accessible computerized databases in medical offices or third parties such as insurance providers or employers. The association between Alzheimer disease and *APOE-ε4* is only one of several genetic associations that have encouraged the establishment of policies to regulate access to genetic information for diseases that occur in old age.

The possibility of discrimination by employers and insurance companies against people who work also could become a serious issue in patients with Alzheimer disease. As more people work past their seventh decade, employers will begin to worry about the risks and costs of their health coverage. Diseases, such as Alzheimer disease, are not transient conditions and could stigmatize an individual, potentially leading to emotional injury and financial harm. Pre-employment genetic screening for common disabling diseases such as Alzheimer disease or cardiovascular disease is not designed to reduce the individual risk of disease associated with exposures in the workplace, but to reduce health-related costs (e.g., health insurance, disability insurance, and lost productivity). Therefore detection of susceptibility to Alzheimer disease by using a genetic test, such as APOE, would not lower the population burden of disease but would protect an employer. One could imagine a scenario in which obtaining even minimal health insurance or long-term care insurance for the elderly would be jeopardized. Third parties could also potentially make decisions for an individual on the basis of genetic information that would, in turn, affect the future employment or insurability for the person's offspring. Access to genetic data by employers and insurers should be limited. Permission from the individual should be required. Protection for workers should also require strict security and enforcement of penalties for unauthorized use of such data.

Conclusions

The discovery of the association between *APOE-ε4* and Alzheimer disease was a major advance in our understanding of the disease and its causes. Further elucidation of the role of *APOE-ε4* in the pathogenesis of this disease will, no doubt, lead to improved therapy. As a genetic variant indicating susceptibility, *APOE* genotyping already contributes to the investigations of other predisposing genes and risk factors. However, *APOE* genotyping does not yet provide the clinician with enough additional information to make routine genotyping necessary for diagnosis. This situation may change if a preventive therapy is identified. Investigations identifying additional chromosome loci indicate that variations in other genes are likely to be found. How these discoveries will be incorporated into clinical practice remains to be determined.

Future research will need to address the complex genetics of Alzheimer disease. The possibility of genetic profiles for research and clinical care are real. However, until a meaningful treatment or preventive measure is established, genetic testing may be limited to research.

Acknowledgments
Support was provided by federal grants AG15473, AG08702, AG07232, the Charles S. Robertson Memorial Gift for Alzheimer's Disease Research from the Banbury Fund, and the Blanchette Hooker Rockefeller Foundation.

References

1. McKhann G, Drachman D, Folstein M, et al. Clinical diagnosis of Alzheimer's disease: report of the NINCDS-ADRDA Work Group under the auspices of Department of Health and Human Services Task Force on Alzheimer's Disease. Neurology 1984;34:939–944.
2. Dickson DW. Neuropathology of Alzheimer's disease and other dementias. Clin Geriatr Med 2001;17:209–228.
3. Winkler J, Thal LJ, Gage FH, Fisher LJ. Cholinergic strategies for Alzheimer's disease. J Mol Med 1998;76:555–567.
4. Evans DA, Funkenstein HH, Albert MS, et al. Prevalence of Alzheimer's disease in a community population of older persons. Higher than previously reported. JAMA 1989;262:2551–2556.
5. Evans DA. Estimated prevalence of Alzheimer's disease in the United States. Milbank Q 1990;68:267–289.
6. Breteler MM, Claus JJ, van Duijn CM, et al. Epidemiology of Alzheimer's disease. Epidemiol Rev 1992;14:59–82.
7. Farrag A, Farwiz HM, Khedr EH, et al. Prevalence of Alzheimer's disease and other dementing disorders: Assiut-Upper Egypt study. Dement Geriatr Cogn Disord 1998;9: 323–328.
8. Chandra V, Ganguli M, Pandav R, et al. Prevalence of Alzheimer's disease and other dementias in rural India: the Indo-US study. Neurology 1998;51:1000–1008.
9. Rocca WA, Hofman A, Brayne C, et al. The prevalence of vascular dementia in Europe: facts and fragments from 1980–1990 studies. EURODEM–Prevalence Research Group. Ann Neurol 1991;30:817–824.
10. Rocca WA, Hofman A, Brayne C, et al. Frequency and distribution of Alzheimer's disease in Europe: a collaborative study of 1980–1990 prevalence findings. The EURODEM-Prevalence Research Group. Ann Neurol 1991;30:381–390.
11. Ott A, Breteler MM, van Harskamp F, et al. Prevalence of Alzheimer's disease and vascular dementia: association with education. The Rotterdam study. BMJ 1995;310: 970–973.
12. White L, Petrovitch H, Ross GW, et al. Prevalence of dementia in older Japanese-American men in Hawaii: The Honolulu-Asia Aging Study. JAMA 1996;276:955–960.
13. Ganguli M, Dodge HH, Chen P, et al. Ten-year incidence of dementia in a rural elderly US community population: the MoVIES Project. Neurology 2000;54:1109–1116.
14. Gurland BJ, Wilder DE, Lantigua R, et al. Rates of dementia in three ethnoracial groups. Int J Geriatr Psychiatry 1999;14:481–493.
15. Perkins P, Annegers JF, Doody RS, et al. Incidence and prevalence of dementia in a multiethnic cohort of municipal retirees. Neurology 1997;49:44–50.
16. Ogunniyi A, Baiyewu O, Gureje O, et al. Epidemiology of dementia in Nigeria: results from the Indianapolis-Ibadan study. Eur J Neurol 2000;7:485–490.
17. Helmer C, Joly P, Letenneur L, Commenges D, et al. Mortality with dementia: results from a French prospective community-based cohort. Am J Epidemiol 2001;154: 642–648.
18. Wolfson C, Wolfson DB, Asgharian M, et al. A reevaluation of the duration of survival after the onset of dementia. N Engl J Med 2001;344:1111–1116.
19. Aguero-Torres H, Fratiglioni L, Guo Z, et al. Mortality from dementia in advanced age: a 5-year follow-up study of incident dementia cases. J Clin Epidemiol 1999;52:737–743.
20. Jagger C, Andersen K, Breteler MM, et al. Prognosis with dementia in Europe: A collaborative study of population-based cohorts. Neurologic Diseases in the Elderly Research Group. Neurology 2000;5411:S16–S20.

21. Lapane KL, Gambassi G, Landi F, et al. Gender differences in predictors of mortality in nursing home residents with AD. Neurology 2001;56:650–654.

22. Ostbye T, Hill G, Steenhuis R. Mortality in elderly Canadians with and without dementia: a 5-year follow-up. Neurology 1999;53:521–526.

23. Gussekloo J, Heeren TJ, Izaks GJ, Ligthart GJ, Rooijmans HG. A community based study of the incidence of dementia in subjects aged 85 years and over. J Neurol Neurosurg Psychiatry 1995;59:507–510.

24. Hebert LE, Scherr PA, Beckett LA, et al. Age-specific incidence of Alzheimer's disease in a community population. JAMA 1995;273:1354–1359.

25. Hendrie HC, Ogunniyi A, Hall KS, et al. Incidence of dementia and Alzheimer disease in 2 communities: Yoruba residing in Ibadan, Nigeria, and African Americans residing in Indianapolis, Indiana. JAMA 2001;285:739–747.

26. Letenneur L, Commenges D, Dartigues JF, et al. Incidence of dementia and Alzheimer's disease in elderly community residents of south-western France. Int J Epidemiol 1994;23:1256–1261.

27. Morgan K, Lilley JM, Arie T, et al. Incidence of dementia in a representative British sample. Br J Psychiatry 1993;163:467–470.

28. Paykel ES, Huppert FA, Brayne C. Incidence of dementia and cognitive decline in over-75s in Cambridge: overview of cohort study. Soc Psychiatry Psychiatr Epidemiol 1998; 33:387–392.

29. Rocca WA, Cha RH, Waring SC, et al. Incidence of dementia and Alzheimer's disease: a reanalysis of data from Rochester, Minnesota, 1975–1984. Am J Epidemiol 1998; 148:51–62.

30. Tang MX, Cross P, Andrews H, et al. Incidence of AD in African-Americans, Caribbean Hispanics, and Caucasians in northern Manhattan. Neurology 2001;56:49–56.

31. Zhang M, Katzman R, Yu E, et al. A preliminary analysis of incidence of dementia in Shanghai, China. Psychiatry Clin Neurosci 1998;52 suppl:S291–S294.

32. Hendrie HC, Hall KS, Hui S, et al. Apolipoprotein E genotypes and Alzheimer's disease in a community study of elderly African Americans. Ann Neurol 1995;37:118–120.

33. Li YS, Meyer JS, Thornby J. Longitudinal follow-up of depressive symptoms among normal versus cognitively impaired elderly. Int J Geriatr Psychiatry 2001;16:718–727.

34. Mehta KM, Ott A, Kalmijn S, et al. Head trauma and risk of dementia and Alzheimer's disease: The Rotterdam Study. Neurology 1999;53:1959–1962.

35. Plassman BL, Havlik RJ, Steffens DC, et al. Documented head injury in early adulthood and risk of Alzheimer's disease and other dementias. Neurology 2000;55: 1158–1166.

36. Guo Z, Cupples LA, Kurz A, et al. Head injury and the risk of AD in the MIRAGE study. Neurology 2000;54:1316–1323.

37. Speck CE, Kukull WA, Brenner DE, et al. History of depression as a risk factor for Alzheimer's disease. Epidemiology 1995;6:366–369.

38. Schofield PW, Tang M, Marder K, et al. Alzheimer's disease after remote head injury: an incidence study. J Neurol Neurosurg Psychiatry 1997;62:119–124.

39. Nemetz PN, Leibson C, Naessens JM, et al. Traumatic brain injury and time to onset of Alzheimer's disease: a population-based study. Am J Epidemiol 1999;149:32–40.

40. Brayne C, Gill C, Huppert FA, et al. Vascular risks and incident dementia: results from a cohort study of the very old. Dement Geriatr Cogn Disord 1998;9:175–180.

41. Breteler MM. Vascular risk factors for Alzheimer's disease: an epidemiologic perspective. Neurobiol Aging 2000;21:153–160.

42. Katzman R, Aronson M, Fuld P, et al. Development of dementing illnesses in an 80-year-old volunteer cohort. Ann Neurol 1989;25:317–324.

43. Launer LJ, Andersen K, Dewey ME, et al. Rates and risk factors for dementia and Alzheimer's disease: results from EURODEM pooled analyses. EURODEM Incidence Research Group and Work Groups. European Studies of Dementia. Neurology 1999;52: 78–84.
44. Zimetbaum P, Frishman W, Aronson M. Lipids, vascular disease, and dementia with advancing age. Epidemiologic considerations. Arch Intern Med 1991;151:240–244.
45. Doll R, Peto R, Boreham J, Sutherland I. Smoking and dementia in male British doctors: prospective study. BMJ 2000;320:1097–1102.
46. Merchant C, Tang MX, Albert S, Manly J, Stern Y, Mayeux R. The influence of smoking on the risk of Alzheimer's disease. Neurology 1999;52:1408–1412.
47. Ott A, Slooter AJ, Hofman A, et al. Smoking and risk of dementia and Alzheimer's disease in a population-based cohort study: the Rotterdam Study. Lancet 1998;351: 1840–1843.
48. Zhang MY, Katzman R, Salmon D, et al. The prevalence of dementia and Alzheimer's disease in Shanghai, China: impact of age, gender, and education. Ann Neurol 1990;27:428–437.
49. Stern Y, Gurland B, Tatemichi TK, Tang MX, Wilder D, Mayeux R. Influence of education and occupation on the incidence of Alzheimer's disease. JAMA 1994;271: 1004–1010.
50. Baldereschi M, Di Carlo A, Lepore V, et al. Estrogen-replacement therapy and Alzheimer's disease in the Italian Longitudinal Study on Aging. Neurology 1998;50:996–1002.
51. Carlson MC, Zandi PP, Plassman BL, et al. Hormone replacement therapy and reduced cognitive decline in older women: The Cache County Study. Neurology 2001;57:2210–2216.
52. Tang MX, Jacobs D, Stern Y, et al. Effect of oestrogen during menopause on risk and age at onset of Alzheimer's disease. Lancet 1996;348:429–432.
53. Yaffe K, Sawaya G, Lieberburg I, Grady D. Estrogen therapy in postmenopausal women: effects on cognitive function and dementia. JAMA 1998;279:688–695.
54. in 't Veld BA, Launer LJ, Hoes AW, et al. NSAIDs and incident Alzheimer's disease. The Rotterdam Study. Neurobiol Aging 1998;19:607–611.
55. in 't Veld BA, Ruitenberg A, Hofman A, et al. Nonsteroidal antiinflammatory drugs and the risk of Alzheimer's disease. N Engl J Med 2001;345:1515–1521.
56. Stewart WF, Kawas C, Corrada M, et al. Risk of Alzheimer's disease and duration of NSAID use. Neurology 1997;48:626–632.
57. Orgogozo JM, Dartigues JF, Lafont S, et al. Wine consumption and dementia in the elderly: a prospective community study in the Bordeaux area. Rev Neurol (Paris) 1997;153:185–192.
58. Friedland RP, Fritsch T, Smyth KA, et al. Patients with Alzheimer's disease have reduced activities in midlife compared with healthy control-group members. Proc Natl Acad Sci U S A 2001;98:3440–3445.
59. Laurin D, Verreault R, Lindsay J, et al. Physical activity and risk of cognitive impairment and dementia in elderly persons. Arch Neurol 2001;58:498–504.
60. Yoshitake T, Kiyohara Y, Kato I, et al. Incidence and risk factors of vascular dementia and Alzheimer's disease in a defined elderly Japanese population: the Hisayama Study. Neurology 1995;45:1161–1168.
61. Hocking LB, Breitner JC. Cumulative risk of Alzheimer-like dementia in relatives of autopsy-confirmed cases of Alzheimer's disease. Dementia 1995;6:355–356.
62. Mayeux R, Sano M, Chen J, et al. Risk of dementia in first-degree relatives of patients with Alzheimer's disease and related disorders. Arch Neurol 1991;48:269–273.
63. Farrer LA, Myers RH, Cupples LA, et al. Transmission and age-at-onset patterns in familial Alzheimer's disease: evidence for heterogeneity. Neurology 1990;40:395–403.

64. Breitner JC, Welsh KA, Gau BA, et al. Alzheimer's disease in the National Academy of Sciences-National Research Council Registry of Aging Twin Veterans. III. Detection of cases, longitudinal results, and observations on twin concordance. Arch Neurol 1995;52:763–771.

65. St George-Hyslop PH. Molecular genetics of Alzheimer's disease. Biol Psychiatry 2000;47:183–199.

66. Rogaeva EA, Fafel KC, Song YQ, et al. Screening for PS1 mutations in a referral-based series of AD cases: 21 novel mutations. Neurology 2001;57:621–625.

67. Corder EH, Saunders AM, Strittmatter WJ, et al. Gene dose of apolipoprotein E type 4 allele and the risk of Alzheimer's disease in late onset families. Science 1993;261:921–923.

68. Levy-Lahad E, Bird TD. Genetic factors in Alzheimer's disease: a review of recent advances. Ann Neurol 1996;40:829–840.

69. Schupf N, Kapell D, Lee JH, et al. Onset of dementia is associated with apolipoprotein E epsilon4 in Down's syndrome. Ann Neurol 1996;40:799–801.

70. Pericak-Vance MA, Bass MP, Yamaoka LH, et al. Complete genomic screen in late-onset familial Alzheimer disease. Evidence for a new locus on chromosome 12. JAMA 1997;278:1237–1241.

71. Mayeux R, Lee JH, Romas SN, et al. Chromosome-12 Mapping of Late-Onset Alzheimer Disease among Caribbean Hispanics. Am J Hum Genet 2002;70:237–243.

72. Pericak-Vance MA, Grubber J, Bailey LR, et al. Identification of novel genes in late-onset Alzheimer's disease. Exp Gerontol 2000;35:1343–1352.

73. Rogaeva E, Premkumar S, Song Y, et al. Evidence for an Alzheimer disease susceptibility locus on chromosome 12 and for further locus heterogeneity. JAMA 1998;280:614–618.

74. Wu WS, Holmans P, Wavrant-DeVrieze F, et al. Genetic studies on chromosome 12 in late-onset Alzheimer disease. JAMA 1998;280:619–622.

75. Ertekin-Taner N, Graff-Radford N, Younkin LH, et al. Linkage of plasma Abeta42 to a quantitative locus on chromosome 10 in late-onset Alzheimer's disease pedigrees. Science 2000;290:2303–2304.

76. Myers A, Holmans P, Marshall H, et al. Susceptibility locus for Alzheimer's disease on chromosome 10. Science 2000;290:2304–2305.

77. Bertram L, Blacker D, Mullin K, et al. Evidence for genetic linkage of Alzheimer's disease to chromosome 10q. Science 2000;290:2302–2303.

78. Warwick Daw E, Payami H, Nemens EJ, et al. The number of trait loci in late-onset Alzheimer disease. Am J Hum Genet 2000;66:196–204.

79. Class CA, Unverzagt FW, Gao S, Sahota A, Hall KS, Hendrie HC. The association between Apo E genotype and depressive symptoms in elderly African-American subjects. Am J Geriatr Psychiatry 1997;5:339–343.

80. Farrer LA, Cupples LA, Haines JL, et al. Effects of age, sex, and ethnicity on the association between apolipoprotein E genotype and Alzheimer disease. A meta-analysis. APOE and Alzheimer Disease Meta Analysis Consortium. JAMA 1997;278:1349–1356.

81. Kalaria RN, Ogeng'o JA, Patel NB, et al. Evaluation of risk factors for Alzheimer's disease in elderly east Africans. Brain Res Bull 1997;44:573–577.

82. Maestre G, Ottman R, Stern Y, et al. Apolipoprotein E and Alzheimer's disease: ethnic variation in genotypic risks. Ann Neurol 1995;37:254–259.

83. Osuntokun BO, Sahota A, Ogunniyi AO, et al. Lack of an association between apolipoprotein E epsilon 4 and Alzheimer's disease in elderly Nigerians. Ann Neurol 1995;38:463–465.

84. Sayi JG, Patel NB, Premkumar DR, et al. Apolipoprotein E polymorphism in elderly east Africans. East Afr Med J 1997;74:668–670.

85. Tang MX, Stern Y, Marder K, et al. The APOE-epsilon4 allele and the risk of Alzheimer disease among African Americans, whites, and Hispanics. JAMA 1998; 279:751–755.

86. Tang MX, Maestre G, Tsai WY, et al. Relative risk of Alzheimer disease and age-at-onset distributions, based on APOE genotypes among elderly African Americans, Caucasians, and Hispanics in New York City. Am J Hum Genet 1996;58:574–584.

87. Sahota A, Yang M, Gao S, et al. Apolipoprotein E-associated risk for Alzheimer's disease in the African-American population is genotype dependent. Ann Neurol 1997; 42:659–661.

88. Mayeux R, Saunders AM, Shea S, et al. Utility of the apolipoprotein E genotype in the diagnosis of Alzheimer's disease. Alzheimer's Disease Centers Consortium on Apolipoprotein E and Alzheimer's Disease. N Engl J Med 1998;338:506–511.

89. Morishima-Kawashima M, Oshima N, Ogata H, et al. Effect of apolipoprotein E allele epsilon4 on the initial phase of amyloid beta-protein accumulation in the human brain. Am J Pathol 2000;157:2093–2099.

90. Schmechel DE, Saunders AM, Strittmatter WJ, et al. Increased amyloid beta-peptide deposition in cerebral cortex as a consequence of apolipoprotein E genotype in late-onset Alzheimer disease. Proc Natl Acad Sci U S A 1993;90:9649–9653.

91. Horsburgh K, Cole GM, Yang F, et al. β-amyloid (Aβ)42(43), aβ42, aβ40 and apoE immunostaining of plaques in fatal head injury. Neuropathol Appl Neurobiol 2000;26: 124–132.

92. Bales KR, Verina T, Cummins DJ, et al. Apolipoprotein E is essential for amyloid deposition in the APP(V717F) transgenic mouse model of Alzheimer's disease. Proc Natl Acad Sci U S A 1999;96:15233–15238.

93. Dodart JC, Mathis C, Bales KR, et al. Behavioral deficits in APP(V717F) transgenic mice deficient for the apolipoprotein E gene. Neuroreport 2000;11:603–607.

94. Gordon I, Ben-Eliyahu S, Rosenne E, et al. Derangement in stress response of apolipoprotein E-deficient mice. Neurosci Lett 1996;206:212–214.

95. Chapman S, Michaelson DM. Specific neurochemical derangements of brain projecting neurons in apolipoprotein E-deficient mice. J Neurochem 1998;70:708–714.

96. Raber J, Wong D, Buttini M, et al. Isoform-specific effects of human apolipoprotein E on brain function revealed in ApoE knockout mice: increased susceptibility of females. Proc Natl Acad Sci U S A 1998;95:10914–10919.

97. Raber J, Wong D, Yu GQ, et al. Apolipoprotein E and cognitive performance. Nature 2000;404:352–354.

98. Buttini M, Orth M, Bellosta S, et al. Expression of human apolipoprotein E3 or E4 in the brains of Apoe-/- mice: isoform-specific effects on neurodegeneration. J Neurosci 1999;19:4867–4880.

99. Mayeux R, Small SA, Tang M, Tycko B, Stern Y. Memory performance in healthy elderly without Alzheimer's disease: effects of time and apolipoprotein-E. Neurobiol Aging 2001;22:683–689.

100. Davignon J, Gregg RE, Sing CF. Apolipoprotein E polymorphism and atherosclerosis. Arteriosclerosis 1988;8:1–21.

101. Haffner SM, Stern MP, Miettinen H, Robbins D, Howard BV. Apolipoprotein E polymorphism and LDL size in a biethnic population. Arterioscler Thromb Vasc Biol 1996;16:1184–1188.

102. Kao JT, Tsai KS, Chang CJ, et al. The effects of apolipoprotein E polymorphism on the distribution of lipids and lipoproteins in the Chinese population. Atherosclerosis 1995;114:55–59.

103. Lee JH, Tang MX, Schupf N, et al. Mortality and apolipoprotein E in Hispanic, African-American, and Caucasian elders. Am J Med Genet 2001;103:121–127.

104. Muros M, Rodriguez-Ferrer C. Apolipoprotein E polymorphism influence on lipids, apolipoproteins and Lp(a) in a Spanish population underexpressing apo E4. Atherosclerosis 1996;121:13–21.

105. Pablos-Mendez A, Mayeux R, Ngai C, Shea S, Berglund L. Association of apo E polymorphism with plasma lipid levels in a multiethnic elderly population. Arterioscler Thromb Vasc Biol 1997;17:3534–3541.

106. Robitaille N, Cormier G, Couture R, Bouthillier D, Davignon J, Perusse L. Apolipoprotein E polymorphism in a French Canadian population of northeastern Quebec: allele frequencies and effects on blood lipid and lipoprotein levels. Hum Biol 1996; 68:357–370.

107. Shriver MD, Boerwinkle E, Hewett-Emmett D, Hanis CL. Frequency and effects of apolipoprotein E polymorphism in Mexican-American NIDDM subjects. Diabetes 1991;40:334–337.

108. Srinivasan SR, Ehnholm C, Elkasabany A, et al. Apolipoprotein E polymorphism modulates the association between obesity and dyslipidemias during young adulthood: The Bogalusa Heart Study. Metabolism 2001;50:696–702.

109. Valdez R, Howard BV, Stern MP, Haffner SM. Apolipoprotein E polymorphism and insulin levels in a biethnic population. Diabetes Care 1995;18:992–1000.

110. Eichner JE, Kuller LH, Orchard TJ, et al. Relation of apolipoprotein E phenotype to myocardial infarction and mortality from coronary artery disease. Am J Cardiol 1993; 71:160–165.

111. Nicoll JA, McCarron MO. APOE gene polymorphism as a risk factor for cerebral amyloid angiopathy-related hemorrhage. Amyloid 2001;8 suppl 1:51–55.

112. Srinivasan SR, Ehnholm C, Wattigney WA, Bao W, Berenson GS. The relation of apolipoprotein E polymorphism to multiple cardiovascular risk in children: the Bogalusa Heart Study. Atherosclerosis 1996;123:33–42.

113. Zekraoui L, Lagarde JP, Raisonnier A, Gerard N, Aouizerate A, Lucotte G. High frequency of the apolipoprotein E ϵ4 allele in African pygmies and most of the African populations in sub-Saharan Africa. Hum Biol 1997;69:575–581.

114. Corbo RM, Scacchi R. Apolipoprotein E (APOE) allele distribution in the world. Is APOEϵ4 a 'thrifty' allele? Ann Hum Genet 1999;63:301–310.

115. Allan CM, Taylor S, Taylor JM. Two hepatic enhancers, HCR.1 and HCR.2, coordinate the liver expression of the entire human apolipoprotein E/C-I/C-IV/C-II gene cluster. J Biol Chem 1997;272:29113–29119.

116. Simonet WS, Bucay N, Lauer SJ, Taylor JM. A far-downstream hepatocyte-specific control region directs expression of the linked human apolipoprotein E and C-I genes in transgenic mice. J Biol Chem 1993;268:8221–8229.

117. Shachter NS, Zhu Y, Walsh A, Breslow JL, Smith JD. Localization of a liver-specific enhancer in the apolipoprotein E/C-I/C-II gene locus. J Lipid Res 1993;34: 1699–1707.

118. Chartier-Harlin MC, Parfitt M, Legrain S, et al. Apolipoprotein E, epsilon 4 allele as a major risk factor for sporadic early and late-onset forms of Alzheimer's disease: analysis of the 19q13.2 chromosomal region. Hum Mol Genet 1994;3:569–574.

119. Lambert JC, Perez-Tur J, Dupire MJ, et al. Distortion of allelic expression of apolipoprotein E in Alzheimer's disease. Hum Mol Genet 1997;6:2151–2154.

120. Lambert JC, Pasquier F, Cottel D, et al. A new polymorphism in the APOE promoter associated with risk of developing Alzheimer's disease. Hum Mol Genet 1998;7: 533–540.

121. Lambert JC, Berr C, Pasquier F, et al. Pronounced impact of Th1/E47cs mutation compared with -491 AT mutation on neural APOE gene expression and risk of developing Alzheimer's disease. Hum Mol Genet 1998;7:1511–1516.

122. Lambert JC, Brousseau T, Defosse V, et al. Independent association of an APOE gene promoter polymorphism with increased risk of myocardial infarction and decreased APOE plasma concentrations-the ECTIM study. Hum Mol Genet 2000;9:57–61.

123. Meyer MR, Tschanz JT, Norton MC, et al. APOE genotype predicts when—not whether—one is predisposed to develop Alzheimer disease. Nat Genet 1998;19:321–2.

124. Nacmias B, Latorraca S, Piersanti P, et al. ApoE genotype and familial Alzheimer's disease: a possible influence on age of onset in APP717 Val → Ile mutated families. Neurosci Lett 1995;183:1–3.

125. Sorbi S, Nacmias B, Forleo P, Piacentini S, Latorraca S, Amaducci L. Epistatic effect of APP717 mutation and apolipoprotein E genotype in familial Alzheimer's disease. Ann Neurol 1995;38:124–127.

126. Juottonen K, Lehtovirta M, Helisalmi S, Riekkinen PJ, Sr., Soininen H. Major decrease in the volume of the entorhinal cortex in patients with Alzheimer's disease carrying the apolipoprotein E epsilon4 allele. J Neurol Neurosurg Psychiatry 1998;65:322–327.

127. Killiany RJ, Gomez-Isla T, Moss M, et al. Use of structural magnetic resonance imaging to predict who will get Alzheimer's disease [see comments]. Ann Neurol 2000;47:430–439.

128. Wahlund LO, Julin P, Lannfelt L, Lindqvist J, Svensson L. Inheritance of the ApoE epsilon4 allele increases the rate of brain atrophy in dementia patients. Dement Geriatr Cogn Disord 1999;10:262–268.

129. Carmelli D, DeCarli C, Swan GE, et al. The joint effect of apolipoprotein E epsilon4 and MRI findings on lower-extremity function and decline in cognitive function. J Gerontol A Biol Sci Med Sci 2000;55:M103–M109.

130. Small BJ, Graves AB, McEvoy CL, Crawford FC, Mullan M, Mortimer JA. Is APOE—epsilon4 a risk factor for cognitive impairment in normal aging? Neurology 2000;54:2082–2088.

131. Cambon K, Davies HA, Stewart MG. Synaptic loss is accompanied by an increase in synaptic area in the dentate gyrus of aged human apolipoprotein E4 transgenic mice. Neuroscience 2000;97:685–692.

132. Teter B, Xu PT, Gilbert JR, et al. Human apolipoprotein E isoform-specific differences in neuronal sprouting in organotypic hippocampal culture. J Neurochem 1999;73:2613–2616.

133. Romas SN, Tang MX, Berglund L, et al. APOE genotype, plasma lipids, lipoproteins, and AD in community elderly. Neurology 1999;53:517–521.

134. Prince M, Lovestone S, Cervilla J, et al. The association between APOE and dementia does not seem to be mediated by vascular factors. Neurology 2000;54:397–402.

135. Hogervorst E, Barnetson L, Jobst KA, et al. Diagnosing dementia: interrater reliability assessment and accuracy of the NINCDS/ADRDA criteria versus CERAD histopathological criteria for Alzheimer's disease. Dement Geriatr Cogn Disord 2000;11:107–13.

136. Massoud F, Devi G, Moroney JT, et al. The role of routine laboratory studies and neuroimaging in the diagnosis of dementia: a clinicopathological study. J Am Geriatr Soc 2000;48:1204–1210.

137. Bianchetti A, Trabucch M. Clinical aspects of Alzheimer's disease. Aging (Milano) 2001;13:221–230.

138. Chertkow H, Bergman H, Schipper HM, et al. Assessment of suspected dementia. Can J Neurol Sci 2001;28 Suppl 1:S28–41.

139. Nagy Z, Esiri MM, Hindley NJ, et al. Accuracy of clinical operational diagnostic criteria for Alzheimer's disease in relation to different pathological diagnostic protocols. Dement Geriatr Cogn Disord 1998;9:219–226.

140. Holmes C, Cairns N, Lantos P, et al. Validity of current clinical criteria for Alzheimer's disease, vascular dementia and dementia with Lewy bodies. Br J Psychiatry 1999;174:45–50.

141. Lim A, Tsuang D, Kukull W, et al. Clinico-neuropathological correlation of Alzheimer's disease in a community-based case series. J Am Geriatr Soc 1999;47:564–569.
142. Chui H, Zhang Q. Evaluation of dementia: a systematic study of the usefulness of the American Academy of Neurology's practice parameters. Neurology 1997;49:925–935.
143. Small GW. Differential diagnosis and early detection of dementia. Am J Geriatr Psychiatry 1998;6:S26–S33.
144. Jagust WJ. Neuroimaging in dementia. Neurol Clin 2000;18:885–902.
145. Larsson M, Semb H, Winblad B, et al. Odor identification in normal aging and early Alzheimer's disease: effects of retrieval support. Neuropsychology 1999;13:47–53.
146. Devanand DP, Michaels-Marston KS, Liu X, et al. Olfactory deficits in patients with mild cognitive impairment predict Alzheimer's disease at follow-up. Am J Psychiatry 2000;157:1399–1405.
147. Rosler N, Wichart I, Jellinger KA. Clinical significance of neurobiochemical profiles in the lumbar cerebrospinal fluid of Alzheimer's disease patients. J Neural Transm 2001; 108:231–246.
148. Andreasen N, Minthon L, Davidsson P, et al. Evaluation of CSF-tau and CSF-Abeta42 as diagnostic markers for Alzheimer disease in clinical practice. Arch Neurol 2001; 58:373–379.
149. Lovestone S. Early diagnosis and the clinical genetics of Alzheimer's disease. J Neurol 1999;246:69–72.
150. Mayeux R, Schupf N. Apolipoprotein E and Alzheimer's disease: the implications of progress in molecular medicine. Am J Public Health 1995;85:1280–1284.
151. McConnell LM, Koenig BA, Greely HT, et al. Genetic testing and Alzheimer disease: recommendations of the Stanford Program in Genomics, Ethics, and Society. Genet Test 1999;3:3–12.
152. Croes EA, Dermaut B, van Der Cammen TJ, et al. Genetic testing should not be advocated as a diagnostic tool in familial forms of dementia. Am J Hum Genet 2000;67: 1033–1035.
153. Genetic testing for late-onset Alzheimer's disease. AGS Ethics Committee. J Am Geriatr Soc 2001;49:225–226.
154. Apolipoprotein E epsilon 4 allele (APOE epsilon 4) and Alzheimer's disease: role of genetic testing for diagnosis and risk assessment. Tecnologica MAP Suppl 1999:4–7.
155. Genetic testing and Alzheimer's disease. Health News 1998;4:5.
156. Statement on use of apolipoprotein E testing for Alzheimer disease. American College of Medical Genetics/American Society of Human Genetics Working Group on ApoE and Alzheimer disease. JAMA 1995;274:1627–1629.
157. Alzheimer's disease and genetic testing. Alzheimer Dis Assoc Disord 1994;8:63–147.

22

Immunogenetic factors in chronic beryllium disease

Erin C. McCanlies, Michael E. Andrew, and Ainsley Weston

Individuals who are exposed to beryllium dust or fumes are at risk of developing a granulomatous lung disease called chronic beryllium disease (CBD). Epidemiologic investigations with an integrated genetic component have linked the development of CBD with the human leukocyte antigen *(HLA)-DPB1* gene (1–6). This case study will focus on the role of *HLA-DPB1* in CBD and the practical application of genetic information for the benefit of all members of the beryllium industry workforce.

Beryllium (Be), atomic number 4 on the periodic table of the elements, is extracted from either beryl ore or bertrandite. It is extremely light and, although brittle, stiffer than steel. Beryllium is fused with copper, nickel, or aluminum to form highly stable flexible alloys that also have tensile strength. These properties make beryllium an ideal element for numerous technological applications and in high demand (Table 22.1). In 2000, more than 500 metric tons of beryllium were produced worldwide for commercial distribution (7). Currently it is not known how many individuals have been exposed to beryllium; though estimates as high as 800,000 have been suggested (8). Exposure to beryllium primarily occurs among workers in beryllium manufacturing plants in which metal fabrication and pressing of beryllia ceramics occurs. For this reason, the beryllium worker population is most often afflicted with CBD.

Though CBD is the principal form of beryllium disease seen today, prior to 1946 acute beryllium disease (ABD) was also a problem. ABD was the result of short, extremely high doses of beryllium. Although removal from exposure and treatment often lead to resolution, in its most severe form ABD was fatal. Of those cases that were not fatal, approximately 17% progressed to CBD (9–11).

CBD was first described among workers in the fluorescent light industry (12). Soon after, it was also recognized in beryllium industry workers. In the United States

Table 22.1 Some Uses of Beryllium

Industry/Uses	Application	Current (C)/Discontinued (D)
Aerospace	Engines	D
	Braking systems	D
	Structures	C
	Rockets	C
	Satellites	C
	Gyroscopes	C
	Precision tools	C
	Mirrors (Hubble telescope)	C
Telecommunications	Undersea repeater housings	C
	Cell phones	C
	Personal computers	C
	Transistor mountings	C
Energy/electrical	Relays and switches	C
	Microelectronics	C
	Microwave devices	C
	Nuclear reactors	C
	Fluorescent lights	D
Biomedical	X-ray tube window	C
	Scanning electron microscope	C
	Dental prostheses	C
	(Crowns, bridges, and partial dentures)	
Defense	Tank mirrors	C
	Springs on submarine hatches	C
	Nuclear triggers	C
	Missile guidance	C
	Mast mounted sights	C
Fire prevention	Non-sparking tools	C
	Sprinkler system springs	C
Other	Plastic molds	C
	Laser tubes	C
	Bellows (CuBe)	C
	Springs	C
	Formula 1 race car brakes	D
	Valve seats (drag racer engines)	C
	Airbag triggers	C
	Jewelry (gems)	C
	Jewelry (rings)	C
	Golf clubs	C
	Bicycle frames[a]	C
	Camera shutters	C
	Pen clips	C

[a]Very limited use.

in 1949, to reduce exposure experienced by beryllium workers, the Atomic Energy Commission, predecessor of the Department of Energy (DOE) proposed a beryllium exposure limit of 2 $\mu g/m^3$ averaged over an 8-hour work period (13). This restriction largely eliminated cases of ABD, but despite overall reduction in workplace exposure, CBD continues to occur (13). This has resulted in a series of investigations aimed at better understanding the natural history of disease, routes of exposure, and the immunogenetics of CBD (1–6,14–29).

In the 1950s, a beryllium case registry was established in the United States to study the pathology and natural history of CBD. Case registry information and cross-sectional surveillance of beryllium worker populations have been utilized to better define the mechanism of CBD. In conjunction with this research, laboratory-based research determined that exposure to beryllium triggered a cell-mediated, type IV delayed hypersensitivity reaction resulting in the proliferation of beryllium-specific T lymphocytes (16,17). Based on an understanding of this mechanism, a specific, cellular hypersensitivity response to beryllium challenge was demonstrated in vitro. This formed the basis for the development of a beryllium lymphocyte proliferation test (BeLPT) (18,19). Utilization of the BeLPT in cross-sectional beryllium worker surveys identified workers who were beryllium sensitized, but who did not have impaired lung function or overt clinical symptoms of CBD (20–25). This suggested that CBD progressed from a subclinical state in which individuals first became sensitized to beryllium prior to developing CBD. It is not yet known if all beryllium-sensitized individuals will eventually develop CBD.

Currently, the BeLPT is used to determine if beryllium workers have become immunologically sensitized to beryllium. Due to interlaboratory and intralaboratory variation, some persons only test positive at one laboratory or the other (20,23–29). For this reason, a diagnosis of beryllium sensitization is dependent on repeated positive BeLPT responses (20–29).

Early detection and diagnosis of CBD is dependent on physical screening of workers that includes the BeLPT. Workers who test positive in successive BeLPT tests are referred for bronchoalveolar lavage to assess the immune status of lymphocytes (BAL-LPT) in the lung and bronchial biopsy to determine if granuloma formation has occurred in the lungs. Radiographic changes consistent with CBD have also been used to diagnose CBD (20,23–29).

Epidemiology of Chronic Beryllium Disease

Cross-sectional surveys have found that between 1% to 12% of beryllium workers are sensitized to beryllium. Of these, 36% to 100% were found to have CBD (20,23–28). The variation in these rates may be due to at least four different factors: the form of beryllium the worker is exposed to; the magnitude or duration of beryllium exposure; temporal variation in immune system response; and differing susceptibilities. As previously discussed, rates of sensitization are also dependent on BeLPT laboratory variation and whether split sampling is conducted. In the case of CBD, prior to the manifestation of physical symptoms, missed granulomas in bronchial biopsies contribute to variable rates of CBD.

Cross-sectional epidemiologic studies that evaluated the prevalence of sensitization and CBD in relation to work processes and beryllium exposure measurements found that risk of sensitization and CBD was dependent on having performed certain jobs or tasks (23,25). In 1992, a study that evaluated 136 beryllium ceramics workers found that eight workers were beryllium sensitized. Of these, six had granulomatous lung disease (CBD). When the specific jobs and tasks performed by these workers were evaluated, machinists were significantly more likely to be beryllium sensitized than the other employees (14.3% vs. 1.2%; OR = 14.3, p = 0.003) (25).

In 1993–1994 cross-sectional surveillance was conducted at a beryllium manufacturing plant in which pure metal, oxides, and alloys are produced. Prior to 1980, this plant had also manufactured beryllium ceramics, but this process had since moved to another facility (23). Fifty-nine workers of 646 participating employees were identified as being beryllium sensitized. Forty-seven of these individuals underwent clinical evaluation that included bronchoscopy. Twenty-four cases of CBD were identified (27). When specific areas, jobs, and tasks were evaluated, individuals who had ever worked in the area referred to as the pebble plant were at the greatest risk of CBD (OR = 23.5; 95% CI = 4.4–125.5) followed by those who had ever worked in ceramics (OR = 4.4; 95% CI = 1.8–10.5). Employees who worked in metal production were more likely to be beryllium sensitized compared to other workers (7.3% vs. 1.3%), and had an increased prevalence of CBD (19.2%) (27).

Air sample measurements taken at the ceramics plant in 1992 indicated that compared to other processes, machining had higher general area and breathing zone beryllium measurements (p = 0.0001). These results demonstrated a quantitative relationship between process-related beryllium exposure and the risk of beryllium sensitization (23). However, air sample measurements and rates of sensitization and CBD have not always correlated. For example, even though rates of CBD were highest in individuals who had ever worked in the pebbles plant, the median average beryllium air exposure since 1984 was estimated to be 1.3 $\mu g/m^3$, not significantly different than other areas (27). Further, a follow-up study of this ceramics plant found that the overall decline in beryllium exposures was not matched by a decline in the prevalence of sensitization or CBD (28). These results are consistent with previous findings in which the number of individuals with CBD did not reflect the level of beryllium exposure (27,30). This discrepancy is even more apparent in light of documented cases of CBD in workers who had very little apparent beryllium exposure (e.g., administrative workers), as well as family and community clusters of cases among individuals who had never worked in the beryllium industry (30). These results are important because they raised concerns about how beryllium exposure is assessed (31). They are also suggestive of the immunogenetic nature of CBD.

Gene Associations with CBD

In vitro studies conducted to evaluate the immunogenetic nature of CBD demonstrated that anti-*HLA-DP* antibodies blocked beryllium-stimulated T lymphocyte proliferation (32). This suggested that MHC class II antigen-bearing cells were involved in a beryllium-specific T-lymphocyte mediated (type II) response in the development of CBD. Thus, genetic variants of the *HLA* (*-DP, -DQ,* and *-DR*) loci were evaluated as potential risk factors for CBD. A map of chromosome 6p12.3 shows the relative locations of the HLA genes (Figure 22.1) (33,34).

Heteroduplex analysis, allele-specific polymerase chain reaction (PCR), restriction fragment length polymorphism (RFLP), oligonucleotide hybridization, and direct and indirect sequencing of PCR products have all been used to investigate the HLA-gene locus. *HLA-DP* variants have been identified as being associated with sensitization and CBD.

The *HLA-DP* molecule is composed of two chains, alpha (*A1*) and beta (*B1*).[1] The *A1* locus lies proximal to the *B1* locus and the gene products form a heterodimer. There are fewer known human *HLA-DPA1* variants than *B1* variants (20 vs. 103), and it appears that there are a limited number of *A/B* haplotypes (35). Structurally, the alpha and beta chains form a groove in the DP molecule that plays an important function in immunologic processes. Initial research focused on the *HLA-DPB1*, but more recent investigations have also evaluated the role of *HLA-DPA1*, alone and in combination with *HLA-DPB1*, in the risk of sensitization and CBD (1–6).

Richeldi et al., 1993 used oligonucleotide hybridization techniques to evaluate *HLA-DPB1* haplotypes characterized by polymorphisms in codons 36,55–57, and

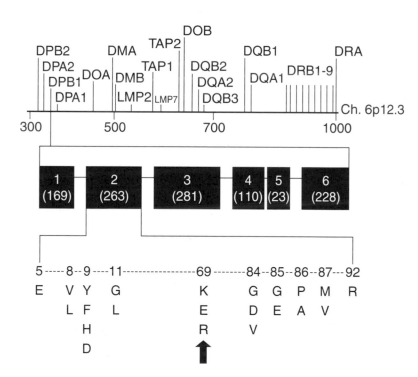

Figure 22.1 Partial map of chromosome 6p12.3 showing relative positions of genes in the HLA-complex. HLA-DPβ1 has been expanded to show the coding region (exons 1–6 and their relative size in base pairs). Some of the key amino acid substitution polymorphisms are indicated.

65–69 in beryllium workers with and without CBD (Table 22.2) (1). The results of this initial study indicated that CBD cases were more likely to have *HLA-DPB1* alleles coding for aspartic acid (D) and glutamic acid (E) in positions 55 and 56, respectively, compared to the controls who where more likely to code for an alanine (A) in those positions (79% vs. 41%; OR = 5.4, 95% CI = 1.7-17.6). Furthermore, alleles characterized by a codon for glutamic acid residue at position 69 (E69) in the amino acid sequence also occurred more often in individuals with CBD than in those without (97% vs. 27%; OR = 85.3, 95% CI, 10.9-3,578.0) (Table 22.2). No significant association was seen between CBD and the polymorphic codon 36. Allele-specific analysis implicated an association between inheritance of the common *HLA-DPB1*0201*, glutamic acid 69 containing allele, and CBD (OR = 4.3, 95% CI = 1.1–20.8). Conversely, *HLA-DPB1*0401*, which does not code for glutamic acid at position 69, occurred less often in cases compared to controls (14% vs. 48%; OR = 0.2, 95% CI = 0.05–0.7).

Wang et al. investigated the presence and absence of both *HLA-DPB1* and *HLA-DPA1* alleles in beryllium-exposed individuals with (n = 20) and without (n = 75) CBD (3). This study was important for two reasons: it verified the association between *HLA-DPB1^{E69}* and CBD; and it evaluated allele-specific relations including the effect of homozygosity versus heterozygosity and disease status. Although there are some methodological concerns regarding the geographic origin of the cases and controls (3), this study was technically superior to those of Richeldi et al., 1993, 1997 (1,2) in its use of allele-specific dideoxy-chain termination-DNA sequencing to characterize HLA alleles.

In this beryllium-exposed population, the odds of disease in the presence of *HLA-DPB1^{E69}* was estimated to be 22.9 (95% CI = 4.8–108.2). Thus, in beryllium industry workers with CBD, 19/20 (95%) were found to carry at least one *HLA-DPB1^{E69}* variant (or putative disease allele) compared to 34/75 (45%) without disease. The one individual with CBD who did not carry at least one *HLA-DPB1^{E69}* variant was found to be homozygous K69/K69 (i.e., lysine homozygote, no putative disease alleles). Although the numbers were small, the data also suggested that individuals homozygous for *HLA-DPB1^{E69}* were at an increased risk of disease compared to heterozygous individuals. Among the 19 individuals with CBD who

Table 22.2 Summary of *HLA-DPB1* Genotyping Results

Allele/Marker	Cases	Controls	OR (95% CI)
D55, E56 (+)	26	18	
D55, E56 (−)	7	26	5.4 (1.7–17.6)
E69 (+)	32	12	
E69 (−)	1	32	85.3 (10.9–3,578.0)
*0201 (+)	10	4	
*0201 (−)	23	40	4.3 (1.1–20.8)
*0401 (+)	5	21	
*0401 (−)	28	23	0.2 (0.05–0.7)
Total (77)	33	44	

Source: From Richeldi et al., 1993 (1).

carried at least one *HLA-DPB1^{E69}* variant, 6 (~32%) were homozygous for *HLA-DPB1^{E69}* compared to only 1 of the 34 individuals without CBD (1.3%) (OR = 15.2; 95% CI = 15.2–721.0) (3).

Wang et al., 1999 (3) also evaluated the distribution of alleles in *HLA-DPB1^{E69}* individuals with and without CBD. These data suggested that variants in positions other than 69 also had a bearing on CBD risk (Fig. 22.1) (3). Alleles coding for amino acids valine, histidine or tyrosine, and leucine (V, H/Y, L) at positions 8, 9, and 11 were found to occur more often in individuals with CBD than in those without (79% vs. 29%; OR = 9.0, 95% CI = 2.6–31.6). Alleles coding for aspartic acid, glutamic acid, alanine, and valine (D, E, A, V) at positions 84–87 were also found to occur more frequently in individuals with CBD than alleles coding for glycine, glycine, proline, methionine (G, G, P, M) (84% vs. 35%; OR = 9.8; 95% CI, 2.6–36.6). Data for absolute characterization of genotypes with detail pertaining to these positions are not presented and cannot be deduced from the literature reports (1–3). However, analysis of the allele-specific data suggests a hierarchy with respect to the risk associated with specific alleles coding for E69. The lowest odds appears to be associated with *HLA-DPB1*0201/2* (OR ~15; 95% CI ~3–85). The highest odds were associated with non-*HLA-DPB1*0201/2* alleles (e.g., *HLA-DPB1*1901* < *HLA-DPB1*1301* < *HLA-DPB1*0901* = *HLA-DPB1*1001* < *HLA-DPB1*0601* < *HLA-DPB1*1701*(OR ~246; 95% CI; ~38–1594)). In the case of individual alleles, a small sample size results in extremely large overlapping confidence intervals. A consensus has not yet been reached in the literature concerning the relative risk potency of the *HLA-DPB1*0201/2* versus the rarer non-**0201/2* *HLA-DPB1^{E69}* alleles. The main differences between these alleles lie in codons 9 and 84–87. This observation forms the basis of a potentially important hypothesis that will be difficult to address because of small numbers and the problem of multiple comparisons.

Based on the *HLA-DPB1* allele specific information, Wang et al. evaluated the distribution of *HLA-DPA1* alleles in *HLA-DPB1*0201* and non-**0201/2* *HLA-DPB1^{E69}* carriers. Among the *HLA-DPB1*0201* individuals 29 of 30 (7/8 CBD vs 22/22 controls) were found to have at least one *HLA-DPA1*0103* allele (3). In contrast, *HLA-DPA1*0201* occurred more often in the non-**0201/2 HLA-DPB1^{E69}* carriers (14/16 CBD vs 12/13 controls). Although these results are preliminary, they indicate that *HLA-DPA1* may also play a role in CBD risk and warrant further study.

In 2001, Wang et al. extended their earlier work to include 25 beryllium-sensitized individuals (BeLPT positive without CBD) (4). They then evaluated the frequency of *HLA-DPB1^{E69}* in individuals with CBD, beryllium sensitization, and beryllium-exposed individuals without either beryllium sensitization or CBD. Interestingly, the frequency of both the high-risk *HLA-DPB1^{E69}* and non-**0201/2* *HLA-DPB1^{E69}* alleles among the sensitized, consistently fell between the frequency observed in the CBD cases and controls. For example, 24% of the beryllium-sensitized individuals were homozygous *HLA-DPB1^{E69}* compared to 30% of the individuals with CBD and 3% of the controls ($p < 0.001$). Similarly, sensitized individuals were more likely to have at least one non-*0201/2 HLA-DPB1^{E69} allele compared to the controls (52% vs. 13%; $p < 0.001$), but this occurred less often than in those with CBD (52% vs. 80%; not significant). Of the non-*0201/2 HLA-DPB1^{E69} alleles examined, HLA-DPB1*1701 occurred most often. Among the beryllium sensitized and individuals with CBD, 16% and 30%, respectively, were HLA-DPB1*1701 positive compared to only 2% of the control group ($p < 0.01$).

These results are of interest for two reasons. First, they extended the work by Richeldi et al (1,2). Second, it was the first study to evaluate the role of HLA-DPB1*0201 and non-*0201/2 HLA-DPB1^{E69} alleles in individuals who were beryllium sensitized. However, due to the small sample size and potential problems with the composition of the CBD cases, sensitized, and control groups, these results must be viewed with caution. For example, 10 of the sensitized did not have signs of respiratory impairment, but none were clinically evaluated for granulomatous lung disease, and 2 of the sensitized individuals in the most recent study were not known to have been occupationally exposed to beryllium. Any of these factors might affect the observed rates.

Saltini and collegues recently conducted a study that analyzed the presence and absence of specific HLA-DPB1 alleles in 22 individuals with CBD, 23 individuals with beryllium sensitivity (BeLPT positive without CBD), and 93 control samples (5). They reported an association between HLA-DPB1^{E69} and disease (OR = 3.7; 95% CI, 1.4–10.0), but not with sensitization (OR = 0.9; 95% CI, 0.3–2.2). They also stated that an increased frequency of *0201 HLA-DPB1^{E69} alleles was associated with CBD; however, this did not appear to be significant (5).

In 2002 Rossman et al. published information on the genetics of beryllium sensitization and CBD. The study population consisted of 137 individuals who had been referred to a tertiary referral hospital for clinical evaluation of CBD. Fifty-five of the participants had a positive BeLPT and were designated as having beryllium hypersensitivity (BH). On clinical exam 25 of the 55 were determined to have CBD and 30 of the 55 were defined as having beryllium hypersensitivity without clinical disease (BHWCD). The control group consisted of 82 beryllium-exposed individuals who had been evaluated for CBD at the hospital. None had BH or positive BeLPT results, though 10 had abnormal chest radiographs. HLA-DPB1 genotyping was conducted on all the samples, and the frequency of alleles compared across the groups with and without disease (6). HLA-DQB1 and HLA-DRB1 were evaluated, but not in conjunction with HLA-DPB1, and so will not be discussed here.

HLA-DPB1^{E69} occurred more often in BHWCD (90%) and individuals with CBD (84%) than those without disease (48%). The highest odds of disease was associated with BHWCD (OR = 9.9; 95% CI, 2.8–35.3). When the frequency of HLA-DPB1^{E69} among individuals with BHWCD was compared to the frequency among individuals with CBD, there was no significant difference. When specific HLA-DPB1^{E69} alleles were evaluated, HLA-DPB1*0601 and HLA-DPB1*1301 occurred more often in BHWCD individuals than controls ($p < 0.05$). These, however, did not remain significant after they were corrected for the number of alleles evaluated.

The presence of lysine at position 11 ($HLA\text{-}DPB1^{L11}$) and aspartic acid at position 55 ($HLA\text{-}DPB1^{D55}$) were significantly associated with BH. $HLA\text{-}DPB1^{L11}$ was also associated with BHWCD. However, Rossman et al. also note that this association only remained significant in the presence of $HLA\text{-}DPB1^{E69}$. Further, there was no difference between the frequency of either the $HLA\text{-}DPB1^{-E69\text{-}L11}$ or the $HLA\text{-}DPB1^{-E69\text{-}D55}$ in the individuals with CBD or BHWCD (6). Rossman et al. concluded that $HLA\text{-}DPB1^{E69}$ was the most important epitope in the development of BH, but could not be used to predict whether someone would develop CBD.

While all of the studies conducted agree that $HLA\text{-}DPB1^{E69}$ is associated with CBD, they differ in the relative importance placed on the role of the *HLA-DPB1*0201* alloforms in CBD (1,3–6). Further, it is of interest that while Wang et al., 2001 (4) and Rossman et al. (6) showed a relationship between $HLA\text{-}DPB1^{E69}$ and beryllium sensitization, Saltini et al., 2001 did not report this relationship (5). This discrepancy might be the result of the different methods used to determine HLA-haplotypes, or differences in the populations under study. Future studies will formally address the differences observed across these studies, and studies with larger sample sizes might help to elucidate specific high-risk *HLA-DP* alleles or other high-risk genes.

Population Frequencies of *HLA-DPB1*

A numbers of studies have examined the frequency of $HLA\text{-}DPB1^{E69}$ in various racial/ethnic groups (35–54). The data in the literature are variable in the amount of haplotype detail given, however, typing for 36 populations that discriminated between *0201/2 $HLA\text{-}DPB1^{E69}$ alleles and non-*0201/2 $HLA\text{-}DPB1^{E69}$ alleles is summarized in Figure 22.2. These data indicate that some populations have a complement of E69 alleles that form the majority of *HLA-DPB1* haplotypes, while others have very low frequencies (in some populations undetectable). Moreover in some populations the *0201* family is more frequent than the *non*0201*, while in others the *non*0201* alleles are more common (Fig. 22.2). These data are consistent with recently published finding based on an RFLP test, where the E69 carrier frequencies among Caucasians, African-Americans, Hispanics, and Chinese fell in the range 0.33–0.59 (55).

Interactions

Richeldi et al. evaluated the risk of CBD in the presence of both a high-risk job and $HLA\text{-}DPB1^{E69}$ (2). As beryllium machining had been found to confer the highest job-related risk, machining and inheritance of the $HLA\text{-}DPB1^{E69}$ were evaluated. Using logistic regression, the odds of CBD associated with $HLA\text{-}DPB1^{E69}$, independent of a machining job history was estimated to be 11.8 (95% CI, 1.3–108.8).

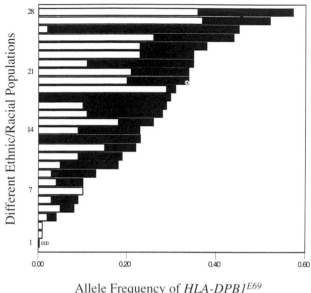

Allele Frequency of *HLA-DPB1*^{E69}

Figure 22.2 Charted frequencies of HLA-DPB1^{E69} alleles by racial/ethnic groups. Open bars represent population frequency of the HLA-DPB1*0200 allele family. Solid bars represent population frequency of the non-*02 family of E69-alleles. Racial/ethnic populations [and their HLA-DPB1*02/non*02 allele frequencies]: (1) Amerindian (Asario, Ijka, Ingano, Kogui, Nukak, Sikuani, Tule, Vaupes, Wayuu) [0/0], (2) Asian/Oceanic (Trobriand, 1) [0.01/0], (3) Amerindian (Coreguaje) [0.01/0], (4) Asian/Oceanic (W. Samoa) [0.02/0.02], (5) Taiwanese [0.05/0.03], (6) African-American (Choco) [0.03/0.06], (7) Asian/Oceanic (New Guinea) [0.10/0], (8) Amerindian (Cayapa) [0.04/0.06], (9) African-American (Cauca) [0.03/0.10], (10) Africa (Liberia) [0.05/0.13], (11) Australian (Caucasian) [0.09/0.10], (12) Caucasian (Slovak) [0.15/0.07], (13) Amerindian (Waunana) [0/0.23], (14) Africa (Nigeria) [0.09/0.14], (15) Africa (Gabonese) [0.18/0.08], (16) African-American (NYC) [0.11/0.17], (17) African American (Columbia Providencia) [0.10/0.19], (18) Amerindian (Embera) [0/0.30], (19) Asian/Oceanic (New Guinea) [0.29/0.02], (20) Australian (Aboriginal-North) [0.20/0.14], (21) Chinese (North) [0.21/0.13], (22) Asian/Oceanic (Java) [0.11/0.24], (23) Caucasian (Czech) [0.23/0.12], (24) Caucasian (Poland) [0.23/0.15], (25) Chinese (South) [0.26/0.18], (26) Asian/Oceanic (Borneo) [0.02/0.43], (27) African (Cameroon/Bantu) [0.37/0.15], (28) Australian (Aboriginal-Central) [0.36/0.21] (Ref. No. 35-54).

The odds of disease associated with machining alone was 10.1 (95% CI, 1.1–93.7) (Table 22.3). On the basis of these results, they reported that genetic and job factors had at least an additive effect for risk of beryllium disease in the industrial environment. However, because of small numbers they were unable to statistically verify this in the regression model. We have included an additional summary of disease prevalence by *HLA-DPB1*^{E69} and machining job history for this study (Table 22.4). While it was not possible to estimate odds ratios referenced to the lowest-risk group because of zero observed cases, it is clear from looking at the prevalence estimates

Table 22.3 Odds of Chronic Beryllium Disease Associated with Machining Beryllium and Presence or Absence of *HLA-DPB1*^{E69}

Potential Risk Factor	Cases	Controls	OR (95% CI)
Machining (+)	5	42	
Machining (−)	1	79	10.1 (1.1–93.7)
E69 (+)	5	36	
E69 (−)	1	85	11.8 (1.3–108.8)

Source: Reference 2.

and confidence intervals that the presence of both *HLA-DPB1*^{E69} and machining job history account for a remarkable proportion of cases.

Saltini et al. (5) utilized a series of 2 × 2 tables extracted from a larger 2 × 4 table to evaluate the risk of beryllium sensitization or CBD in the presence of either one, or a combination of the genes: *HLA-DPB1*^{E69}; tumor necrosis factor *(TNF)*-α-308*2; and *HLA–DR*^{R74}.[2] *HLA-DR*^{R74} was found to be independently associated with sensitization (OR = 4.0, 95% CI, 1.5–10.1), but not with CBD (OR = 0.9; 95% CI, 0.3–2.6) (5). An association was also reported between BeLPT-positive individuals (sensitized and CBD) and *TNF-α-308*2* (OR = 7.8; 95% CI, 3.2–19.1).

Gene–gene analyses identified an association between sensitization and the presence of *TNF-α-308*2* and *HLA-DR*^{R74} (5). An association between the risk of sensitization in individuals who were *HLA-DPB1*^{E69} positive, *HLA-DR*^{R74} negative was also reported. However, scrutiny of their tabulated data shows that sensitization was associated with *HLA-DR*^{R74}-positive, *HLA-DPB1*^{E69}-negative individuals. Interestingly, neither *HLA-DPB1*^{E69} nor *HLA-DR*^{R74} alone, or in combination, were associated with CBD. This may be an effect of the construction of the 2 × 4 tables, resulting in small cell counts. *TNF-α-308*2* was independently associated with CBD, but in the presence of *HLA-DPB1*^{E69}, this risk was even greater. The risk associated with *TNF-α-308*2* alone was estimated to be 4.6. In individuals with both a *TNF-α-308*2* and *HLA-DPB1*^{E69} compared to those with neither, the odds of disease increased to 9.7. The extent to which these analyses may have been affected by the use of different laboratory methods to determine *TNF-α-308*2*, *HLA-DPB1*^{E69}, and *HLA-DR*^{R74} alleles is unknown. Although exposure to beryllium is a necessary component in the development of CBD, these results suggest that genes other than, or in conjunction with, *HLA-DPB1*^{E69} may play a role in the risk of both sensitization and disease.

Table 22.4 Disease Prevalence by *HLA-DPB1*^{E69}-Machining

E69/Machining	Cases (n = 6)*	Controls (n = 121)	Prevalence (95%CI)†
−/−	0	55	0.00 (0.00–0.06)
−/+	1	30	0.03 (0.00–0.17)
+/−	1	24	0.04 (0.00–0.20)
+/+	4	12	0.25 (0.07–0.52)

Source: Reference 2.

Availability of *HLA-DPB1* Gene Testing

One of the leading beryllium manufacturing facilities, in collaboration with a tertiary referral hospital, is offering prospective employees the opportunity to confidentially obtain *HLA-DPB1*^{E69} genotyping. Individuals applying for work at the plant are referred to the hospital, which is responsible both for conducting the genetic analysis and counseling the prospective employees about their risk of beryllium sensitization and disease. The plant does not receive individual results, rather they receive a summary report including the number of individuals who requested the test and the number of individuals who accepted employment at the plant. This report is only conveyed after enough individuals have participated to prevent the plant from being able to identify individual participants.

The decision by the beryllium industry to offer applicants *HLA-DPB1*^{E69} genotyping through an independent third party was reinforced by recommendations both by current workers and the Beryllium Industry Scientific Advisory Committee. Their goal is twofold; first, it is to enable applicants to better be able to assess their risks associated with working with beryllium. Secondly, it is to potentially lower the risk of CBD among new hires.

Besides this program, as part of ongoing molecular epidemiologic research studies, experimental laboratory methods are being conducted to characterize *HLA-DPB1* sequence motifs in individuals with CBD, those with beryllium sensitization, and individuals without CBD. Individuals who participate in these research studies can obtain their genetic information upon request.

Currently there are no commercial kits available, nor are there Clinical Laboratory Improvement Advisory Committee (CLIAC)-approved laboratories offering *HLA-DPB1*^{E69} genetic testing. Thus persons interested in obtaining their *HLA-DPB1*^{E69} genotype, but who are not seeking employment at the beryllium plant, nor participating in the epidemiologic research are required to seek out an independent laboratory capable of providing such a test. There are a number of potential problems associated with this. The cost of the test could be prohibitive and depending on the reliability of the laboratory, the results may or may not be accurate. Further, it is not evident that a test sought in this manner would include counseling. What the results mean and how this genetic information can be used to benefit, or negatively impact current, former, and prospective workers is of particular concern to beryllium researchers, the beryllium industry, and to the individuals who obtain their genetic information. The benefits and risks are discussed below.

Potential Benefits and Risks Associated with Genetic Information

HLA-DPB1^{E69} genetic information is valuable for better understanding the molecular mechanism of CBD and beryllium sensitization. It is a necessary component in the development of animal models and may lead to better treatments and modes of intervention to prevent disease for all workers. In addition, for some individuals, knowing their *HLA-DPB1^{E69}* status may be important in assessing whether or not they want to work in the beryllium industry.

An animal model for beryllium sensitization or CBD is not available. For example, when mice are exposed to beryllium they can become sensitized, but they do not develop granulomas like humans. A major difference between the mouse and human histocompatibility antigens is that the mouse does not have an *HLA-DPB1* homologue (56). Therefore, if the correct *HLA-DP* haplotypes could be introduced, it is possible that a mouse model could be developed that would become sensitized to beryllium and develop CBD. A transgenic mouse model susceptible to a disease condition similar to humans would be invaluable for studying, (*1*) the pathobiology of sensitization and CBD, (*2*) questions concerning dose and route of exposure, (*3*) the development of better diagnostic tools, and (*4*) the development of postexposure intervention strategies.

Currently, treatment options for chronic beryllium disease are limited to antiinflammatory and immunosuppressive agents (e.g., prednisone). Thus, treatment is generally palliative and can have serious adverse effects (i.e., electrolyte imbalance, diabetes, congestive heart failure, hypertension, and others) (57). In the absence of complete abrogation of exposure, research on the genetic underpinnings of sensitization and disease may lead to better diagnostic tests, effective postexposure interventions, and implementation of exposure limits that would protect all workers.

For individuals, knowing their *HLA-DPB1^{E69}* status is beneficial if they are interested in knowing more about their own risk. To help these individuals understand what their results might mean, and how the information may or may not benefit them, it is important that they receive extensive written and oral information describing the risks and benefits associated with receiving their genetic results. This includes information about potential insurance and employment discrimination, the predictive value for *HLA-DPB1^{E69}* and the odds of disease associated with *HLA-DPB1^{E69}*.

The most obvious risks prospective, current, and former beryllium workers face is the potential for insurance and employment discrimination. There are no recorded cases to date in which beryllium workers have been refused health insurance or employment based on their genotype status. However, current laws vary as to who is protected and the extent of protection, individuals who decide to obtain their genetic information must be made aware of this, and that insurance or employment discrimination is a potential risk.

Another issue prospective, current, and former beryllium workers must be made aware of is the predictive value associated with *HLA-DPB1^{E69}* and beryllium disease. The positive predictive value of a diagnostic test provides information about how well the presence of a positive diagnostic test outcome will detect the presence of disease. Typically the positive predictive value has been defined as the probability that an individual will have the disease given that the diagnostic test is positive. It is a function of test sensitivity, test specificity, and disease prevalence (58).

This definition of positive predictive value is cross-sectional in nature and does not involve information about disease incidence.

Positive predictive value can also be defined in terms of the probability that individuals will develop disease subsequent to screening (58,59). This definition of positive predictive value is based on test sensitivity, test specificity and disease incidence rather than disease prevalence. This modified definition of positive predictive value is interpreted as the probability that an individual will develop the disease subsequent to screening, given that they have a positive screening test.

While both definitions of positive predictive value have utility in disease screening and prevention, it is essential to preserve the distinction between the cross-sectional and longitudinal definitions when interpreting estimates of positive predictive value. For situations where both disease prevalence and incidence are available then it may be useful to calculate both the cross-sectional and longitudinal positive predictive value. Since disease incidence data are not always readily available, then the cross-sectional positive predictive value provides useful information about the probability that an individual has developed disease given a positive-test result.

Alternatively, negative predictive value is defined longitudinally as the probability that an individual will not develop the disease subsequent to screening given that a negative screening test has occurred. Cross-sectionally negative predictive value is defined as the probability that an individual does not have the disease of interest given a negative outcome on a given screening test.

With these definitions in mind we used the model developed by Khoury et al. (59) and cross-sectional data from Richeldi et al. (2) to estimate the positive and negative predictive value and the sensitivity and specificity of a genetic test based on determination of the supratypic marker $HLA\text{-}DPB1^{E69}$ (Fig. 22.3). This estimate spans a wide range of $HLA\text{-}DPB1^{E69}$ carrier frequencies, and assumes that risk is independent of other genetic risk factors that are not in disequilibrium with the $HLA\text{-}DPB1$ locus. It is based on a disease prevalence of 5% for beryllium industry workers and 15% for high-exposure tasks (machinists and lappers). Figure 22.3 shows that, regardless of the prevalence of disease (5% or 15%), as the frequency of $HLA\text{-}DPB1$ increases in a population, the positive predictive value decreases well below 50%. What this means, and what must be stressed to the workers is that although the odds of beryllium sensitization and disease is greater in individuals with $HLA\text{-}DPB1^{E69}$, the absence of $HLA\text{-}DPB1^{E69}$ does not protect them from beryllium sensitization and CBD (1–6).

Furthermore, while prospective employees might benefit from knowing their genotype (if it is confidentially provided), the utility of risk information for people already exposed to beryllium is less clear, since CBD risk continues even with exposure cessation. Currently, it is not known whether risk can be lowered by leaving the industry or whether genetic characterization of sensitized or CBD cases has prognostic implications. These issues must also be discussed with prospective, current, and former beryllium workers, as do the potential risks associated with insurance and employment discrimination.

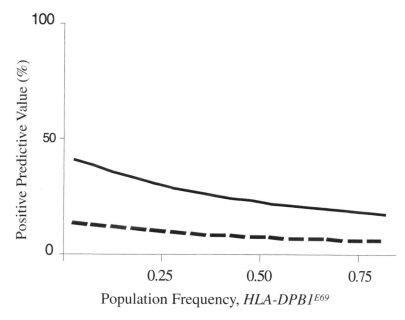

Figure 22.3 Curves showing positive predictive value of HLA-DPB1^{E69} testing for suscepti-bility to beryllium disease over a range of carrier frequencies for a workforce with a 5% prevalence of disease (dashed line), and a workforce at high exposure risk (15% disease prevalence; solid line).

Conclusions

Occupational exposure to beryllium presents a clear risk of adverse health outcome. Despite the implementation of the 2 μg/m^3 exposure limit and implementation of control technology (e.g., respirators), CBD and beryllium sensitization continue to be problematic (20, 23–28). In addition to beryllium exposure, a strong association between certain genetic factors and the risk of disease has also been identified (1–6). In light of the high odds of disease associated with specific genes, it would seem prudent to implement a genetic testing program that would potentially reduce the numbers of high-risk people from being exposed to beryllium. However, where this has proven effective for other diseases that have a strong genetic component, (e.g., hemochromotosis and *HFE* gene) (60), in the case of CBD, because of potential employment and insurance discrimination, the ethical, legal, and social implications must be stressed.

Here also, we focused on the scientific issue of the positive predictive value of such a genetic test and have found that *HLA-DPB1^{E69}* in the beryllium worker pop-ulation results in a low positive predictive value. However, because the association between *HLA-DPB1^{E69}* and CBD is unequivocal, if confidentially provided, prospective employees may find having their genetic information useful for decid-ing if they want to work in the beryllium industry. Despite the level of uncertainty, for some, knowing only that their risk was higher is sufficient, regardless of the predictive value. Similarly for current workers, though they have already been exposed, the realization that they are at an increased risk may also affect their decision about remaining in the industry.

Unfortunately the scientific evidence is not yet available to advise beryllium workers on a definite course of action based on their genotype. Rather, only the potential risks and benefits can be discussed. However, as integrated genetic and epidemiologic research studies continue, many of these issues may be resolved. Research will allow specific questions associated with the natural history of CBD to be addressed. The identification of other high-risk genes, gene-exposure, and gene-gene interactions will also improve personal risk assessment and help determine if specific genes or alleles are more valuable as prognostic indicators. Furthermore, these and similar studies will be able to clarify the pathology of CBD, opening the door to more effective treatment and interventions.

Notes

[1] Both the Greek letters α and β as well as the Roman letters A and B are used to describe the HLA chains, however, this usage is inconsistent in the literature. For consistency we use A and B throughout this text in reference to the alpha and beta chains.
[2] There are more than 220 *HLA-DRB* alleles that are classified by 9 sub-groupings (*B1–B9*) of which *HLA-DRB1* is the largest (more than 180 alleles). An arginine residue is found at position 74 of the mature protein, which is encoded by 13 alleles designated *HLA-DRB1* (e.g., *HLA-DRB1*03011*) and 6 alleles designated *HLA-DRB3* (e.g., *HLA-DRB3*0302*) (39). We have adopted the nomenclature used by Saltini et al. (5) in their original report.

References

1. Richeldi L, Sorrentino R, Saltini C. HLA-DPB1 glutamate 69: a genetic marker of beryllium disease. Science 1993;262:242–244.
2. Richeldi L, Kreiss K, Mroz M, et al. Interaction of genetic and exposure factors in the prevalence of berylliosis. Am J Ind Med 1997;32:337–340.
3. Wang Z, White PS, Petrovic M, et al. Differential susceptibilities to chronic beryllium disease contributed by different Glu69 HLA-DPB1 and -DPA1 alleles. J Immunol 1999; 163:1647–1653.
4. Wang Z, Farris GM, Newman LS, et al. Beryllium sensitivity is linked to *HLA-DP* genotype. Toxicology 2001;165:27–38.
5. Saltini C, Richeldi L, Losi M, et al. Major histocompatibility locus genetic markers of beryllium sensitization and disease. Eur Respir J 2001;18:677–684.
6. Rossman M, Stubbs J, Lee CW, et al. Human leukocyte antigen class II amino acid epitopes. Susceptibility and progression markers for beryllium hypersensitivity. Am J Respir Crit Care Med 2002;165:788–794.
7. Cunningham L. Beryllium. In: US Geological Survey Minerals yearbook–2000, pp. 12.1–12.5.
8. Cullen MR, Kominsky JR, Rossman MD, et al. Chronic beryllium disease in a precious metal refinery. Am Rev Resp Dis 1987;135:201–208.

9. Van Ordstrand HS. Chemical pneumonia in worker extracting beryllium oxide. The Cleveland Clinic Quarterly 1943;10.

10. Van Ordstrand HS Hughes R, De Nardi JM. Beryllium poisoning. JAMA 1945;129:1084.

11. Hardy HL. Beryllium poisoning: lessons in control of man-made disease. N Engl J Med 1965; 273:1188–1199.

12. Hardy HL, Tabershaw IR. Delayed chemical pnuemonitis occurring in workers exposed to beryllium compounds. J Ind Hyg Toxicol 1946;28:197–211.

13. Eisenbud M. Origin of the standards for control of beryllium disease (1947–1949). Environ Res 1982;27:79–88.

14. Tinkle SS, Antonini JM, Abrigo BA, et al. 2000. Particle penetration of the skin as a route of sensitization in occupational lung disease (abstract). Society of Toxicology Meetings 54:149.

15. Lademann J, Weigmann H, Rickmeyer C. Penetration of titanium dioxide microparticles in a sunscreen formulation into the horny layer and the follicular orifice. Skin Pharmacol Appl Skin Physiol 1999;12:247–256.

16. Newman LS. Lloyd J. Daniloff E. The natural history of beryllium sensitization and chronic beryllium disease. Environl Health Perspec 1996;104S:937–943.

17. Curtis GH. The diagnosis of beryllium disease with special reference to the patch test. Arch Ind. Health 1959;19:150–153.

18. Hanifin JM. Epstein WL. Cline MJ. In vitro studies on granulomatous hypersensitivity to beryllium. Journal of Invest Dermatol 1970;55:284–288.

19. Deodhar SD, Barna B, Van Ordstrand HS. A study of the immunologic aspects of chronic berylliosis. Chest 1973;63:309–313.

20. Kreiss K, Newman LS, Mroz MM, et al. Screening blood test identifies subclinical beryllium disease. J Occup Med 1989;31:603–608.

21. Williams WR, Williams WJ. Development of beryllium lymphocyte transformation tests in chronic beryllium disease. Int Arch Allergy Appl Immunol 1982;67:175–180.

22. Rossman MD, Kern JA, Elias JA. Proliferative response of bronchioalveolar lymphocytes to beryllium: a test for chronic beryllium disease. Ann Intern Med 1988;108:687–693.

23. Kreiss K, Wasserman S, Mroz MM, et al. Beryllium disease screening in the ceramics industry. Blood lymphocyte test performance and exposure-disease relations. J Occup Med 1993;35:267–274.

24. Kreiss K, Mroz MM, Zhen B, et al. Epidemiology of beryllium sensitization and disease in nuclear workers. Am Rev Respir Dis 1993;148:985–991.

25. Kreiss K, Mroz M, Newman LS, et al. Machining risk of beryllium disease and sensitization with median exposures below 2 micrograms/m^3. Am J Ind Med 1996;30:16–25.

26. Stange AW, Furman FJ, Hilmas DE. Rocky Flats Beryllium Health Surveillance. Environ Health Perspect 1996;104S:981–986.

27. Kreiss K, Mroz MM, Zhen B, et al. Risks of beryllium disease related to work processes at a metal, alloy, and oxide production plant. Occup Environ Med 1997;54:605–612.

28. Henneberger PK, Cumro D, Deubner D, et al. Beryllium sensitization and disease among long-term and short-term workers in a beryllium ceramics plant. Int Arch Occup Environ Health 2001;74:167–176.

29. Deubner DC, Goodman M, Iannuzzi J. Variability, predictive value, and uses of the beryllium blood lymphocyte proliferation test (BLPT): preliminary analysis of the ongoing workforce survey. Appl Occ Env Hyg 2001;16:521–526.

30. Eisenbud M and Lisson J. Epidemiological aspects of beryllium induced nonmalignant lung disease: A 30 year update. J Occ Med 1983;25:196–202.

31. Kent MS, Robins TG, Madl AK. Is total mass or mass of alveolar-deposited airborne particles of beryllium a better predictor of the prevalence of disease? A preliminary study of a beryllium processing facility. Appl Occ and Envir Hyg 2001;16:539–558.

32. Saltini C. Weinstock K. Kirby M, et al. Maintenance of alveolitis in patients with chronic beryllium disease by beryllium specific helper T cells. N Engl J Med 1989;320:1103–1109.

33. International Human Genome Sequencing Consortium. Initial sequencing and analysis of the human genome. Nature 2001;409:860–921.

34. www.ebi.ac.uk/imgt/hla/allele

35. al-Daccak R, Wang FQ, Theophille D, et al. Gene polymorphism of HLA-DPB1 and DPA1 loci in caucasoid population: frequencies and DPB1-DPA1 associations. Hum Immunol 1991;31:277–285.

36. Cariappa A, Sands B, Forcione D, et al. Analysis of MHC class II DP, DQ and DR alleles in Crohn's disease. Gut 1998;43:210–215.

37. Loudova M, Sramkova I, Cukrova V, et al. Frequencies of HLA-DRB1, -DQB1 and -DPB1 alleles in Czech population. Folia Biologica 1999;45:27–30.

38. Lui Z, Lin J, Chen W, et al. Sequence of complete exon 2 and partial intron 2 of HLA-DPB1*8001 allele. GenBank accession # AF336231. 2001.

39. MacHulla HK, Schonermarck U, Schaaf A, et al. HLA-A, B, C and DRB1, DRB3/4/5, DQB1, DPB1 frequencies in German immunoglobulin A-deficient individuals. Scand J Immun 2000;52:207–211.

40. Magzoub MM, Stephens HA, Sachs JA, et al. HLA-DP polymorphism in Sudanese controls and patients with insulin-dependent diabetes mellitus. Tissue Antigens 1992; 40: 64–68.

41. Marsh SG. 2000. Nomenclature for factors of the HLA system, update September 2000, WHO Nomenclature Committee for Factors of the HLA System. Tissue Antigens 2000; 56:565–566.

42. Marsh SG. 1998. Nomenclature for factors of the HLA system, update March 1998. WHO Nomenclature Committee for Factors of the HLA System. Tissue Antigens 1998; 51: 681–683.

43. May J, Mockenhaupt FP, Loliger CC, et al. HLA DPA1/DPB1 genotype and haplotype frequencies, and linkage disequilibria in Nigeria, Liberia, and Gabon. Tissue Antigens 1998;52:199–207.

44. McTernan CL, Mijovic CH, Cockram CS, et al. The nucleotide sequence of two new DP alleles, DPA1*02015 and DPB1*8401, identified in a Chinese subject. GenBank accession # AF077015, 2001.

45. Moonsamy PV, Suraj VC, Bugawan TL, et al. Genetic diversity within the HLA class II region: ten new DPB1 alleles and their population distribution. Tissue Antigens 1992;40: 153–157.

46. Nishimaki K, Kawamura T, Inada H, et al. HLA DPB1*0201 gene confers disease susceptibility in Japanese with childhood onset type I diabetes, independent of HLA-DR and DQ genotypes. Diabetes Res Clin Pract 2000;47:49–55.

47. Poulton KV, Kennedy LJ, Ross J, et al. A study of HLA-DPB1 phenotypes reveals DPB1*6301 in a rural population from Cameroon. Eur J Immunogenet 1998;25:375–377.

48. Rani R, Sood A, Lazaro AM, et al. Associations of MHC class II alleles with insulin-dependent diabetes mellitus (IDDM) in patients from North India. Hum Immunol 1999; 60:524–531.

49. Ravikumar M, Dheenadhayalan V, Rajaram K, et al. Associations of HLA-DRB1, DQB1 and DPB1 alleles with pulmonary tuberculosis in south India. Tuber Lung Dis 1999; 79:309–317.

50. Rozmuller EH, van der Zwan AW, Voorter CE, et al. DPB1*8501, a novel variant in the US Black population. Tissue Antigens 2000;56:282–284.

51. Steiner LL, Wu J, Noreen HJ, et al. Four new alleles identified in a study of 500 unrelated bone marrow donor-recipient pairs. Tissue Antigens 1999;53:201–206.

52. Varney MD, Tait BD. Identification of a novel DPB1* allele (DPB1*6901) and the occurrence of HLA haplotypes that extend to DPB1. Tissue Antigens 2000;55:188–190.
53. Voorter CE, Tilanus MG, van den Berg-Loonen EM. Two new HLA DPB1 alleles identified by sequence-based typing: DPB1*8201 and DPB1*8301. Tissue Antigens 2000; 56:560–562.
54. Wang FQ, al-Daccak R, Ju LY, et al. HLA-DP distribution in Shanghai Chinese—a study by polymerase chain reaction—restriction fragment length polymorphism. Hum Immunol 1992;33:129–132.
55. Weston A, Ensey J, Kreiss K, et al. Racial differences in prevalence of a supratypic HLA-genetic marker, immaterial to pre-employment testing for chronic beryllium disease. Am J Ind Med 2002; 41:457–465.
56. Mouse Genome Informatics. The Jackson Laboratory. Http://www.informatics.jax.org/searches/homology form.shtml
57. Medical Economics Company. 1998. Physicians Desk Reference, 52nd Edition, Medical Economics Company Inc. pp. 1797–1799.
58. Van Damme K, Casteleyn L, Heseltine E, et al. Individuals susceptibility and prevention of occupational diseases: Scientific and ethical issues. JOEM 1995; 37:91–99.
59. Khoury MJ, Newill CA, Chase GA. Epidemiologic evaluation of screening for risk factors: application to genetic screening. Am J Pub Health 1985; 75:1204–1208.
60. Hanson EH, Imperatore G, Burke W. HFE gene and hereditary hemochromatosis: A HuGE review. Am J Epi 2001;154:193–206.

23

Fragile X syndrome: from gene identification to
clinical diagnosis and population screening

Dana C. Crawford and Stephanie L. Sherman

Clinical Overview

The fragile X syndrome, an X-linked dominant disorder with reduced penetrance, is one of the most common forms of inherited mental retardation. The disorder-causing mutation is the amplification of a cytosine-guanine-guanine (CGG) repeat in the 5′ untranslated region of *FMR1* located at Xq27.3 (1). The fragile X mutation in affected persons results in the loss of the *FMR1* gene product fragile X mental retardation protein (FMRP), an RNA-binding protein (2). Although its precise function is not yet understood, FMRP is thought to regulate the translation of other proteins (3). Thus, the loss of FMRP and possibly the loss of regulation of other as-yet-unidentified proteins result in the clinical phenotype of the fragile X syndrome.

   The clinical phenotype associated with the fragile X syndrome is wide and includes a variety of cognitive, physical, and behavioral characteristics (4). With regard to cognitive function, affected males often exhibit developmental delay early in childhood. By the age of 3 years, most males will test in the mentally retarded range (5). Ultimately, mental retardation is diagnosed in almost all males with the fragile X syndrome, with severity ranging from profound (IQ <20) to mild mental retardation (IQ 50–70), with most being moderately retarded (IQ 40–54) (6). Physically, adult males often have a long narrow face, prominent ears, a prominent jaw, and increased testicular volume (6). Other common physical features include a high arched palate, hyper-extensible finger joints, double-jointed thumbs, single palmar crease, hand calluses, velvet-like skin, flat feet, and mitral valve prolapse (7). Males with the fragile X syndrome also tend to exhibit behavioral features such as hyperactivity, social anxiety, tactile defensiveness, stereotypies (e.g., hand-flapping), and hand biting (6). Language delay and perseverative speech are also commonly observed among males with the fragile X syndrome. Compared with other children who have developmental delay without the fragile X syndrome, males with the fragile X syndrome exhibit some autistic-like features, such as social avoidance, at a very young age (8). As many as 25% of males with the fragile X syndrome meet the diagnostic criteria for autism (9). The association of fragile X with autism, however, is not clear because the proportion of males with the fragile X syndrome meeting the diagnostic criteria for autism seems to diminish with age (9).

Females with the full mutation are often less affected than males, presumably because of X-inactivation (10,11). Approximately 30% to 50% of females with the full mutation have an IQ of <70, and 50% to 70% of females with the full mutation have an IQ of <85 (12). With regard to the specific cognitive and neuropsychological profile, deficits in specific IQ subtests (such as the arithmetic score), executive function, and visual information processing also have been noted for females with the full mutation (4,13–15). The pattern of autistic-like behavior among males also occurs among a proportion of females with the full mutation (16). Interestingly, autistic-like behavior is not correlated with IQ in females with the full mutation, which suggests that the degree of mental retardation is not the underlying cause of this behavior (16).

Epidemiology of the Fragile X Syndrome

The fragile X syndrome is found in all world populations tested. Its prevalence among populations of northern European descent is approximately 1 in 6000 to 1 in 4000 males in the general population (17–19). A more detailed description of the prevalence of the fragile X syndrome within a general population is given below (Epidemiologic findings). The prevalence of the fragile X syndrome among phenotypically defined populations is difficult to describe, as studies published in the literature employed different criteria for ascertaining persons for testing. For example, von Koskull et al. (20) examined clinically referred persons with mental retardation, while Mazzocco et al. (21) examined preschool children referred for language delay. Although both of these studies identified persons with the fragile X syndrome, the fraction of males identified with fragile X differed between these two phenotypically defined populations (4.9% vs. 0.3%, respectively). Despite these difficulties in comparing across studies, the prevalence of the fragile X syndrome is higher for populations with developmental delay or mental retardation than in the general population (17).

Genetics of the Fragile X Syndrome

The genetics of the fragile X syndrome are unique in that they represent a new class of disorder-causing mutations known as trinucleotide or dynamic repeats. One characteristic of these disorders is anticipation, which is the increased disease occurrence or increased severity of the disease observed in pedigrees of affected families in succeeding generations (22). The understanding of this phenomenon came with the cloning of the fragile X mental retardation (*FMR*)-1 gene in 1991. A repeated sequence of CGG was found within the 5' UTR of the gene. These repeats were unstable when transmitted from parent to offspring. The polymorphic fragile X CGG repeat can now be categorized in at least four forms on the basis of the size of the repeat: full mutation (>200–230 repeats), premutation (61–200 repeats), intermediate (41–60 repeats), and common (6–40 repeats). Among the general population, the common repeats are usually transmitted from parent to offspring in a stable manner. In contrast, premutation alleles are unstable when transmitted from parent to offspring and usually expand in the next generation. The size of the repeat expansion is positively correlated with maternal CGG repeat size, with >90 repeats almost always expanding to the full mutation in the next generation (23). Intermediate alleles are larger repeats that may or may not be transmitted stably from parent to offspring. Thus, these alleles overlap the boundary between common and premutation alleles (24). The full mutation allele is the form associated with the fragile X syndrome phenotype. All full mutations identified have been derived from premutation or full mutation alleles from the previous generation. In contrast to female transmission, paternal transmission of the full mutation to the offspring is rare. The end result of the full mutation allele is the hypermethylation (25) and deacetylation (26) of the promoter region of *FMR1*, which effectively prevents transcription of the gene (27).

In summary, the fragile X syndrome is transmitted as an X-linked dominant mutation with reduced penetrance. The reduced penetrance is related to the form of the mutation; that is, only full mutation alleles lead to the syndrome. The increased occurrence of the syndrome in succeeding generations of an identified family, anticipation, results from the instability of the repeat mutation and the bias for expansion. There is a parent-of-origin effect in that only premutation alleles carried by females are at risk for expansion to the full mutation in the next generation. Premutation alleles carried by males are relatively stable.

Epidemiologic Findings

Prevalence: Full Mutation
Both the full mutation and premutation genotypes are directly related to the expression of the fragile X syndrome phenotype (Tables 23.1 and 23.2). For the full mutation, all general population prevalence estimates have tested target populations with some type of cognitive impairment and extrapolated these findings to the general population. The general assumption among these studies is that males affected with the fragile X syndrome will be found only among the targeted population being tested (e.g., special education). According to the results of these studies, point estimates for the fragile X syndrome range from 1 in 3717 (28) to 1 in 8918 (29) males in the general white population (Table 23.1). The confidence limits, available for only four of these studies, vary widely, with a lower boundary of 1 in 1333 (30) to an upper boundary of 1 in 8922 (18).

Table 23.1 Population-Based Prevalence of the Fragile X Syndrome among Males

			Estimated Prevalence	
Country (reference)	Target Population	No. Positive/ No. Tested	Target Population (%)	General Population (95%CI)
U.K. (Wessex) (18,35)	SpEd population (aged 5–18 years), unknown etiology	20/3738	SpEd: 0.5	1/5530 (1/8992–1/4007)
U.S. (Atlanta, Georgia) (28,34)	SpEd population (aged 7–10 years), regardless of etiology	White: 4/1,572	White SpEd: 0.3	White: 1/3717 (1/7692–1/1869)
		African-American: 3/753	African-American SpEd: 0.4	African-American: 1/2545 (1/5208–1/1293)
		Total: 7/2471	Total SpEd: 0.3	Total: 1/3623 (1/6024–1/2212)
Southwest Netherlands (19)	Schools and institutes for MR, unknown etiology	9/866	Mild MR: 2.0 Moderate/severe MR: 2.4	1/6045 (1/9981–1/3851)
Hellenic population of Greece and Cyprus (30)	Referred clinical population of idiopathic MR	8/611	MR: 1.3	1/4246 (1/16440–1/1333)
Australia (Sydney) (95,96)	Children with MR in SpEd	10/472	MR: 2.1 Mild MR: 0.6 Moderate/severe MR: 5.4	1/4350[a]
Southern Häme, Finland (97)	Adult males (>16 years) registered in the Southern Häme Care Organization with MR, unknown etiology	6/344	MR: 1.7	1/4400[b]
U.K. (Coventry) (96,98,99)	Children with MR in institutions or SpEd	6/219	MR: 2.7 Mild MR: 1.3 Moderate/severe MR: 6.7	1/4090[a]
Poland (Warsaw) (100)	Males in institutions or SpEd with MR	6/201	MR: 3.0	1/2857–1/5882[c]
U.K. (Wessex) (29)	SpEd population (aged 5–18 years), unknown etiology	4/180	SpEd: 2.2	1/8918[d]

(continued)

Table 23.1 Population-Based Prevalence of the Fragile X Syndrome among Males (*continued*)

Country (reference)	Target Population	No. Positive/ No. Tested	Target Population (%)	General Population (95%CI)[e]
			ESTIMATED PREVALENCE	
Spain (101)	Children in SpEd or clinically referred with MR of unknown etiology; no known family history of MR	5/180	MR: 2.7	1/6200–1/8200[e]
Guadeloupe, French West Indies (31)	SpEd population, unknown etiology	11/163	SpEd: 6.7	1/2359 (1/4484–1/276)
U.K. (Oxfordshire) (102)	Children in schools for moderate to severe learning difficulties, unknown etiology	4/103	MR: 3.9	1/4130[f]

Special education or special schools (SpEd), mental retardation (MR).

[a]Turner et al. (96) provided only a point estimate.
[b]Arvio et al. (97) provided only a range on the basis of past cytogenetic and DNA-based diagnoses.
[c]Mazurczak et al. (100) provided only a range, not a point estimate.
[d]Jacobs et al. (29) provided only a point estimate.
[e]Millan et al. (101) provided a range, not a point estimate. Millan et al. (101) also acknowledged that persons with mild MR might have been missed, so the range could be as high as 1/5000–1/6800.
[f]Slaney et al. (102) only provided a lower boundary, not a point estimate.

Table 23.2 Prevalence of Premutation among Females in the General Population

Country (reference)	Target Population	Number Tested	Prevalence (95% CI)[a] (61–200 repeats)	Prevalence (95% CI)[a] (55–200 repeats)
Israel (33,89)	Women of reproductive age with no family history of fragile X or MR	14,334	1/231 (1/299–1/182)	1/116 (1/138–1/97)
Canada (Quebec) (38)	Unselected female blood donors	10,624	1/379 (1/560–1/267)	1/259 (1/373–1/198)
Israel (37)	Women of reproductive age or pregnant women	9660	—	1/159 (1/205–1/124)
Israel (32)	Women of reproductive age with no history of fragile X or MR	8426	1/468 (1/766–1/303)	1/145 (1/189–1/113)
Finland (87)	Pregnant women with no known history of fragile X	1477	1/246 (1/605–1/119)	—
United States (Fairfax, Virginia) (85)	Screening egg donors or pregnant women with no history of MR or LD	745	1/248[b] (1/961–1/93)	1/149[c] (1/404–1/68)
United States (Atlanta, Georgia) (28,34)	SpEd population (ages 7–10 years)	White: 670 African-American: 321	White: 1/335 (1/1934–1/84) African-American: 0/321	

MR, mental retardation; LD, learning disability; SpEd, special education.

[a]Because the frequency of premutations is close to zero, 95% confidence intervals were calculated using the equations recommended by Fleiss (103).

[b]Spence et al. (85) reported one premutation as 60 ± 3 repeats.

[c]Spence et al. (85) reported one premutation at 52 ± 3 repeats and one as 55 ± 3 repeats.

In contrast to the populations of northern European descent, few estimates exist for other racial/ethnic groups. There have been only two other population-based estimates, both for African-derived populations. Elbaz et al. (31) examined an Afro-Caribbean population in the French West Indies, and Crawford et al. (28) examined an African-American population in metropolitan Atlanta, Georgia, USA. Surprisingly, both studies suggested that the point estimate in these admixed, African-derived populations is approximately 1 in 2500 males, which is higher than that observed in white populations (Table 23.1). Further studies are needed to explore this possible higher general population prevalence as the confidence intervals for both studies overlap with estimates in populations of northern European descent.

Even less is known about the prevalence of the full mutation among females in the general population. On the basis of the prevalence of the full mutation among white males in the general population (approximately 1 in 4000) and the transmission only by females of the full mutation to their offspring, the expected prevalence of the full mutation allele among females is 1 in 4000. Assuming that 50% of females with the full mutation will be affected with the fragile X syndrome phenotype, the expected prevalence of females affected by the fragile X syndrome is approximately 1 in 8000 females. Among 8462 women in Israel who had no family history of mental retardation, Pesso et al. (32) identified one woman with the full mutation. In a separate study of 14,334 normal healthy women in Israel, Toledano-Alhadef et al. (33) identified three women with a full mutation, resulting in a higher prevalence (1 in 4778 females) compared with Pesso et al (32). Neither study was population-based and both were biased toward testing unaffected females. Thus, conclusions cannot be made about the prevalence of the fragile X syndrome among females based on these studies. Large population-based studies are needed to better understand the prevalence of the full mutation and its expression among females.

Prevalence: Premutations

Similar to the data for the prevalence of the full mutation, most of the data on the prevalence of premutations were collected from white populations. Results from two large, population-based studies published in the literature suggest that the prevalence of premutations (61–200 repeats) in males is probably 1 in 1000 (28,34) to 1 in 2000 (18,35) males in the general white population.

For this review, we defined premutations in females as either 61–200 repeats or 55–200 repeats. The former definition of premutations characterizes repeats that are always unstable and expand to the full mutation. The latter definition of premutations characterizes premutations of smaller size that can be found in families with the fragile X syndrome (24,36). The smallest premutation among families with the fragile X syndrome to expand to the full mutation in a single generation is 59 repeats (24,36). In the general population, however, these smaller premutations may or may not be unstable (37). Thus, the former estimate of premutations (Table 23.2) represents the lower limits of premutation carriers whereas the latter estimate of premutations possibly represents the upper limits of premutation carriers. Results from recent, large studies suggest that the lower limits of the prevalence of the premutation range from 1 in 231 (33) to 1 in 468 (32) females, and the upper limits range from 1 in 116 (33) to 1 in 259 (38) females (Table 23.2).

Interactions

No interactions with environmental factors or other genes have been identified to explain the variation in the phenotype of the fragile X syndrome. However, such interactions are expected for two reasons. First, the range in clinical severity of fully methylated, full-mutation males and females, even among monozygotic twins, has been observed and is significant (39–41). Among repeat-size or methylation mosaic males and females, variability in IQ can be partially explained by the variability in FMRP levels (42). However, neither features of *FMR1* or its gene product FMRP can account for the majority of the variability among fully methylated, full-mutation males. Although the low but measurable level of FMRP seems to be related to the level of development among affected males, it does not seem to be related to the rate of development or the expression of autism (43,44). In fact, the co-occurrence of autistic behavior and the fragile X syndrome more accurately predicts developmental status than does the level of FMRP (44), possibly suggesting the existence of additional factors involved in the fragile X phenotype.

The second reason to expect gene–gene interactions is that the normal function of FMRP is to regulate translation of other proteins. FMRP is an RNA-binding protein (2,45) that is capable of binding to itself as well as other proteins (46). In addition to its RNA-binding capabilities, FMRP has both a nuclear localization signal (NLS) and a nuclear export signal (NES) (47). The current model suggests that oligomerized FMRP, in conjunction with other proteins, shuttles specific mRNAs from the nucleus to the cytoplasm for translation. The identification of the specific mRNAs and their corresponding genes is necessary for the complete understanding of the molecular consequences related to the fragile X syndrome phenotype (3).

Laboratory Tests

The Quality Assurance Subcommittee of the American College of Medical Genetics Laboratory Practice Committee has recently published technical standards and guidelines for fragile X syndrome testing (36). According to the subcommittee, DNA-based tests that determine the size of the fragile X CGG repeat are considered diagnostic and are 99% sensitive and 100% specific (36). These DNA-based tests are also applicable for prenatal diagnosis in both amniotic fluid cells and chorionic villus samples (CVS). However, all DNA-based tests for the fragile X syndrome have important caveats that impact the interpretation of the test, most of which are reviewed below. For a more comprehensive checklist, refer to the standards and guidelines published by the Quality Assurance Subcommittee (36).

The most popular and accepted method for DNA-based testing for the expanded CGG repeat is the Southern blot. Many different restriction enzymes can be used in combination to determine both expansion (*Eco*RI, *Pst*I, *Bgl*II, *Hind*III, *Bcl*I) and methylation (*Sac*II, *Bss*HII, *Eag*I, *Bst*ZI) status for an individual (7,48). Methylation status is particularly useful for distinguishing between borderline premutation and full mutation alleles (200–230 repeats) (36). Methylation-sensitive enzymes can also describe the degree of methylation of the full mutation allele for both males and females as well as the X-inactivation pattern for females. However, neither of these measures can be used to predict the degree of mental retardation status for either sex (36). The main disadvantage of the Southern blot is that it requires a large amount of DNA and is laborious, both of which prevent the rapid and inexpensive screening of large populations. Because of these limitations, other diagnostic tests based on DNA and protein properties of the fragile X syndrome have been developed for fragile X screening.

New DNA diagnostic tests have concentrated on the use of the polymerase chain reaction (PCR). Many different PCR protocols have been developed for the fragile X CGG repeat, with different degrees of amplification abilities and sizing accuracies. Regardless of the variations in protocol, compared with Southern blots, the PCR test is inexpensive, automated, and fast. Also, PCR can be performed on very small amounts of DNA, making collection of the samples relatively painless and convenient for the patients. The disadvantage of PCR is that the test results may not be straightforward for several reasons. The amplification of large repeats is difficult, especially in the presence of a second, smaller repeat. For many PCR protocols, the DNA fragment with the expanded repeat does not amplify. This is especially problematic for females and persons with repeat-size mosaicism, who could appear to have a single, normal repeat size. To avoid these false negatives, many screening programs follow up by Southern blot any sample that fails to amplify by PCR and any female who appears to be homozygous. This strategy could potentially produce a false-negative result for persons who are normal/full mutation mosaic; however, few data exist to suggest that this occurs frequently.

Most DNA-based methods can distinguish between premutations and full mutations. Because of ethical issues in identifying asymptomatic carriers, some proposed screening strategies are designed to identify only affected persons or those with the full mutation. Also, affected persons with point mutations and deletions that result in the loss of FMRP would not be routinely identified with the use of the above methods that examine repeat size, and potential sequencing of *FMR1* is not routinely practiced for screening strategies or even for clinical diagnosis (49). For 99% of persons with the fragile X syndrome, affected status depends not only on the expansion of the repeat but on the subsequent lack of FMRP as well. The development of antibodies against FMRP has made screening possible on the basis of affected status only (45,50). In this protein-based assay, the percentage of FMRP detected in lymphocytes from blood smears is used to determine affected status (51–54). Typically, fewer than 40% of the lymphocytes from males with the fragile X syndrome have detectable amounts of FMRP (52). This protein-based test has recently been adapted for hair root (55,56) and prenatal (57,58) samples. Although promising, this technique cannot accurately identify affected females (42,54) and may not be appropriate for males who are normal/full or premutation/full mosaic.

Potential Contribution of Genetic Information to Improve Health Outcomes

One potential contribution resulting from the cloning of *FMR1* could be the prompt identification of children eligible for early intervention services. Although no cure exists for the fragile X syndrome, all infants and toddlers identified with the fragile X syndrome in the United States are eligible for early intervention services as described in Part C of the Individuals with Disabilities Education Act (IDEA, PL 101–476) (59). Early intervention programs in the United States vary among the states and are generally meant to facilitate access to existing services and programs (59). These programs can also provide direct services as a supplement to existing programs. Only one report in the literature describes intervention services specifically for children with the fragile X syndrome. Given the fact that services vary from state to state, this report from North Carolina may not accurately describe services available to children with fragile X who reside in other states. In this report, Hatton et al. (60) described in detail the types of services and therapies received by young boys with fragile X, including special education, speech-language therapy, occupational therapy, and physical therapy. The report further describes the amount of services and therapies provided, and demonstrates that the amount of services provided is positively correlated with age (60). Hatton et al. (60) also identify the need of interventionists to learn about the fragile X syndrome and its specific behavioral features, a finding consistent with other surveys of special education teachers in the United States (61) and the United Kingdom (62).

Although the medical and education communities recommend participation in programs and services after diagnosis of developmental delay or the fragile X syndrome, the effectiveness of early intervention services has not been yet demonstrated. This gap in research, however, is difficult to fill because it requires diagnosis with the fragile X syndrome soon after birth (63). In many countries, including the United States, the fragile X syndrome is commonly diagnosed through a referral for fragile X testing. A referral is indicated if a person has unexplained mental retardation, developmental delay, or autism, especially if physical or behavioral characteristics commonly associated with the fragile X syndrome are evident (64). Despite these recommendations, studies of school-aged children receiving special education in the Netherlands (19), the United States (28), and the United Kingdom (18) suggest that the current referral system can fail to identify toddlers and school-aged children who do not yet manifest the physical hallmarks of the syndrome (65,66).

A second potential contribution resulting from the cloning of *FMR1* could be the identification of families at risk of having a child with the fragile X syndrome. Based on the genetics of the fragile X syndrome, strategies for identifying at risk families usually focus on identifying female premutation and full mutation carriers. Ideally, women at risk would be identified before pregnancy, which would give these women the most options including egg donation, adoption, and prenatal diagnosis. Also, these women would be advised of their risk for premature ovarian failure (POF). Whereas the mean age of menopause is 51 years, women with POF experience the cessation of menses before age 40 years. Twenty-one percent (95% CI: 15%–27%) of premutation carriers develop POF, compared with 1% of women in the general population (67). Women identified as premutation carriers could be advised to plan their families earlier rather than later in their reproductive lives because of their risk for POF.

Although knowledge of genotype status for the fragile X syndrome offers benefits to many families, this knowledge may also solicit poorly understood harms. One such harm is the impact carrier status may have on the psychology of an individual. Retrospective interviews of parents with children in whom the fragile X syndrome had been diagnosed indicate that many carrier parents feel guilty for passing the mutation to their children and are worried about the implications fragile X test results have on their families (68,69). Furthermore, a cross-sectional study of obligate carriers found that knowledge of carrier status was upsetting and caused a proportion of women to change their views about themselves (70). Finally, an important study by McConkie-Rosell and colleagues describes the emotions and attitudes of women without children with fragile X but who know before testing that they have a 50% chance of carrying the premutation allele. In this longitudinal study, white women over age 18 years who were at risk of carrying the premutation allele completed a structured interview and standardized measures both at the time of testing and 6 months after learning the results of the test (71). Results of the study suggest that the idea of being "at risk" before testing was upsetting and continued to be so only for the women found to be carriers (72). Surprisingly, noncarrier women surveyed after genetic testing viewed the fragile X syndrome as a more serious problem than they did the first time they were surveyed and than did carrier women (72). Results also suggest that women's responses regarding feelings of self were not related to global self-concept but to the implications the positive carrier test would have on themselves and their families (71). These responses could be categorized into five areas of specific concerns: implications their positive test had for their children, reproductive options, possible expression of the fragile X syndrome phenotype in themselves, genetic identity, and regret of having not known sooner (71). Most noncarrier women and a small proportion of carrier women expressed relief or a positive emotion after learning carrier status (71,72). Although the only study of its kind in the literature, the results of the study may be limited to educated and married white women. Additional studies in this area are needed so that the impact of carrier testing can be fully understood in terms of race/ethnicity, education, economics, and marital status. Also, studies must be designed to explore the impact of carrier testing on individuals from the general population who have a very low a priori risk of carrying the fragile X premutation or full mutation.

A second potential harm associated with knowledge of genotype is related to insurance coverage. The focus of much concern is the use of genetic information that reveals susceptibility or carrier status of a presently healthy individual in determining health insurance coverage. In the United States, several states have enacted laws that prohibit the use of genetic information in pricing, issuing, or structuring of health insurance (73,74). At the federal level, the Health Insurance Portability and Accountability Act enacted in 1996 (HIPAA: Stat 1936, Pub L, No 110: 104–191) prohibits group health insurers from applying preexisting condition exclusions to genetic conditions that are identified solely by genetic tests. Although fear of genetic discrimination has generated much attention, a recent survey of insurers, agents, and professionals in medical genetics (e.g., genetic counselors and clinicians) suggests that discrimination based solely on a genetic test is not common with regard to health insurance coverage (73,74).

Consistent with these findings is one report about the effects of fragile X testing on health insurance coverage for 39 families living in Colorado (75). All families surveyed by Wingrove et al. (75) had a child affected with the fragile X syndrome. None of the families reported having their health insurance coverage cancelled as a result of genetic testing. Six families reported that carriers of the fragile X syndrome were declined for coverage; however, all six included a child affected with fragile X with their applications (75). Also, six families reported a member who refused testing because of fear of genetic discrimination (75). This result highlights the need to dispel the myths associated with genetic testing and health insurance that may otherwise discourage a person from seeking genetic testing.

Future Directions

Clinical Practice
Recent research findings dictate two trends in clinical practice related to the fragile X syndrome: (*1*) effective medical treatments and (*2*) earlier diagnosis. With regard to effective medical treatments, most current treatments available target the behavioral problems often observed in children with fragile X. Treatments include interventions such as occupational therapy with an emphasis in sensory integration and speech and language therapy. This multidisciplinary approach, detailed in Hagerman and Cronister (76), will help children with fragile X develop to their full potential by minimizing some of the behavioral problems that would impede their developmental progress. In addition to intervention, stimulation medications used to treat attention deficit hyperactivity disorder (ADHD) are commonly prescribed to children with fragile X syndrome to alleviate some of the behavioral problems that would interfere with learning and socialization. Other problems such as aggression and anxiety have also been successfully treated (77,78).

Beyond treatment of the symptoms of the syndrome, a new emphasis in research has been directed toward correcting the fragile X defect. Basic research suggests that reduced acetylation of histones H3 and H4 at the 5' end of *FMR1* leads to the condensation of chromatin and subsequent inhibition of transcription (26). The deacetylation is mediated through a methylcytosine binding protein, MeCP2, at the abnormally methylated CpG island observed in individuals with fragile X (79). Much interest lies in reactivating *FMR1* by demethylating the CpG island. Although research groups have successfully reactivated *FMR1* using 5 azadeoxycytidine (5 azadC), the chemical is too toxic for human use (26,80). Also, 5 azadC requires cell division, which makes it an unlikely candidate for neurons, the cells presumably most affected by the fragile X mutation. Chemical therapy of the primary defect is not yet a reality, but researchers remain hopeful that accumulating basic research and knowledge of the fragile X defect will translate into an effective treatment in the future.

With regard to earlier diagnosis, as mentioned above, the referral system often used by physicians does not identify all young children with the fragile X syndrome. The current referral system relies, in part, on the classic physical features present in adults with the fragile X syndrome, which are not usually present in young children with fragile X (66). Oftentimes the parents or guardians of young children with the fragile X syndrome will notice developmental delay or behavioral problems within the first few months of the children's lives (81). In fact, an analysis of retrospective interviews of North Carolina mothers with children in whom fragile X was diagnosed reported that, on average, parents noted developmental delay at 9 months (81). Despite the early signs of developmental delay among this study group, the diagnosis of developmental delay averaged 24 months and the diagnosis of fragile X averaged 35 months (81). The lag time between the first signs of fragile X and a diagnosis could be related to the physician's reluctance to test the child at such an early age. In the North Carolina study, 28 parents (68%) of affected children voiced frustration with their pediatricians or other health-care providers for dismissing the parents' concern for their children's development (81). This frustration has also been documented in a similar survey of parents with children with fragile X in the United Kingdom (82).

The importance of a prompt diagnosis for the fragile X syndrome cannot be understated. The benefits of an accurate diagnosis radiate beyond the person diagnosed because both immediate and extended family members will be affected by its consequences. The lag time between the birth of the first child with fragile X and a diagnosis must be short as possible to ensure that couples at risk have ample time to learn about the syndrome and to explore their reproductive options. As a result of this movement for a more timely diagnosis, researchers and parents have begun the discussion of general population screening programs for the fragile X syndrome.

Population Testing

For the fragile X syndrome, development of many different types of screening programs could be based on the timing of the test and the persons to whom the test is offered. Conceivably, the fragile X test could be offered at four different periods: preconceptional, prenatal, newborn, and symptomatic. Because of the unique genetics of the fragile X syndrome, only women transmit the full mutation to their offspring. Thus, either during the preconceptional period or during pregnancy, women could be offered carrier testing for the fragile X syndrome. Prenatal screening of the fetus could then be offered to women identified as premutation or full mutation carriers. If the goal is to diagnose the fragile X syndrome, systematic testing of newborns could be considered. The fragile X test could also be offered to toddlers or school-aged children who have unexplained developmental delay (symptomatic screening). However, because the goal is to shorten the lag time between birth and diagnosis, this option will not be discussed here.

With regard to screening preconceptional or pregnant women in the United States, the policy for fragile X syndrome testing has not changed since a working group for the American College of Medical Genetics published their recommendations in 1994 (64). That is, in the United States, the fragile X syndrome test is offered on a referral basis and not routinely to the general population. Examples of carrier screening offered to U.S. women with histories of fragile X or mental retardation (83,84) have been published in the literature. Only one program in the United States has reported screening reproductive-age women without a history of mental retardation. This program at the Genetics & IVF Institute in Fairfax, Virginia, reported offering fragile X carrier screening to women on a self-pay basis (85,86). Most women were referred to the clinic because of their advanced reproductive age. From December 1993 through June 1995, 3345 women were offered testing, and 668 (21%) accepted. Most (69%) of these women did not have family histories of mental retardation. Among these women, three premutation carriers (60–199 repeats) were identified (85).

Unlike the screening studies performed in the United States, investigators in Finland have implemented a large population-based carrier-screening program. The program implemented by the Kuopio University Hospital in Finland offered an *FMR1* gene test free of charge to all pregnant women seeking prenatal care from July 1995 until December 1996 (87). According to Ryynanen et al. (87), almost all pregnant women in Finland seek prenatal care and are registered in antenatal clinics during the 6th through 10th weeks of pregnancy because this registration is required for maternity allowance provided by the state. Among women without family histories of the fragile X syndrome, 1477 (85%) elected genetic testing. Of these women, six were identified as premutation carriers (60–199 repeats) and all six women elected prenatal testing. The program has since screened an additional 1358 women through December 1997. Six more women with the premutation allele were identified, and all six elected prenatal diagnosis (88). The program has also expanded to offer an *FMR1* genetic test to pregnant women undergoing invasive prenatal testing because of advanced maternal age or history of trisomy pregnancy. In this expanded program, 241 (80%) of the 302 women offered the test consented, and one woman was identified as a premutation carrier (88).

As in Finland, carrier testing is also widely employed and accepted in Israel. At least three groups have published results obtained from their large screening programs. Unlike the program in Finland, these programs offer the test on a self-pay basis and rely on either a self-referral or physician referral for testing. In one published report, the Rabin Medical Center screened 14,334 preconceptional or pregnant women who were self-referred and had no family histories of mental retardation between January 1992 and October 2000 (33,89). Women identified as carriers were offered prenatal testing free of charge as instructed by the Israeli Ministry of Health. A total of 204 women were identified as premutation carriers (51–200 repeats), and three women were identified as full mutation carriers (33). Of the pregnant women identified as premutation/full mutation carriers (n = 173), only 14 women refused prenatal diagnosis (33). In a second report, the Genetic Institute of the Tel Aviv Sourasky Medical Center offered an *FMR1* genetic test to 9660 women during September 1994 through October 1998 (37). A total of 38 premutation carriers (60–200 repeats) were identified, and all of these women consented to prenatal diagnosis (37). Finally, during January 1994 through March 1999, the Danek-Gertner Institute of Human Genetics screened 8426 women of reproductive age who had no family histories of the fragile X syndrome or mental retardation (32). Among these women, 18 were identified as premutation carriers (61–199 repeats), and one was identified as a full mutation carrier (>200 repeats).

Newborn screening for the fragile X syndrome has not yet been implemented in any country. Compared against the criteria published by the National Academy of Sciences (NAS) (90), the Institute of Medicine (IOM) (91), and Task Force on Genetic Testing (92), the fragile X syndrome meets at least two criteria essential for a successful program in newborn screening. Specifically, based on its morbidity and prevalence, the fragile X syndrome is an important public health problem. Also, approximately 99% of cases diagnosed thus far are caused by a single, inherited mutation, making the fragile X syndrome particularly amenable to DNA-based testing for an accurate diagnosis.

Despite meeting these criteria, the fragile X syndrome does not meet several other key criteria for newborn screening. One crucial gap in research is the lack of a cure or effective treatment available for persons with the disorder, as mentioned previously. A second gap is the lack of knowledge of the potential harms associated with the diagnosis in an apparently healthy child. Many researchers and parents worry that a diagnosis at the newborn period would disrupt the parent–child bond. However, little evidence supports this, and many parents contend that a diagnosis would strengthen the parent–child bond through a greater understanding of the child's special needs. Nevertheless, because the fragile X genetic test cannot predict the severity of mental retardation (especially regarding females with the full mutation), the effects of a diagnosis on the family's perception of the child's prognosis and future development are important to consider. Finally, a third gap is the lack of general consensus for the appropriate time to screen that would maximize the benefits of a diagnosis (93). Newborn screening will identify affected infants who are eligible for early intervention services. Newborn screening would also provide parents with information about their children's future development and methods to optimize this development. However, identification of an affected person will also identify at risk families. Although these identified families could benefit from genetic counseling, newborn screening is neither ideal nor designed for identifying most at risk families for the fragile X syndrome in the general population (68,94). Also, many parents may want to know before pregnancy or birth about the fragile X syndrome. Indeed, a proportion of parents with fragile X children surveyed in the United Kingdom believed that a diagnosis based on newborn screening would be "too late (82)."

The debate concerning screening for the fragile X syndrome will no doubt continue. In the meantime, more information about the risk for expansion based on premutation size should be collected to better assess a women's risk of conceiving a child with the fragile X syndrome. Also, the psychological impact of genetic testing should be thoroughly explored because these results have implications that reach far beyond the fragile X syndrome in this new genetic age. Finally, effective treatments need to be developed and properly evaluated so that persons affected by the fragile X syndrome and their families can live life to the fullest potential.

References

1. Fu Y-H, Kuhl DPA, Pizzuti A, et al. Variation of the CGG repeat at the fragile X site results in genetic instability: resolution of the Sherman paradox. Cell 1991;67: 1047–1058.
2. Ashley CT, Sutcliffe JS, Kunst CB, et al. Human and murine FMR-1: alternative splicing and translational initiation downstream of the CGG repeat. Nat Genet 1993; 4:244–251.
3. Jin P, Warren ST. Understanding the molecular basis of fragile X syndrome. Hum Mol Genet 2000;9:901–908.
4. Mazzocco MMM. Advances in research on the fragile X syndrome. Ment Retard Dev Disabil Res Rev 2000;6:96–106.
5. Bailey DB, Hatton DD, Skinner M. Early developmental trajectories of males with fragile X syndrome. Am J Ment Retard 1998;1:29–39.
6. Hagerman RJ (1996). Physical and behavioral phenotype. in: Hagerman RJ, Cronister A, Fragile X Syndrome: Diagnosis, Treatment, and Research. Baltimore: Johns Hopkins University Press, pp. 3–88.
7. Warren ST, Sherman SL. The fragile X syndrome, in The Metabolic and Molecular Basis of Inherited Disease, Scriver CR, Beaudet AL, Sly WS, eds. New York: McGraw-Hill, pp. 1257–1289.
8. Kau ASM, Reider EE, Payne L, et al. Early behavioral signs of psychiatric phenotypes in fragile X syndrome. Am J Ment Retard 2000;105:266–299.
9. Bailey DB, Mesibov GB, Hatton DD, et al. Autistic behavior in young boys with fragile X syndrome. J Autism Dev Dis 1998;28:499–508.
10. Wolff PH, Gardner J, Lappen J, et al. Variable expression of the fragile X syndrome in heterozygous females of normal intelligence. Am J Med Genet 1988;30:213–225.
11. Rousseau F, Heitz D, Tarleton J, et al. A multicenter study on genotype-phenotype correlations in fragile X syndrome, using direct diagnosis with probe StB128: the first 2,253 cases. Am J Hum Genet 1994;55:225–237.
12. de Vries BBA, Wiegers AM, Smits APT, et al. Mental status of females with an *FMR1* gene full mutation. Am J Hum Genet 1996;58:1025–1032.

13. Franke P, Leboyer M, Hardt J, et al. Neuropsychological profiles of FMR-1 premutation and full-mutation carrier females. Psychiatry Res 1999;87:223–231.

14. Block SS, Brusca-Vega R, Pizzi WJ, et al. Cognitive and visual processing skills and their relationship to mutation size in full and premutation female fragile X carriers. Optom Vis Sci 2000;77:592–599.

15. Bennetto L, Pennington BF, Porter D, et al. Profile of cognitive functioning in women with the fragile X mutation. Neuropsychology 2001;15:290–299.

16. Mazzocco MMM, Kates WR, Baumgardner TL, et al. Autistic behaviors among girls with fragile X syndrome. J Autism Dev Disord 1997;27:415–435.

17. Crawford, DC, Acuña JM, Sherman SL. FMR1 and the fragile X syndrome: Human genome epidemiology review. Genet Med 2001; 3:359–371.

18. Youings SA, Murray A, Dennis N, et al. FRAXA and FRAXE: the results of a five-year survey. J Med Genet 2000;37:415–421.

19. de Vries BBA, van den Ouweland AMW, Mohkamsing S, et al. Collaborative Fragile X Study Group. Screening and diagnosis for the fragile X syndrome among the mentally retarded: an epidemiological and psychological survey. Am J Hum Genet 1997;61:660–667.

20. von Koskull H, Gahmberg N, Salonen R, et al. FRAXA locus in fragile X diagnosis: family studies, prenatal diagnosis, and diagnosis of sporadic cases of mental retardation. Am J Med Genet 1994;51:486–489.

21. Mazzocco MMM, Sonna NL, Teisl JT, et al. The FMR1 and FMR2 mutations are not common etiologies of academic difficulty among school-age children. J Dev Behav Pediatr 1997;18:392–398.

22. Cummings CJ, Zoghbi HY. Fourteen and counting: unraveling trinucleotide repeat diseases. Hum Mol Genet 2000;9:909–916.

23. Ashley-Koch AE, Robinson H, Glicksman AE, et al. Examination of factors associated with instability of the FMR1 CGG repeat. Am J Hum Genet 1998;63:776–785.

24. Nolin SL, Lewis FA, Ye LL, et al. Familial transmission of the FMR1 CGG repeat. Am J Hum Genet 1996;59:1252–1261.

25. Sutcliffe JS, Nelson DL, Zhang F, et al. DNA methylation represses FMR-1 transcription in fragile X syndrome. Hum Mol Genet 1992;1:397–400.

26. Coffee B, Zhang F, Warren ST, et al. Acetylated histones are associated with FMR1 in normal but not fragile X syndrome cells. Nat Genet 1999;22:98–101.

27. Pieretti M, Zhang F, Fu YH, et al. Absences of expression of the FMR-1 gene in fragile X syndrome. Cell 1991;66:817–822.

28. Crawford DC, Meadows KL, Newman JL, et al. Prevalence of the fragile X syndrome in African-Americans. Am J Med Genet 2002;110:226–233.

29. Jacobs PA, Bullman H, Macpherson J, et al. Population studies of the fragile X: a molecular approach. J Med Genet 1993;30:454–459.

30. Patsalis PC, Sismani C, Hettinger JA, et al. Molecular screening of fragile X (FRAXA) and FRAXE mental retardation syndromes in the Hellenic population of Greece and Cyprus: incidence, genetic variation, and stability. Am J Med Genet 1999;84:184–190.

31. Elbaz A, Suedois J, Duquesnoy M, et al. Prevalence of fragile X syndrome and FRAXE among children with intellectual disability in a Caribbean island, Guadeloupe, French West Indies. J Intellect Disabil Res 1998;42:81–89.

32. Pesso R, Berkenstadt M, Cuckle H, et al. Screening for fragile X syndrome in women of reproductive age. Prenat Diagn 2000;20:611–614.

33. Toledano-Alhadef H, Basel-Vanagaite L, Magal N, et al. Fragile-X carrier screening and the prevalence of premutation and full-mutation carriers in Israel. Am J Hum Genet 2001;69:351–360.

34. Crawford DC, Meadows KL, Newman JL, et al. Prevalence and phenotype consequence of FRAXA and FRAXE alleles in a large, ethnically diverse, special education-needs population. Am J Hum Genet 1999;64:495–507.

35. Murray A, Youings SA, Dennis N, et al. Population screening at the FRAXA and FRAXE loci: molecular analyses of boys with learning difficulties and their mothers. Hum Mol Genet 1996;5:727–735.

36. Maddalena A, Richards CS, McGinniss MJ, et al. Technical standards and guidelines for fragile X: The first of a series of disease-specific supplements to the standards and guidelines for clinical genetics laboratories of the American College of Medical Genetics. Genet Med 2001;3:200–205.

37. Geva E, Yaron Y, Shomrat R, et al. The risk of fragile X premutation expansion is lower in carriers detected by general prenatal screening than in carriers from known fragile X families. Genet Testing 2000;4:289–292.

38. Rousseau F, Rouillard P, Morel M-L, et al. Prevalence of carriers of premutation-size alleles of the FMR1 gene—and implications for the population genetics of the fragile X syndrome. Am J Hum Genet 1995;57:1006–1018.

39. Helderman-van den Enden AT, Maaswinkel-Mooij PD, Hoogendoorn E, et al. Monozygotic twin brothers with the fragile X syndrome: different CGG repeat and different mental capacities. J Med Genet 1999;36:253–257.

40. Willemsen R, Olmer R, De Diego Otero Y, et al. Twin sisters, monozygotic with the fragile X mutations, but with a different phenotype. J Med Genet 2000;37:603–604.

41. Sheldon L, Turk J. Monozygotic boys with fragile X syndrome. Dev Med Child Neuro 2000;42:768–774.

42. Tassone F, Hagerman RJ, Ikle DN, et al. FMRP expression as a potential prognostic indicator in fragile X syndrome. Am J Med Genet 1999;84:250–261.

43. Bailey DB, Hatton DD, Tassone F, et al. Variability in FMRP and early development in males with fragile X syndrome. Am J Ment Retard 2001;106:16–27.

44. Bailey DB, Hatton DD, Skinner M, et al. Autistic behavior, FMR1 protein, and developmental trajectories in young males with fragile X syndrome. J Autism Dev Disord 2001;31:165–174.

45. Siomi H, Siomi MC, Nussbaum RL, et al. The protein product of the fragile X gene, *FMR1*, has characteristics of an RNA-binding protein. Cell 1993;74:291–298.

46. Siomi MC, Zhang Y, Siomi H, et al. Specific sequences in the fragile X syndrome protein FMR1 and the FXR proteins mediate their binding to 60S ribosomal subunits and the interactions among them. Mol Cell Biol 1996;16:3825–3832.

47. Eberhart DE, Malter HE, Feng Y, et al. The fragile X mental retardation protein is a ribonucleoprotein containing both nuclear localization and nuclear export signals. Hum Mol Genet 1996;5:1083–1091.

48. Murray J, Cuckle H, Taylor G, et al. Screening for fragile X syndrome: information needs for health planners. J Med Screen 1997;4:60–94.

49. Gronskov K, Hallberg A, Brondum-Nielsen K. Mutational analysis of the FMR1 gene in 118 mentally retarded males suspected of fragile X syndrome: absence of prevalent mutations. Hum Genet 1998;102:440–445.

50. Verheij C, Bakker CE, de Graaff E, et al. Characterization and localization of the *FMR-1* gene product associated with fragile X syndrome. Nature 1993;363:722–724.

51. Willemsen R, Mohkamsing S, de Vries BA, et al. Rapid antibody test for fragile X syndrome. Lancet 1995;345:1147–1148.

52. Willemsen R, Smits A, Mohkamsing S, et al. Rapid antibody test for diagnosing fragile X syndrome: a validation of the technique. Hum Genet 1997;99:308–311.

53. de Vries BBA, Halley DJJ, Oostra BA, et al. The fragile X syndrome. J Med Genet 1998;35:579–589.

54. Willemsen R, Oostra BA. FMRP detection assay for the diagnosis of the fragile X syndrome. Am J Med Genet 2000;97:183–188.

55. Willemsen R, Anar B, De Diego Otero Y, et al. Noninvasive test for fragile X syndrome, using hair root analysis. Am J Hum Genet 1999;65:98–103.

56. Tuncbilek E, Alikasifoglu M, Aktas D, et al. Screening for the fragile X syndrome among mentally retarded males by hair root analysis. Am J Med Genet 2000;95:105–107.

57. Jenkins EC, Wen GY, Kim KS, et al. Prenatal fragile X detection using cytoplasmic and nuclear-specific monoclonal antibodies. Am J Med Genet 1999;83:342–346.

58. Lambiris N, Peters H, Bollmann R, et al. Rapid FMR1-protein analysis of fetal blood: an enhancement of prenatal diagnostics. Hum Genet 1999;105:258–260.

59. Bailey DB, Aytch LS, Odom SL, et al. Early intervention as we know it. Ment Retard Dev Disabil Res Rev 1999;5:11–20.

60. Hatton DD, Bailey DB, Roberts JP, et al. Early intervention services for young boys with fragile X syndrome. J Early Intervention 2000;23:235–251.

61. Wilson PG, Mazzocco MMM. Awareness and knowledge of fragile X syndrome among special educators. Ment Retard 1993;31:221–227.

62. York A, von Fraunhofer N, Turk J, et al. Fragile-X syndrome, Down's syndrome and autism: awareness and knowledge amongst special educators. J Intellect Disabil Res 1999;43:314–324.

63. Bailey DB, Roberts JE, Mirrett P, et al. Identifying infants and toddlers with fragile X syndrome: issues and recommendations. Inf Young Children 2001;14:24–33.

64. American College of Medical Genetics. Fragile X syndrome: diagnostic and carrier testing. Am J Med Genet 1994;53:380–381.

65. Stoll C. Problems in the diagnosis of fragile X syndrome in young children are still present. Am J Med Genet 2001;100:110–115.

66. Lachiewicz AM, Dawson DV, Spiridigliozzi GA. Physical characteristics of young boys with fragile X syndrome: reasons for difficulties in making a diagnosis in young males. Am J Med Genet 2000;92:229–236.

67. Sherman SL. Premature ovarian failure in the fragile X syndrome. Am J Med Genet 2000;97:189–194.

68. van Rijn MA, de Vries BBA, Tibben A, et al. DNA testing for fragile X syndrome: implications for parents and family. J Med Genet 1997;34:907–911.

69. McConkie-Rosell A, Spiridigliozzi GA, Rounds K, et al. Parental attitudes regarding carrier testing in children at risk for fragile X syndrome. Am J Med Genet 1999; 82:206–211.

70. McConkie-Rosell A, Spiridigliozzi GA, Iafolla T, et al. Carrier testing in the fragile X syndrome. Am J Med Genet 1997;68:62–69.

71. McConkie-Rosell A, Spiridigliozzi GA, Sullivan JA, Dawson DV, Lachiewicz AM. Carrier testing in fragile X syndrome: effect on self-concept. Am J Med Genet 2000;92: 336–342.

72. McConkie-Rosell A, Spiridigliozzi GA, Sullivan JA, Dawson DV, Lachiewicz AM. Longitudinal study of the carrier testing process for fragile X syndrome: perceptions and coping. Am J Med Genet 2001;98:37–45.

73. Hall MA and Rich SS. Laws restricting health insurers' use of genetic information: impact on genetic discrimination. Am J Hum Genet 2000;66:293–307.

74. Hall MA and Rich SS. Patients' fear of genetic discrimination by health insurers: the impact of legal protections. Genet Med 2000;2:214–221.

75. Wingrove KJ, Norris J, Barton PL, Hagerman R. Experiences and attitudes concerning genetic testing and insurance in a Colorado population: a survey of families diagnosed with fragile X syndrome. Am J Med Genet 1996;64:378–381.

76. Hagerman RJ, Cronister A (1996). Fragile X syndrome: diagnosis, treatment, and research, in Hagerman RJ, Cronister A (eds.). 2nd ed. Baltimore, Johns Hopkins University Press.

77. Hagerman RJ. Fragile X syndrome. In: Neurodevelopmental Disorders: Diagnosis and Treatment. New York: Oxford University Press, 1999:61–132.

78. Hagerman RJ, Stackhouse T, Scharfenaker S, Hickman L. Fragile X 2000: Current research and treatment. Boulder, Belle Curve Records, Inc, 2000.
79. Robertson KD and Wolffe AP. DNA methylation in health and disease. Nat Rev 2000;1:11–19.
80. Chiurazzi P, Pomponi MG, Pietrobono R, Bakker CE, Neri G, Oostra BA. Synergistic effect of histone hyperacetylation and DNA demethylation in the reactivation of the FMR1 gene. Hum Mol Genet 1999;8:2317–2323.
81. Bailey DB, Skinner D, Hatton D, Roberts J. Family experiences and factors associated with the diagnosis of fragile X syndrome. J Dev Behav Pediatr 2000;21:315–321.
82. Carmichael B, Pembrey M, Turner G, Barnicoat A. Diagnosis of fragile-X syndrome: the experiences of parents. J Intellect Disabil Res 1999;43:47–53.
83. Brown WT, Nolin SL, Houck GE, Ding X, Glicksman AE, Li S-Y, Stark-Houck S, Brophy P, Duncan C, Dobkin C, Jenkins EC. Prenatal diagnosis and carrier screening for fragile X by PCR. Am J Med Genet 1996;64:191–195.
84. Wenstrom KD, Descartes M, Franklin J, et al. A five-year experience with fragile X screening of high-risk gravid women. Am J Obstet Gynecol 1999;181:789–792.
85. Spence WC, Black SH, Fallon L, et al. Molecular fragile X screening in normal populations. Am J Med Genet 1996;64:181–183.
86. Howard-Peebles PN, Maddalena A, Black SH, et al. Fragile X screening in pediatric and obstetrical patients. Dev Brain Dysfunct 1995;8:408–410.
87. Ryynanen M, Heinonen S, Makkonen M, et al. Feasibility and acceptance of screening for fragile X mutations in low-risk pregnancies. Eur J Hum Genet 1999;7:212–216.
88. Kallinen J, Heinonen S, Mannermaa A, et al. Prenatal diagnosis of fragile X syndrome and the risk of expansion of a premutation. Clin Genet 2000;58:111–115.
89. Drasinover V, Ehrlich S, Magal N, et al. Increased transmission of intermediate alleles of the FMR1 gene compared with normal alleles among female heterozygotes. Am J Med Genet 2000;93:155–157.
90. National Research Council and Committee for the Study of Inborn Errors of Metabolism. Genetic Screening: Programs, Prinicples, and Research. Washington, DC, National Academy of Sciences, 1975.
91. Assessing genetic risks. Implications for health and social policy. Andrews LB, Fullarton JE, Holtzman NA, Motulsky AG. Washington, DC, National Academy of Sciences, 1994.
92. Holtzman NA and Waston MS. Promoting safe and effective genetic testing in the United States: Final report of the Task Force on Genetic Testing. Bethesda, MD, National Institutes of Health, 1997.
93. Pembrey ME, Barnicoat AJ, Carmichael B, Bobrow M, Turner G. An assessment of screening strategies for fragile X syndrome in the UK. Health Technol Assess 2001; 5:1–95.
94. Wildhagen MF, van Os TAM, Polder JJ, ten Kate LP, Habbema JDF. Efficacy of cascade testing for fragile X syndrome. J Med Screen 1999;6:70–76.
95. Turner G, Robinson H, Laing S, Purvis-Smith S. Preventive screening for the fragile X syndrome. N Engl J Med 1986;315:607–609.
96. Turner G, Webb T, Robinson H. Prevalence of fragile X syndrome. Am J Med Genet 1996;64:196–197.
97. Arvio M, Peippo M, Simola KOJ. Applicability of a checklist for clinical screening of the fragile X syndrome. Clin Genet 1997;52:211–215.
98. Webb T, Bundey S, Thake J, et al. The frequency of the fragile X chromosome among schoolchildren in Coventry. J Med Genet 1986;23:396–399.
99. Morton JE, Bundey S, Webb TP, et al. Fragile X syndrome is less common than previously estimated. J Med Genet 1997;34:1–5.

100. Mazurczak T, Bocian E, Milewski M, et al. Frequency of fraX syndrome among institutionalized mentally retarded males in Poland. Am J Med Genet 1996;64:184–186.
101. Millan JM, Martinez F, Cadroy A, Gandia J, Casquero M, Beneyto M, Badia L, Prieto F. Screening for FMR1 mutations among the mentally retarded: prevalence of the fragile X syndrome in Spain. Clin Genet 1999;56:98–99.
102. Slaney SF, Wilkie AOM, Hirst MC, Charlton R, McKinley M, Pointon J, Christodoulou Z, Huson SM, Davies KE. DNA testing for fragile X syndrome in schools for learning difficulties. Arch Dis Child 1995;72:33–37.
103. Fleiss JL. An introduction to applied probability, in Statistical Methods for Rates and Proportions. New York: John Wiley and Sons, 1981, pp. 1–18.

24

The connexin connection: from epidemiology to clinical practice

Aileen Kenneson and Coleen Boyle

Individuals who are deaf or hard-of-hearing account for a significant portion of the population. Currently about one-sixth of the U.S. population (i.e., about 40 million individuals) have some degree of hearing loss, a half of a million of which fall into the severe-to-profound range (1). Hearing loss may occur late in life, or may be present at birth, as is the case for one to three in 1000 newborns (2). Formerly, these children were not diagnosed until they were 2 to 3 years of age (3), resulting in developmental delay, particularly in the arena of language development (4). However, recent technological advances and increased public health attention have resulted in the development of Early Hearing Detection and Intervention (EHDI) programs in most of the United States. EHDI programs seek the early identification of infants with hearing loss via universal newborn hearing screening programs followed by linkage with appropriate intervention options, of which there are many, resulting in the avoidance of developmental delays (4–6).

Genetic evaluations are often included as part of the medical care of individuals with hearing loss. Based on family history and physical examination, clinical geneticists categorize hearing loss into syndromic and nonsyndromic cases, a distinction that is useful both for appropriate clinical care and estimation of recurrence rates (7). Recent advances in our understanding of single-gene causes of both syndromic and nonsyndromic hearing loss are likely to increase the role of specific genetic tests in the evaluation of individuals with hearing loss.

As genetic research progresses and the demand for use of genetic testing increases, genetic tests often move rapidly from research interest into clinical use, sometimes before the clinical utility of the test is fully defined (8,9). This is the case for *GJB2*, a gene recently implicated in up to 50% of cases of nonsydromic hearing loss in some populations. Thus, *GJB2* testing may serve as a model by which we can examine the translation of research advances into clinical and public health use.

Background

Translation of sound waves into integrated neuronal signals in the brain is an amazingly complex process, so it is not surprising that hundreds of genes are involved in the development and operation of this machinery (10). Variation in any one of these genes can result in hearing loss. The extent of locus heterogeneity of hearing loss is illustrated by the more than 400 hereditary syndromes which claim hearing loss as a component (11), and which account for about 30% of cases of hearing loss (12–14). Heterogeneity is also observed in nonsyndromic cases, which are typically sensorineural in nature, and for which there is linkage evidence for almost 70 loci, including autosomal recessive variants in 75% to 80% of nonsyndromic cases (designated with the prefix DFNB), autosomal dominant variants in 20% to 25% (DFNA), and X-linked variants in 1–1.5% (DFN) (15).

The estimated numbers and types of loci for nonsyndromic hearing loss based on linkage analysis are: 30 autosomal recessive, 29 autosomal dominant, 8 X-linked, and two mitochondrial (16). The *Gap Junction Beta 2, GJB2,* gene was recently identified as the source of both the DFNB1 (autosomal recessive) and the DFNA3 (autosomal dominant) loci (17–20). The *GJB2* gene encodes for connexin 26, a beta class gap junction protein expressed in the cochlea and in the epidermis. Connexin 26 hexamers form channels between cells that, when open, allow cell-to-cell diffusion of small molecules. This function is necessary for the recycling of potassium in the cochlea that is critical for sensorineural hearing function (21). More than 90 variants of the *GJB2* gene have been reported, and many are rare; recessive alleles, dominant alleles, and polymorphisms have all been described (22).

Epidemiological Findings

Contribution of GJB2 Variants to Hearing Loss
A large number of studies related to the association between hearing loss and *GJB2* variants in a broad range of populations have been published in the past few years. In general, the conclusions are limited by factors such as small sample sizes, lack of population-based ascertainment methods, and lack of population descriptions. For example, a common source of ascertainment is hearing loss clinics; potential biases in this scenario include self-selection for clinic attendance, and under-ascertainment of mild or unilateral hearing loss. Ascertainment details are often not published, and reports often lack population descriptions including age, sex, and race/ethnicity. Comparison between studies is also limited by the differences in inclusion criteria, which have included all familial cases, recessive cases, sporadic cases, or all cases combined. However, despite the limitations, enough data have been amassed that we can begin to develop a picture of the relationship between *GJB2* and hearing loss.

Given the extraordinary genetic heterogeneity of nonsyndromic hearing loss, it was believed that no single gene would play a significant role in its etiology. So it was surprising to discover that sequence variations at the *GJB2* locus account for up to 50% of cases of nonsyndromic prelingual sensorineural hearing loss in some populations. While more than 90 alleles have been described in the literature, three account for the majority of *GJB2*-related hearing loss in studied populations: 167ΔT, 35ΔG, and 235ΔC, the most common variant alleles in the Ashkenazi Jewish population (23,24), populations of northern European descent (25–32), and in the Korean (33) and Japanese (34–36) populations, respectively.

Figure 24.1 summarizes data from several sources and depicts the contribution of the *GJB2* variants to hearing loss in several different populations around the world. The available population-specific epidemiologic data consistently indicate population differences in two key measures that are important to determine the clinical validity of genetic testing: (*1*) the percent of cases of nonsyndromic hearing loss that is associated with *GJB2* variants, and (*2*) the population frequency of the different *GJB2* alleles. For example, nonsyndromic sensorineural hearing loss is associated with *GJB2* variants in almost 50% of cases in Israel, but in only 8% in Korea and 20% in Japan. Likewise, the 35ΔG allele accounts for about 10% to 20% of cases of nonsyndromic hearing loss in persons of northern European descent, but about 30% to 40% of cases in Mediterranean regions.

Table 24.1 presents the contribution of *GJB2* variants to nonsyndromic hearing loss in several populations. The table presents both the percent of cases of hearing loss associated with *GJB2* variants in general, as well as the percent of cases of hearing loss that are associated with the most common allele in that population. For example, in European and North American whites, 38% of individuals with hearing loss carry one or more *GJB2* variant allele, but 34% of individuals with hearing loss carry one or more copies of 35ΔG. Thus, the 35ΔG allele accounts for the majority of variant alleles in this population. In contrast, 50% of Ashkenazi Jewish individuals with hearing loss carry a *GJB2* variant allele, but only 31% carry at least one copy of the 167ΔT variant. Thus, there is more allelic heterogeneity in the Ashkenazi Jewish population than there is among European and North American whites. Table 24.1 also demonstrates that these alleles are not uncommon in the general population. The 235ΔC allele is carried by 1% of individuals in Korea and Japan, and the 35ΔG allele is carried by 1 in 50 whites. In the Ashkenazi Jewish population, the 167ΔT allele may be carried by up to 8% of the general population.

In general, there is a lack of phenotype data related to the contribution of the less common *GJB2* variants to hearing loss. In addition, because of the small numbers of individuals with mild, unilateral, and late-onset hearing loss included in studies, the potential involvement of *GJB2* variants with the full spectrum of forms of hearing loss has not been fully assessed, particularly in the case of the less common variants.

GJB2 Variants and Age of Onset
To fully assess the relationship between *GJB2* variants and age of onset, genotype data is needed on individuals with congenital (present at birth), noncongenital prelin-

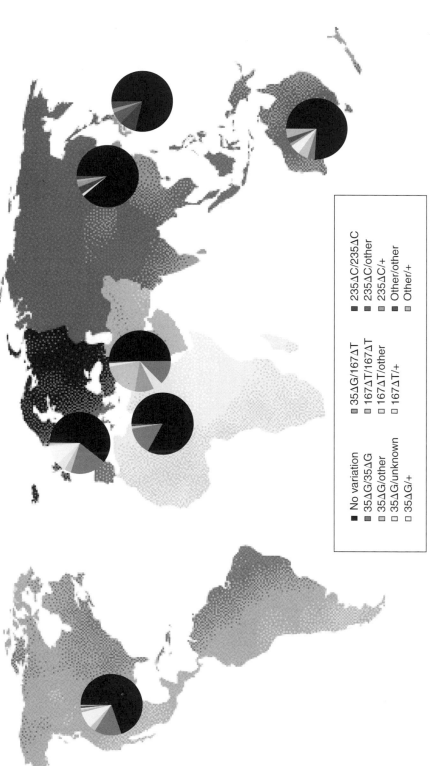

Figure 24.1 Population-specific epidemiological data pertaining to GJB2 and nonsyndromic hearing loss. Contribution of GJB2 variants (35ΔG, 167ΔT, and 235ΔC) to nonsyndromic hearing loss in the United States (White Americans) (37,55,57,59,66), Europe (29–32), Israel (23,24), Tunisia (42), Korea (33), Japan (34–36), and Australia (60).

Table 24.1 The Prevalence of *GJB2* Variants and Specific Common Alleles among Individuals with Nonsyndromic Prelingual Hearing Loss and in General Population Controls in European, White American, Ashkenazi Jewish, and Korean and Japanese Populations

Population	Cases with one or two GJB2 Variant Alleles (%)[a]	Most Common Allele in Population	Cases with one or two Copies of the Most Common Allele (%)	References for Case Data	Controls with one Copy Common Allele (%)	References for Control Data
Ashkenazi Jews	50	167ΔT	31	23,70	8	23,24,69,70
European and North American whites	38	35ΔG	34	19,26–28, 30,31,40,41, 47,55,57	2	24,27,28,57, 63,64,66–68
Japan	21	235ΔC	17	34–36	1	34–36
Korea	11	235ΔC	7	33	1	33

[a]Excluding the V27I, E114G, I203T polymorphisms.

gual, postlingual, and late-onset (after age 30) hearing loss. In the absence of newborn hearing screening, hearing loss is usually not diagnosed until late infancy or early childhood. Thus, in most published studies, it is not possible to distinguish between congenital and noncongenital prelingual hearing loss.

Only one published study has examined the contribution of *GJB2* variants to congenital hearing loss. Allele-specific methods were used to determine the prevalence of the 35ΔG and 167ΔT genotypes in 42 infants identified with hearing loss through the Rhode Island universal newborn hearing screen. The study identified three 35ΔG homozygotes, two 35ΔG/167ΔT compound heterozygotes, and one 35ΔG carrier. The two compound heterozygotes were reported as having Ashkenazi Jewish ancestry. The remaining newborns were of mixed European background. Thus, the 35ΔG and 167ΔT genotypes in this newborn population with hearing loss did not differ from other American populations with hearing loss who were ascertained in childhood, and who were of similar race/ethnicity (Table 24.1) (37).

More studies of this type, as well as studies including documented noncongenital prelingual hearing loss, are needed to assess the relationship between *GJB2* variants and congenital hearing loss. In this regard, it is important to note that there have been case reports of newborns who passed the newborn hearing screen but were diagnosed with *GJB2*-related hearing loss later in infancy (38,39). It is not clear whether these cases represent false-negative results of the newborn hearing screening programs or are indicative of a noncongenital and/or progressive nature of some *GJB2*-related cases of hearing loss.

Only four studies have addressed the possibility of an association between *GJB2* variants and postlingual hearing loss. Three of these studies did not detect any *GJB2* variants among individuals with postlingual hearing loss, including 11 individuals in Israel (age of onset undefined) (24), 16 in France (onset before age 20) (26), and 39 in Japan (onset between 3 and 30 years) (36). The fourth study, conducted in

Austria, found 4 carriers of *GJB2* variants among 16 individuals with postlingual (undefined) hearing loss (40). The genotypes were L90P/I20T (onset in first decade), L90P/35ΔG (onset in first decade), 35ΔG/+ (onset in first decade), and G160S/+ (onset in fourth decade). The L90P allele is of interest in this population because it is seen in 2 of 16 postlingual cases, and 3 of 53 prelingual cases of hearing loss. Thus, this allele may contribute to postlingual, as well as prelingual, hearing loss. The failure to detect *GJB2* variants in the other three postlingual studies may be due to a higher prevalence of the L90P allele in the Austrian population, as this allele was detected only rarely in individuals with hearing loss in France (2 of 88) (26) and Italy (3 of 147) (31,41), but not at all in Israel (0 of 102) (23,24), Japan (0 of 94) (34–36), Korea (0 of 147) (33), Tunisia (0 of 70) (42), or the United Kingdom (0 of 210) (30).

Another allele, C202F, has also been implicated in postlingual hearing loss, as it was observed to co-segregate with hearing loss (age of onset between 10 and 20 years) over five generations in a French family (43). This allele was not detected in 95 French control individuals (43), nor has it been reported in other studies that provided sequence data on controls, including 100 Korean newborns (33), 209 Japanese individuals (34–36), and 119 additional French individuals (27).

No published studies have assessed the possible relationship between *GJB2* variants and late-onset (after age 30) hearing loss. Thus, additional population-based studies involving individuals with congenital, non-congenital prelingual, postlingual, and late-onset hearing loss will be needed to fully assess the relationship between *GJB2* variants and age of onset.

Factors Influencing Phenotypic Outcome

Many study groups have reported that the degree of hearing loss of individuals with the same *GJB2* genotype varies in severity, even within sibships (23,24,26,31). This suggests that other factors, genetic and/or environmental, may be acting to modify the phenotypic outcome of *GJB2* variant alleles. Hearing loss is typically described as 50% genetic and 50% environmental in nature, involving a wide range of both genetic and environmental factors (12); any number of these factors could potentially influence the phenotypic expression of *GJB2* variant genotypes.

Influence of Environmental Factors

The possible contribution of environmental factors to *GJB2*-related hearing loss has not been assessed. Most of the studies pertaining to the contribution of *GJB2* variants to hearing loss excluded cases with known risk factors from the genetic analysis. These factors include infections (e.g., meningitis, rubella), low birth weight, ventilator use, ototoxic medications (e.g., aminoglycosides), and hyperbilirubinemia. However, a case report of two individuals with hearing loss originally attributed to rubella infection that were later found to be homozygous for the 167ΔT variant (39) suggests that the presence of known risk factors should not necessarily preclude genetic analysis. Indeed, the proportion of *GJB2* cases that have been attributed to other causes has not been elucidated, and the possibility of gene–environment interactions has not been examined. Studies pertaining to the relationship between *GJB2* variants, environmental factors, and hearing loss may identify factors that modify the *GJB2* phenotype, and may implicate *GJB2* variants in the susceptibility to known ototoxic factors. In addition to lending clues about the developmental etiology of hearing function, studies of this nature are important for accurate genetic counseling. For example, in the above case report, the couple would have originally been counseled that the chance of having a child with hearing loss was low due to the environmental nature of their hearing loss. However in retrospect, it actually was 100% due to the recessive nature of their alleles.

Contribution of GJB2 Variants to Syndromic Hearing Loss

The currently published *GJB2* studies have generally excluded cases of syndromic hearing loss from analysis, thus precluding the analysis of possible gene–gene interactions in the phenotypic expression of these syndromes. One study in the United Kingdom included DNA analysis of seven families with syndromic hearing loss. The DNA analysis looked only for the 35ΔG allele, which was not detected in any of these families (44). The small number of participants and the allele-specific DNA analysis limit any conclusions about the role of *GJB2* variants in syndromic hearing loss.

Over 400 different recognizable syndromes have hearing loss as a component, varying in degree of loss, age of onset, and penetrance (11). The aforementioned variation in degree of hearing loss in siblings with identical *GJB2* genotypes indicates the importance of genetic and/or environmental backgrounds in the expression of *GJB2*-related hearing loss. Hence, it is also possible that variants in genes such as *GJB2* influence the penetrance and expressivity of hearing loss associated with syndromes. This possibility remains to be explored.

Laboratory Detection Techniques

The known genetic variants in the *GJB2* gene are amenable to detection by standard molecular genetic laboratory techniques. The majority of *GJB2* variants fall in the 680-base pair coding region in exon 2, and the rest fall in the 3′ untranslated region in exon 1. Detection methods include allele-specific PCR-based methods, scanning technologies such as SSCP, and sequencing. As some common alleles account for the majority of variants in some populations, allele-specific methods are often used, either alone or in conjunction with sequencing methods. A recent survey of laboratory practices pertaining to clinical use of *GJB2* testing indicated that U.S. laboratories vary in their chosen methodology. Most of the laboratories used sequencing, either alone or as a follow-up to allele-specific methods. Of the laboratories that employed sequencing, most analyzed exon 2 only, while a few sequenced both exons 1 and 2 (45). Most of the published studies that have utilized sequencing methods have included analysis of exon 2 only. Therefore, information is lacking to accurately determine the relative clinical validity and utility of these two methods.

Potential Contribution of Genetic Information to Improved Health Outcomes

The American College of Medical Genetics recommends the provision of genetic services to individuals with hearing loss "to establish the etiology whenever possible, (46)" and *GJB2* testing may be one potential option in this process. Several potential clinical uses of *GJB2* testing in children with hearing loss have been proposed, including (*1*) ruling-out risk of syndromic complications, (*2*) predicting moderate-to-profound hearing loss requiring aggressive language intervention, (*3*) indicating potential candidacy for cochlear implant use, and (*4*) allowing genetic counseling regarding recurrence rates (37,47,48). However, there is little evidence in support of some of these proposed uses, and there are many factors to be weighed in the decision to include *GJB2* testing.

Much of the information regarding a child's phenotype can be obtained through physical examination by audiologists, otolaryngologists, and clinical geneticists, so a child's course of intervention may or may not be significantly altered by the knowledge of *GJB2* genotype. It has been argued a *GJB2* diagnosis may reduce the burden of additional tests that are traditionally used to rule-out syndromic complications (e.g., ophthalmologic, cardiac, and vestibular evaluations) (48). This may be particularly relevant in the case of infants with hearing loss, as the medical tests used to distinguish syndromic and nonsyndromic cases may not have as much predictive power during infancy as they do later in childhood. However, the potential role of *GJB2* variants in the penetrance and expressivity of hearing loss in syndromes has not been assessed, so while the presence of *GJB2* variants in a newborn with hearing loss will most likely be associated with nonsydromic hearing loss, more data need to be collected to determine the sensitivity and specificity of this use of *GJB2* testing. Likewise, the use of *GJB2* testing in the prediction of the success of various intervention options has not been assessed.

Determination of a genetic etiology also allows for the provision of recurrence information for the family. However, organizations of professional geneticists, including the American Society of Human Genetics and the American College of Medical Genetics, generally discourage the genetic testing of minors in the absence of direct intervention benefits for the child (49). Given the Deaf community's concerns about genetic testing (50), this point is particularly germane in regards to hearing loss, because a child may prefer not to know this information as an adult. (The "Deaf" community with a capital "D" refers to a community that shares a specific linguistic [American Sign Language] and cultural identity.)

Another factor to be considered in the decision to use *GJB2* testing is the lack of epidemiologic data pertaining to the less common variants. This paucity of information must be considered when counseling families about *GJB2* test options and results. Consider, for example, the M34T allele. In 1997, the M34T variant was found to co-segregate with three generations of hearing loss in one family in an apparently dominant fashion (20). This was the first evidence implicating *GJB2* in nonsyndromic hearing loss. But several years later, a second variant was characterized in this family, found *in trans* with the M34T allele in the individuals with hearing loss (51). This suggests that the M34T allele is recessive in nature. Since then, several groups have documented the failure of M34T to co-segregate with hearing loss in several families, raising the possibility that M34T may be a benign polymorphism (52–55). More recent evidence that supports the recessive allele model includes the observation of M34T compound heterozygotes and homozygotes among individuals with hearing loss but not among control populations (51–60). This example cautions researchers and clinicians to interpret the role of less common, and hence less well-characterized, variants with care, particularly in regards to family genetic counseling issues.

These issues will continue to affect an increasing number of families as an increasing number of states are screening newborns for audiologic function, so that infants with hearing loss are identified and referred for intervention services very early in life. The role that *GJB2* testing will play in conjunction with EHDI programs is still in the process of being defined.

Conclusions

Several areas of current research are aimed at the definition of the clinical utility of *GJB2* genetic testing. One such area of research focuses on the potential role of *GJB2* genotyping in the prediction of success of various intervention options, such as cochlear implants. A second area of research involves the contribution of *GJB2* variants to hearing loss in diverse populations. Given the interpopulation variability of the prevalence of *GJB2* variants and their apparent contribution to hearing loss, it would be helpful to define these measures in all potential target populations. In the United States, for example, there is a lack of data pertaining to non-white American populations. Epidemiologic data specific to these populations are necessary to provide population-specific determinations of clinical validity and utility.

While *GJB2* variants have been shown to be associated with a large fraction of cases of nonsyndromic moderate-to-profound prelingual hearing loss, the potential contribution of *GJB2* variants to mild, unilateral, late-onset, syndromic, or environmentally acquired cases of hearing loss has not yet been determined. Research into potential associations such as these may help to unravel the interactions between genetic and environmental influences in the phenotypic expression of hearing function and hearing loss.

The emergence of EHDI programs presents an excellent opportunity for population-based ascertainment of cases of congenital hearing loss. Similar population-based strategies are needed for complete ascertainment of cases of hearing loss arising sometime after the newborn period. Inclusion of all cases of hearing loss, regardless of etiology (syndromic vs. nonsyndromic, other known genetic factors), presence of known risk factors, degree of hearing loss, and age at onset are required to fully assess the contribution of the *GJB2* gene to the spectrum of hearing loss phenotypes. In addition, as greater than 5% of the general population have a hearing loss of some kind (61), the ascertainment of control populations should also be carefully considered in this type of analysis and should include individuals known not to have a hearing loss. Confounding variables including age, sex, race/ethnicity, and presence of known risk factors are important considerations in case-control analyses of *GJB2* variants and hearing loss.

A final note about the process of defining the potential role of genetic testing in medical practice relates to the consideration of the viewpoints of all stakeholders. The case of genetic testing pertaining to hearing loss has raised some important issues in this regard. The goal of the medical community to eliminate disease and disability is at odds with the viewpoint of the Deaf community that hearing loss is not a disability. This viewpoint challenges society to reconsider the definitions of disease and disability. The Deaf community has also expressed concerns that genetic testing will do more harm than good and will devalue individuals with hearing loss (50). The issues raised by the Deaf community provide a unique opportunity by challenging scientists and society to find culturally-sensitive methods for genetic research and testing that are acceptable to all cultural groups.

References

1. Blanchfield BB, Feldman JJ, Dunbar JL, Gardner EN. The severely to profoundly hearing-impaired population in the United States: prevalence estimates and demographics. J Am Acad Audiol 2001;12:183–189.
2. Davidson J, Hyde ML, Alberti PW. Epidemiologic patterns in childhood hearing loss: a review. Int J Pediatr Otorhinolaryngol 1989;17:239–266.
3. Elssman SA, Matkin ND, Sabo MP. Early identification of congenital sensorineural hearing impairment. The Hearing J 1987;40:13–17.
4. Yoshinaga-Itano C, Sedey AL, Coulter DK, Mehl AL. Language of early- and later-identified children with hearing loss. Pediatrics 1998;102:1161–1171.
5. Moeller MP. Early intervention and language development in children who are deaf and hard of hearing. Pediatrics 2000;106:E43.
6. Yoshinaga-Itano C. Benefits of early intervention for children with hearing loss. Otolaryngol Clin North Am 1999;32:1089–1102.
7. Tomaski SM, Grundfast KM. A stepwise approach to the diagnosis and treatment of hereditary hearing loss. Pediatr Clin North Am 1999;46:35–48
8. Holtzman NA, Watson MS, eds. Final Report of the Task Force on Genetic Testing: Promoting Safe and Effective Genetic Testing in the United States. Washington DC: National Human Genome Research Institute, 1997.
9. Secretary's Advisory Committee on Genetic Testing. Enhancing the oversight of genetic tests: Recommendations of the SACGT. Bethesda, MD: National Institutes of Health, 2000.
10. Hudspeth AJ. How the ear's works work. Nature 1989;341:397–404.
11. Gorlin RJ, Toriello HV, Cohen MM, eds. Hereditary Hearing Loss and Its Syndromes. New York: Oxford University Press, 1995.

12. Marazita ML, Ploughman LM, Rawlings B, et al. Genetic epidemiological studies of early-onset deafness in the U.S. school-age population. Am J Med Genet 1993;46:486–491.
13. Morton NE. Genetic epidemiology of hearing impairment. Ann N Y Acad Sci 1991; 630:16–31.
14. Reardon W. Genetic deafness. J Med Genet 1992;29:521.
15. Van Camp G, Willems PJ, Smith RJ. Nonsyndromic hearing impairment: unparalleled heterogeneity. Am J Hum Genet 1997;60:758–764.
16. Hereditary Hearing Loss Homepage http://dnalab-www.uia.ac.be/dnalab/hhh/index.html
17. Guilford P, Ben Arab S, Blanchard S, Levilliers J, Weissenbach J, Belkahia A, Petit C. A non-syndrome form of neurosensory, recessive deafness maps to the pericentromeric region of chromosome 13q. Nat Genet 1994;6:24–28.
18. Chaib H, Lina-Granade G, Guilford P, Plauchu H, Levilliers J, Morgon A, Petit C. A gene responsible for a dominant form of neurosensory non-syndromic deafness maps to the NSRD1 recessive deafness gene interval. Hum Mol Genet 1994;3:2219–2222.
19. Denoyelle F, Lina-Granade G, Plauchu H, Bruzzone R, Chaib H, Levi-Acobas F, Weil D, Petit C. Connexin 26 gene linked to a dominant deafness. Nature 1998;393:319–320.
20. Kelsell DP, Dunlop J, Stevens HP, Lench NJ, Liang JN, Parry G, Mueller RF, Leigh IM. Connexin 26 mutations in hereditary non-syndromic sensorineural deafness. Nature 1997;387:80–83.
21. Richard G. Connexins: a connection with the skin. Exp Dermatol 2000;9:77–96.
22. Kenneson A, Boyle C. GJB2 (Connexin 26) Variants and nonsyndromic sensorineural hearing loss. Genet Med 2002;4:258–274.
23. Lerer I, Sagi M, Malamud E, Levi H, Raas-Rothschild A, Abeliovich D. Contribution of connexin 26 mutations to nonsyndromic deafness in Ashkenazi patients and the variable phenotypic effect of the mutation 167delT. Am J Med Genet 2000;95:53–56.
24. Sobe T, Erlich P, Berry A, et al. High frequency of the deafness-associated 167delT mutation in the connexin 26 (GJB2) gene in Israeli Ashkenazim. Am J Med Genet 1999; 86:499–500.
25. Casademont I, Bizet C, Chevrier D, et al. Rapid detection of campylobacter fetus by polymerase chain reaction combined with non-radioactive hybridization using an oligonucleotide covalently bound to microwells. Mol Cell Probes 2000;14:233–240.
26. Denoyelle F, Marlin S, Weil D, et al. Clinical features of the prevalent form of childhood deafness, DFNB1, due to a connexin-26 gene defect: implications for genetic counselling. Lancet 1999;353:1298–1303.
27. Denoyelle F, Weil D, Maw MA, et al. Prelingual deafness: high prevalence of a 30delG mutation in the connexin 26 gene. Hum Mol Genet 1997;6:2173–2177.
28. Estivill X, Fortina P, Surrey S, et al. Connexin-26 mutations in sporadic and inherited sensorineural deafness. Lancet 1998;351:394–398.
29. Lench N, Houseman M, Newton V, et al. Connexin-26 mutations in sporadic non-syndromal sensorineural deafness. Lancet 1998;351:415.
30. Mueller RF, Nehammer A, Middleton A, et al. Congenital non-syndromal sensorineural hearing impairment due to connexin 26 gene mutations—molecular and audiological findings. Int J Pediatr Otorhinolaryngol 1999;50:3–13.
31. Murgia A, Orzan E, Polli R, et al. Cx26 deafness: mutation analysis and clinical variability. J Med Genet 1999;36:829–832.
32. Tessa A, Patrono C, Santorelli FM, et al. Rapid detection of the 35delG mutation in the GJB2 gene in childhood deafness. J Med Screen 2000;7:167.
33. Park HJ, Hahn SH, Chun YM, et al. Connexin26 mutations associated with nonsyndromic hearing loss. Laryngoscope 2000;110:1535–1538.
34. Abe S, Usami S, Shinkawa H, et al. Prevalent connexin 26 gene (GJB2) mutations in Japanese. J Med Genet 2000;37:41–43.

35. Fuse Y, Doi K, Hasegawa T, Sugii A, et al. Three novel connexin26 gene mutations in autosomal recessive non-syndromic deafness. Neuroreport 1999;10:1853–1857.
36. Kudo T, Ikeda K, Kure S, et al. Novel mutations in the connexin 26 gene (GJB2) responsible for childhood deafness in the Japanese population. Am J Med Genet 2000;90: 141–145.
37. Milunsky JM, Maher TA, Yosunkaya E, et al. Connexin-26 gene analysis in hearing-impaired newborns. Genet Test 2000;4:345–349.
38. Green GE, Smith RJ, Bent JP, et al. Genetic testing to identify deaf newborns. JAMA 2000;284:1245.
39. Salvador MQ, Fox MA, Schimmenti LA, et al. Homozygosity for the connexin 26 167delT mutation in an Ashkenazi Jewish family. Am J Hum Genet 2000;67:202.
40. Loffler J, Nekahm D, Hirst-Stadlmann A, et al. Sensorineural hearing loss and the incidence of Cx26 mutations in Austria. Eur J Hum Genet 2001;9:226–230.
41. Orzan E, Polli R, Martella M, et al. Molecular genetics applied to clinical practice: the Cx26 hearing impairment. Br J Audiol 1999;33:291–295.
42. Masmoudi S, Elgaied-Boulila A, Kassab I, et al. Determination of the frequency of connexin26 mutations in inherited sensorineural deafness and carrier rates in the tunisian population using DGGE. J Med Genet 2000;37:E39.
43. Morle L, Bozon M, Alloisio N, et al. A novel C202F mutation in the connexin26 gene (GJB2) associated with autosomal dominant isolated hearing loss. J Med Genet 2000;37: 368–370.
44. Parker MJ, Fortnum HM, Young ID, Davis AC, Mueller RF. Population-based genetic study of childhood hearing impairment in the Trent Region of the United Kingdom. Audiology 2000;39:226–231.
45. Kenneson A, Myers MF, Lubin IM, Boyle C. Genetic laboratory practices related to testing of the GJB2 (connexin 26) gene in the United States in 1999 and 2000. Genet Test (in press).
46. Nance WE, Cunningham GC, Davis JG, et al. Statement of the American College of Medical Genetics on Universal Newborn Hearing Screening. Genet Med 2000;2:149–150.
47. Cohn ES, Kelley PM, Fowler TW, et al. Clinical studies of families with hearing loss attributable to mutations in the connexin 26 gene. Pediatrics 1999;103:546–550.
48. Cohn ES, Kelley PM. Clinical phenotype and mutations in connexin 26 (DFNB1/GJB2), the most common cause of childhood hearing loss. Am J Med Genet 1999;89:130–136.
49. American Society of Human Genetics and the American College of Medical Genetics (1995) Points to consider: Ethical, legal, and psychological implications of genetic testing in children and adolescents. Am J Hum Genet 1995;57:1233–1241.
50. Middleton A, Hewison J, Mueller RF. Attitudes of deaf adults toward genetic testing for hereditary deafness. Am J Hum Genet 1998;63:1175–1180.
51. Kelsell DP, Wilgoss AL, Richard G, et al. Connexin mutations associated with palmoplantar keratoderma and profound deafness in a single family. Eur J Hum Genet 2000; 8:469–472.
52. Cucci RA, Prasad S, Kelley PM, et al. The M34T allele variant of connexin 26. Genet Test 2000;4:335–344.
53. Griffith AJ, Chowdhry AA, Kurima K, et al. Autosomal recessive nonsyndromic neurosensory deafness at DFNB1 not associated with the compound-heterozygous GJB2 (connexin 26) genotype M34T/167delT. Am J Hum Genet 2000;67:745–749.
54. Houseman MJ, Ellis LA, Pagnamenta A, et al. Genetic analysis of the connexin-26 M34T variant: identification of genotype M34T/M34T segregating with mild-moderate nonsyndromic sensorineural hearing loss. J Med Genet 2001;38:20–25.
55. Kelley PM, Harris DJ, Comer BC, et al. Novel mutations in the connexin 26 gene (GJB2) that cause autosomal recessive (DFNB1) hearing loss. Am J Hum Genet 1998;62:792–799.

56. Bason LD, Shah UK, Potsic WP, et al. Diagnostic evaluation and counseling of 97 families with hearing loss. Am J Hum Genet 2000;67:240.
57. Green GE, Scott DA, McDonald JM, et al. Carrier rates in the midwestern United States for GJB2 mutations causing inherited deafness. JAMA 1999;281:2211–2216.
58. Pandya A, Oelrich K, Arnos KS, et al. Connexin (Cx) testing in a nationwide repository of samples from deaf probands: relevance to clinical practice. Am J Hum Genet 2000; 67:240.
59. Prasad S, Cucci RA, Green GE, et al. Genetic testing for hereditary hearing loss: Connexin 26 (GJB2) allele variants and two novel deafness-causing mutations (R32C and 645-648delTAGA). Hum Mutat 2000;16:502–508.
60. Wilcox SA, Saunders K, Osborn AH, et al. High frequency hearing loss correlated with mutations in the GJB2 gene. Hum Genet 2000;106:399–405.
61. Niskar AS, Kieszak SM, Holmes A, et al. Prevalence of hearing loss among children 6 to 19 years of age: the Third National Health and Nutrition Examination Survey. JAMA 1998;279:1071–1075.
62. Antoniadi T, Gronskov K, Sand A, et al. Mutation analysis of the GJB2 (connexin 26) gene by DGGE in Greek patients with sensorineural deafness. Hum Mutat 2000;16:7–12.
63. Antoniadi T, Rabionet R, Kroupis C, et al. High prevalence in the Greek population of the 35delG mutation in the connexin 26 gene causing prelingual deafness. Clin Genet 1999;55:381–382.
64. Gasparini P, Rabionet R, Barbujani G, Melchionda S, Petersen M, Brondum-Nielsen K, Metspalu A, Oitmaa E, Pisano M, Fortina P, Zelante L, Estivill X. High carrier frequency of the 35delG deafness mutation in European populations. Genetic Analysis Consortium of GJB2 35delG. Eur J Hum Genet 2000;8:19–23.
65. Lucotte G, Bathelier C, Champenois T. PCR test for diagnosis of the common GJB2 (connexin 26) 35delG mutation on dried blood spots and determination of the carrier frequency in France. Mol Cell Probes 2001;15:57–59.
66. Storm K, Willocx S, Flothmann K, et al. Determination of the carrier frequency of the common GJB2 (connexin-26) 35delG mutation in the Belgian population using an easy and reliable screening method. Hum Mutat 1999;14:263–266.
67. Morell RJ, Kim HJ, Hood LJ, et al. Mutations in the connexin 26 gene (GJB2) among Ashkenazi Jews with nonsyndromic recessive deafness. N Engl J Med 1998;339: 1500–1505.
68. Scott DA, Kraft ML, Carmi R, et al. Identification of mutations in the connexin 26 gene that cause autosomal recessive nonsyndromic hearing loss. Hum Mutat 1998;11:387–394.
69. Dong J, Katz DR, Eng CM, et al. Nonradioactive Detection of the Common Connexin 26 167delT and 35delG Mutations and Frequencies among Ashkenazi Jews. Mol Genet Metab 2001;73:160–163.
70. Sobe T, Vreugde S, Shahin H, et al. The prevalence and expression of inherited connexin 26 mutations associated with nonsyndromic hearing loss in the Israeli population. Hum Genet 2000;106:50–57.

25

Genetic and environmental factors in cardiovascular disease

Molly S. Bray

In Western societies, cardiovascular disease (CVD) is the leading cause of morbidity and mortality in adults. For example, an estimated 61 million Americans currently have some form of CVD. CVD refers to a variety of diseases and conditions affecting the heart and blood vessels, including stroke, atherosclerosis, hypertension, arteriosclerosis, and coronary heart disease (CHD). In 1999 in the United States, almost one million men and women died of CVD (1). African Americans continue to bear a disproportionate burden of CVD, with the age-adjusted death rates from CHD currently greater than 35% higher in African-American men and almost 70% higher in African-American women than in their non-Hispanic, white counterparts. In 2002, an estimated $329 billion in health care costs will be directly attributable to CVD, including both direct costs associated with medical care and indirect costs resulting from lost productivity from both CVD morbidity and mortality (1). Extensive research has demonstrated that the development of CVD begins in childhood in both African Americans and whites (2), and detecting those at increased risk earlier in life so they can seek healthier lifestyles is an important goal of CVD primary prevention programs.

Of all common chronic diseases, none is so clearly influenced by the contribution of both genes and environments as CVD. Years of epidemiologic research have provided evidence for numerous behavioral, physiologic, and environmental factors that increase risk for the development of CVD. Yet, in no case do these risk factors, either singly or combined, explain all the variation observed in the onset and progression of CVD. In fact, many instances exist of persons who are resistant to CVD despite having a number of established risk factors, as well as those who develop CVD while living a healthy and low-risk lifestyle. Sir Winston Churchill, renowned leader of Great Britain during World War II, died at age 91 years despite a daily regimen of alcohol drinking, cigar smoking, and heavy eating throughout

436

his adult life, while Jim Fixx, America's first running and health guru, died from a massive heart attack at age 52. Clearly, a family history of longevity in Churchill's case and of heart disease in Fixx's were important factors in determining the health outcomes of these individuals.

That CVD has a multifactorial etiology involving both genetic and environmental factors is widely accepted. A model for such gene–environment contribution to disease is depicted in Figure 25.1. In this model, both genetic and environmental factors can influence the disease state directly or through their interaction with other factors. Susceptibility to CVD is continuously distributed among individuals and is determined in large part by variation in levels of quantitative intermediate traits that influence the onset, development, and severity of disease (3,4). Such trait variation can result from both genomic and environmental variation, as well as from the interaction between the two. In addition to risk factors such as age and smoking, the central role of plasma lipids in the etiology of CVD has been established by more than 50 years of research (5). Elevated total- and low-density lipoprotein (LDL) cholesterol levels are associated with higher CVD risk, as are low levels of high-density lipoprotein (HDL) cholesterol (6). Elevated blood pressure also has long been established as a major risk factor for premature CHD, cerebrovascular disease, and renal disease (7). In addition, obesity and increased intra-abdominal fat are associated with increased risk for CVD, as well as for other CVD risk factors including type 2 diabetes, dyslipidemia, and insulin resistance (8). Each of these risk factors has, in turn, been shown to have both a genetic and environmental basis for its determination.

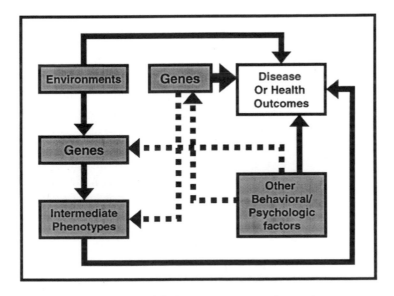

Figure 25.1 Model of gene–environment interaction.

The consensus from multiple reviews of the literature is that the contribution of genetic variation to quantitative CVD risk factor variation is statistically significant and greater than that of shared familial environmental factors (9–11). Positive family history, particularly in early-onset CHD, is a well-established risk factor for CVD, and familial clustering of atherosclerotic and thrombotic cardiovascular disease has been reported in numerous studies (12). Nora et al. estimated the heritability of early-onset CHD (before age 55 years) to be as high as 0.56 in families in which ischemic heart disease had occurred in at least one family member before age 55 years, even after excluding cases with genetic-based hyperlipidemias (13). Significantly higher concordance for death from ischemic heart disease has also been reported for monozygotic versus dizygotic twins from the Norwegian Twin Panel (14).

In addition to disease states, substantial heritability for quantitative CVD risk factors has been reported. Approximately 50% to 83% of the interindividual variance in total, HDL, and LDL serum cholesterol levels has been attributed to genetic variation in several studies of white families (15,16). In Mexican Americans, genetic variation accounted for 30% to 45% of the variation in lipid and lipoprotein levels and from 15% to 30% of the phenotypic variation in measures of glucose, hormones, adiposity, and blood pressure (17). Estimates of heritability of blood pressure ranged from 22% (pedigrees) to 63% (twins) in families from Utah (18), and several determinants of blood pressure (urinary kallikrein, high fat pattern index, intraerythrocytic sodium, sodium-lithium countertransport, and ouabain binding sites) have also been demonstrated to have a highly heritable component (19).

Studies of monogenic cases of hyperlipidemias have provided much insight into the etiology of atherosclerotic CVD and support for the importance of genetic variation in the determination of disease phenotypes. Familial hypercholesterolemia (FH), one of the best-studied of these conditions, results primarily from mutations in the receptor for low-density lipoprotein (*LDLR*). In a classic series of experiments, Goldstein and Brown demonstrated that genetic variation in the *LDLR* gene can produce receptor binding defects, receptor internalization defects, and/or a nonfunctional receptor that ultimately results in the severe atherosclerotic plaque accumulation and premature mortality associated with FH (20). These studies helped to elucidate the critical role of cholesterol metabolism in the etiology of atherosclerosis and thrombotic heart disease. Other monogenic disorders, such as glucocorticoid remediable aldosteronism (GRA) and Liddle's syndrome, both of which produce severe early hypertension and stroke (albeit through different pathways), also have provided insight into the role of hypertension in prevalent forms of CVD (21,22). Nevertheless, although the genetic basis of these diseases is known (abnormal splicing of the aldosterone synthase and steroid 11-β-hydroxylase genes produces GRA, and mutations in the β subunit of the epithelial sodium channel gene result in Liddle's syndrome), these rare single-gene syndromes account for a tiny fraction of CVD, and most common forms of CVD neither result from mutations in these genes nor follow any clear Mendelian form of inheritance. Thus, the most fruitful searches for genetic factors that predict common forms of CVD have come from studies of candidate genes based on the known physiology underlying CVD etiology rather than on studies of familial inheritance.

Literally hundreds of genes have been proposed as candidates for cardiovascular-related diseases. Most studies of candidate genes have focused on components of lipid metabolism, blood pressure regulation, and hemostasis and thrombosis. Other candidate genes for CVD may include those related to obesity, insulin signaling and diabetes, and inflammation. A brief list of potential candidate genes is presented in Table 25.1. This list is necessarily incomplete because new candidate genes are being continuously proposed and tested, and the functional classification of these candidate genes in the etiology of CVD is not always distinct.

Table 25.1 Summary of Candidate Genes for Cardiovascular Disease

Function	Candidate Genes
Lipid metabolism	Apolipoproteins (*APOA1, APOA2, APOA4, APOB, APOC1, APOC2, APOC3, APOC4, APOD, APOE, LPA*), cholesteryl ester transfer protein (*CETP*), CD36 antigen (*CD36*), HMG Co-A reductase (*HMGCR*), lecithin:cholesterol acyltransferase (*LCAT*), LDL receptor (*LDLR*), hepatic lipase (HL), hormone-sensitive lipase (*LIPE*), lipoprotein lipase (*LPL*), paraoxonase (*PON1*)
Blood pressure regulation	Adducin (*ADDA*), adrenergic receptors (*ADRA2A, ADRB1, ADRB2, ADRB3*), angiotensinogen (*AGT*), angiotensin receptors (*AT1R, AT2R*), arginine vasopressin (*AVP*), AVP receptor 2 (*AVPR2*), bradykinin receptor (*BDKRB2*), angiotensin converting enzyme (*ACE*), dopamine receptor D1 (*DRD1*), guanine nucleotide binding protein (*GNB3*), renin (REN), renin binding protein (*RENBP*), very-low-density lipoprotein receptor (*VLDLR*)
Hemostasis and thrombosis	Antithrombin (*AT3*), endothelin (*EDN1, EDN2*), endothelin receptor (*EDNRA, EDNRB*), factor X (*F10*), factor XI (*F11*), factor XIII (*F13A1, F13B*), prothrombin (*F2*), thrombin receptor (*F2R*), factor III (*F3*), factor V (*F5*), factor VII (*F7*), factor IX (*F9*), fibrinogen (*FGA, FGB, FGG*), haptoglobin (*HP*), intercellular adhesion molecule (*ICAM1, ICAM2*), platelet glycoprotein (*ITGA2B, ITGB3*), atrial naturietic peptide (*NPPA*), nitric oxide synthase (*NOS3*), brain naturietic peptide (*NPPB*), plasminogen activator inhibitor (*PAI1, PAI2*), plate-derived growth factor (*PDGFA*), selectins (*SELE, SELL, SELP*), thrombomodulin (*THBD*), vascular cell adhesion molecule (*VCAM*)
Insulin signaling	Glucagon (*GCG*), glucagon receptor (*GCGR*), glycogen synthase (*GYS1*), islet amyloid polpeptide (*IAPP*), insulin-like growth factor (*IGF1, IGF2*), insulin (*INS*), insulin receptor (*INSR*), insulin promoter factor (*IPF1*), insulin receptor substrate (*IRS2*), glucose transporters (*SLC2A1, SLC2A2, SLC2A3, SLC2A4*)
Obesity	Leptin (*LEP*), leptin receptor (*LEPR*), neuropeptide Y (*NPY*), NPY receptor (*NPYY1, NPYY5*), pro-opiomelanocortin (*POMC*), melanocortin receptors (*MC4R, MC3R*), agout-related protein (*AGRP*), peroxisome-proliferator activated receptors (*PPARG, PPARA*), uncoupling proteins (*UCP2, UCP3*)
Inflammation	Interleukins (*IL1A, IL6, IL10, IL18*), tumor necrosis factor-α (*TNFA*), interferon (*IFN*)

Although a few candidate genes have consistently been associated with several intermediate risk factors for CVD, no single gene has been shown or is expected to be responsible for the majority of the variation in CVD risk (23). Rather, the combined effects of multiple genes, each with polymorphic alleles having moderate trait effects, and multiple environments produce most of the genetic variation in CVD risk (4,10). Thus, determination of the genetic components influencing CVD and related morbidities is complicated by the multiplicity of contributing factors. Elucidating genes for CVD is particularly challenging due to the fact that genes do not act independently of behavior and environment, and detecting genetic effects may be critically dependent upon the environment men in which they are being manifested.

Case Studies of Selected CVD Candidate Genes

Of the many potential candidate gene polymorphisms associated with CVD outcomes, two genes that have consistently been associated with both risk factors and disease states include the cholesteryl ester transfer protein (*CETP*) and the guanine nucleotide binding protein β3 subunit (*GNB3*) genes. These genes are presented here as representative examples of candidate genes for CVD.

High levels of total cholesterol, in particular high levels of LDL cholesterol and low levels of HDL cholesterol, are associated with the development of atherosclerosis (24). Deposits of cholesterol in arterial plaques may be reduced by reverse cholesterol transport, wherein excess cellular cholesterol is carried by HDL, transferred to triglycerides, and removed by the liver for secretion in bile. Reverse cholesterol transport rate depends on the activity of a plasma protein, cholesteryl ester transfer protein (CETP), which facilitates transport of cholesteryl esters from HDL-cholesterol to triglyceride-rich lipoproteins (25). The protein CETP regulates the rate of cholesterol transport toward the liver for excretion (26). Genetic variation that results in decreased activity of CETP has been associated with marked increases in HDL cholesterol. Therefore, CETP has been viewed as potentially both a protective factor against and a risk-raising factor in atherosclerosis.

The *CETP* gene has been localized to chromosome 16q21 and encompasses 16 exons. Several rare mutations in the *CETP* gene have been identified that result in the absence of detectable CETP mass or activity, reducing the rate of reverse cholesterol transport and producing elevated levels of HDL cholesterol in affected individuals (27). Common variants in the *CETP* gene have been studied to determine the complex association between CETP, HDL cholesterol, and atherosclerotic diseases. One specific polymorphism in the *CETP* gene, referred to as *Taq*IB and located in intron 1 of the gene, has been associated with altered lipid-transfer activity (27–31) and HDL cholesterol levels (32–34). The *Taq*IB variant influences CETP plasma concentrations, which in turn modify the reverse cholesterol transport process and HDL cholesterol levels.

We recently investigated the *CETP Taq*IB variant and its relation to incident CHD and HDL cholesterol in a sample of white men and women from the Atherosclerosis Risk in Communities (ARIC) study. The genotype distribution for the study sample was 31.9% for B1/B1 (n = 315), 50.2% for B1/B2 (n = 495), and 17.8% for B2/B2 (n = 176), and this is consistent with other reports from the literature. In this sample of 375 incident CHD cases and 611 controls, the frequency of the B2 allele was significantly lower (p = 0.023) in the incident CHD cases (39.9%) than in the noncases (46.1%). In a Cox proportional hazards model including age and gender as covariates, the *CETP* variant was significantly associated with incident CHD case status (p = 0.0106, hazard ratio (HR) = 0.58; 95% confidence interval (CI), 0.38–0.88). When other established cardiovascular risk factors, such as body mass index (BMI), smoking status, hypertension, total cholesterol, diabetes, alcohol drinking status, and leisure-time physical activity, were included in the model, the *CETP* variant remained significantly associated with incident CHD (p = 0.0239, HR = 0.49; 95% CI, 0.26–0.91), and the risk for CHD decreased by almost half for B2 homozygotes, compared with those with one or two B1 alleles (35). In addition, in a multivariate regression model including age, gender, smoking status, BMI, and leisure-time physical activity, the *CETP* B1/B2 (p = 0.0007) and B2/B2 (p < 0.0001) genotypes were significant predictors of HDL cholesterol levels after adjusting for covariates.

The results from our work concur with previous reports (36,37). In the Framingham population, presence of the *CETP* B2 allele decreased the risk for CHD by 30% in men after adjusting for age, BMI, systolic blood pressure, diabetes, smoking, and alcohol consumption (37). A recent study published by Durlach et al. also suggested a sex-dependent association of the *CETP* polymorphism for CHD in type 2 diabetic patients. Male diabetic patients with the B2/B2 genotype had a lower risk for coronary artery pathology than did patients with one or two B1 alleles (38). This sex-dependent association was not confirmed in a similar study of Japanese diabetics, but the *CETP* polymorphism was associated with CHD when genders were combined (36). In general, results from the literature support the conclusion that the *CETP* B2/B2 genotype protects against CHD, although further investigations are needed to determine the mechanism of the effect of *CETP* on atherosclerosis.

As another example of a candidate gene for CVD, the protein product of the *GNB3* gene plays a critical role in sodium processing and total body fluid homeostasis by directing the trafficking of sodium/potassium channels in and out of the cell membrane. In addition, intracellular signaling by G proteins is an essential step in the formation of mature adipocytes (39). Thus, *GNB3* has been proposed as a candidate gene for both essential hypertension and obesity. The *GNB3* gene, located on chromosome 12p13, was first identified as a potentially important gene for human hypertension when cell lines derived from hypertensive subjects demonstrated an increase in G-protein signal transduction and activity of the sodium–proton exchanger compared with cells from nonhypertensive patients (40). Exhaustive sequence analysis subsequently resulted in the discovery of a polymorphism in the *GNB3* gene, C825T, which results in a 123 base pair deletion in exon 9 of the heterotrimeric G protein (41). This discovery was particularly exciting because the variant was located outside of the known consensus splice site regions and suggested that much of variation in DNA sequence might have the potential to be functional (42). Interestingly, allele frequencies for this polymorphism differ dramatically between white and African populations (43). In non-Hispanic white populations, the 825C and 825T alleles have a frequency of approximately 0.70 and 0.30, respectively, while in black Africans (BA) and African Americans (AA), these allele frequencies are almost completely reversed, with frequencies ranging from 0.82 (BA) to 0.74 (AA) for the 825T allele. Allele frequencies in Asian populations are intermedidate between these two extremes (43). Detection of population-specific alleles or differential allele frequencies between ethnicities in which disease prevalence also differs is one strategy that researchers have used to identify putatively disease-causing alleles. Such allele frequency differences in the *GNB3* gene may represent, at least in part, the underpinnings of hypertension prevalence differences between these populations.

In initial investigations, the 825T allele was observed in only 44% of normotensive subjects compared with 53.1% of hypertensive subjects (41). This discovery led to several studies that reported an association of hypertension and the 825T allele in a number of populations including Australians, Germans, and Canadian Oji-Cree (41,44–46). The *GNB3* 825T allele also has been associated with left ventricular hypertrophy (47), enhanced coronary vasoconstriction (48), and impaired left ventricular diastolic filling (49). In Japanese, the 825T allele was associated with serum potassium, total cholesterol, and hypertension (50,51). Additional researchers have demonstrated an increase in sodium–proton exchanger activity in obese, but not in lean, persons (52), and stimulation of adipogenesis by increased signaling of pertussis toxin-sensitive G-proteins (53). Obesity is a well-known risk factor for hypertension, and studies have shown that obesity is associated with the 825T allele in hypertensive (39) and normotensive subjects (45). Subsequently, the 825T allele has been associated with obesity in such diverse ethnic populations as Germans, Chinese, Black African, and Canadian Inuits (39,45).

Of the many candidate genes that have been identified for CVD-related traits, we have focused only on two, *CETP* and *GNB3*, which have demonstrated consistent associations with both disease states and quantitative risk factors in multiple studies and in multiple populations. Other genes that have shown similarly consistent findings include the angiotensinogen (AGT) gene, the angiotensin converting enzyme (ACE) gene, and the apolipoprotein E (APOE) gene.

Gene–Environment Interaction and CVD Candidate Genes

Although several studies have consistently demonstrated the effects of both the *CETP* and *GNB3* genes on both qualitative and quantitative traits related to CVD, and functional experiments have provided support for a physiologic mechanism by which these genetic variants may produce change in disease phenotypes, of essential importance is the determination of environmental factors that may modify these effects. Detection of gene–environment interactions that influence the development of CVD-related morbidities may be difficult because of a number of factors. First, the disease states are highly heterogeneous and can develop via multiple mechanisms (e.g., one person may develop CHD consequent to chronic inflammation, while another may eat a very high salt diet that contributes to hypertension and atherosclerosis). Second, several to many genes are likely to influence the development of CVD, each with small to moderate effects. Third, expression of the genetic defect producing CVD may be context dependent (e.g., a defect in estrogen signaling that contributes to alterations in lipid levels may be evident only in females and not in males). Fourth, complex physiologic functions such as the regulation of blood pressure or vasoconstriction will have multiple pathways and redundancy to ensure survival of the organism. Many association studies of candidate CVD gene polymorphisms have been performed, often with conflicting results due, at least in part, to the context dependency of the genetic effects.

We were particularly interested in investigating how environments known to influence HDL cholesterol might modify or be modified by variation in *CETP*. Thus, we tested the interaction between physical activity and alcohol consumption and this gene variant in predicting HDL cholesterol levels. Physical activity and alcohol consumption have both been linked to higher HDL cholesterol levels (54), and physical activity is considered protective against CHD. In our study of incident CHD cases and controls, leisure-time physical activity was an independent predictor of HDL cholesterol levels. Although the *CETP* variant by activity level interaction was not significant, physically active individuals had HDL levels that averaged 5.15 mg/dL higher than those of inactive individuals, which according to Gordon et al. would equate to a decrease in CHD risk of 10% for active men and 15% for active women (55). HDL cholesterol levels are also influenced by alcohol intake, with a moderate alcohol intake associated with a slight increase in HDL cholesterol levels. In a study of Japanese men and women, Hata et al. quantified this association and reported that drinking approximately 23 g of alcohol increased HDL cholesterol by 2.5 mg/dL (56). Furthermore, Fumeron et al. showed that alcohol intake influences the *CETP* polymorphism effects and that increases in HDL cholesterol resulting from the B2 allele are evident only in moderate drinkers (57). In our study, current drinkers had a HDL level of approximately 3 mg/dL higher than former or nondrinkers across all *CETP* genotypes, but this finding was not statistically significant. Our findings show that both variation in the *CETP* gene and environmental factors related to HDL cholesterol levels are associated with decreased risk for CHD.

One major obstacle in detecting gene–environment effects is the large sample size needed in order to detect small effects. Thus, we analyzed the interaction between variation in the *GNB3* gene and both obesity and physical activity in predicting hypertension in a sample of almost 4,000 AAs from the ARIC study. In our study of

444   CASE STUDIES

the *GNB3* gene, we found a significant interaction between obesity status and the
GNB3 C825T polymorphism in predicting hypertension ($p = 0.007$). When the sample was stratified by obesity status (obese = BMI ≥ 30 kg/m^2), nonobese individuals who were homozygous for the 825T allele had significantly lower risk for
hypertension (odds ratio (OR) = 0.66; 95% CI, 0.46–0.95), while obese 825T
homozygotes experienced higher risk for hypertension (OR = 1.46; 95% CI,
0.94–2.26) compared to 825C/825C homozygotes. Similarly, we observed a significant interaction ($p = 0.008$) between *GNB3* variation and physical activity in predicting obesity status. Subjects in the highest tertile of physical activity (based on
the Baecke Leisure Activity Index) and homozygous for the 825T allele had significantly lower risk for obesity (OR = 0.56; 95% CI, 0.35 = 0.88), while low active, 825T/825T subjects had higher risk for obesity (OR = 1.59; 95% CI, 0.95–2.65)
than their respective counterparts who were homozygous for the 825C allele. Further analyses designed to test how the combination of physical activity and obesity
might modify the effect of the *GNB3* C825T variant in predicting hypertension resulted in a significant interaction in which increasing levels of obesity and physical
inactivity in individuals homozygous for the 825T allele resulted in significantly increased risk for hypertension compared with subjects with one or two copies of
825C. Though one would expect obesity and physical inactivity to be associated
with increased risk for hypertension in all individuals, in fact, when the sample was
stratified by *GNB3* genotype, subjects with two copies of the 825C allele experienced essentially *no increase* in disease risk, while those with one or two copies of
the 825T allele had dramatically increased risk for hypertension as they became
more obese and inactive (Table 25.2). Because the 825C allele is the least frequent
in AAs, these findings suggest that 825C homozygosity may confer some level of
protection from hypertension, even in the face of obesity and a sedentary lifestyle.

Other studies have provided evidence that *GNB3* variation may produce a type
of hypertension that may be exacerbated or precluded by obesity (43), and a recent
report suggested that hypertension in Japanese might result from a high prevalence
of the *GNB3* 825T allele combined with a high salt diet (51). These findings support the concept that genes and environments can interact to produce differential
disease outcomes in different physiologic and behavioral environments. Further stud-

Table 25.2 Stratification by *GNB3* C825T Genotype and the Association with Obesity and
Physical Activity in Predicting Hypertension

Genotype	Nonobese/Active (n = 718)	Nonobese/Inactive (n = 518)	Obese/Active (n = 537)	Obese/Inactive (n = 566)
825C/825C	1.00	1.18 0.61–2.30	1.26 0.61–2.60	1.15 0.61–2.58
825C/825T	1.00	1.15 0.86–1.55	1.57* 1.13–2.17	2.03* 1.40–2.85
825T/825T	1.00	1.47* 1.15–1.87	2.07* 1.58–2.72	2.37* 1.82–3.09

*$p < 0.05$.

ies in large samples are needed to enhance our understanding of how and which genes interact with behavioral, nutritional, and other environments to produce CVD and related co-morbidities.

Gene–Gene Interactions and CVD Risk

In addition to gene–environment interaction, the interaction between genetic variation at multiple loci may be of significant importance in influencing the onset of CVD. No single gene is expected to be completely predictive of CVD, and it is likely that multiple genetic alterations, in combination with environmental factors, are required for overt disease to be manifested. Detecting such gene–gene interactions that may contribute to CVD is difficult due to the multitude of putative factors involved both in the development of disease and the determination of underlying quantitative traits that contribute to CVD. Nevertheless, one example of the impact of variation within multiple genes on HDL cholesterol is presented below.

To investigate the role of gene–gene interaction in determining HDL cholesterol levels, genetic variants within three genes known to be associated with HDL cholesterol, *CETP* (*Taq*IB variant), paraoxonase (*PON1*, Gln192Arg), and hepatic lipase (HL, $-514C > T$ variant), were genotyped in a sample of approximately 800 white individuals. In order to analyze the collective effect of the multiple loci, each variant genotype was collapsed into two categories, indicating presence or absence of the less common allele within each gene, and combined to formulate a multilocus genotype consisting of presence/absence of the less common allele at each of the three loci. We then tested this multilocus genotype in multivariate models in which the common allele at each of the three loci was used as the referent category and age, gender, BMI, and diabetes status were included as covariates. As shown in Table 25.3, the mean HDL cholesterol value for the referent class (multilocus genotype = *CETP* B2/B2, HL $-514C/-514C$, *PON1* Arg192/Arg192) was 55.81 ± 3.6 mg/dL. While HDL cholesterol levels were not significantly different from the

Table 25.3 Multi-locus Genotypes for CETP, HL, and PON in Predicting HDL Cholesterol Levels

Presence of Common Allele	*Presence of Alternate Allele*[a]	*n*	*HDL Cholesterol*
CETP, HL, PON1		90	55.81
CETP, PON1	*HL*	158	56.46
CETP, HL	*PON1*	100	55.96
CETP	*HL, PON1*	174	53.89
HL, PON1	*CETP*	42	45.34[b]
PON1	*CETP, HL*	70	44.32[b]
HL	*CETP, PON1*	40	47.47[b]
	CETP, HL, PON1	71	52.16

[a]Referent genotype = *CETP* B2/B2, HL $-514C/-514C$, *PON1* Arg192/Arg192; alternate alleles = *CETP* B1, HL $-514T$, and *PON1* Gln192.
[b]$p < 0.01$.

referent for other multilocus genotypes that included the B2 allele at *CETP*, mean values for HDL cholesterol were significantly lower in the presence of the B1 allele within *CETP*. HDL cholesterol was an average of 10.5 mg/dl lower for all multilocus genotypes that included the CETP B1 allele, regardless of genotype at the other two loci, *except* when the alternate alleles were present at *both* HL and PON. The B1 allele in *CETP* has been strongly associated with lower HDL cholesterol and increased risk for CHD, while the alternate alleles at both HL (-514T) and *PON1* (Gln192) have been associated with higher HDL cholesterol levels and decreased risk for CHD (58–64). These results suggest that the risk-raising effect of the *CETP* B1 allele may be compensated for by the combined effect of the risk-lowering alleles at both HL and *PON1* but that alterations in either HL or *PON1* alone are not sufficient to counteract the effect of the *CETP* B1 allele. Such analyses, while simplistic, provide evidence that selecting and testing the effects of candidate gene variation based on known associations with quantitative risk factors is one approach to dissecting complex interactions between genetic variation within multiple loci in influencing the underlying factors that contribute to CVD. More sophisticated analyses, such as combinatorics, neural networks, decision trees, and other approaches are also effective in dissecting complex interactions between multiple genes and environments in determining risk factor levels and disease outcomes.

Limitations of Genetic Information in Predicting Cardiovascular Disease

The discovery of gene–environment and gene–gene interactions that influence CVD risk is important not only to identify individuals at greatest risk for disease but also to identify persons *most likely to respond* to treatment or intervention. What is the future of genetic testing for CVD risk? For rare, monogenetic disorders such as familial hypercholesterolemia and Liddle's syndrome, screening of individuals from affected families may be more efficacious in detection and diagnosis of new cases than screening of the population at large (65). Because these syndromes are highly treatable, such detection procedures should be used, particularly once an affected individual has been identified. To this end, researchers at the University of Utah have created a database of FH pedigrees (MEDPED) in an effort to identify new cases and to provide information to affected individuals and their family members and physicans (65).

Numerous association studies of CVD positional and biologic candidate genes have been conducted to date, and there are a substantial number of conflicting results for these genes, depending on both the gene(s) and population(s) under study. For more common forms of CVD, with the exception of rare Mendelian disorders, no *single* gene is likely to be completely deterministic of CVD outcomes. Thus, unless an association is strong or a putative risk-raising allele is common, population screening of single-gene variation may not be an effective strategy in reducing overall CVD prevalence. With the completion of the Human Genome Project, our knowledge of genetic variation and its effects on human disease is growing at an exponential rate. Nevertheless, we are currently limited by the need for genotype information on large, population-based samples and for statistical methods designed to analyze large data sets of variables that are inter-related in complex ways. As genotyping costs continue to decrease and analytical strategies continue to be developed, we can envision a time when panels of multiple risk-raising gene variants can be easily characterized and provide information not only about who is at increased risk for CVD but also about what types of environmental change or intervention will most effectively reduce risk.

Conclusions

Although many critical factors exist in the development of CVD, the contribution of genes and the interaction between genes and environments play a substantial role in its etiology. Understanding the interplay between behavior, nutrition, physiology, psychology, other environments, and genes in producing CVD may lead to more efficacious strategies for its treatment and prevention. CVD is directly associated with a number of chronic diseases, and preventing its onset may help to alleviate these substantial public health concerns.

Acknowledgments

Support for this work was provided by the Centers for Disease Control and Prevention contract #UR6/CCU617218. Special thanks are given to Mitzi Laughlin and Megan Grove for their help in preparing the manuscript.

References

1. American Heart Association. 2002 Heart and Stroke Statistical Update. Dallas, TX: American Heart Association, 2001.
2. PDAY Research Group. Natural history of aortic and coronary atherosclerotic lesions in youth: Findings from the PDAY Study. Arterioscler Thromb 1993;13:1291–1298.
3. Rose G (1987). Implications of genetic research for control measures: Ciba Foundation Symposium 130, in Molecular Approaches to Human Polygenic Disease. Bock G, Collins G (eds). Chichester, UK: John Wiley & Sons, Ltd., pp. 247–256.
4. Sing C, Haviland M, Reilly S (1996). Genetic architecture of common multifactorial diseases. in Variation in the Human Genome. Chadwick D, Cardew G (eds.). Chichester, UK: John Wiley & Sons, pp. 211–232.
5. Myant N. The Biology of Cholesterol and Related Steroids. London: William Heinemann Medical Books Ltd., 1981.
6. Castelli WP. Lipids, risk factors and ischaemic heart disease. Atherosclerosis 1996;124 suppl:S1–S9.
7. Kannel WB. Framingham study insights into hypertensive risk of cardiovascular disease. Hypertens Res 1995;18:181–196.
8. Thompson D, Edelsberg J, Colditz GA, et al. Lifetime health and economic consequences of obesity. Arch Intern Med 1999;159:2177–2183.

9. Winkelmann BR, Hager J. Genetic variation in coronary heart disease and myocardial infarction: methodological overview and clinical evidence. Pharmacogenomics 2000;1: 73–94.

10. Boerwinkle E, Hixson J. Genes and normal lipid variation. Curr Opinion Lipidol 1990; 1:151–159.

11. Sing C, Moll P. Genetics of atherosclerosis. Annu Rev Genet 1990;24:171–188.

12. Grundy S, Balady G, Criqui M, et al. Primary prevention of coronary heart disease: Guidance from Framingham. Circulation 1998;97:1876–1887.

13. Nora JJ, Lortscher RH, Spangler RD, et al. Genetic—epidemiologic study of early-onset ischemic heart disease. Circulation 1980;61:503–508.

14. Berg K. Twin studies of coronary heart disease and its risk factors. Acta Genet Med Gemellol (Roma) 1984;33:349–361.

15. Perusse L, Rice T, Bouchard C, et al. Cardiovascular risk factors in a French-Canadian population: resolution of genetic and familial environmental effects on blood pressure by using extensive information on environmental correlates. Am J Hum Genet 1989; 45:240–251.

16. Berg K. Role of genetic factors in atherosclerotic disease. Am J Clin Nutr 1989;49: 1025–1029.

17. Mitchell BD, Kammerer CM, Blangero J, et al. Genetic and environmental contributions to cardiovascular risk factors in Mexican Americans. The San Antonio Family Heart Study. Circulation 1996;94:2159–2170.

18. Hunt SC, Williams RR (1993). Genetic factors, family history, and blood pressure. In: Hypertension Primer. Isso JL, Black HR, eds. Dallas, TX: American Heart Association, pp. 155–158.

19. Williams RR, Hasstedt SJ, Hunt SC, et al. Genetic traits related to hypertension and electrolyte metabolism. Hypertension 1991;17:169–173.

20. Goldstein JL, Brown MS. Lipoprotein receptors and the control of plasma LDL cholesterol levels. Eur Heart J 1992;13 suppl B:34–36.

21. Shimkets RA, Lifton RP. Recent advances in the molecular genetics of hypertension. Curr Opin Nephrol Hypertens 1996;5:162–165.

22. Lifton RP, Dluhy RG, Powers M, et al. A chimaeric 11 beta-hydroxylase/aldosterone synthase gene causes glucocorticoid-remediable aldosteronism and human hypertension. Nature 1992;355:262–265.

23. Goldstein J, Hobbs H, Brown M (1995). Familial hypercholesterolemia. in The Metabolic and Molecular Bases of Inherited Disease. Scriver C, Beaudet A, Sly W, et al. (eds). New York, NY: McGraw-Hill, Inc., pp. 1981–2030.

24. Gordon T, Castelli WP, Hjortland MC, et al. High density lipoprotein as a protective factor against coronary heart disease. The Framingham Study. Am J Med 1977;62:707–714.

25. Bruce C, Sharp DS, Tall AR. Relationship of HDL and coronary heart disease to a common amino acid polymorphism in the cholesteryl ester transfer protein in men with and without hypertriglyceridemia. J Lipid Res 1998;39:1071–1078.

26. Barter P. CETP and atherosclerosis. Arterioscler Thromb Vasc Biol 2000;20:2029–2031.

27. Corbex M, Poirier O, Fumeron F, et al. Extensive association analysis between the CETP gene and coronary heart disease phenotypes reveals several putative functional polymorphisms and gene-environment interaction. Genet Epidemiol 2000;19:64–80.

28. Stein O, Stein Y. Atheroprotective mechanisms of HDL. Atherosclerosis 1999;144: 285–301.

29. Kondo I, Berg K, Drayna D, et al. DNA polymorphism at the locus for human cholesteryl ester transfer protein (CETP) is associated with high density lipoprotein cholesterol and apolipoprotein levels. Clin Genet 1989;35:49–56.

30. Berg K, Kondo I, Drayna D, et al. "Variability gene" effect of cholesteryl ester transfer protein (CETP) genes. Clin Genet 1989;35:437–445.
31. Corella D, Saiz C, Guillen M, et al. Association of TaqIB polymorphism in the cholesteryl ester transfer protein gene with plasma lipid levels in a healthy Spanish population. Atherosclerosis 2000;152:367–376.
32. Kuivenhoven JA, Jukema JW, Zwinderman AH, et al. The role of a common variant of the cholesteryl ester transfer protein gene in the progression of coronary atherosclerosis. The Regression Growth Evaluation Statin Study Group. N Engl J Med 1998;338:86–93.
33. Inazu A, Jiang XC, Haraki T, et al. Genetic cholesteryl ester transfer protein deficiency caused by two prevalent mutations as a major determinant of increased levels of high density lipoprotein cholesterol. J Clin Invest 1994;94:1872–1882.
34. Inazu A, Brown ML, Hesler CB, et al. Increased high-density lipoprotein levels caused by a common cholesteryl-ester transfer protein gene mutation. N Engl J Med 1990; 323:1234–1238.
35. Laughlin M, Jackson A, Boerwinkle E, et al. Association of the cholesteryl ester transfer protein TaqIB polymorphism with coronary heart disease risk and plasma lipoprotein levels. Athersclerosis. In press.
36. Meguro S, Takei I, Murata M, et al. Cholesteryl ester transfer protein polymorphism associated with macroangiopathy in Japanese patients with type 2 diabetes. Atherosclerosis 2001;156:151–156.
37. Ordovas JM, Cupples LA, Corella D, et al. Association of cholesteryl ester transfer protein-TaqIB polymorphism with variations in lipoprotein subclasses and coronary heart disease risk: the Framingham study. Arterioscler Thromb Vasc Biol 2000;20:1323–1329.
38. Durlach A, Clavel C, Girard-Globa A, et al. Sex-dependent association of a genetic polymorphism of cholesteryl ester transfer protein with high-density lipoprotein cholesterol and macrovascular pathology in type II diabetic patients. J Clin Endocrinol Metab 1999;84:3656–3659.
39. Siffert W, Naber C, Walla M, et al. G protein beta3 subunit 825T allele and its potential association with obesity in hypertensive individuals. J Hypertens 1999;17:1095–1098.
40. Siffert W, Rosskopf D, Moritz A, et al. Enhanced G protein activation in immortalized lymphoblasts from patients with essential hypertension. J Clin Invest 1995;96:759–766.
41. Siffert W, Rosskopf D, Siffert G, et al. Association of a human G-protein beta3 subunit variant with hypertension. Nat Genet 1998;18:45–48.
42. Iiri T, Bourne H. G proteins propel surprise. Nat Genet 1998;18:8–10.
43. Siffert W, Forster P, Jockel KH, et al. Worldwide ethnic distribution of the G protein beta3 subunit 825T allele and its association with obesity in Caucasian, Chinese, and Black African individuals. J Am Soc Nephrol 1999;10:1921–1930.
44. Benjafield A, Jeyasingam C, Nyholt D, et al. G-protein beta3 subunit gene (GNB3) variant in causation of essential hypertension. Hypertension 1998;32:1094–1097.
45. Hegele R, Harris S, Hanley A, et al. G protein beta3 subunit gene variant and blood pressure variation in Canadian Oji-Cree. Hypertension 1998;32:688–692.
46. Schunkert H, Hense H, Doring A, et al. Association between a polymorphism in the G protein beta3 subunit gene and lower renin and elevated diastolic blood pressure levels. Hypertension 1998;32:510–513.
47. Poch E, Gonzalez D, Gomez-Angelats E, et al. G-Protein beta(3) subunit gene variant and left ventricular hypertrophy in essential hypertension. Hypertension 2000;35: 214–218.
48. Meirhaeghe A, Bauters C, Helbecque N, et al. The human G-protein beta3 subunit C825T polymorphism is associated with coronary artery vasoconstriction. Eur Heart J 2001;22:845–848.

49. Jacobi J, Hilgers KF, Schlaich MP, et al. 825T allele of the G-protein beta3 subunit gene (GNB3) is associated with impaired left ventricular diastolic filling in essential hypertension. J Hypertens 1999;17:1457–1462.

50. Ishikawa K, Imai Y, Katsuya T, et al. Human G-protein beta3 subunit variant is associated with serum potassium and total cholesterol levels but not with blood pressure. Am J Hypertens 2000;13:140–145.

51. Tozawa Y. G protein beta3 subunit variant: tendency of increasing susceptibility to hypertension in Japanese. Blood Press 2001;10:131–134.

52. Delva P, Pastori C, Provoli E, et al. Erythrocyte Na(+)-H+ exchange activity in essential hypertensive and obese patients: role of excess body weight. J Hypertens 1993; 11:823–830.

53. Moxham CM, Hod Y, Malbon CC. Gi alpha 2 mediates the inhibitory regulation of adenylylcyclase in vivo: analysis in transgenic mice with Gi alpha 2 suppressed by inducible antisense RNA. Dev Genet 1993;14:266–273.

54. Hashimoto Y, Futamura A, Nakarai H, Nakahara K. Effects of the frequency of alcohol intake on risk factors for coronary heart disease. Eur J Epidemiol 2001;17:307–312.

55. Gordon DJ, Probstfield JL, Garrison RJ, et al. High-density lipoprotein cholesterol and cardiovascular disease. Four prospective American studies. Circulation 1989;79:8–15.

56. Hata Y, Nakajima K. Life-style and serum lipids and lipoproteins. J Atheroscler Thromb 2000;7:177–197.

57. Fumeron F, Betoulle D, Luc G, et al. Alcohol intake modulates the effect of a polymorphism of the cholesteryl ester transfer protein gene plasma high density lipoprotein and the risk of myocardial infarction. J Clin Invest 1995;96:1664–1671.

58. Deeb SS, Peng R. The C-514T polymorphism in the human hepatic lipase gene promoter diminishes its activity. J Lipid Res 2000;41:155–158.

59. Couture P, Otvos JD, Cupples LA, et al. Association of the C-514T polymorphism in the hepatic lipase gene with variations in lipoprotein subclass profiles: The Framingham Offspring Study. Arterioscler Thromb Vasc Biol 2000;20:815–822.

60. Jansen H, Chu G, Ehnholm C, et al. The T allele of the hepatic lipase promoter variant C-480T is associated with increased fasting lipids and HDL and increased preprandial and postprandial LpCIII:B : European Atherosclerosis Research Study (EARS) II. Arterioscler Thromb Vasc Biol 1999;19:303–308.

61. Zambon A, Deeb SS, Hokanson JE, et al. Common variants in the promoter of the hepatic lipase gene are associated with lower levels of hepatic lipase activity, buoyant LDL, and higher HDL2 cholesterol. Arterioscler Thromb Vasc Biol 1998;18:1723–1729.

62. Tomas M, Senti M, Elosua R, et al. Interaction between the Gln-Arg 192 variants of the paraoxonase gene and oleic acid intake as a determinant of high-density lipoprotein cholesterol and paraoxonase activity. Eur J Pharmacol 2001;432:121–128.

63. Aubo C, Senti M, Marrugat J, et al. Risk of myocardial infarction associated with Gln/Arg 192 polymorphism in the human paraoxonase gene and diabetes mellitus. The REGICOR Investigators. Eur Heart J 2000;21:33–38.

64. Hegele RA, Brunt JH, Connelly PW. A polymorphism of the paraoxonase gene associated with variation in plasma lipoproteins in a genetic isolate. Arterioscler Thromb Vasc Biol 1995;15:89–95.

65. Williams RR, Hopkins PN, Stephenson S, et al. Primordial prevention of cardiovascular disease through applied genetics. Prev Med 1999;29:S41–S49.

26

BRCA1/2 and the prevention of breast cancer

Jenny Chang-Claude

With one million new cases diagnosed in the world each year, breast cancer is responsible for a substantial burden of disease. In the United States, breast cancer accounts for an estimated 29% of all new cancers and 16% of all cancer-related deaths in women during 1999. Lifetime risk for breast cancer for an individual woman varies considerably between countries: 12% in the United States, 10% in the United Kingdom, and less than 5% in Japan. Despite decades of intensive research, incidence of breast cancer continues to increase in most countries. The number of deaths due to breast cancer would be further reduced if effective preventive strategies could be added to the screening and treatment options that already exist and have contributed to reduced mortality (1). The focus in recent years has been directed towards preventive intervention to persons at increased risk of disease due to inherited susceptibility. The identification of women at particularly high risk of developing breast cancer will provide a group for whom expensive and rigorous screening programs are cost-effective and a cohort of women who may benefit from trials of preventive interventions. As women become aware of their inherited susceptibility, they will need to have options for preventing the disease. Scientists expected that the discovery of breast cancer genes would also shed light on the origins of breast cancer, not only for women with inherited gene mutations but also for the many more women with noninherited disease. Understanding the functions of the gene products could lead to strategies to compensate for the loss or impairment of function.

Risk Factors

Breast cancer is a complex disease attributed to the strong joint effects of genetic and environmental factors. Compelling evidence from epidemiologic and experimental data implicates endogenous estrogen levels as critical determinants of breast cancer risk (2,3). Most of the established risk factors with respect to menstrual and reproductive history, such as early age at menarche, late age at menopause, nulliparity, and late age at first birth, increase the number of menstrual cycles and thereby increase cumulative exposure of the breast epithelium to estrogen. Early age at ovarectomy minimizes the number of ovulatory cycles experienced and can thus reduce risk by 50% (4). Prolonged lactation and physical activity can reduce the number of ovulatory cycles and protect against breast cancer (5,6). Physical activity in adolescence may delay the age of onset of ovulatory cycles and reduce the cycle frequency and circulating ovarian hormone levels (7). High levels of occupational and recreational activity in the lifetime have also been reported to be associated with a 12% to 60% decrease in risk, although a dose-response trend was not evident in most studies (8). Higher levels of alcohol consumption have been associated with two- to threefold increased breast cancer risk. Although some authors suggest a linear increase in risk up to 60 g of alcohol per day (five drinks) (9,10), low to moderate levels of consumption may have negligible effect on breast cancer risk (11). Alcohol may influence risk for breast cancer by affecting hormone metabolism, thus increasing serum hormones and insulin-like growth factor levels (12–14), as well as affecting alcohol-derived reactive oxygen species that induce DNA modifications and carcinogenesis (15). Obesity has consistently been found to be positively associated with postmenopausal but not premenopausal breast cancer risk (16). This is attributed to the conversion of androstenedione to estrone in adipose tissue, which is the primary source of estrogens in postmenopausal women. Obesity is also associated with increased proportions of free and albumin-bound estrogens and decreased sex-hormone binding globulin production. Indeed, studies of serum hormone levels have consistently shown that Asian women at lower breast cancer risk have lower serum and urinary estrogen levels than do Caucasian women at higher risk for breast cancer (17–19). Cohort studies have shown that postmenopausal women who subsequently developed breast cancer have a 15% higher mean serum estradiol concentration than do unaffected women (20,21).

Genetic Factors

Clinical reports of families with multiple cases of breast cancer in successive generations and the consistent finding of increased risk in women with a family history of breast cancer in the mother or a sister suggested a role of genetic predisposition in determining the risk for the disease. A positive family history in first-degree relatives is reported by 10% to 12% of patients in most studies. Both younger age at diagnosis of breast cancer in the relative and a larger number of affected relatives are associated with a greater risk elevation (reviewed in ref. [22]). The existence of susceptibility genes is supported by results of pedigree and segregation analysis indicating that the familial clustering in some families is accounted for by rare high-penetrance autosomal dominant genes, which explain about 5% of all breast cancer in the general population (22,23). The widely accepted model of breast cancer sus-

Table 26.1 Genes Conferring High Risk for Breast Cancer

Associated Syndrome	Clinical Manifestation	Gene
Hereditary breast/ovarian cancer	Breast cancer, ovarian cancer	*BRCA1, BRCA2*
Li-Fraumeni syndrome	Sarcoma, brain and breast cancer	*PT53*
Cowden disease	Multiple hamartomatous lesions of skin, mucous membrane, cancer of breast and thyroid	*PTEN/NMAC1*
Peutz-Jeghers syndrome	Melanocytic macules of lips, multiple polyps, tumors of intestinal tract, breast, ovaries, etc.	*STK11*
Hereditary nonpolyposis colorectal cancer	Colorectal cancer, predominantly also tumors of endometrium, ovaries, intestinal tract, and breast	*MSH2, MLH1, PMS1, PMS2*
Ataxia telangiectasia	Progressive cerebral ataxia, hypersensitivity to radiation, increased cancer risk	*ATM*

ceptibility now is that it is due to a small number of highly penetrant genes (such as *BRCA1* and *BRCA2*) and a much larger number of low-penetrance variants (24,25).

Several major genes for breast cancer susceptibility have been cloned as predisposing genes for different cancer syndromes in which breast cancer is a constituent tumor (Table 26.1). Mutation analysis has confirmed that the genes identified can largely explain the cancer occurrence in these syndromes. However, a proportion of such families remain in which the disease-associated allele has not been identified. This can be attributed to both the imperfect sensitivity of the mutation detection method and the existence of additional as yet unidentified susceptibility genes.

The two major breast cancer susceptibility genes, *BRCA1* and *BRCA2*, were identified by using linkage analysis in large extended breast and/or ovarian cancer pedigrees and subsequent molecular cloning (26–28). Germline mutations in *BRCA1* have been identified in 15% to 45% of women with familial breast cancer, depending on the population and extent of family history (29–32), and in 45% to 80% of women with a family history of both breast and ovarian cancer (29–31,33–35). Women with an inherited germline mutation in *BRCA1* have a 10- to 20-fold increased risk for breast cancer in early adult life. The median age at diagnosis in mutation carriers is around 40 years, about 20 years earlier than that for unselected women in Western countries (36). *BRCA1* carriers are also at highly increased risk for ovarian cancer. *BRCA2* mutations predispose individuals to a high lifetime risk for breast cancer, similar to that associated with *BRCA1* mutations, together with a lower, although still significantly increased risk for ovarian cancer (30). *BRCA2* mutations also confer a very high risk for breast cancer to male carriers (37). In addition to the risks for breast and ovarian cancers, *BRCA1* and *BRCA2* mutations can be associated with an increased risk for a variety of other cancers, such as cancers of the colon, prostate, pancreas, gallbladder, bile duct, stomach, and skin (38,39). *BRCA1* mutations have also been found in excess in breast cancer patients with multiple primary cancers who also have a family history of breast cancer (40).

BRCA1 and *BRCA2* are both large genes, encoding a nuclear protein of 220 and 384 kilodaltons, respectively. *BRCA1* contains a zinc-binding RING domain at the amino terminus and a conserved acidic carboxyl terminus that function in transcriptional co-activation (26). Binding of *BRCA1* to *BRCA2*, p53, *RAD51* and many other proteins involved in cell cycling and DNA-damage response has been observed. *BRCA1* is part of the RAD50–MRE11–p95 complex, which is an essential component of homologous recombination repair of DNA double-stranded breaks (reviewed in refs. 15 and 41). The *BRCA2* protein binds to *RAD51* and to *BRCA1* and is thus also involved in recombination-mediated repair of double-stranded breaks and the maintenance of chromosome integrity. *BRCA1* and *BRCA2* are clearly involved in DNA-damage response. However, the reasons cancer risks in *BRCA1* and *BRCA2* mutation carriers are particularly increased for breast and ovarian cancer are not yet understood.

The p53-responsive genes have been identified as the main downstream targets of *BRCA1*, and both *BRCA1* and *BRCA2* also act as transcriptional co-regulators. Cells that lack *BRCA1* or *BRCA2* accumulate chromosomal abnormalities (42). It has been proposed that the unusually high densities of repetitive DNA elements in the genomic regions of both *BRCA1* and *BRCA2* can mediate genomic rearrangements and may thus contribute to genomic instability (43). This genetic instability may predispose to additional mutations, including mutations in checkpoint genes, which in accumulation lead to tumorigenesis. Various genetic mechanisms may be responsible for the somatic inactivation of the multiple functions of the BRCA genes, and they have yet to be resolved.

Inherited susceptibility is conferred also by mutations in several of the highly penetrant genes of the cancer syndromes associated with hereditary breast cancer. Germline mutations in the *TP53* cause Li-Fraumeni syndrome involving childhood leukemias, brain tumors, adrenal carcinomas, and soft tissue sarcomas. The breast cancer penetrance approaches 100% in mutation carriers surviving childhood (44,45). *STK11/LKB1* is a serine-threonine kinase and mutations in this gene cause the Peutz-Jeghers syndrome, which is characterized by hamartomatous polyps in the small bowel and pigmented macules of the buccal mucosa, lips, fingers, and toes. Mutation carriers have a 20-fold higher risk for breast cancer than noncarriers (46,47). Cowden syndrome (adenomas and follicular cell carcinomas of the thyroid gland, polyps and adenocarcinomas of the gastrointestinal tract, and ovarian cancer) is caused by mutations in *PTEN*, a dual specificity phosphatase, which is also associated with a 20% to 30% lifetime risk for breast cancer (48). However, all these susceptibility alleles are extremely rare in the general population and are considered to account for less than 1% of inherited susceptibility for breast cancer (49,50).

Cancer risk, especially breast cancer risk, has been reported in epidemiologic studies to be highly increased (fourfold to eightfold) in blood relatives (presumptive heterozygous gene carriers) of patients with the recessive disease ataxia telangiectasia (51,52). Subsequent studies based on mutation screening and haplotype analysis yielded lower estimates of about a threefold increase in the risk for breast cancer risk among heterozygous carriers (53,54). Contrary to previous estimates, heterozygosity for germline *ATM* mutations appears to be observed rarely in unilateral breast cancer but may be more prevalent among patients with nonfamilial early-onset bilateral breast cancer (55–59). Most previous studies have screened only for truncating *ATM* mutations, so that studies aimed at identifying germline missense mutations and rare allelic variants of the *ATM* gene may provide better estimates of the contribution of *ATM* variants to early-onset breast cancer (60).

The search for genes with predisposing alleles has not been restricted to highly penetrant susceptibility genes, such as the *BRCA1* and *BRCA2*, which are associated with multiple occurrences of disease in a family, often in a Mendelian pattern of an autosomal dominant trait. Population-based data suggest that a number of common, low-penetrance genes with additive effects may account for the residual non-*BRCA1/2* familial aggregation of breast cancer (24). Candidate genes studied included genes involved in steroid hormone metabolism which may modulate the levels of bioavailable steroid hormones, such as the catechol-O-methyltransferase (*COMT*), cytochrome p450 genes *CYP19* (aromatase), *CYP17* (steroid 17 alpha-hydroxylase/17,20 lyase), *CYP1B1*, and the steroid hormone receptor genes, such as estrogen receptor (ER), progesterone receptor (PR), androgen receptor (AR) as well as vitamin D receptor (VDR), reviewed by Dunning et al. (61). In addition, common alleles of high-penetrance genes such as *TP53*, *BRCA1*, *BRCA2*, and *ATM* can affect the integrity of cell cycle checkpoint and DNA repair, and thus modify cancer risk. For example, homozygous carriers of a common variant in *BRCA2* (N372H) were found to confer a 30% increased risk for breast cancer compared with noncarriers (62). Effects of single common alleles are being continuously reported: these require replication and confirmation in large population-based studies and gene–gene effects need to be taken into account.

Joint effects of several genes and environmental factors could also explain the majority of breast cancer that occurs in the absence of a strong family history and with later age at onset. Inherited susceptibility to breast cancer can thus also arise from common polymorphisms of genes governing the metabolism of exogenous and endogenous substances and genes for specific DNA repair mechanisms. These include genes coding for phase I enzymes, such as *CYP1A1*, *CYP1A2*, and *CYP2D6*, which act on tobacco smoke associated carcinogens; *CYP2E1*, which metabolizes ethanol; as well as genes for phase II enzymes, such as the glutathione-S-transferases μ (*GSTM1*), π (*GSTP1*), θ; (*GSTT1*); and the N-acetyl transferases *NAT1* and *NAT2*. A large number of metabolic polymorphisms have been studied, but they are not the focus of this chapter. As an example, the family of N-acetyltransferase (NAT) enzymes affects metabolism of mutagenic heterocyclic amines found in well-cooked red meat. There are data suggesting that consumption of well-done meat may be a risk factor for breast cancer and colorectal cancer in individuals with rapid-metabolizing variants of the *NAT1* and *NAT2* enzymes (57,63). Although earlier association studies suffer from both small and/or poorly selected samples (61), increasingly larger studies are being conducted to elucidate the complex interplay between many genes and environmental factors. This type of susceptibility confers increased risk only in combination with exposure to the particular substances, which is less effectively metabolized by the genetic variant. The genetic variant alone is generally not predictive of disease risk. Prevention of disease in these individuals will be feasible and effective by reducing or eliminating the environmental exposures.

Epidemiologic Findings on BRCA1 and BRCA2 Prevalence

Mutations in *BRCA1* account for 15% to 45% of hereditary breast cancers families, depending on the population and types of families under study. Higher estimates were obtained from large families with four or more breast cancer cases diagnosed before 60 years of age (29,30) and lower estimates from more recent studies which have included a broader spectrum of families seen at genetic clinics, including a larger proportion of families with only two or three breast cancer cases (31,33–35,64,65). A large proportion of multicase families including ovarian cancer was attributed to *BRCA1*, although most studies of smaller families in different countries have consistently reported that about 45% of families with both breast and ovarian cancer were due to mutations in the *BRCA1* gene (31,33,35,64–66).

BRCA2 mutations account for 20% to 56% of hereditary breast cancer families and 5% to 25% of families including ovarian cancer, depending on the study population and number of affected members in the family (30,34,35,67,68). However, breast cancer in families that have a male breast cancer patient appears to be due to *BRCA2* mutations (30).

Recent population-based studies of incident breast cancer patients suggest that germline mutation *BRCA1* and *BRCA2* may be responsible for only about 6% to 10% of early-onset breast cancer cases in the general population (32,69–71). Furthermore, only about 10% to 20% of unselected patients with a first-degree family history of breast or ovarian cancer harbored a germline mutation in *BRCA1* or *BRCA2*. On the basis of mutation screening in ovarian cancer families and unselected ovarian cancer patients in the United Kingdom, the allele frequency of mutant *BRCA1* and *BRCA2* alleles has been estimated to be 0.0013 and 0.0017, indicating that these disease-causing alleles are relatively rare in the general population (72). Therefore, despite their strong effect on individual risk, *BRCA1* and *BRCA2* probably account for only a modest proportion of familial aggregation in most outbred Western populations. On the other hand, where founder (or ancestral) mutations exist, disease allele frequencies can be much higher, such as 0.6% for the *BRCA2* 999del5 mutation in the Icelandic population (73,74) and 1.0% to 1.5% for the two recurrent *BRCA1* mutations, 185delAG and 5382insC, and about 1.2% for one recurrent *BRCA2* 6174delT mutation among Ashkenazi Jews (75,76).

Although the risk for breast cancer in men is about 80 times higher than in the general population, *BRCA2* mutations account for an estimated 10% of breast cancer in the male population (37). The majority of studies in non-Icelandic populations supports these estimates (77–80). The higher prevalence of *BRCA2* mutations in Iceland is consistent with the high frequency of the 999del5 mutation.

Penetrance

Based on large extended pedigrees, female carriers of *BRCA1* mutations were initially estimated to have 82% (95% confidence interval 64%–91%) risk of developing breast cancer by age 70 years (81). Mutation testing in these families led to revised estimates of 71% (53%–82%) (30). Affected mutation carriers were also estimated to have a 65% lifetime risk for developing a second breast cancer (38). Estimated lifetime risk of developing ovarian cancer was 44%, but unlike breast cancer, age-specific penetrances were not skewed toward early onset of disease (82). For *BRCA2* mutation carriers, the estimated lifetime breast cancer risk of 84% (43%–95%) is as high as for *BRCA1* mutations, whereas the estimated lifetime ovarian cancer risk of 27% (0%–47%) is lower (30). In male carriers of *BRCA2* mutations, the estimated lifetime risk for breast cancer by age 80 years is estimated at 7%, which is 80-fold higher than the risk in the male population (37).

On the other hand, population-based studies have generally found smaller estimates of cumulative risks. The penetrance estimates were 47% (5%–82%) up to age 70 for *BRCA1* and 56% (5%–80%) for *BRCA2* in the United Kingdom (83), 40% (15%–65%) for *BRCA1* and *BRCA2* combined in Australia (69), and 68% up to age 80 for *BRCA1* and no detectable increased risk for *BRCA2* among ovarian cancer patients (84). For the *BRCA2* 999del5 mutation in Iceland, reported breast cancer risk estimates were 37% (22%–54%) (85), and among Ashkenazi Jewish patients, 46% (31%–80%) for *BRCA1* and 26% (14%–50%) for *BRCA2* for the three Ashkenazi ancestral mutations (86).

The difference in the reported estimates are not surprising because the different studies involved different populations as well as different types of mutations. The higher risk estimates derived from studies of multiple case families may indicate that the detected mutations have an inherently higher risk. In addition, mutation carriers in the family may share factors which enhance their genetically determined risk, and these can be both other genes as well as lifestyle and environmental factors.

Genotype–Phenotype Correlation

Estimation of risks associated with specific mutations has been possible only for common founder mutations in the Ashkenazi and the Icelandic population. Some variation in cancer risk by mutation position has been reported for both *BRCA1* and *BRCA2* mutation carriers. Mutations in a central portion of the *BRCA2* gene, named the "ovarian cancer cluster region" (OCCR) (87), were associated with a significantly higher ratio of cases of ovarian/breast cancer in female carriers than were mutations 5′ or 3′ of this region, which lies in exon 11 (37). In 164 families with breast/ovarian cancer and germline *BRCA2* mutations, relative risks associated with mutations in the OCCR region were 0.63 for breast cancer and 1.88 for ovarian cancer. A similar reduction in breast cancer risk associated with mutations in a central region of the *BRCA1* gene has been observed (88). The OCCR in *BRCA2* contains six BRC repeat motifs shown to bind *RAD51*, a gene involved in homologous repair of double-stranded DNA breaks (89). *BRCA1* also binds *RAD51* through a domain in exon 11 that appears to coincide with the *BRCA1* central region (90). These observations suggest an involvement of *RAD51* binding, although the biologic mechanism underlying the genotype–phenotype correlation is not yet known. If confirmed in future studies, variance in penetrance for breast cancer by mutation location may affect treatment decisions for mutation carriers (84).

Modifiers of BRCA1/2

The age of onset and the site of cancer occurrence in mutation carriers of *BRCA1* and *BRCA2* vary substantially between as well as within families. Variability is also observed for founder mutations (74,91). Therefore, suggestions that different variants may be associated with different disease severity or may predispose differentially by cancer site cannot fully explain the variability. *BRCA1* and *BRCA2* are considered to be tumor-suppressor genes and, therefore, changes in both alleles are required for the complete loss of normal gene function. Even in individuals with an inherited mutation in one gene copy, loss or aberration of the normal gene copy later in life will be required for disease to develop. Other factors, both genetic and environmental, may therefore modify cancer risk.

Several other gene loci have been reported to modify the penetrance of *BRCA1* mutations, although the findings have seldom been confirmed. Mutation carriers harboring rare variable number of tandem repeats (VNTR) alleles of the HRAS proto-oncogene had a 2.11 higher risk for ovarian cancer than did carriers with only common alleles, but breast cancer risk was not increased (92). Genotypes with long CAG repeats at the androgen receptor gene were reported to be associated with earlier age at diagnosis of breast cancer in *BRCA1* mutation carriers in the United States (93). However, neither CAG nor GGC repeats modified breast cancer risk in *BRCA1* and *BRCA2* carriers and noncarriers in another study, even if Ashkenazi descent was accounted for (94). Thus, any effect of the AR repeat length on *BRCA1* penetrance, if real, is likely to be weak. Two studies have found a single nucleotide polymorphism in *RAD51* to increase the risk for breast cancer among *BRCA2* mutation carriers; however, the biochemical basis of the risk modification is unknown (95,96). Only recently has replicating initial findings in relatively large replicating sets of *BRCA1/2* mutation carriers become possible. Therefore, future studies will focus on the identification of genetic modifiers specific for *BRCA1* and/or *BRCA2*.

A few studies have addressed the question of modifying effects of known reproductive and lifestyle risk factors on the *BRCA1*-associated risk. Some indicate similar effects of reproductive and hormonal risk factors on breast or ovarian cancer risk, and others suggest that effects in mutation carriers may differ from that seen in breast and ovarian cancer in the general population (29,97–99). Most of these studies suffer from a very small sample size and/or a survival bias, because prevalent cases were used. To address these questions appropriately, prospective cohort studies of *BRCA1/2* mutation carriers in international collaboration are being carried out (100). Knowledge gained about risk modifiers in mutation carriers may be useful for refining risk estimation of the individual and may provide further insight into the relevant pathways of breast cancer tumorigenesis.

Gene Testing

The *BRCA1* gene comprises of 24 exons (22 coding exons), which translate into a protein containing 1863 amino acids, covering more than 100 kb genomic DNA, and the *BRCA2* gene is equally large, with 27 exons (exon 1 untranslated) extending over 70 kb genomic DNA. Due to the time sequence in the cloning of the *BRCA1* and *BRCA2* genes, most laboratories initially established *BRCA1* mutation screening, thus the mutational spectrum of *BRCA2* is less well established than that of *BRCA1*. Hundreds of different mutations in *BRCA1/2* have been found and are available at the Breast Cancer Information Core (BIC) Web site, http:/www. nhgri.nih.gov/intramural_research/lab_transfer/bic/member/index.html. Only in a few populations (e.g., Icelandic, Ashkenazi Jewish) does breast cancer predisposition primarily result from a limited number of founder (or ancestral) mutations, although such mutations have been reported in different countries, such as Russia, the Netherlands, Norway, Scotland, and Finland. Therefore, screening for mutation is generally still laborious and costly because the entire coding sequence and intron/exon junctions are screened.

The commonly used methods of mutation screening aim to detect point mutations and small rearrangements. Most of the detected mutations are frameshift mutations, but several missense mutations alter *BRCA1* protein function, and mutations at splice acceptor/donor sites have also been found. Generally, because of the costs involved in direct sequencing of the complete genes, prescreening is performed initially, and the variants identified are then sequenced to define the mutation. The DNA-based methods employed include single strand conformation polymorphism (SSCP), combined SSCP and heteroduplex analysis (SSCP/HA), conformation strand gel electrophoresis (CSGE), denaturing gradient gel electrophoresis (DGGE), fluorescent assisted missense analysis (FAMA), and denaturing high power liquid chromatography (DHPLC). RNA-based protein truncation test (PTT) has been frequently used for detecting truncating mutation in the larger exons, for example exon 11 in both genes, and allele-specific oligonucleotides (ASO) for specific common founder mutations.

With the use of conventional methods, the estimated probability of detecting a point mutation in large families showing linkage to *BRCA1* was 63% (30). Mutations involving several kilobases of genomic sequence in *BRCA1*, which are not detected by conventional screening methods, partially accounted for this relatively low sensitivity. Genomic rearrangements within *BRCA1* and its regulatory regions could account for about 10% of families with breast and ovarian cancer (101–103). The *BRCA1* exon 13 duplication and three other genomic rearrangements are significant founder mutations in the United Kingdom and Dutch populations, respectively (101). Only one genomic rearrangement in *BRCA2* deleting exon 3 has been described

(104). Since *BRCA2* intronic sequence contains fewer *Alu* repeats than does *BRCA1*, genomic rearrangements involving *Alu*-mediated recombination events are likely to be less frequent.

   Sensitivity of the different methods for mutation screening of point mutations and small rearrangements (when optimally implemented) has been estimated for other disease genes to range between 70% and 90% and should apply to *BRCA1/2*. However, for *BRCA1*, detection of large genomic rearrangements using Southern blot analysis would increase the sensitivity of mutation screening. RNA studies to study changes in size of transcript or loss of transcript have seldom been conducted and may help to account for some large families with both breast and ovarian cancer (105).

Implications for Breast Cancer Risk Assessment, Management, and Prevention

The initially widely quoted lifetime penetrance figures of over 80% for mutations in the *BRCA1/2* genes have caused women with a family history of breast cancer to be particularly concerned about their own risk. Geneticists and clinicians are encouraged by the possibility of screening for germline mutations in high-risk families and thus providing targeted care to family members differentiated as carriers or noncarriers. The identification of women at particularly high risk of developing breast cancer was considered also to provide a group for whom expensive and rigorous screening programs are cost-effective and a cohort of women in whom it might be particularly appropriate to undertake trials of chemoprevention.

Risk Assessment—Eligibility Criteria
Services at most breast cancer genetics clinics are targeted to those at greatest risk and will accept referrals of women between the age of 35 and 50 years if their lifetime risk for breast cancer is at least 20%. The computation of risk generally follows one of the widely accepted algorithms based on a genetic model of rare highly penetrant genes for susceptibility to breast cancer, and precise figures can be generated by using complete pedigree structure (106–108). However, even many multicase breast cancer families may not present with a Mendelian pattern of inheritance because of incomplete penetrance, inheritance of the mutant gene through paternal lineage generally not apparent because of unaffected male carriers, or the occurrence of one or more sporadic cases, breast cancer being a relatively common disease. In addition, not all women at substantially increased genetic risk for breast cancer will have an affected close relative. Studies of unselected women with early-onset breast cancer have found that less than a half of such patients who have inherited a germline mutation in *BRCA1* and *BRCA2* have a family history of breast and ovarian cancer, and less than 10% have two or more affected first- or second-degree relatives (109). One of the most difficult tasks of the geneticist and counselor is therefore to explain the degree of uncertainty that accompanies translation from the general to the individual. Unfortunately, different methods currently used

in clinical practice can result in a significant degree of variability of the breast cancer risk estimates, which has serious implications for individual patient management, service provision, and study evaluation (110).

As discussed previously, the estimate of penetrance of the cancer trait is undergoing continuous revision, with the focus shifted from families with substantial numbers of affected individuals to those derived primarily from cancer genetics practices and population-based studies. Although this affects the calculation of the chance that a mutant gene exists in the family, it is crucial for the specification of risk for a particular family member carrying the mutation. Furthermore, penetrance and the pattern of cancer may be affected by nature and position of the mutation and whether these findings should influence advice remains controversial. For most women currently attending genetics clinics, no mutation in the known breast cancer susceptibility genes has yet been identified in their families, so that molecular testing may not improve risk assessment. They will remain at the a priori risk level calculated on the basis of their family history because the involvement of unknown predisposing genes cannot be excluded and the sensitivity of the mutation analysis is below 100% (111). Thus, even for the minority of those seeking advice who can be offered molecular testing, uncertainties remain about the advantages and disadvantages of proceeding with the test, about the age-specific cancer risk associated with any given germline mutation, and about the environmental or genetic factors that may influence the development of breast, ovarian, or other cancers.

Surveillance and Management

At present, recommendations for management of women with inherited susceptibility to breast cancer vary, and familial cancer programs in some countries are undergoing prospective audit of the outcome. Proposed guidelines include offering genetic counseling, education in "breast awareness," and annual mammography and clinical expert examination starting at 30 or 35 years of age or 5 years earlier than the age at cancer diagnosis of the youngest affected relative (112,113). Indeed, surveys in several countries have found that women attending breast cancer genetics clinics want, above all, access to mammographic screening (114,115). Although the imaging modality of choice for radiologic screening has been mammography, sensitivity of mammography is reduced in young women who have dense breast tissue (116–118). Women with *BRCA1* and *BRCA2* germline mutations typically present with early-onset tumors of higher grade (grade 3) (119,120), and little is known about the effects of mammography screening in this high-risk group. Nevertheless, preliminary data suggest that increased surveillance does detect very small and node-negative tumors in high-risk women, although the impact on breast cancer mortality remains to be established (121–123).

Addition of breast magnetic resonance imaging (MRI) to the more commonly available triad of mammography, ultrasound, and breast examination appears to improve detection of small, node-negative invasive tumors as observed in small prospective and retrospective studies (124–127). Although breast MRI showed higher sensitivity than mammography, some cancers may still be detected as interval cancer (126). The preliminary encouraging findings will require confirmation by large prospective studies on the cost-effectiveness of breast MRI presently under way in the United Kingdom and the Netherlands (128). The appropriateness of breast cancer screening schedules and ovarian cancer screening also needs to be further evaluated in larger prospective series of women with proven *BRCA1* or *BRCA2* mutations or a strong family history of breast or ovarian cancer.

The question of breast-conserving therapy should also be addressed in view of the high risk for second primary breast cancer, as well as a possibly higher risk for ipsilateral breast tumor recurrence and an increased radiation sensitivity in *BRCA1* and *BRCA2* mutation carriers (129,130). Risk for contralateral breast cancer is higher for younger age at onset in *BRCA1*-associated breast cancer, and this must be made clear when discussing treatment choices (131,132).

Strategies for Primary Prevention

Few recognized lifestyle risk factors for breast cancer are easily amenable for prevention of breast cancer. Modifiable lifestyle characteristics most studied for their relation to breast cancer are dietary fat, obesity, and physical activity levels. Dietary fat is assumed to affect breast cancer risk by increasing the availability of estrogen, and reduction of serum estradiol levels have been observed in both premenopausal and postmenopausal women randomized to a low-fat diet (133). The Women's Health Initiative will evaluate the possibility of a direct effect of dietary fat intake and breast cancer risk in postmenopausal women, including a randomized trial to evaluate a low-fat dietary pattern (134). Dietary intake is also closely related to obesity, which is associated with increased breast cancer risk only in postmenopausal women. Half of the *BRCA1* mutation carriers with breast cancer are diagnosed by the age of 50 and therefore predominantly premenopausal. Therefore, lifestyle changes to prevent disease after the menopause may not be effective for these women. Regular strenuous physical activity has been shown to reduce breast cancer risk in most studies, however, some recent studies suggest that the protective effect may be more evident among the postmenopausal women (8,135). Intensive sport during adolescence, which delays the age at onset, may be another option to reduce the cyclic exposure to circulating ovarian hormones. However, it is yet unclear whether the recognized lifestyle and reproductive risk factors for breast cancer also apply to the inherited form or are modified by genetic predisposition due to *BRCA1/2*. There is only preliminary evidence for modifiers of the penetrance of *BRCA1/2* (as discussed earlier). Data from prospective series of *BRCA1/2* mutations carriers will be required to provide unbiased estimation of potential risk modification.

Preventive options discussed for high-risk women include prophylactic surgery and chemoprevention. Prophylactic mastectomy is usually considered by a minority of women with the strongest family histories or personal experience of the disease, although there are cultural differences in acceptability (114). Recent reports from retrospective follow-up suggest that prophylactic mastectomy can confer a high degree of long-term protection. The cohort of high-risk women who underwent prophylactic mastectomy at the Mayo Clinic experienced an 81% reduction in incident breast cancer compared with their sisters (136). Supportive data are now emerging from prospective follow-up studies, although the follow-up period is still short (137). At the Rotterdam Family Cancer Clinic, 76 women with *BRCA1/2* mutations who underwent prophylactic bilateral mastectomy remained unaffected while 8 of 63 other carriers, followed according to a surveillance protocol of a monthly self-examination, a semiannual clinical breast cancer examination, and annual mammography, developed breast cancer after 3 years of follow-up (138). Prophylactic oophorectomy, which reduces breast cancer risk, was more common among women who underwent prophylactic mastectomy than those only under surveillance, and women in the study were allowed to choose their treatment, so that the reduced incidence may not be entirely attributable to prophylactic mastectomy. Longer follow-up will be required to establish the impact, if any, on mortality of breast cancer. Bilateral prophylactic mastectomy has also been found to reduce psychological morbidity and anxiety (139). This may be explained by the strong conviction in women who chose surgery that the removal of breast tissue will reduce their risk significantly. Long-term prospective follow-up on an extended group of women will be necessary to truly address the risk of subsequent breast cancer and the psychological sequela.

Bilateral prophylactic oophorectomy has been reported in a study of 122 *BRCA1* mutation carriers to reduce breast cancer risk by 50% or more with longer duration of follow-up and protection was not abrogated by hormone replacement therapy (140). Since most options for ovarian cancer prevention are not highly efficacious, bilateral prophylactic oophorectomy is an option to also reduce the high risk of ovarian cancer particularly in *BRCA1* carriers, although it is clear that the subsequent development of ovarian cancers cannot be completely prevented because of reports of peritoneal carcinoma and cancer of the fallopian tube subsequent to surgery (141,142). Removal of ovaries and fallopian tubes in their entirety has thus been recommended when prophylactic bilateral oophorectomy is performed (143).

The use of compounds that ablate the production of ovarian hormones may provide a nonsurgical alternative to prophylactic surgery so that effective chemoprevention would reduce the need for prophylactic surgery. However, if chemoprevention is to be equivalent to (the gold standard of) prophylactic mastectomy, it must provide an 80% reduction. The results of approximately 50% reduction in the risk of developing breast cancer by tamoxifen in the US trial have not so far been reproduced in two smaller European trials (144–146). Tamoxifen reduced the occurrence of estrogen receptor-positive but not estrogen receptor-negative tumors and therefore may not be effective in preventing *BRCA1*-associated tumors. In the Multiple Outcomes of Raloxifene Evaluation (MORE) multicenter trial, which included 7705 women, raloxifene, another selective estrogen receptor modulator (SERM), also decreased the risk for estrogen receptor-positive postmenopausal breast cancer by 90% but not estrogen receptor-negative invasive breast cancer (147). However, the efficacy of SERMs as chemopreventive agents for women with a *BRCA1* mutation has also been questioned because most tumors arising in these women do not express estrogen receptors (148). Fenretinide, a synthetic retinoic acid derivative, has recently been shown to decrease the second breast malignancies in premenopausal women, although these findings require confirmation (149). The evaluation of chemoprevention, however, particularly in *BRCA1/2* mutation carriers, may be hampered by poor recruitment, as was recently observed in the United Kingdom, where the low level of recruitment of highest risk women into chemopreventive trials seems to be associated with a reluctance to be randomly allocated to placebo (150).

Social, Ethical, and Economic issues

Familial cancer clinics have been established to meet the new demands, and several studies of patients who have attended clinical services for familial cancer have shown that genetic counseling fulfills a useful function to alleviate anxiety. Individual self-estimates of breast cancer risk among members of multicase families are often unrealistic; and accuracy of risk perception improves somewhat after attendance, and levels of anxiety tend to decline, at least in the short term, regardless of changes in risk perception. However, considerable time must be allocated to dealing with these issues (151–153). Furthermore, only half of eligible women choose to undergo clinical *BRCA1/2* testing after appropriate counseling, and they are the women who have the highest risk of carrying a mutation and thus the greatest probability of gaining useful information from the test results (154,155).

Highly educated professional women are clearly overrepresented and the most socially deprived groups underrepresented among the clientele of familial breast cancer clinics (156). This social imbalance needs to be corrected if health benefits from existing programs are confirmed and are to be applied in general. Education of both professionals (general practitioners and nurses) and laypersons is required to improve the delivery of genetic services as well as the perception of familial cancer risks and limitations of molecular testing (157–159).

Not unique to familial breast cancer is the difficulty of balancing the individual's right to privacy against the duty to share relevant information with the wider family. Confirmation and extension of family histories are necessary to provide a best estimate of genetic risk. Nevertheless, questions exist about ownership of the information, particularly when its disclosure may increase risks for "genetic discrimination" in education, employment, insurance coverage, and access to health care (156,160). Restricting access to medical records may compromise the conduct of both genetic and epidemiologic studies of profound importance for improving future health. Fortunately, most women appear to believe that genetic information should be shared within families, unless that violates a patient's wishes (161).

The actual costs of familial cancer clinical services are high, regardless of whether they are provided through national, insurance-based, or privately funded health-care systems. Reduction in morbidity and mortality from breast cancer in this case particularly involves young women who are still in the prime of life. The potential economic and social gain can thus be enormous (162). Further studies are warranted to determine factors to increase cost-effectiveness, such as better identifying women who would most benefit from specialist genetics services and providing management and surveillance with greatest efficacy.

Conclusions

Clinical Practice

The awareness of genetic predisposition to breast cancer has increased tremendously since the identification of *BRCA1* and *BRCA2* and demand for risk assessment and genetic counseling as well as for genetic testing has been fueled by publicity in both the popular media and the professional literature. Breast cancer genetics clinics have been established in many countries to meet this new demand. Although progress is continuously being made, the unresolved questions are still "Can we identify those at high risk?" and "Can we do anything for them?" Molecular screening should theoretically contribute to defining the target group at highest risk. In practice, this is only feasible in limited settings, chiefly where founder mutations are relatively common. In most populations, we must rely on family history to obtain an empirical estimate of genetic risk, in particular when molecular data are not available or a germline mutation was not identified in one of the known susceptibility genes. The variability in risk estimates from different methods of computation, the use of often incomplete and /or inaccurate information on family history, and the reliance on molecular data for only a minority of affected families are reflected in the imprecision in selection of family members for special surveillance and decisions about management. A breast cancer family program must be provided to every woman who meets the eligibility criteria, regardless of mutation screening.

Currently available surveillance includes annual mammography, clinical breast examination, and breast self-examination. Despite the recognized reduced sensitivity of the mammography in young women, accumulated experience to date suggests that increased surveillance detects breast cancer at an early stage, although longer follow-up will be required to establish the real impact on breast cancer morbidity and mortality. Breast MRI appears to have higher sensitivity, but the higher costs involved may limit its application to a defined group at highest risk. Other developments in breast imaging and technologic advances could have a major impact in this field. Bilateral prophylactic oophorectomy is an option to reduce the high risk of both breast and ovarian cancer particularly in *BRCA1* carriers and may be recommended for women with *BRCA1* and *BRCA2* mutations after childbearing has been completed. The more effective but also more daunting preventive option is prophylactic mastectomy and is the right choice for some women. While it is justified to demand that prevention should not be more extreme than the cure, less invasive preventive measures such as chemoprevention has not yet been shown to be effective for women with *BRCA1* and *BRCA2* mutations.

Population Testing

The previous discussions pointed out that *BRCA1* and *BRCA2* mutations are extremely rare in most populations, the upper limit of the estimated combined population frequencies being generally about 0.3%. Mutation screening has remained time consuming and costly because of the large size of the gene and the broad spectrum of mutations. The sensitivity of molecular testing with the use of conventional methods is also well below 100%, because large deletions of the genes are bound to remain undetected. In addition, the available surveillance tools are imperfect, and the choices for preventive surgery difficult. More importantly, clear evidence is lacking about reduction in cancer morbidity and mortality from long-term follow-up of women participating in breast cancer genetics programs. In general, population testing for *BRCA1* and *BRCA2* is out of the question in most populations, except for example, the populations in Iceland and in Israel, where founder mutations are relatively common.

Agenda for Future Human Genome Epidemiology Research

Individual risk estimates need to be refined, taking into account of major germ-line predisposing mutations, polymorphisms at potential modifying genes, and the influence of environmental factors. Uncertainties in risk assessment could be partly reduced given more precise estimates of gene frequencies and age-specific penentrances for defined populations. Large population-based studies are thus still required to determine the contribution of the BRCA genes to breast cancer in the general population. A substantial contribution to the genetic component of breast cancer etiology comes, almost certainly, from common low-penetrant mutations or polymorphisms in genes that have yet to be identified. The discovery of additional major breast cancer genes will need to make use of more homogeneous population to increase the power of genetic mapping. Many large population-based case-control or cohort studies are required to detect and replicate associations observed with common low-penetrant polymorphisms in candidate genes, taking into account possible effects of environmental exposures.

Another important research goal is to develop management strategies according to genetic profile and categories of risk. Ongoing and novel prospective studies to define the impact of prophylactic surgery and optimal screening strategies must be supported. Many of the issues surrounding the provision and evaluation of services for women at risk for familial cancers are common to the different countries. An extended network of scientists and professionals in the field, as has been developed in the United States and in Europe, should therefore be supported on the international level. High priority must be given to the collation of prospective data on the efficacy of surveillance and early detection for breast and ovarian cancer to better define risks and benefits according to age, family history, molecular findings, and population background.

Chemoprevention is an option still in its infancy. New chemopreventive agents will need to be developed for younger women because breast cancer risk increases most steeply between 40 and 50 years of age. The efficacy of existing or novel agents for chemoprevention has to be carefully assessed in properly designed clinical trials of women with defined increased risk for breast cancer. In the process, other modifying factors of penetrance in mutation carriers need to be accounted for to evaluate the true effect of the chemopreventive agents.

Health-care resources are limited everywhere. Therefore, clinical services for familial cancers should undergo auditing and comprehensive economic evaluation, so that these can evolve effectively in parallel with developments in molecular technology.

Acknowledgments
I would like to thank Neva Haites, Michael Steel, Gareth Evans, Diana Eccles, Shirley Hodgson, Pål Møller, Dominique Stoppa-Lyonnet, Patrick Morrison, and Hans Vasen, from whose diverse expertise in the field of genetic oncology I have learned tremendously, and for their continuing enthusiasm and efforts for collaborative work.

References

1. Blanks R, Moss S, McGahan C, et al. Effect of NHS breast screening programme on mortality from breast cancer in England and Wales, 1990–8: comparison of observed with predicted mortality. BMJ 2000;321:665–669.
2. Pike M, Spicer D, Dahmoush L, et al. Estrogens, progestogens, normal breast cell proliferation, and breast cancer risk. Epidemiol Rev 1993;15:17–35.
3. Bernstein L, Ross R. Endogenous hormones and breast cancer risk. Epidemiol Rev 1993;15:48–65.
4. Kelsey JL, Gammon MD, John EM. Reproductive factors and breast cancer. Epidemiol Rev 1993;15:36–47.
5. Chang Claude J, Eby N, Kiechle M, et al. Breastfeeding and breast cancer risk by age 50 among women in Germany. Cancer Causes Control 2000;11:687–695.
6. Bernstein L, Henderson BE, Hanisch R, et al. Physical exercise and reduced risk of breast cancer in young women. J Natl Cancer Inst 1994;86:1403–1408.
7. Broocks A, Pirke K, Schweiger U, et al. Cyclic ovarian function in recreational athletes. J Appl Physiol 1990;68:2083–2086.
8. Gammon MD, Schoenberg JB, Britton JA, et al. Recreational physical activity and breast cancer risk among women under age 45 years. Am J Epidemiol 1998;147:273–280.
9. Longnecker MP. Alcoholic beverage consumption in relation to risk of breast cancer: meta-analysis and review. Cancer Causes Control 1994;5:73–82.
10. Smith Warner S, Spiegelman D, Yaun S, et al. Alcohol and breast cancer in women: a pooled analysis of cohort studies. JAMA 1998;279:535–540.
11. Kropp S, Becher H, Nieters A, et al. Low-to-moderate alcohol consumption and breast cancer risk by age 50 years among women in Germany. Am J Epidemiol 2001;154:624–634.
12. Ginsburg E. Estrogen, alcohol and breast cancer risk. J Steroid Biochem Mol Biol 1999;69:299–306.
13. Yu H, Berkel J. Do insulin-like growth factors mediate the effect of alcohol on breast cancer risk? Med Hypotheses 1999;52:491–496.
14. Dorgan J, Baer D, Albert P, et al. Serum hormones and the alcohol-breast cancer association in postmenopausal women. J Natl Cancer Inst 2001;93:710–715.
15. Wright R, McManaman J, Repine J. Alcohol-induced breast cancer: a proposed mechanism. Free Radic Biol Med 1999;26:348–354.

16. Hunter D, Willett W. Diet, body size, and breast cancer. Epidemiol Rev 1993;15: 110–132.

17. MacMahon B, Cole P, Brown J, et al. Urine estrogens, frequency of ovulation, and breast cancer risk: case-control study in premenopausal women. J Natl Cancer Inst 1983; 70:247–250.

18. Bernstein L, Yuan JM, Ross RK, et al. Serum hormone levels in pre-menopausal Chinese women in Shanghai and white women in Los Angeles: results from two breast cancer case-control studies. Cancer Causes Control 1990;1:51–58.

19. Shimizu H, Ross RK, Bernstein L, et al. Serum oestrogen levels in postmenopausal women: comparison of American whites and Japanese in Japan. Br J Cancer 1990; 62:451–453.

20. Thomas HV, Key TJ, Allen DS, et al. A prospective study of endogenous serum hormone concentrations and breast cancer risk in premenopausal women on the island of Guernsey. Br J Cancer 1997;75:1075–1079.

21. Hankinson SE, Willett WC, Colditz GA, et al. Circulating concentrations of insulin-like growth factor-I and risk of breast cancer. Lancet 1998;351:1393–1396.

22. Eby N, Chang Claude J, Bishop DT. Familial risk and genetic susceptibility for breast cancer. Cancer Causes Control 1994;5:458–470.

23. Claus EB, Risch N, Thompson WD. Genetic analysis of breast cancer in the cancer and steroid hormone study. Am J Hum Genet 1991;48:232–242.

24. Antoniou A, Pharoah P, McMullan G, et al. Evidence for further breast cancer susceptibility genes in addition to BRCA1 and BRCA2 in a population-based study. Genet Epidemiol 2001;21:1–18.

25. Cui J, Antoniou A, Dite G, et al. After BRCA1 and BRCA2-what next? Multifactorial segregation analyses of three-generation, population-based Australian families affected by female breast cancer. Am J Hum Genet 2001;68:420–431.

26. Miki Y, Swensen J, Shattuck Eidens D, et al. A strong candidate for the breast and ovarian cancer susceptibility gene BRCA1. Science 1994;266:66–71.

27. Wooster R, Bignell G, Lancaster J, et al. Identification of the breast cancer susceptibility gene BRCA2. Nature 1995;378:789–792.

28. Tavtigian SV, Simard J, Rommens J, et al. The complete BRCA2 gene and mutations in chromosome 13q-linked kindreds. Nat Genet 1996;12:333–337.

29. Narod SA, Ford D, Devilee P, et al. An evaluation of genetic heterogeneity in 145 breast-ovarian cancer families. Breast Cancer Linkage Consortium. Am J Hum Genet 1995; 56:254–264.

30. Ford D, Easton DF, Stratton M, et al. Genetic heterogeneity and penetrance analysis of the BRCA1 and BRCA2 genes in breast cancer families. The Breast Cancer Linkage Consortium. Am J Hum Genet 1998;62:676–689.

31. Couch FJ, Deshano ML, Blackwood MA, et al. BRCA1 mutations in women attending clinics that evaluate the risk of breast cancer. N Engl J Med 1997;336:1409–1415.

32. Peto J, Collins N, Barfoot R, et al. Prevalence of BRCA1 and BRCA2 gene mutations in patients with early-onset breast cancer. J Natl Cancer Inst 1999;91:943–949.

33. Stoppa Lyonnet D, Laurent Puig P, Essioux L, et al. BRCA1 sequence variations in 160 individuals referred to a breast/ovarian family cancer clinic. Institut Curie Breast Cancer Group. Am J Hum Genet 1997;60:1021–1030.

34. Frank TS, Manley SA, Olopade OI, et al. Sequence analysis of BRCA1 and BRCA2: correlation of mutations with family history and ovarian cancer risk. J Clin Oncol 1998;16:2417–2425.

35. Martin A, Blackwood M, Antin Ozerkis D, et al. Germline mutations in BRCA1 and BRCA2 in breast-ovarian families from a breast cancer risk evaluation clinic. J Clin Oncol 2001;19:2247–2253.

36. Easton DF, Narod SA, Ford D, et al. The genetic epidemiology of BRCA1. Breast Cancer Linkage Consortium [letter]. Lancet 1994;344:761.

37. Thompson D, Easton D. Variation in cancer risks, by mutation position, in BRCA2 mutation carriers. Am J Hum Genet 2001;68:410–419.

38. Ford D, Easton DF, Bishop DT, et al. Risks of cancer in BRCA1-mutation carriers. Breast Cancer Linkage Consortium. Lancet 1994;343:692–695.

39. Anonymous. Cancer risks in BRCA2 mutation carriers.The Breast Cancer Linkage Consortium. J Natl Cancer Inst 1999;91:1310–1316.

40. Shih H, Nathanson K, Seal S, et al. BRCA1 and BRCA2 mutations in breast cancer families with multiple primary cancers. Clin Cancer Res 2000;6:4259–4264.

41. Nathanson K, Wooster R, Weber B. Breast cancer genetics: what we know and what we need. Nat Med 2001;7:552–556.

42. Deng C, Scott F. Role of the tumor suppressor gene Brca1 in genetic stability and mammary gland tumor formation. Oncogene 2000;19:1059–1064.

43. Welcsh P, King M. BRCA1 and BRCA2 and the genetics of breast and ovarian cancer. Hum Mol Genet 2001;10:705–713.

44. Malkin D, Li FP, Strong LC, et al. Germ line p53 mutations in a familial syndrome of breast cancer, sarcomas, and other neoplasms. Science 1990;250:1233–1238.

45. Birch J, Hartley AL, Tricker KJ, et al. Prevalence and diversity of constitutional mutations in the p53 gene among 21 Li-Fraumeni families. Cancer Res 1994;54:1298–1304.

46. Jenne DE, Reimann H, Nezu J, et al. Peutz-Jeghers syndrome is caused by mutations in a novel serine threonine kinase. Nat Genet 1998;18:38–43.

47. Boardman L, Thibodeau S, Schaid D, et al. Increased risk for cancer in patients with the Peutz-Jeghers syndrome. Ann Intern Med 1998;128:896–899.

48. Liaw D, Marsh D, Li J, et al. Germline mutations of the PTEN gene in Cowden disease, an inherited breast and thyroid cancer syndrome. Nat Genet 1997;16:64–67.

49. Borresen AL. Role of genetic factors in breast cancer susceptibility. Acta Oncol 1992; 31:151–155.

50. Rapakko K, Allinen M, Syrjakoski K, et al. Germline TP53 alterations in Finnish breast cancer families are rare and occur at conserved mutation-prone sites. Br J Cancer 2001;84:116–119.

51. Swift M, Reitnauer PJ, Morrell D, et al. Breast and other cancers in families with ataxia-telangiectasia. N Engl J Med 1987;316:1289–1294.

52. Easton DF. Cancer risks in A-T heterozygotes. Int J Radiat Biol 1994;66:S177–S182.

53. Athma P, Rappaport R, Swift M. Molecular genotyping shows that ataxia-telangiectasia heterozygotes are predisposed to breast cancer. Cancer Genet Cytogenet 1996;92:130–134.

54. Janin N, Andrieu N, Ossian K, et al. Breast cancer risk in ataxia telangiectasia (AT) heterozygotes: haplotype study in French AT families. Br J Cancer 1999;80:1042–1045.

55. FitzGerald MG, Bean JM, Hegde SR, et al. Heterozygous ATM mutations do not contribute to early onset of breast cancer. Nat Genet 1997;15:307–310.

56. Vorechovsy I, Rasio D, Luo L, et al. The ATM gene and susceptibility to breast cancer: analysis of 38 breast tumors reveals no evidence for mutation. Cancer Res 1996;56: 2726–2732.

57. Chen J, Stampfer M, Hough H, et al. A prospective study of N-acetyltransferase genotype, red meat intake, and risk of colorectal cancer. Cancer Res 1998;58:3307–3311.

58. Laake K, Vu P, Andersen T, et al. Screening breast cancer patients for Norwegian ATM mutations. Br J Cancer 2000;83:1650–1653.

59. Broeks A, Urbanus JH, Floore AN, et al. ATM-heterozygous germline mutations contribute to breast cancer-susceptibility. Am J Hum Genet 2000;66:494–500.

60. Dork T, Bendix R, Bremer M, et al. Spectrum of ATM gene mutations in a hospital-based series of unselected breast cancer patients. Cancer Res 2001;61:7608–7615.

61. Dunning AM, Healey CS, Pharoah PD, et al. A systematic review of genetic polymorphisms and breast cancer risk. Cancer Epidemiol Biomarkers Prev 1999;8:843–854.

62. Healey C, Dunning A, Teare M, et al. A common variant in BRCA2 is associated with both breast cancer risk and prenatal viability. Nat Genet 2000;26:362–364.

63. Deitz A, Zheng W, Leff M, et al. N-Acetyltransferase-2 genetic polymorphism, well-done meat intake, and breast cancer risk among postmenopausal women. Cancer Epidemiol Biomarkers Prev 2000;9:905–910.

64. Hakansson S, Johannsson O, Johansson U, et al. Moderate frequency of BRCA1 and BRCA2 germ-line mutations in Scandinavian familial breast cancer. Am J Hum Genet 1997;60:1068–1078.

65. Dong J, Chang Claude J, Wu Y, et al. A high proportion of mutations in the BRCA1 gene in German breast/ovarian cancer families with clustering of mutations in the 3′ third of the gene. Hum Genet 1998;103:154–161.

66. Shattuck Eidens D, McClure M, Simard J, et al. A collaborative survey of 80 mutations in the BRCA1 breast and ovarian cancer susceptibility gene. Implications for presymptomatic testing and screening. JAMA 1995;273:535–541.

67. Serova Sinilnikova OM, Boutrand L, Stoppa Lyonnet D, et al. BRCA2 mutations in hereditary breast and ovarian cancer in France [letter]. Am J Hum Genet 1997;60:1236–1239.

68. Ikeda N, Miyoshi Y, Yoneda K, et al. Frequency of BRCA1 and BRCA2 germline mutations in Japanese breast cancer families. Int J Cancer 2001;91:83–88.

69. Hopper JL, Southey MC, Dite GS, et al. Population-based estimate of the average age-specific cumulative risk of breast cancer for a defined set of protein-truncating mutations in BRCA1 and BRCA2. Cancer Epidemiology, Biomarkers & Prevention 1999;8:741–747.

70. Syrjakoski K, Vahteristo P, Eerola H, et al. Population-based study of BRCA1 and BRCA2 mutations in 1035 unselected Finnish breast cancer patients. J Natl Cancer Inst 2000;92:1529–1531.

71. Loman N, Johannsson O, Kristoffersson U, et al. Family history of breast and ovarian cancers and brca1 and brca2 mutations in a population-based series of early-onset breast cancer. J Natl Cancer Inst 2001;93:1215–1223.

72. Antoniou AC, Gayther SA, Stratton JF, et al. Risk models for familial ovarian and breast cancer. Genet Epidemiol 2000;18:173–190.

73. Johannesdottir G, Gudmundsson J, Bergthorsson JT, et al. High prevalence of the 999del5 mutation in icelandic breast and ovarian cancer patients. Cancer Res 1996;56:3663–3665.

74. Thorlacius S, Sigurdsson S, Bjarnadottir H, et al. Study of a single BRCA2 mutation with high carrier frequency in a small population. Am J Hum Genet 1997;60:1079–1084.

75. Struewing JP, Hartge P, Wacholder S, et al. The risk of cancer associated with specific mutations of BRCA1 and BRCA2 among Ashkenazi Jews. N Engl J Med 1997;336:1401–1408.

76. Bahar A, Taylor P, Andrews L, et al. The frequency of founder mutations in the BRCA1, BRCA2, and APC genes in australian Ashkenazi Jews. Cancer 2001;92:440–445.

77. Couch FJ, Farid LM, Deshano ML, et al. BRCA2 germline mutations in male breast cancer cases and breast cancer families. Nat Genet 1996;13:123–125.

78. Friedman LS, Gayther SA, Kurosaki T, et al. Mutation analysis of BRCA1 and BRCA2 in a male breast cancer population. Am J Hum Genet 1997;60:313–319.

79. Haraldsson K, Loman N, Zhang Q, et al. BRCA2 germ-line mutations are frequent in male breast cancer patients without a family history of the disease. Cancer Res 1998;58:1367–1371.

80. Csokay B, Udvarhelyi N, Sulyok Z, et al. High frequency of germ-line BRCA2 mutations among Hungarian male breast cancer patients without family history. Cancer Res 1999;59:995–998.

81. Easton DF, Bishop DT, Ford D, et al. Genetic linkage analysis in familial breast and ovarian cancer: results from 214 families. The Breast Cancer Linkage Consortium. Am J Hum Genet 1993;52:678–701.

82. Easton DF, Ford D, Bishop DT. Breast and ovarian cancer incidence in BRCA1-mutation carriers. Breast Cancer Linkage Consortium. Am J Hum Genet 1995;56:265–271.

83. Anglian Breast Cancer Study Group. Prevalence and penetrance of BRCA1 and BRCA2 mutations in a population-based series of breast cancer cases. Br J Cancer 2000;83:1301–1308.

84. Risch H, McLaughlin J, Cole D, et al. Prevalence and penetrance of germline BRCA1 and BRCA2 mutations in a population series of 649 women with ovarian cancer. Am J Hum Genet 2001;68:700–710.

85. Thorlacius S, Struewing JP, Hartge P, et al. Population-based study of risk of breast cancer in carriers of BRCA2 mutation. Lancet 1998;352:1337–1339.

86. Satagopan J, Offit K, Foulkes W, et al. The lifetime risks of breast cancer in Ashkenazi Jewish carriers of BRCA1 and BRCA2 mutations. Cancer Epidemiol Biomarkers Prev 2001;10:467–473.

87. Gayther SA, Mangion J, Russell P, et al. Variation of risks of breast and ovarian cancer associated with different germline mutations of the BRCA2 gene. Nat Genet 1997;15:103–105.

88. Gayther SA, Warren W, Mazoyer S, et al. Germline mutations of the BRCA1 gene in breast and ovarian cancer families provide evidence for a genotype-phenotype correlation. Nat Genet 1995;11:428–433.

89. Bork P, Blomberg N, Nilges M. Internal repeats in the BRCA2 protein sequence [letter]. Nat Genet 1996;13:22–23.

90. Scully R, Chen J, Plug A, et al. Association of BRCA1 with Rad51 in mitotic and meiotic cells. Cell 1997;88:265–275.

91. Levy Lahad E, Catane R, Eisenberg S, et al. Founder BRCA1 and BRCA2 mutations in Ashkenazi Jews in Israel: frequency and differential penetrance in ovarian cancer and in breast-ovarian cancer families. Am J Hum Genet 1997;60:1059–1067.

92. Phelan CM, Rebbeck TR, Weber BL, et al. Ovarian cancer risk in BRCA1 carriers is modified by the HRAS1 variable number of tandem repeat (VNTR) locus. Nat Genet 1996;12:309–311.

93. Rebbeck TR, Kantoff PW, Krithivas K, et al. Modification of BRCA1-associated breast cancer risk by the polymorphic androgen-receptor CAG repeat. Am J Hum Genet 1999;64:1371–1377.

94. Kadouri L, Easton D, Edwards S, et al. CAG and GGC repeat polymorphisms in the androgen receptor gene and breast cancer susceptibility in BRCA1/2 carriers and non-carriers. Br J Cancer 2001;85:36–40.

95. Levy Lahad E, Lahad A, Eisenberg S, et al. A single nucleotide polymorphism in the RAD51 gene modifies cancer risk in BRCA2 but not BRCA1 carriers. Proc Natl Acad Sci U S A 2001;98:3232–3236.

96. Wang W, Spurdle A, Kolachana P, et al. A Single Nucleotide Polymorphism in the 5′ Untranslated Region of RAD51 and Risk of Cancer among BRCA1/2 Mutation Carriers. Cancer Epidemiol Biomarkers Prev 2001;10:955–960.

97. Chang Claude J, Becher H, Eby N, et al. Modifying effect of reproductive risk factors on the age at onset of breast cancer for German BRCA1 mutation carriers. J Cancer Res Clin Oncol 1997;123:272–279.

98. Ursin G, Henderson BE, Haile RW, et al. Does oral contraceptive use increase the risk of breast cancer in women with BRCA1/BRCA2 mutations more than in other women? Cancer Res 1997;57:3678–3681.

99. Jernstrom H, Lerman C, Ghadirian P, et al. Pregnancy and risk of early breast cancer in carriers of BRCA1 and BRCA2. Lancet 1999;354:1846–1850.

100. Goldgar D, Bonnardel C, Renard H, et al. The International BRCA1/2 Carrier Cohort Study: Purpose, Rationale, and Study Design. Breast Cancer Res 2000;2:E010

101. Petrij Bosch A, Peelen T, van Vliet M, et al. BRCA1 genomic deletions are major founder mutations in Dutch breast cancer patients. Nat Genet 1997;17:341–345.
102. Puget N, Stoppa Lyonnet D, Sinilnikova O, et al. Screening for germ-line rearrangements and regulatory mutations in BRCA1 led to the identification of four new deletions. Cancer Res 1999;59:455–461.
103. Unger M, Nathanson K, Calzone K, et al. Screening for genomic rearrangements in families with breast and ovarian cancer identifies BRCA1 mutations previously missed by conformation-sensitive gel electrophoresis or sequencing. Am J Hum Genet 2000;67:841–850.
104. Nordling M, Karlsson P, Wahlstrom J, et al. A large deletion disrupts the exon 3 transcription activation domain of the BRCA2 gene in a breast/ovarian cancer family. Cancer Res 1998;58:1372–1375.
105. Serova O, Montagna M, Torchard D, et al. A high incidence of BRCA1 mutations in 20 breast-ovarian cancer families. Am J Hum Genet 1996;58:42–51.
106. Claus EB, Risch N, Thompson WD. Autosomal dominant inheritance of early-onset breast cancer. Implications for risk prediction. Cancer 1994;73:643–651.
107. Schmidt S, Becher H, Chang Claude J. Breast cancer risk assessment: use of complete pedigree information and the effect of misspecified ages at diagnosis of affected relatives. Hum Genet 1998;102:348–356.
108. Parmigiani G, Berry D, Aguilar O. Determining carrier probabilities for breast cancer-susceptibility genes BRCA1 and BRCA2. Am J Hum Genet 1998;62:145–158.
109. Cui J, Hopper J. Why are the majority of hereditary cases of early-onset breast cancer sporadic? A simulation study. Cancer Epidemiol Biomarkers Prev 2000;9:805–812.
110. Tischkowitz M, Wheeler D, France E, et al. A comparison of methods currently used in clinical practice to estimate familial breast cancer risks. Ann Oncol 2000;11:451–454.
111. Stoppa Lyonnet D, Caligo M, Eccles D, et al. Genetic testing for breast cancer predisposition in 1999: which molecular strategy and which family criteria? Dis Markers 1999;15:67–68.
112. Moller P, Evans G, Haites N, et al. Guidelines for follow-up of women at high risk for inherited breast cancer: consensus statement from the Biomed 2 Demonstration Programme on Inherited Breast Cancer. Dis Markers 1999;15:207–211.
113. Eisinger F, Alby N, Bremond A, et al. Recommendations for medical management of hereditary breast and ovarian cancer: the French National Ad Hoc Committee. Ann Oncol 1998;9:939–950.
114. Julian Reynier, Bouchard, Evans, et al. Women's attitudes toward preventive strategies for hereditary breast or ovarian carcinoma differ from one country to another: differences among English, French, and Canadian women. Cancer 2001;92:959–968.
115. Lalloo F, Boggis C, Evans D, et al. Screening by mammography, women with a family history of breast cancer. Eur J Cancer 1998;34:937–940.
116. National Institutes of Health Consensus Development Panel. National Institutes of Health Consensus Development Conference Statement: Breast Cancer Screening for Women Ages 40–49, January 21–23, 1997. National Institutes of Health Consensus Development Panel. J Natl Cancer Inst 1997;89:1015–1026.
117. van Gils C, Otten J, Hendriks J, et al. High mammographic breast density and its implications for the early detection of breast cancer. J Med Screen 1999;6:200–204.
118. Nixon R, Pharoah P, Tabar L, et al. Mammographic screening in women with a family history of breast cancer: some results from the Swedish two-county trial. Rev Epidemiol Sante Publique 2000;48:325–331.
119. Eisinger F, Stoppa Lyonnet D, Longy M, et al. Germ line mutation at BRCA1 affects the histoprognostic grade in hereditary breast cancer. Cancer Res 1996;56:471–474.
120. Lakhani S, Jacquemier J, Sloane J, et al. Multifactorial analysis of differences between sporadic breast cancers and cancers involving BRCA1 and BRCA2 mutations. J Natl Cancer Inst 1998;90:1138–1145.

121. Moller P, Reis M, Evans G, et al. Efficacy of early diagnosis and treatment in women with a family history of breast cancer. European Familial Breast Cancer Collaborative Group. Dis Markers 1999;15:179–186.

122. Tilanus Linthorst MM, Bartels CC, Obdeijn AI, et al. Earlier detection of breast cancer by surveillance of women at familial risk. Eur J Cancer 2000;36:514–519.

123. Macmillan RD. Screening women with a family history of breast cancer—results from the British Familial Breast Cancer Group. Eur J Surg Oncol 2000;26:149–152.

124. Kuhl C, Schmutzler R, Leutner C, et al. Breast MR imaging screening in 192 women proved or suspected to be carriers of a breast cancer susceptibility gene: preliminary results. Radiology 2000;215:267–279.

125. Warner E, Plewes D, Shumak R, et al. Comparison of breast magnetic resonance imaging, mammography, and ultrasound for surveillance of women at high risk for hereditary breast cancer. J Clin Oncol 2001;19:3524–3531.

126. Brekelmans C, Seynaeve C, Bartels C, et al. Effectiveness of breast cancer surveillance in BRCA1/2 gene mutation carriers and women with high familial risk. J Clin Oncol 2001;19:924–930.

127. Stoutjesdijk M, Boetes C, Jager G, et al. Magnetic resonance imaging and mammography in women with a hereditary risk of breast cancer. J Natl Cancer Inst 2001;93: 1095–1102.

128. Tilanus Linthorst M, Obdeijn I, Bartels K, et al. First experiences in screening women at high risk for breast cancer with MR imaging. Breast Cancer Res Treat 2000;63:53–60.

129. Robson M, Levin D, Federici M, et al. Breast conservation therapy for invasive breast cancer in Ashkenazi women with BRCA gene founder mutations. J Natl Cancer Inst 1999;91:2112–2117.

130. Turner BC, Harrold E, Matloff E, et al. BRCA1/BRCA2 germline mutations in locally recurrent breast cancer patients after lumpectomy and radiation therapy: implications for breast-conserving management in patients with BRCA1/BRCA2 mutations. J Clin Oncol 1999;17:3017–3024.

131. Verhoog LC, Brekelmans CT, Seynaeve C, et al. Contralateral breast cancer risk is influenced by the age at onset in BRCA1-associated breast cancer. Br J Cancer 2000; 83:384–386.

132. Eccles D, Simmonds P, Goddard J, et al. Management of hereditary breast cancer. European Familial Breast Cancer Collaborative Group. Dis Markers 1999;15:187–189.

133. Wu AH, Pike MC, Stram DO. Meta-analysis: dietary fat intake, serum estrogen levels, and the risk of breast cancer. J Natl Cancer Inst 1999;91:529–534.

134. The Women's Health Initiative Study Group. Design of the Women's Health Initiative clinical trial and observational study Control Clin Trials 1998;19:61–109.

135. Friedenreich C, Bryant H, Courneya K. Case-control study of lifetime physical activity and breast cancer risk. Am J Epidemiol 2001;154:336–347.

136. Hartmann LC, Schaid DJ, Woods JE, et al. Efficacy of bilateral prophylactic mastectomy in women with a family history of breast cancer. N Engl J Med 1999;340:77–84.

137. Evans D, Anderson E, Lalloo F, et al. Utilisation of prophylactic mastectomy in 10 European centres. Dis Markers 1999;15:148–151.

138. Meijers Heijboer, van Geel, van Putten, et al. Breast cancer after prophylactic bilateral mastectomy in women with a BRCA1 or BRCA2 mutation. N Engl J Med 2001;345:159–164.

139. Hatcher, Fallowfield, A'Hern. The psychosocial impact of bilateral prophylactic mastectomy: prospective study using questionnaires and semistructured interviews. BMJ 2001;322:76

140. Rebbeck TR, Levin AM, Eisen A, et al. Breast cancer risk after bilateral prophylactic oophorectomy in BRCA1 mutation carriers. J Natl Cancer Inst 1999;91:1475–1479.

141. Struewing J, Watson P, Easton D, et al. Prophylactic oophorectomy in inherited breast/ovarian cancer families. J Natl Cancer Inst Monogr 1995;33–35.

142. Paley P, Swisher E, Garcia R, et al. Occult cancer of the fallopian tube in BRCA-1 germline mutation carriers at prophylactic oophorectomy: a case for recommending hysterectomy at surgical prophylaxis. Gynecol Oncol 2001;80:176–180.

143. Lu KH, Garber JE, Cramer DW, et al. Occult ovarian tumors in women with BRCA1 or BRCA2 mutations undergoing prophylactic oophorectomy. J Clin Oncol 2000;18: 2728–2732.

144. Fisher B, Costantino J, Wickerham D, et al. Tamoxifen for prevention of breast cancer: report of the National Surgical Adjuvant Breast and Bowel Project P-1 Study. J Natl Cancer Inst 1998;90:1371–1388.

145. Powles T, Eeles R, Ashley S, et al. Interim analysis of the incidence of breast cancer in the Royal Marsden Hospital tamoxifen randomised chemoprevention trial. Lancet 1998;352:98–101.

146. Veronesi U, Maisonneuve P, Costa A, et al. Prevention of breast cancer with tamoxifen: preliminary findings from the Italian randomised trial among hysterectomised women. Italian Tamoxifen Prevention Study. Lancet 1998;352:93–97.

147. Cummings S, Eckert, Krueger K, et al. The effect of raloxifene on risk of breast cancer in postmenopausal women: results from the MORE randomized trial. Multiple Outcomes of Raloxifene Evaluation. JAMA 1999;281:2189–2197.

148. Lidereau R, Eisinger F, Champeme M, et al. Major improvement in the efficacy of BRCA1 mutation screening using morphoclinical features of breast cancer. Cancer Res 2000;60:1206–1210.

149. Veronesi U, De Palo G, Marubini E, et al. Randomized trial of fenretinide to prevent second breast malignancy in women with early breast cancer. J Natl Cancer Inst 1999;91:1847–1856.

150. Evans D, Lalloo F, Shenton A, et al. Uptake of screening and prevention in women at very high risk of breast cancer. Lancet 2001;358:889–890.

151. Evans DG, Blair V, Greenhalgh R, et al. The impact of genetic counselling on risk perception in women with a family history of breast cancer. Br J Cancer 1994;70:934–938.

152. Lerman C, Daly M, Masny A, et al. Attitudes about genetic testing for breast-ovarian cancer susceptibility. J Clin Oncol 1994;12:843–850.

153. Cull A, Anderson E, Campbell S, et al. The impact of genetic counselling about breast cancer risk on women's risk perceptions and levels of distress. Br J Cancer 1999;79:501–508.

154. Armstrong K, Calzone K, Stopfer J, et al. Factors associated with decisions about clinical BRCA1/2 testing. Cancer Epidemiol Biomarkers Prev 2000;9:1251–1254.

155. Hofferbert S, Worringen U, Backe J, et al. Simultaneous interdisciplinary counseling in German breast/ovarian cancer families: first experiences with patient perceptions, surveillance behavior and acceptance of genetic testing. Genet Couns 2000;11:127–146.

156. Steel M, Smyth E, Vasen H, et al. Ethical, social and economic issues in familial breast cancer: a compilation of views from the E.C. Biomed II Demonstration Project. Dis Markers 1999;15:125–131.

157. Bankhead C, Emery J, Qureshi N, et al. New developments in genetics-knowledge, attitudes and information needs of practice nurses. Fam Pract 2001;18:475–486.

158. Schwartz M, Benkendorf J, Lerman C, et al. Impact of educational print materials on knowledge, attitudes, and interest in BRCA1/BRCA2: testing among Ashkenazi Jewish women. Cancer 2001;92:932–940.

159. Watson E, Austoker J, Lucassen A. A study of GP referrals to a family cancer clinic for breast/ovarian cancer. Fam Pract 2001;18:131–134.

160. Morrison P, Steel C, Vasen H, et al. Insurance implications for individuals with a high risk of breast and ovarian cancer in Europe. Dis Markers 1999;15:159–165.

161. Lehmann L, Weeks J, Klar N, et al. Disclosure of familial genetic information: perceptions of the duty to inform. Am J Med 2000;109:705–711.

162. Heimdal K, Maehle L, Moller P. Costs and benefits of diagnosing familial breast cancer. Dis Markers 1999;15:167–173.

27

The role of chemokine and chemokine receptor genes in HIV-1 infection

Thomas R. O'Brien

HIV-1 Pathogenesis

Human immunodeficiency virus type 1 (HIV-1) is an RNA virus that replicates through a double-stranded DNA intermediate by use of the viral enzyme reverse transcriptase (i.e., a retrovirus). HIV-1 infects key components of the immune system, especially macrophages and CD4$^+$ lymphocytes (T helper cells). Depletion of CD4$^+$ lymphocytes disrupts various aspects of host defense, which leads to the opportunistic infections, cancers, and other conditions that comprise the acquired immunodeficiency syndrome (AIDS).

HIV-1 infection consists of acute and chronic stages (1–3). Most patients who contract acute (primary) HIV-1 infection are symptomatic. Signs and symptoms begin several weeks after the initial infection and can be mononucleosis-like, with fever, rash, and swollen lymph nodes. Very high levels of circulating HIV-1 and a decreased CD4$^+$ lymphocyte count mark acute HIV-1 infection. About 6 to 12 weeks after the initial infection, the patient's immunologic response results in decreased viremia and a partial rebound in the CD4$^+$ lymphocyte count. By that point, however, HIV-1 has become widely disseminated to lymphoid organs, and chronic infection with HIV-1 has been established.

Most patients are asymptomatic during early chronic HIV-1 infection, but ongoing viral replication steadily damages the immune system (2). The rate of HIV-1 replication, as reflected by the blood HIV-1 RNA level, is the most important determinant of the rate of loss of CD4$^+$ lymphocytes, which in turn determines the clinical prognosis (4,5). A variety of factors, including the patient's genetic makeup, affect the HIV-1 RNA level and the rate of progression to AIDS. Without effective treatment, HIV-1-infected patients become severely immunosuppressed and develop AIDS within an average of 8 to 11 years after initial infection (6).

Epidemiology and Prevention of HIV-1 and AIDS

HIV-1 originated in Central Africa, apparently through transmission of a chimpanzee retrovirus to humans sometime in the first half of the 20th century (7,8). HIV-1 appears to have been restricted to Africa for a number of years and the virus was not recognized until it spread to North America. In 1981, the HIV-1 pandemic was heralded by cases of *Pneumocystis carinii* pneumonia and Kaposi's sarcoma among U.S. homosexual men (9,10). It was several more years before the HIV-1 was discovered and antibody tests were developed that allowed investigators to determine the extent of the epidemic. The level of basic, clinical, epidemiologic, and prevention research that has been conducted on HIV-1 is unprecedented, but the virus continues to spread despite these efforts.

Worldwide, HIV-1 sexual transmission is by far the most common mode of acquisiton. The virus is also transmitted parenterally, through transfusion or drug injection, and perinatally during pregnancy, childbirth, or breastfeeding. It is impossible to determine the incidence or prevalence of HIV-1 precisely, but the World Health Organization (WHO) estimates that 20 million people have died of AIDS since the start of the epidemic and that 36 million people are now living with HIV-1 infection (11).

The epidemic is most severe in sub-Saharan Africa, where about three-fourths of worldwide AIDS deaths have occurred thus far and where ~25 million adults and children are currently infected (11). The overall prevalence of HIV-1 among African adults is ~9%, but rates vary widely. Countries in southern African are thought to have the . highest rates of HIV-1 infection in the world. In Botswana, for example, over one-third of adults are thought to be infected (11). Heterosexual intercourse is the most common mode of transmission in Africa, but perinatal transmission is also extensive. Blood transfusions (12) and the use of unsterilized medical equipment may be additional sources of HIV-1 transmission in some countries. Severe poverty undermines efforts to stop the spread of HIV-1 and, in turn, morbidity and mortality from the epidemic has damaged the social and economic structure of many African countries.

HIV-1 is also a severe problem in some Asian countries. HIV-1 reached Asia relatively recently, but it has spread rapidly (via sexual transmission or injection drug use) in some areas. About 7 million Asians are living with HIV-1 infection (11), most of whom reside in Southeast Asia or India, where about 1 in 200 adults is infected. There is the potential for a much larger Asian epidemic as HIV-1 is becoming more common in Chinese injection drug users and reportedly has been introduced to some rural areas of China via contaminated blood collection equipment.

HIV-1 infection is less common in the developed world, but about 1 million North Americans and about 500,000 Western Europeans are currently infected (11). In these regions, the HIV-1 epidemic arose initially among homosexual or bisexual men and then spread to injection drug users and heterosexuals. HIV-1 infection is also a major problem in parts of the Caribbean and Latin America, where about 1.8 million HIV-1-infected people reside.

Prevention of HIV-1 infection is challenging, and it is especially difficult where resources are limited. In developed countries, blood screening has virtually eliminated transmission of HIV-1 to transfusion recipients, but transfusion transmission of HIV-1 remains a risk in some parts of Africa (12). Needle exchange programs reduce HIV-1 transmission among drug users (13), but these programs have not been broadly implemented. Anti-retroviral treatment of HIV-1 infected pregnant women and their newborns can prevent perinatal transmission of HIV-1, but this approach is currently too expensive for implementation in regions that need it the most (14). Less expensive prophylactic regimens to prevent perinatal transmission are under consideration for implementation in developing countries.

Prevention of sexual transmission of HIV-1 requires changes in a fundamental human behavior, and the success of such programs has been mixed. Homosexual men in the United States and Europe changed their behaviors in response to the AIDS epidemic, and Thailand mounted a successful campaign to reduce HIV-1 transmission among heterosexuals (15). Efforts have been less successful in Africa, where limited resources (and sometimes denial of the problem) are impediments. A vaccine may be the only hope for controlling the HIV-1 epidemic, but developing an anti-HIV-1 vaccine has proven challenging and none is likely to be available in the near future (16).

Effective therapy can prevent or delay immune system damage and resultant disease (17). Widespread use of combination treatment regimens for HIV-1 infection has markedly reduced death from AIDS in developed countries (18–20). In the United States, for example, mortality among HIV-1-infected patients plummeted after combination antiretroviral regimens became available in 1996 (Fig. 27.1). Treat-

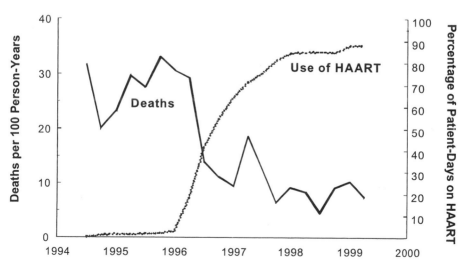

Figure 27.1 Mortality and frequency of use of combination antiretroviral therapy including a protease inhibitor among HIV-1-infected patients with <100 CD4+ lymphocytes/mm³, according to calendar quarter, from 1994 through 1999. (Data updated from Pallela et al. [18], courtesy of Scott Holmberg.)

ment is, therefore, an important and effective strategy for preventing AIDS and other consequences of HIV-1 infection.

Therapy for HIV-1 Infection

The goal of antiretroviral therapy is to minimize HIV-1 replication and, thereby, halt or reverse immunologic damage (17,21). Zidovudine (AZT), an inhibitor of reverse transcriptase, was the first treatment for HIV-1, but effective treatment was not available until new drugs that interfere with the HIV-1 protease gene (protease inhibitors) became available to offer a second major treatment target. Used singly, antiretroviral agents have only short-term benefits because HIV-1 mutates rapidly and drug resistance usually develops within several months. At least three drugs are usually needed to suppress viral replication for extended periods of time. Suppression of HIV-1 replication can prevent selection for drug-resistant HIV-1 strains, reconstitute the immune system, and decrease the risk of AIDS. The HIV-1 RNA level is used to monitor the effectiveness of therapy (17). Hopes that combination regimens might cure HIV-1 were dashed by findings that the virus remains integrated within resting CD4$^+$ lymphocytes despite years of therapy and that the virus reactivates when treatment is stopped (22–24). It is likely, therefore, that HIV-1-infected patients will require lifelong treatment.

Although current antiretroviral drugs have markedly changed the prognosis for HIV-1-infected patients, these agents often produce adverse effects that diminish the quality of life. Adverse effects can lead to poor compliance, which increases the likelihood that drug resistant HIV-1 strains will develop. Furthermore, for many patients these regimens do not yield lasting viral suppression. Because current therapies are imperfect, new treatments that act through novel mechanisms are needed. The role of chemokine genes in HIV-1 infection has sparked studies of new approaches to therapy.

Chemokine Receptors as HIV-1 Coreceptors

To enter a cell, HIV-1 needs both the CD4 molecule and a coreceptor. These HIV-1 coreceptors are chemokine receptors (25,26) that are members of the super family of seven transmembrane, G protein–coupled receptors. Chemokines (chemoattractant cytokines) and their receptors form a complex, redundant, and incompletely understood network that regulates leukocyte migration by recruiting specific cells to particular tissue compartments in response to local signals (27).

HIV-1 uses certain chemokine receptors to gain entry into cells. CC-chemokine receptor 5 (CCR5) is the major coreceptor for HIV-1 strains that predominate during early infection (26) and CXC-chemokine receptor 4 (CXCR4) is the major coreceptor for the more pathogenic, syncytium-inducing HIV-1 strains that often emerge in late infection (25). HIV-1 strains are now termed R5, X4, or R5/X4 depending on whether they use one or both of these major coreceptors. CCR5 is the natural receptor for the chemokines macrophage inflammatory protein (MIP)-1α, MIP-1β, and regulated-on-activation-normal-T-cell-expressed-and-secreted (RANTES) (27). In contrast, CXCR4 appears to have a specific relationship with a single chemokine ligand named stromal cell derived factor-1 (SDF-1). A number of other chemokine receptors can act as HIV-1 coreceptors (e.g., CCR2), but CCR5 and CXCR4 are by far the most important HIV-1 coreceptors.

Role of Human Genetics in HIV-1 Infection

Chemokine Receptor Genes

CCR5-Δ32. The gene that encodes CCR5 is located on the short arm of chromosome 3. The coding region consists of a single open reading frame of 1055 base pairs (28). Soon after the key role of CCR5 in HIV-1 cell entry was reported, several groups found a 32 base-pair deletion polymorphism (*Δ32*) of the *CCR5* gene (29–31). A frameshift in this mutant allele results in a truncated protein that does not bind chemokines or HIV-1. The *CCR5-Δ32* allele is common in persons of northern European descent (~10% to 15% allele frequency), but very rare or absent in Asians and Africans (Table 27.1). The restricted racial/ethnic distribution and high

Table 27.1 Chemokine and Chemokine Receptor Alleles (or Haplotypes) that Have Been Associated with Differences in Response to HIV-1

Allele or Haplotype	ESTIMATED FREQUENCY (BY ANCESTRY) (%)			Genotype	Effect
	African	Asian	European		
CCR5-Δ32	0[a]	0[a]	10[b]	Homozygous	Resistant to HIV-1 infection
				Heterozygous	Resistant to HIV-1 infection (in some studies); Slower progression to AIDS
CCR5 Promoter 59029A (P1)	43	47	56	Homozygous	Faster progression to AIDS
CCR2-64I	15	25	10	Homozygous or heterozygous	Slower progression to AIDS
RANTES-28G	0	6–17	4	Homozygous or heterozygous	Initial report of better prognosis not confirmed
RANTES−403A	36–52	27	19	Heterozygous	More susceptible to HIV-1 infection; slower progression to AIDS;
SDF-1 3'A	6	26	21	Homozygous	Initial report of slower progression to AIDS not confirmed
CX₃CR1-I249 M280	0*	0*	17	Homozygous	Initial report of increased susceptibility to HIV-1 infection and faster progression to AIDS not confirmed

[a]Allele appears to be generally absent in these groups except through population admixture.
[b]Allele frequency is higher in northern Europeans than in southern Europeans.

frequency of the *CCR5-Δ32* allele suggest that it results from a relatively recent mutation that underwent strong selective pressure, perhaps because it protects against a lethal infectious agent (other than the recently emergent HIV-1) that has not yet been identified (32).

Epidemiologic investigations revealed a markedly increased frequency of the *CCR5-Δ32* homozygous genotype among HIV-1-uninfected persons who were at high risk of acquiring the virus (29–31). To date, only nine HIV-1-infected *CCR5-Δ32* homozygotes have been reported worldwide (16). These findings indicate that *CCR5-Δ32* homozygotes strongly resist HIV-1 infection and that CCR5 plays a key role in initial HIV-1infection. Because *CCR5-Δ32* heterozygotes express less CCR5 on lymphocytes than wild-type subjects, it is feasible that the heterozygous genotype might protect against initial HIV-1 infection. In some studies, *CCR5-Δ32* heterozygotes have been found to be at lower risk of acquiring HIV-1 infection (31,33), but not in others (29). It is clear, however, that *CCR5-Δ32* heterozygosity benefits HIV-1-infected patients. Compared to wild-type patients, *CCR5-Δ32* heterozygotes have lower HIV-1 RNA levels and develop AIDS ~25% more slowly (29,34–38). The protective effects of *CCR5-Δ32* genotypes correspond with the amount of CCR5 protein that is expressed on lymphocytes. *CCR5-Δ32* homozygotes express no CCR5, and *CCR5-Δ32* heterozygotes generally express less than people with two functional alleles (39).

CCR5 Promoter. The protective role of *CCR5-Δ32* genotypes in HIV-1 infection led investigators to examine other chemokine receptor gene alleles. Because the amount of CCR5 protein expressed on the cell surface of lymphocytes correlates with clinical prognosis and in vitro infectability by HIV-1, several groups searched for polymorphisms in the *CCR5* promoter region that might alter *CCR5* transcription levels. An allele (*59029 A*) (40) and a haplotype (*P1*) (41) in the *CCR5* promoter 5′ untranslated region were reported to accelerate the rate of HIV-1 disease progression. It was subsequently shown that *59029A* and *P1* define the same haplotype (42), which is present at a frequency of 43% to 68%, depending on race/ethnic group (41). *59029A* (*P1*) does not appear to alter susceptibility to infection, but HIV-1-infected patients who carry neither *CCR5-Δ32* nor *CCR2-64I* and who are homozygous for *59029A* (7% to 13% of all people) have a poorer prognosis than patients with none of these markers.

CCR2-64I. The minor HIV-1 coreceptor CCR2 can be used by some HIV-1 strains that are typically found late in the course of HIV-1 infection (16). Although CCR2 appears to play a much less important role in HIV-1 infection than either CCR5 or CXCR4, a common *CCR2* allele affects the course of HIV-1 infection. Smith and colleagues (43) screened the *CCR2* gene and found a conservative valine to isoleucine change at amino acid position 64 (*CCR2-64I*), which lies in a transmembrane domain of the receptor. *CCR2-64I* has an allele frequency of about 10% in Caucasians, 15% in African Americans, 17% in Hispanics, and 25% in

Asians. This allele does not reduce susceptibility to HIV-1 infection, but Smith found that *CCR2-64I* carriers (heterozygotes and homozygotes combined) had slower progression to AIDS. Some investigators could not confirm this effect (44), but it was verified in an international meta-analysis of individual patient data that included most of the relevant data from cohorts around the world. This meta-analysis also showed that *CCR2-64I* carriers, like *CCR5-Δ32* heterozygotes, have lower levels of viremia than HIV-1-infected patients with two normal copies of this gene (38).

Given that CCR2 is a minor HIV-1 coreceptor, it is unlikely that the effect of *CCR2-64I* reflects the role of CCR2 as a coreceptor. *CCR2-64I* is in linkage disequilibrium with an allele in the closely proximate *CCR5* promoter region and could, therefore, result in decreased expression of CCR5 through that linkage (45). Alternatively, the protein produced by the *CCR2-64I* variant might interact with CCR5 to decrease its expression. In vitro studies have failed to demonstrate that lymphocytes from *CCR2-64I* carriers express less CCR5 or are more resistant to R5 HIV-1 strains than cells from wild-type subjects (46,47), but the effect of *CCR2-64I* may be too subtle to be detected in such studies. Chemokine receptors internalize after they are bound by a chemokine. Preliminary evidence suggests that cells carrying the *CCR2-64I* allele may be slower to re-express CCR5 after internalization, which could result in less CCR5 on the cell surface (48). The *CCR2-64I* allele may also interact with CXCR4, the other major HIV-1 coreceptor. CCR2 from persons with the *CCR2-64I* genotype can form dimers with CXCR4, while the normal CCR2 protein cannot (49). It has been suggested that formation of CCR2/CXCR4 'heterodimers' could reduce the amount of CXCR4 that is expressed on peripheral blood lymphocytes of *CCR2-64I* carriers and reduce susceptibility to X4 HIV-1 strains (49). This explanation is inconsistent, however, with the findings that *CCR2-64I* exerts its effect early in the course of HIV-1 infection (50), that *CCR2-64I* carriers develop X4 HIV-1 strains more rapidly than wild-type subjects (51–53), and that the *CCR2-64I* protection is lost after X4 strains emerge (54). The question of how the *CCR2-64I* allele protects patients from HIV-1 disease progression remains unanswered.

CCR5-Related Haplotypes

The *CCR5-Δ32, CCR5* promoter, and *CCR2-64I* polymorphisms lie in close proximity on chromosome 3p21. Linkage disequilibrium exists between *CCR5 59029A* and both the *CCR5-Δ32* allele and the *CCR2-64I* allele. *CCR5-Δ32* and *CCR2-64I* are found exclusively with *CCR5 59029A*, but not all *CCR5 59029A*-containing haplotypes include *CCR5-Δ32* or *CCR2-64I* (40,41). To further evaluate this region, Ahuja and colleagues developed an evolutionary-based classification (Fig. 27.2) consisting of nine *CCR5*-related haplotypes (designated *HHA* through *HHG*2). The haplotypes are defined by *CCR5 59029, CCR2-64I*, and 6 other common single-nucleotide polymorphisms, as well as by the *CCR5-Δ32* allele (55). Haplotypes *HHE, HHF*1, and *HHG*1 include *59029A*, but not *CCR5-Δ32* or *CCR2-64I*; haplotype *HHF*2 includes *CCR2-64I*; and haplotype *HHG*2 includes *CCR5-Δ32*.

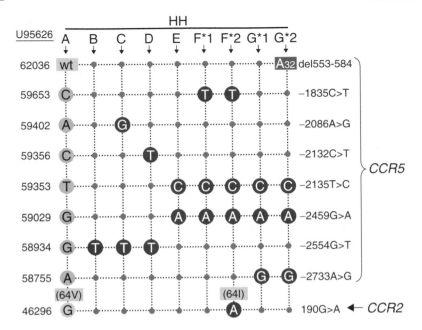

Figure 27.2 CCR2-CCR5 haplotypes. Eight single-nucleotide polymorphisms (SNPs) and the CCR5-Δ32 allele define nine major human haplotypes (HH) designated as HHA through HHG*2 (top). Sequences identical to these shown in HHA are indicated by dots. The 5-digit numbers (left) are based on GenBank sequence U95626, while three- or four-digit numbers (right) refer to positions relative to the ARG translation start site in the transcribed CCR2 and CCR5 sequences. (Courtesy of James Tang and Richard Kaslow; Tang J, Kaslow RA. Polymorphic chemokine receptor and ligand genes in HIV infection, in Susceptibility to Infectious Diseases: The Importance of Host Genetics, *Richard Bellamy (eds.) Cambridge: Cambridge University Press, in press.)*

Analyses based on this system have produced some results that are generally consistent with the aforementioned studies. Among 341 white subjects, mean plasma HIV-1 RNA level during the initial 42 months after infection (a strong predictor of subsequent clinical course) was higher in carriers of the HHE/HHE (which contains *CCR5 59029*) genotype and lower in those carrying HHF*2 (*CCR2-64I*) or HHG*2 (*CCR5-Δ32*) (56). Haplotype analyses have also yielded some novel findings. In a large U.S. Air Force cohort of HIV-1-infected patients, it was reported that the haplotypes associated with disease acceleration or retardation differed between African Americans and whites, and that there was an interactive effect between different haplotypes (55). This haplotype classification may add precision to the evaluation of *CCR5*-related haplotypes, but there are a number of unresolved issues. As discussed elsewhere in this volume (Chapters 6 and 10), methodologic issues relating to haplotype analysis are still under development. Analyses based on even a modest number of possible haplotypes face issues of sparse data for some haplotypes (possible false-negative results) and multiple statistical comparisons (possible false-positive results) (26,27). The number of comparisons increases when subpopulations (e.g., racial ethnic groups) or potential gene–gene interactions are explored. Further research is needed to confirm the findings based on the *CCR5*-related haplotypes.

CX_3CR1. CX_3CR1 is the receptor for the chemokine fractalkine. In vitro studies have shown that this molecule can act as a coreceptor for certain strains of HIV-1, although its role in HIV-1 infection in vivo appears to be limited. The *CX_3CR1* gene is located on chromosome 3p21 in the vicinity of the *CCR5* and *CCR2* genes. Faure et al. (57) examined *CX_3CR1* for polymorphisms associated with HIV-1 infection. They found that a common variant haplotype affecting two amino acids (isoleucine-249 and methionine-280) in the coding region of *CX_3CR1*. Homozygotes for *CX_3CR1-I249 M280* had reduced binding for fractalkine in in vitro studies (57). In an epidemiologic analysis of French patients, homozygosity for *CX_3CR1-I249 M280* was associated with increased risk of becoming infected with HIV-1 and with accelerated disease progression after infection had occurred (57). McDermott et al (58) attempted to confirm these associations in a study of HIV-1-infected men from North America. In contrast to the initial study, there was no evidence that *CX_3CR1* genotype was associated with susceptibility to HIV-1 infection, and homozygotes for *CX_3CR1-I249 M280* did not progress to AIDS or death faster than subjects who were homozygous for the more common *CX_3CR1* haplotype. The results of the initial study may have been due to chance.

CXCR4. The important role of CXCR4 as a coreceptor for more pathogenic X4 HIV-1 strains led to an examination of the gene that codes for this receptor. *CXCR4* is highly conserved and the rare variants that have been identified have no apparent functional consequence (28). Knockout mutations of *CXCR4* are lethal in mice (59–61), which suggests that the gene is essential to normal development.

Chemokine Genes

RANTES. The CCR5 chemokine RANTES can block infection by R5 HIV-1 strains (62). Lymphocytes from individuals vary in their secretion of RANTES and higher levels of RANTES may protect against infection or slow disease progression if infection occurs (28,63–65). These findings may be mediated through polymorphisms of the *RANTES* gene. Liu et al (66) found that the *RANTES* promoter polymorphisms *−403G/A* and *−28C/G* did not alter the incidence of HIV-1 infection, but the *−28G* allele, which had a frequency of 17% in a Japanese population, was associated with reduced CD4$^+$ lymphocyte depletion rates in HIV-1-infected individuals. Functional analyses indicated that this allele increased transcription of the *RANTES* gene.

McDermott and his colleagues (67) examined the effect of these *RANTES* promoter polymorphisms among participants in the Multicenter AIDS Cohort Study of

U.S. homosexual men. The *RANTES -28G* allele was much rarer in this cohort than in the Japanese study, with allele frequencies ranging from 5.7% in Asian Americans to 0.0% in African Americans. The original report could not be confirmed, but these investigators found that those with the *−403G/A*, *−28C/C* compound genotype were more likely to acquire HIV, but, paradoxically, less likely to progress to AIDS rapidly (when compared to those with the *−403G/G*, *−28C/C* compound genotype). Because of these inconsistencies, additional work is required to determine the role of *RANTES* promoter polymorphisms in HIV-1 infection.

SDF-1 3′A. Stromal cell derived factor-1 (SDF-1) is the chemokine ligand of CXCR4 (68;69). Homozygosity for the *SDF-1 3′A* allele, which is found in the 3′ untranslated region of the gene, has been reported to slow disease progression in some studies (70) but not in others (54,71–75). The international meta-analysis (38) found no evidence that *SDF-1 3′A* homozygotes had an altered risk of AIDS, death, or death after AIDS. It is possible, therefore, that the original report represents a chance finding.

Genes Outside the Chemokine System That May Affect HIV-1 Infection

In addition to the chemokines that function as ligands for HIV-1 coreceptors CCR5 or CXCR4, other cytokines may play a role in the response to HIV-1 by stimulating or inhibiting viral replication. A TH2 cytokine profile, determined predominantly by interleukin-4 (IL-4) and IL-10, is associated with a higher rate of HIV-1 replication (76) and more advanced HIV-1 infection (35,77). Furthermore, IL-4 may also modify viral replication by down-regulating CCR5 and up-regulating CXCR4 (78). The *IL-4* gene is located at 5q31. The *IL-4* promoter *−589T/T* genotype was associated with increased rates of X4 strain acquisition in HIV-1-infected Japanese patients (53). The *−589T* allele was also found to be associated with slower progression to AIDS in a French HIV-1 cohort (53), but that association was not be confirmed in a study based on five HIV-1 cohorts from the United States (79). The *IL10* gene lies on the long arm of chromosome 1. Individuals carrying a single nucleotide polymorphism in the *IL10* promoter region (*IL10-5′-592A*) progressed to AIDS more rapidly than homozygotes for the alternative (*IL10-5′-592 C*) genotype, particularly in the later stages of HIV-1 infection (80).

The highly polymorphic *HLA* class I and class II loci lie on chromosome 6. HLA molecules initiate the immune response by presenting antigenic peptides to T cells. HIV-1-infected individuals who are heterozygous at *HLA* class I loci appear to have a more favorable clinical course than homozygotes (81,82). This "heterozygote advantage" presumably results because the immune system of persons who are heterozygous at *HLA* class I loci can present a greater variety of antigenic peptides. Some specific *HLA* alleles appear to play a role in determining prognosis after HIV-1 infection, although the numerous HLA alleles make it difficult to distinguish true associations from chance findings. A strong effect of the class I *HLA-B*35* haplotype on rapid disease progression was observed among 474 HIV-1-infected subjects (81). The *B*35* haplotype appears to have a co-dominant effect in that homozygotes progress more rapidly than heterozygotes, who in turn progress more rapidly than individuals without *B*35*. In a closer examination of these subjects, *HLA-B*35* subtypes were divided into two groups according to peptide-binding specificity (83). The influence of *HLA-B*35* in accelerating progression to AIDS was completely attributable to *HLA-B*35-Px* alleles, which can differ from other *HLA-B*35* alleles by a single amino acid residue. This finding demonstrates that small differences in closely related molecules can affect immune defense against HIV-1. Other specific alleles, especially *B*27* and *B*57* (84–88), also appear to affect the course of HIV infection. A scoring system based on a combination of CCR and HLA genotypes has been shown to be predictive of viremia and immunodeficiency In HIV-infected patients (89).

Genetic polymorphisms in the innate immune system, which provides a line of immunologic defense prior to the acquired (cellular and humoral) immune system, may also play a role in HIV-1 infection. Mannose-binding lectin (MBL) activates complement in response to various infectious agents. The *MBL* gene is located on chromosome 10q (90,91), and low serum levels of MBL are associated with impaired immunity and increased risk of various infections. Three *MBL* coding variants (*B*, *C*, and *D*) are associated with lower levels of MBL in the blood (92–94). Epidemiologic studies suggest that a person who is homozygous for *MBL* variants is more susceptible to HIV-1 infection and more likely to develop AIDS rapidly (95–97).

Chemokine Receptor Gene Polymorphisms as Potential Predictors of Prognosis and Therapeutic Response

Although current HIV-1 treatment regimens have dramatically reduced morbidity and mortality, these therapies can cause significant adverse effects. For that reason, treatment is often delayed for patients whose prognosis is thought to be relatively good on the basis of HIV-1 RNA and CD4$^+$ lymphocyte measurements (98). Additional prognostic markers might improve these predictions. Genetic alleles that predict the course of HIV-1 infection are potential candidates for this role, but because the protective effects of the *CCR5-Δ32* and *CCR2-64I* alleles appear to be reflected in lower HIV-1 RNA levels (38), genotyping for these alleles probably will not provide additional information. Further studies are needed to determine whether other genetic factors act independently of HIV-1 RNA and might, therefore, provide useful prognostic information.

There is evidence that chemokine receptor gene polymorphisms may explain some of the individual differences in response to antiretroviral treatment. Valdez et al. (99) found that combination therapy reduced the HIV-1 RNA level to <400 copies/ml in 81% of *CCR5-Δ32* heterozygotes compared to 57% of wild type subjects. Among HIV-1-infected, protease-inhibitor naive patients enrolled in a French epidemiologic study, *CCR5-Δ32* heterozygotes had a better virologic and immuno-logic response to combination therapy at both 6 and 12 months compared to wild-type patients (100). Results from a study of HIV1-infected children suggest that the *CCR5-Δ32* allele also slows disease progression among children receiving anti-retroviral treatment (101). One study examined the predictive value of several chemokine receptor gene polymorphisms (102). These subjects were enrolled in a clinical trial of combination therapy for patients with relatively high $CD4^+$ lym-phocyte counts (≥ 200 cells/mm^3). Consistent with other studies, there was a trend toward better response among *CCR5-Δ32* heterozygotes, but evidence was found among those who were homozygous for the *CCR5* promoter allele *59029 A* (*P1* hap-lotype). Consistent with the natural history studies, *59029 A* homozygotes were less likely to adequately suppress the virus in response to treatment. In all, these stud-ies suggest that genotyping for chemokine receptor gene polymorphisms might have a role in predicting response to antiretroviral regimens.

Targeting Chemokine Receptor Expression for Treatment and Prevention

The key role of CCR5 in HIV-1 cell entry, together with the fact that the *CCR5-Δ32* allele confers resistance to infection (29) without major deleterious effects (103) led to exploration of potential new avenues of treatment (104). Therapies that pre-vent HIV-1 cell entry by blocking chemokine receptors (or down regulating their expression) could help reduce HIV-1 replication. As noted above, new therapies are needed because HIV-1 can develop resistance to all currently available drugs. Pre-clinical and clinical evaluations of chemokine receptor inhibitors are underway.

Several RANTES derivatives have been developed, such as aminooxypentane (AOP)-RANTES, which was created by chemical modification of the amino termi-nus of RANTES. AOP-RANTES inhibits R5 HIV-1 strains in vitro (105), as do other RANTES derivatives. TAK-779 is a nonpeptide compound with a small mo-lecular weight that inhibits replication by R5, but not X4, HIV-1 strains (106). In vitro studies of TAK-779 revealed no cytotoxicity to the host cells. PRO 140 is an anti-CCR5 monoclonal antibody that potently inhibits cell entry by R5 viruses (107). PRO 140 also inhibits some dual R5/X4 tropic HIV-1 strains. CCR5 may also be targeted by gene therapy. Several groups have developed ribozymes that reduce cel-lular CCR5 mRNA and cell surface CCR5 when introduced into cells by a viral vector (108–110). These ribozymes protect cells from infection by R5 HIV-1 strains and could be the basis for agents that prevent HIV-1 infection or slow disease pro-gression.

Enthusiasm for blockade of CCR5 is tempered by concerns of selection for more pathogenic X4 HIV-1 strains. For that reason, several potential CXCR4 blockers are also under investigation. Met-SDF-1β, a modified form of the chemokine ligand of CXCR4, reduces susceptibility to X4 strains in vitro (111). The effect of Met-SDF-1β is additive to that of some currently used antiretroviral therapies. T22, T134,

ALX-40-4C, and AMD3100 are small peptides that block entry of X4 strains (112–116).

CCR5 is also a target for HIV-1 vaccine strategies. Rhesus macaques were immunized with recombinant simian immunodeficiency virus (SIV) proteins, recombinant granulocyte-macrophage colony-stimulating factor (GM-CSF), and either IL-2 or IL-4 (117). The immunized monkeys had significant increases in the concentrations of CCR5 chemokines (RANTES, MIP-1α, and MIP-1β), and decreases in the proportion of cells expressing CCR5. Upon challenge with SIV, the monkeys were not protected against infection, but the chemokine concentrations were inversely correlated with the plasma SIV RNA levels and positively correlated with the proportion of cells that expressed CCR5. Zuber et al (118) immunized cynomolgus macaques with CCR5 DNA and then boosted with CCR5 peptides. This approach induced CCR5-specific antibodies that neutralized R5 HIV-1 strains in vitro, but did not protect the monkeys against SIV infection. A monoclonal antibody against a cyclic closed-chain dodecapeptide that mimics the conformation-specific domain of CCR5 almost completely inhibited infection by R5 strains in vitro (119) and this approach could be useful in developing an HIV-1 vaccine.

Conclusions

Chemokine receptor gene polymorphisms, such as *CCR5-Δ32*, *CCR2-64I*, and *CCR5* promoter *59029 A*, help determine susceptibility to HIV-1 infection and the clinical course of those who have become infected. Preliminary studies suggest that chemokine receptor gene alleles may also predict which patients are more likely respond to HIV-1 treatment regimens. If these findings are confirmed, genetic testing may contribute to the prognostic evaluation of HIV-1-infected patients. A number of agents that are designed to interfere with chemokine receptor expression or function are currently under development. If this developmental work proves successful, agents that mimic the effects of chemokine receptor gene polymorphisms may become important additions to the treatment of HIV-1-infected patients or even form the basis of a much-needed prophylactic vaccine against HIV-1.

References

1. Phair JP. Determinants of the natural history of human immunodeficiency virus type 1 infection. J Infect Dis 1999;179(suppl 2):S384–S386.
2. Rizzardi GP, Pantaleo G (2002). Pathogenesis of HIV-1 infection, in Chemokine Receptors and AIDS, O'Brien TR (ed.). New York: Marcel Dekker, pp. 51–74.
3. Rizzardi GP, Pantaleo G. Therapeutic perspectives in HIV-1 infection from recent advances in HIV-1 pathogenesis: it is time to move on. J Biol Regul Homeost Agents 1999;13(3):151–157.
4. Mellors JW, Rinaldo CR, Jr., Gupta P, et al. Prognosis in HIV-1 infection predicted by the quantity of virus in plasma. Science 1996;272(5265):1167–1170.

5. O'Brien TR, Blattner WA, Waters D, et al. Serum HIV-1 RNA levels and time to development of AIDS in the Multicenter Hemophilia Cohort Study. JAMA 1996;276 (2):105–110.

6. Collaborative Group on AIDS Incubation and HIV Survival including the CASCADE EU Concerted Action.Concerted Action on SeroConversion to AIDS and Death in Europe. Time from HIV-1 seroconversion to AIDS and death before widespread use of highly-active antiretroviral therapy: a collaborative re-analysis. Lancet 2000;355(9210): 1131–1137.

7. Gao F, Bailes E, Robertson DL, et al. Origin of HIV-1 in the chimpanzee Pan troglodytes troglodytes. Nature 1999;397(6718):436–441.

8. Korber B, Muldoon M, Theiler J, et al. Timing the ancestor of the HIV-1 pandemic strains. Science 2000;288(5472):1789–1796.

9. Centers for Disease Control. Pneumocystis pneumonia—Los Angeles. MMWR Morb Mortal Wkly Rep 1981;30(21):250–252.

10. Centers for Disease Control. Kaposi's sarcoma and Pneumocystis pneumonia among homosexual men—New York City and California. MMWR Morb Mortal Wkly Rep 1981;30(25):305–308.

11. Piot P, Bartos M, Ghys PD, et al. The global impact of HIV/AIDS. Nature 2001;410 (6831):968–973.

12. Moore A, Herrera G, Nyamongo J, et al. Estimated risk of HIV transmission by blood transfusion in Kenya. Lancet 2001;358(9282):657–660.

13. Des Jarlais DC, Marmor M, Paone D, et al. HIV incidence among injecting drug users in New York City syringe-exchange programmes. Lancet 1996;348(9033):987–991.

14. Mofenson LM, McIntyre JA. Advances and research directions in the prevention of mother-to-child HIV-1 transmission. Lancet 2000;355(9222):2237–2244.

15. Kilmarx PH, Supawitkul S, Wankrairoj M, et al. Explosive spread and effective control of human immunodeficiency virus in northernmost Thailand: the epidemic in Chiang Rai province, 1988–99. AIDS 2000;14(17):2731–2740.

16. O'Brien TR, Michael NL, Sheppard HW, Buchbinder S. HIV-1 infection in patients with the CCR5-delta32 homozygous genotype. In: O'Brien TR, editor. Chemokine receptors and AIDS. New York: Marcel Dekker, 2002:215–224.

17. NIH panel to define principles of therapy of HIV infection. Report of the NIH Panel To Define Principles of Therapy of HIV Infection. Ann Intern Med 1998;128(12 Pt 2):1057–1078.

18. Palella FJ, Jr., Delaney KM, Moorman AC, et al. Declining morbidity and mortality among patients with advanced human immunodeficiency virus infection. HIV Outpatient Study Investigators. N Engl J Med 1998;338(13):853–860.

19. Gebhardt M, Rickenbach M, Egger M. Impact of antiretroviral combination therapies on AIDS surveillance reports in Switzerland. Swiss HIV Cohort Study. AIDS 1998;12(10):1195–1201.

20. Mocroft A, Sabin CA, Youle M, et al. Changes in AIDS-defining illnesses in a London Clinic, 1987–1998. J Acquir Immune Defic Syndr 1999;21(5):401–407.

21. Centers for Disease Control. Guidelines for using antiretroviral agents among HIV-infected adults and adolescents. Recommendations of the Panel on Clinical Practices for Treatment of HIV. MMWR Recomm Rep 2002;51(RR-7):1–55.

22. Finzi D, Blankson J, Siliciano JD, et al. Latent infection of CD4$^+$ T cells provides a mechanism for lifelong persistence of HIV-1, even in patients on effective combination therapy. Nat Med 1999;5(5):512–517.

23. Siliciano RF. Latency and reservoirs for HIV-1. AIDS 1999;13(suppl A):S49–S58.

24. Ramratnam B, Mittler JE, Zhang L, et al. The decay of the latent reservoir of replica-

tion-competent HIV-1 is inversely correlated with the extent of residual viral replication during prolonged anti-retroviral therapy. Nat Med 2000;6(1):82–85.

25. Feng Y, Broder CC, Kennedy PE, Berger EA. HIV-1 entry cofactor: functional cDNA cloning of a seven-transmembrane, G protein-coupled receptor. Science 1996;272 (5263):872–877.

26. Dragic T, Litwin V, Allaway GP, et al. HIV-1 entry into CD4$^+$ cells is mediated by the chemokine receptor CC-CKR-5. Nature 1996;381(6584):667–673.

27. Shields PL, Adams DH (2002). Chemokines and chemokine receptor interactions and functions. In: O'Brien TR, editor. Chemokine receptors and AIDS. New York: Marcel Dekker, 2002:1–30.

28. Martin MP, Carrington M. The role of human genetics in HIV-1 infection, in Chemokine receptors and AIDS. O'Brien TR (ed.). New York: Marcel Dekker, pp. 133–162.

29. Dean M, Carrington M, Winkler C, et al. Genetic restriction of HIV-1 infection and progression to AIDS by a deletion allele of the CKR5 structural gene. Hemophilia Growth and Development Study, Multicenter AIDS Cohort Study, Multicenter Hemophilia Cohort Study, San Francisco City Cohort, ALIVE Study. Science 1996;273(5283):1856–1862.

30. Liu R, Paxton WA, Choe S, et al. Homozygous defect in HIV-1 coreceptor accounts for resistance of some multiply-exposed individuals to HIV-1 infection. Cell 1996;86 (3):367–377.

31. Samson M, Libert F, Doranz BJ, et al. Resistance to HIV-1 infection in caucasian individuals bearing mutant alleles of the CCR-5 chemokine receptor gene. Nature 1996; 382(6593):722–725.

32. Stephens JC, Reich DE, Goldstein DB, et al. Dating the origin of the CCR5-Delta32 AIDS-resistance allele by the coalescence of haplotypes. Am J Hum Genet 1998; 62(6):1507–1515.

33. Marmor M, Sheppard HW, Donnell D, et al. Homozygous and heterozygous CCR5-Delta32 genotypes are associated with resistance to HIV infection. J Acquir Immune Defic Syndr 2001;27(5):472–481.

34. Katzenstein TL, Eugen-Olsen J, Hofmann B, et al. HIV-infected individuals with the CCR delta32/CCR5 genotype have lower HIV RNA levels and higher CD4 cell counts in the early years of the infection than do patients with the wild type. Copenhagen AIDS Cohort Study Group. J Acquir Immune Defic Syndr Hum Retrovirol 1997;16(1):10–14.

35. Klein SA, Dobmeyer JM, Dobmeyer TS, et al. Demonstration of the Th1 to Th2 cytokine shift during the course of HIV-1 infection using cytoplasmic cytokine detection on single cell level by flow cytometry. AIDS 1997;11(9):1111–1118.

36. Zimmerman PA, Buckler-White A, Alkhatib G, et al. Inherited resistance to HIV-1 conferred by an inactivating mutation in CC chemokine receptor 5: studies in populations with contrasting clinical phenotypes, defined racial background, and quantified risk. Mol Med 1997;3(1):23–36.

37. Michael NL, Chang G, Louie LG, et al. The role of viral phenotype and CCR-5 gene defects in HIV-1 transmission and disease progression. Nat Med 1997;3(3):338–340.

38. Ioannidis JP, Rosenberg PS, Goedert JJ, et al. Effects of CCR5-Delta32, CCR2-64I, and SDF-1 3'A alleles on HIV-1 disease progression: An international meta-analysis of individual-patient data. Ann Intern Med 2001;135(9):782–795.

39. Wu L, Paxton WA, Kassam N, et al. CCR5 levels and expression pattern correlate with infectability by macrophage-tropic HIV-1, in vitro. J Exp Med 1997;185(9):1681–1691.

40. McDermott DH, Zimmerman PA, Guignard F, Kleeberger CA, Leitman SF, Murphy PM. CCR5 promoter polymorphism and HIV-1 disease progression. Multicenter AIDS Cohort Study (MACS). Lancet 1998;352(9131):866–870.

41. Martin MP, Dean M, Smith MW, et al. Genetic acceleration of AIDS progression by a promoter variant of CCR5. Science 1998;282(5395):1907–1911.
42. An P, Martin MP, Nelson GW, et al. Influence of CCR5 promoter haplotypes on AIDS progression in African-Americans. AIDS 2000;14(14):2117–2122.
43. Smith MW, Dean M, Carrington M, et al. Contrasting genetic influence of CCR2 and CCR5 variants on HIV-1 infection and disease progression. Hemophilia Growth and Development Study (HGDS), Multicenter AIDS Cohort Study (MACS), Multicenter Hemophilia Cohort Study (MHCS), San Francisco City Cohort (SFCC), ALIVE Study. Science 1997;277(5328):959–965.
44. Michael NL, Louie LG, Rohrbaugh AL, et al. The role of CCR5 and CCR2 polymorphisms in HIV-1 transmission and disease progression. Nat Med 1997;3(10):1160–1162.
45. Kostrikis LG, Huang Y, Moore JP, et al. A chemokine receptor CCR2 allele delays HIV-1 disease progression and is associated with a CCR5 promoter mutation. Nat Med 1998;4(3):350–353.
46. Lee B, Doranz BJ, Rana S, et al. Influence of the CCR2-V64I polymorphism on human immunodeficiency virus type 1 coreceptor activity and on chemokine receptor function of CCR2b, CCR3, CCR5, and CXCR4. J Virol 1998;72(9):7450–7458.
47. Mariani R, Wong S, Mulder LC, et al. CCR2-64I polymorphism is not associated with altered CCR5 expression or coreceptor function. J Virol 1999;73(3):2450–2459.
48. Sabbe R, Picchio GR, Pastore C, et al. Donor- and ligand-dependent differences in C-C chemokine receptor 5 reexpression. J Virol 2001;75(2):661–671.
49. Mellado M, Rodriguez-Frade JM, Vila-Coro AJ, de Ana AM, Martinez A. Chemokine control of HIV-1 infection. Letter. Nature 1999;400(6746):723–724.
50. Mulherin SA, O'Brien TR, Ioannidis JP, et al. Effects of CCR5-delta32 and CCR2-64I alleles on HIV-1 disease progression: the protection varies with duration of infection. AIDS 2003.
51. van Rij RP, de Roda Husman AM, Brouwer M, Goudsmit J, Coutinho RA, Schuitemaker H. Role of CCR2 genotype in the clinical course of syncytium-inducing (SI) or non-SI human immunodeficiency virus type 1 infection and in the time to conversion to SI virus variants. J Infect Dis 1998;178(6):1806–1811.
52. Vicenzi E, Ghezzi S, Brambilla A, et al. CCR2-64I polymorphism, syncytium-inducing human immunodeficiency virus strains, and disease progression. J Infect Dis 2000; 182(5):1579–1580.
53. Nakayama EE, Hoshino Y, Xin X, et al. Polymorphism in the interleukin-4 promoter affects acquisition of human immunodeficiency virus type 1 syncytium-inducing phenotype. J Virol 2000;74(12):5452–5459.
54. van Rij RP, Broersen S, Goudsmit J, Coutinho RA, Schuitemaker H. The role of a stromal cell-derived factor-1 chemokine gene variant in the clinical course of HIV-1 infection. AIDS 1998;12(9):F85–F90.
55. Gonzalez E, Bamshad M, Sato N, et al. Race-specific HIV-1 disease-modifying effects associated with CCR5 haplotypes. Proc Natl Acad Sci U S A 1999;96(21):12004–12009.
56. Tang J, Shelton B, Makhatadze NJ, et al. Distribution of chemokine receptor CCR2 and CCR5 genotypes and their relative contribution to human immunodeficiency virus type 1 (HIV-1) seroconversion, early HIV-1 RNA concentration in plasma, and later disease progression. J Virol 2002;76(2):662–672.
57. Faure S, Meyer L, Costagliola D, et al. Rapid progression to AIDS in HIV+ individuals with a structural variant of the chemokine receptor CX3CR1. Science 2000; 287(5461):2274–2277.
58. McDermott DH, Colla JS, Kleeberger CA, et al. Genetic polymorphism in CX3CR1 and risk of HIV disease. Science 2000;290(5499):2031.
59. Nagasawa T, Hirota S, Tachibana K, et al. Defects of B-cell lymphopoiesis and bone-

marrow myelopoiesis in mice lacking the CXC chemokine PBSF/SDF-1. Nature 1996; 382(6592):635–638.

60. Zou YR, Kottmann AH, Kuroda M, Taniuchi I, Littman DR. Function of the chemokine receptor CXCR4 in haematopoiesis and in cerebellar development. Nature 1998;393 (6685):595–599.

61. Ma Q, Jones D, Borghesani PR, T et al. Impaired B-lymphopoiesis, myelopoiesis, and derailed cerebellar neuron migration in C. Proc Natl Acad Sci U S A 1998;95(16): 9448–9453.

62. Cocchi F, Devico AL, Garzino-Demo A, et al. Identification of RANTES, MIP-1α, and MIP-1β as the major HIV-suppressive factors produced by CD8$^+$ T cells. Science 1995;270(5243):1811–1815.

63. Paxton WA, Liu R, Kang S, et al. Reduced HIV-1 infectability of CD4$^+$ lymphocytes from exposed-uninfected individuals: association with low expression of CCR5 and high production of beta-chemokines. Virology 1998;244(1):66–73.

64. Xiao L, Rudolph DL, Owen SM, et al. Adaptation to promiscuous usage of CC and CXC-chemokine coreceptors in vivo correlates with HIV-1 disease progression. AIDS 1998;12(13):F137–F143.

65. Saha K, Bentsman G, Chess L, Volsky DJ. Endogenous production of beta-chemokines by CD4$^+$, but not CD8$^+$, T-cell clones correlates with the clinical state of human immunodeficiency virus type 1 (HIV-1)-infected individuals and may be responsible for blocking infection with non-syncytium-inducing HIV-1 in vitro. J Virol 1998;72(1): 876–881.

66. Liu H, Chao D, Nakayama EE, et al. Polymorphism in RANTES chemokine promoter affects HIV-1 disease progression. Proc Natl Acad Sci U S A 1999;96(8):4581–4585.

67. McDermott DH, Beecroft MJ, Kleeberger CA, et al. Chemokine RANTES promoter polymorphism affects risk of both HIV infection and disease progression in the Multicenter AIDS Cohort Study. AIDS 2000;14(17):2671–2678.

68. Lu Z, Berson JF, Chen Y, et al. Evolution of HIV-1 coreceptor usage through interactions with distinct CCR5 and CXCR4 domains. Proc Natl Acad Sci U S A 1997; 94(12):6426–6431.

69. Bjorndal A, Deng H, Jansson M, et al. Coreceptor usage of primary human immunodeficiency virus type 1 isolates varies according to biological phenotype. J Virol 1997;71(10):7478–7487.

70. Winkler C, Modi W, Smith MW, et al. Genetic restriction of AIDS pathogenesis by an SDF-1 chemokine gene variant. Science 1998;279:389–393.

71. Hendel H, Henon N, Lebuanec H, et al. Distinctive effects of CCR5, CCR2, and SDF1 genetic polymorphisms in AIDS progression. J Acquir Immune Defic Syndr Hum Retrovirol 1998;19(4):381–386.

72. Mummidi S, Ahuja SS, Gonzalez E, et al. Genealogy of the CCR5 locus and chemokine system gene variants associated with altered rates of HIV-1 disease progression. Nat Med 1998;4(7):786–793.

73. Magierowska M, Theodorou I, Debre P, et al. Combined genotypes of CCR5, CCR2, SDF1, and HLA genes can predict the long-term nonprogressor status in human immunodeficiency virus-1-infected individuals. Blood 1999;93(3):936–941.

74. Meyer L, Magierowska M, Hubert JB, et al. CC-chemokine receptor variants, SDF-1 polymorphism, and disease progression in 720 HIV-infected patients. SEROCO Cohort. Amsterdam Cohort Studies on AIDS. AIDS 1999;13(5):624–626.

75. Brambilla A, Villa C, Rizzardi G, et al. Shorter survival of SDF1-3'A/3'A homozygotes linked to CD4$^+$ T cell decrease in advanced human immunodeficiency virus type 1 infection. J Infect Dis 2000;182(1):311–315.

76. Galli G, Annunziato F, Mavilia C, et al. Enhanced HIV expression during Th2-oriented

responses explained by the opposite regulatory effect of IL-4 and IFN-gamma of fusin/CXCR4. Eur J Immunol 1998;28(10):3280–3290.

77. Wasik TJ, Jagodzinski PP, Hyjek EM, et al. Diminished HIV-specific CTL activity is associated with lower type 1 and enhanced type 2 responses to HIV-specific peptides during perinatal HIV infection. J Immunol 1997;158(12):6029–6036.

78. Valentin A, Lu W, Rosati M, et al. Dual effect of interleukin 4 on HIV-1 expression: implications for viral phenotypic switch and disease progression. Proc Natl Acad Sci U S A 1998;95(15):8886–8891.

79. Modi WS, Winkler C, O'Brien TR, et al. Haplotype diversity in the interleukin-4 gene and association analyses with HIV-1 transmission and AIDS progression. Submitted 2004.

80. Shin HD, Winkler C, Stephens JC, et al. Genetic restriction of HIV-1 pathogenesis to AIDS by promoter alleles of IL10. Proc Natl Acad Sci U S A 2000;97 (26):14467–14472.

81. Carrington M, Nelson GW, Martin MP, et al. HLA and HIV-1: heterozygote advantage and B*35-Cw*04 disadvantage. Science 1999;283(5408):1748–1752.

82. Tang J, Costello C, Keet IP, et al. HLA class I homozygosity accelerates disease progression in human immunodeficiency virus type 1 infection. AIDS Res Hum Retroviruses 1999;15(4):317–324.

83. Gao X, Nelson GW, Karacki P, et al. Effect of a single amino acid change in MHC class I molecules on the rate of progression to AIDS. N Engl J Med 2001;344(22): 1668–1675.

84. Kaslow RA, Carrington M, Apple R, et al. Influence of combinations of human major histocompatibility complex genes on the course of HIV-1 infection [see comments]. Nat Med 1996;2(4):405–411.

85. McNeil AJ, Yap PL, Gore SM, et al. Association of HLA types A1-B8-DR3 and B27 with rapid and slow progression of HIV disease. QJM 1996;89(3):177–185.

86. Migueles SA, Sabbaghian MS, Shupert WL, et al. HLA B*5701 is highly associated with restriction of virus replication in a subgroup of HIV-infected long term nonprogressors. Proc Natl Acad Sci U S A 2000;97(6):2709–2714.

87. Hogan CM, Hammer SM. Host determinants in HIV infection and disease. Part 2: genetic factors and implications for antiretroviral therapeutics. Ann Intern Med 2001; 134(10):978–996.

88. Tang J, Tang S, Lobashevsky E, et al. Favorable and unfavorable HLA class I alleles and haplotypes in Zambians predominantly infected with clade C human immunodeficiency virus type 1. J Virol 2002;76(16):8276–8284.

89. Tang J, Wilson CM, Meleth S, et al. Host genetic profiles predict virological and immunological control of HIV-1 infection in adolescents. AIDS 2002;16(17):2275–2284.

90. Taylor ME, Brickell PM, Craig RK, Summerfield JA. Structure and evolutionary origin of the gene encoding a human serum mannose-binding protein. Biochem J 1989; 262(3):763–771.

91. Sastry K, Herman GA, Day L, et al. The human mannose-binding protein gene. Exon structure reveals its evolutionary relationship to a human pulmonary surfactant gene and localization to chromosome 10. J Exp Med 1989;170(4):1175–1189.

92. Lipscombe RJ, Sumiya M, Hill AV, et al. High frequencies in African and non-African populations of independent mutations in the mannose binding protein gene. Hum Mol Genet 1992;1(9):709–715.

93. Madsen HO, Garred P, Kurtzhals JA, et al. A new frequent allele is the missing link in the structural polymorphism of the human mannan-binding protein. Immunogenetics 1994;40(1):37–44.

94. Madsen HO, Garred P, Thiel S, et al. Interplay between promoter and structural gene

variants control basal serum level of mannan-binding protein. J Immunol 1995;155(6): 3013–3020.

95. Garred P, Madsen HO, Balslev U, et al. Susceptibility to HIV infection and progression of AIDS in relation to variant alleles of mannose-binding lectin. Lancet 1997;349(9047): 236–240.

96. Pastinen T, Liitsola K, Niini P, Salminen M, Syvanen AC. Contribution of the CCR5 and MBL genes to susceptibility to HIV type 1 infection in the Finnish population. AIDS Res Hum Retroviruses 1998;14(8):695–698.

97. Maas J, Roda Husman AM, Brouwer M, et al. Presence of the variant mannose-binding lectin alleles associated with slower progression to AIDS. Amsterdam Cohort Study. AIDS 1998;12(17):2275–2280.

98. Carpenter CC, Cooper DA, Fischl MA, et al. Antiretroviral therapy in adults: updated recommendations of the International AIDS Society-USA Panel. JAMA 2000;283(3): 381–390.

99. Valdez H, Purvis SF, Lederman MM, Fillingame M, Zimmerman PA. Association of the CCR5delta32 mutation with improved response to antiretroviral therapy. JAMA 1999;282(8):734.

100. Guerin S, Meyer L, Theodorou I, et al. CCR5 delta32 deletion and response to highly active antiretroviral therapy in HIV-1-infected patients. AIDS 2000;14(17):2788–2790.

101. Barroga CF, Raskino C, Fangon MC, et al. The CCR5Delta32 allele slows disease progression of human immunodeficiency virus-1-infected children receiving antiretroviral treatment. J Infect Dis 2000;182(2):413–419.

102. O'Brien TR, McDermott DH, Ioannidis JP, Carrington M, Murphy PM, Havlir DV et al. Effect of chemokine receptor gene polymorphisms on the response to potent antiretroviral therapy. AIDS 2000;14(7):821–826.

103. Nguyen GT, Carrington M, Beeler JA, et al. Phenotypic expressions of CCR5-delta32/delta32 homozygosity. J Acquir Immune Defic Syndr 1999;22(1):75–82.

104. D'Souza MP, Cairns JS, Plaeger SF. Current evidence and future directions for targeting HIV entry: therapeutic and prophylactic strategies. JAMA 2000;284(2):215–222.

105. Simmons G, Clapham PR, Picard L, et al. Potent inhibition of HIV-1 infectivity in macrophages and lymphocytes by a novel CCR5 antagonist. Science 1997;276(5310):276–279.

106. Baba M, Nishimura O, Kanzaki N, et al. A small-molecule, nonpeptide CCR5 antagonist with highly potent and selective anti-HIV-1 activity. Proc Natl Acad Sci U S A 1999;96(10):5698–5703.

107. Trkola A, Ketas TJ, Nagashima KA, et al. Potent, broad-spectrum inhibition of human immunodeficiency virus type 1 by the CCR5 monoclonal antibody PRO 140. J Virol 2001;75(2):579–588.

108. Feng Y, Leavitt M, Tritz R, et al. Inhibition of CCR5-dependent HIV-1 infection by hairpin ribozyme gene therapy against CC-chemokine receptor 5. Virology 2000;276(2): 271–278.

109. Cagnon L, Rossi JJ. Downregulation of the CCR5 beta-chemokine receptor and inhibition of HIV-1 infection by stable VA1-ribozyme chimeric transcripts. Antisense Nucleic Acid Drug Dev 2000;10(4):251–261.

110. Bai J, Rossi J, Akkina R. Multivalent anti-CCR ribozymes for stem cell-based HIV type 1 gene therapy. AIDS Res Hum Retroviruses 2001;17(5):385–399.

111. Rusconi S, Merrill DP, La Seta CS, et al. In vitro inhibition of HIV-1 by Met-SDF-1beta alone or in combination with antiretroviral drugs. Antivir Ther 2000;5(3):199–204.

112. Murakami T, Nakajima T, Koyanagi Y, et al. A small molecule CXCR4 inhibitor that blocks T cell line-tropic HIV-1 infection. J Exp Med 1997;186(8):1389–1393.

113. Arakaki R, Tamamura H, Premanathan M, et al. T134, a small-molecule CXCR4 in-

hibitor, has no cross-drug resistance with AMD3100, a CXCR4 antagonist with a different structure. J Virol 1999;73(2):1719–1723.

114. Doranz BJ, Grovit-Ferbas K, Sharron MP, et al. A small-molecule inhibitor directed against the chemokine receptor CXCR4 prevents its use as an HIV-1 coreceptor. J Exp Med 1997;186(8):1395–1400.

115. Schols D, Este JA, Henson G, De Clercq E. Bicyclams, a class of potent anti-HIV agents, are targeted at the HIV coreceptor fusin/CXCR-4. Antiviral Res 1997;35(3):147–156.

116. Schols D, Struyf S, Van Damme J, Este JA, Henson G, De Clercq E. Inhibition of T-tropic HIV strains by selective antagonization of the chemokine receptor CXCR4. J Exp Med 1997;186(8):1383–1388.

117. Lehner T, Wang Y, Cranage M, et al. Up-regulation of beta-chemokines and down-modulation of CCR5 co-receptors inhibit simian immunodeficiency virus transmission in non-human primates. Immunology 2000;99(4):569–577.

118. Zuber B, Hinkula J, Vodros D, et al. Induction of immune responses and break of tolerance by DNA against the HIV-1 coreceptor CCR5 but no protection from SIVsm challenge. Virology 2000;278(2):400–411.

119. Misumi S, Nakajima R, Takamune N, Shoji S. A cyclic dodecapeptide-multiple-antigen peptide conjugate from the undecapeptidyl arch (from Arg(168) to Cys(178)) of extracellular loop 2 in CCR5 as a novel human immunodeficiency virus type 1 vaccine. J Virol 2001;75(23):11614–11620.

28

Hereditary hemochromatosis

Giuseppina Imperatore, Rodolfo Valdez, and Wylie Burke

Disease

Hereditary hemochromatosis (HHC, OMIM 235200) is an inherited disorder of iron metabolism characterized by an increased absorption of iron from the diet. Over time the excess iron accumulates in body tissues, a condition known as iron overload, and can lead to organ damage. Iron accumulation occurs primarily in the liver, pancreas, heart, joints, and pituitary gland. This may result in organ failure including liver cirrhosis, primary liver cancer, impotence, arthritis, diabetes, or cardiomyopathy. The disease onset is insidious and often characterized by common nonspecific symptoms such as fatigue, arthralgia, and abdominal pain. For this reason, it can be undetected for a long time and it is usually diagnosed when advanced organ damage has already occurred. Two methods of screening for the detection of the early stage of HHC are available: serum iron measures and molecular testing to detect mutations in the HHC gene, called *HFE*. Iron overload due to HHC can be detected before the appearance of organ damage. The most commonly used test to identify persons at risk of developing iron overload disease is the percent transferrin saturation (TS) (1). An elevated TS usually occurs well before HHC clinical symptoms. The first step for ascertaining HHC is measuring a nonfasting TS. If this result is elevated (usually >45%), the test should be repeated after an overnight fast (2). If fasting TS is also elevated, then more tests need to be performed to check for the presence of increased iron stores. First, serum ferritin levels should be measured. Serum ferritin levels above 300 μg/L in men and postmenopausal women and $\geq$ 200 μg/L in premenopausal women indicate primary iron overload. Liver biopsy or quantitative phlebotomy confirms the diagnosis of HHC and quantifies the degree of iron overload (2). Treatment of iron overload consists in removing the excess iron through repeated phlebotomy, which improves survival in symptomatic persons (3–6). If phlebotomy is initiated before the development of cirrhosis, survival rate of individuals with HHC is similar to that of the general population (3,7,8).

Epidemiology

The lack of a standardized disease definition makes the estimate of the prevalence of HHC complicated. HHC can be defined by the presence of genetic mutations in the *HFE* gene, or biochemical markers of iron metabolism, or the presence of clinical symptoms. The genetic analysis identifies persons carrying one or two copies of the two known mutations in the *HFE* gene, C282Y and H63D. In the United States, a recent population-based study estimated that among whites the frequency of *HFE* genotypes containing two mutations (C2982Y/C282Y, C282Y/H63D, and H63D/H63D) is about 5% (9). Among these genotypes, however, the highest risk for iron overload disease occurs with the C282Y homozygous genotype (10). The preponderance of clinically diagnosed HHC cases with C282Y/C282Y genotype, despite the fact that it is much rarer than other *HFE* genotypes, is evidence of the higher risk associated with this genotype. In the United States, for example, among whites the prevalence of homozygosity for C282Y mutation is 0.30% (95% CI, 0.12–0.82), about 1 in 333 individuals (9). Similar estimates were reported among members of a health maintenance organization where the prevalence of C282/C282Y genotype in whites was 0.4% (11). If we assume that about 81% of affected individuals in the United States are homozygous for C282Y (12), based on mutation analysis the estimate of HHC in the white population of the United States ranges between 37 in 10,000 (30/0.81) to 62 in 10,000 (50/0.81).

In population-based intervention trials, the estimated prevalence of homozygosity based on phenotype, defined as biochemical evidence of iron overload, is 50 per 10,000 (95% confidence interval, 17–84) for men and 62 per 10,000 (95% confidence interval, 27–97) for women (13). In primary care settings among whites, the estimated prevalence of clinically proven or liver biopsy proven HHC is 54 per 10,000 (14). A higher prevalence (80 per 10,000) was obtained in one study when elevated TS alone was used for defining HHC (15). This may simply reflect the fact that a significant proportion of unaffected or heterozygous individuals have TS levels above the cutoff, especially when TS thresholds of 50% are used (13). Lower estimates (3 to 19 per 10,000) are derived from autopsy studies and review of death records. In 1992, the HHC-associated mortality rate in the American population was reported at 1.8 deaths per million (16), far lower than the estimated prevalence of HHC. Similarly, a study using data from the National Hospital Discharge Survey from 1979 through 1997 estimated that the rate of hemochromatosis-associated hospitalizations was 2.3 per 100,000 persons in the United States (17). There are, therefore, fewer people requiring treatment or dying from HHC than is predicted by the frequency of the HHC mutations. This may reflect the fact that the disease is underdiagnosed, or that the penetrance (the likelihood that a person carrying a given genotype will develop clinical disease) of the genotype is low, or both.

Genetics

More than 20 years ago, Simon et al. (18) described HHC as an autosomal recessive disorder linked to the HLA-A3 complex on the short arm of chromosome 6. In 1996, Feder et al. mapped the *HFE* gene on the short arm of chromosome 6 (6p21.3) and described two missense mutations of this gene (C282Y and H63D) that accounted for the majority of HHC patients in their study (19).

HHC due to mutations of the *HFE* gene occurs commonly among whites, especially those of northern European descent (20). *HFE* mutations so far identified, however, do not account for all cases of hemochromatosis (12). For example, in Southern Europe, non-*HFE* related iron overload disorders due to mutations of the transferrin receptor 2 (*TFR2*) and the ferroportion gene (*SLC11A3*) (21–22) have been described. Therefore, the genetics of HHC is complex.

Allelic Variants

The *HFE* gene codes for a 343 residue type I transmembrane protein that associates with class I light chain beta$_2$-microglobulin (19). This protein binds to the transferrin receptor and reduces its affinity for iron-loaded transferrin by 5- to 10-fold (23). The localization of the *HFE* protein in the crypt cells of the duodenum (the site of dietary iron absorption) and its association with the transferrin receptor in those cells are consistent with a role in regulating iron absorption (24–25). The observation that *HFE*-deficient mice (*HFE* gene knockout model) develop iron overload similar to that seen in humans with HHC provides further evidence that the *HFE* protein is involved in iron homeostasis (26). The C282Y mutation results from a G-to-A transition at nucleotide 845 of the *HFE* gene (845G → A) that produces a substitution of cysteine for a tyrosine at the amino acid position 282 in the protein product. This substitution alters the *HFE* protein structure and beta$_2$-microglobulin association, disrupting its transport to and presentation on the cell surface (27). In the H63D mutation, a G replaces C at nucleotide 187 of the gene (187C → G), causing aspartate to substitute for histidine at the amino acid position 63 in the *HFE* protein. The H63D mutation does not seem to prevent beta$_2$-microglobulin association or cell surface expression (24), indicating that the C282Y mutation results in a greater loss of protein function than does H63D (28).

In addition to C282Y and H63D, nine other missense mutations causing amino acid substitutions have been documented. In one, a substitution of a cysteine for serine at the amino acid position 65 (S65C) has been implicated in a mild form of HHC (29). A number of intronic polymorphisms have also been found (30). One polymorphism occurs within the intron 4 (5569G-A) of the *HFE* gene in the binding region of the primer originally described by Feder et al. (19). One laboratory reported that when a polymerase chain reaction (PCR)-based restriction endonuclease digestion assay is used, the presence of this polymorphism might cause C282Y heterozygosity to be misdiagnosed as C282Y homozygosity (31–32). However, three groups could not replicate this finding (33–34). Beutler et al. reported that a mutation in intron 3 (IVS3-48c) can also lead to misdiagnosis of heterozygotes for the C282Y mutation as homozygotes (35).

Genotype Prevalence

A number of studies have reported on both the general population and the proband frequencies of the *HFE* genotype. Recently, a review has summarized the results of these studies according to the geographic origin of the populations studied (12). In the general population, a total of 6203 samples from European countries revealed on average a C282Y homozygous and heterozygous prevalence of 0.4% and 9.2%, respectively. However, C282Y homozygosity has not been reported in the general population of Southern or Eastern Europe. The frequency of the C282Y heterozygosity varies from 1% to 3% in Southern and Eastern Europe to as high as 24.8% in Ireland. In North America, among 3752 samples, the *HFE* genotype distribution had a similar pattern: *C282Y/C282Y* genotype was the rarest with a frequency of 0.5% and C282Y heterozygosity was present in 9.0% of the samples. In the Asian, Indian subcontinent, African/Middle Eastern, and Australasian populations, C282Y homozygotes were not found, and the frequency of C282Y heterozygosity was very low (range: 0% to 0.5%). C282Y/H63D compound and H63D homozygosity each accounted for 2% of the European general population and 2.5% and 2.1% in the North American populations, respectively. The heterozygous frequency of the H63D mutation was 22% in Europe and 23% in North America.

Hanson et al. (12) recently reviewed 17 studies reporting the frequency of the HFE genotypes among patients with clinically diagnosed HHC. Most of the studies used case definitions that included diagnostic evidence of iron overload from either liver biopsy or quantitative phlebotomy. The exceptions were a French (29) and a U.S. study (36), which used a case definition of persistently elevated TS or elevated serum ferritin. In all case series, the majority of patients had the homozygous C282Y genotype. However, there was some variability across studies. For example, among 2229 European HHC patients, the estimated prevalence of homozygosity for the C282Y genotype ranged from 52% (37) to 96% (38). In North America, among 588 patients, the C282Y homozygosity ranged between 67 to 95%. Heterozygosity for the H63D mutation and compound heterozygosity (C282Y/H63D) each accounted for 6% of European cases and 4% of cases in North America. Overall, 3.6% (95% CI, 2.9–4.3) of the patients had the C282Y/wild genotype, and 1.5% (95% CI, 1.1–2.1) had the H63D/H63D genotype. Worldwide, among 2929 patients, 6.9% (95% CI, 6.0–7.9) were homozygous for the wild allele. These findings suggest that nongenetic influences, additional HFE mutations, or variation at additional genes affecting iron metabolism, as recently reported, may also cause or modulate iron overload (21,22,39).

Gene–Gene and Gene–Environment Interactions

The clinical expression of HHC is influenced by a variety of factors, both genetic and environmental. In *HFE* knockout mice, mutations of other genes involved in iron metabolism, such as beta$_2$-microglobulin, transferrin receptor, and transmembrane iron import molecule (DTM1), strongly modify the amount of liver iron (40). It is, therefore, conceivable, that similar gene–gene interactions may influence the course of HHC in humans. The finding of the C282Y heterozygote genotype among some persons classified as being affected with HHC is also suggestive of the influence of other yet-to-be-identified HFE mutations; or of the combination of the C282Y heterozygous state with environmental modifying factors (e.g., high iron intake, viral hepatitis, or alcohol abuse); or of a second genetic disorder (e.g., β-thalassemia trait, iron loading anemia) that could account for clinical disease (41–44).

There is also evidence that sex plays a primary role in the clinical manifestation of HHC. Family studies based on HLA-typing indicate that the frequency of affected brothers and sisters is similar, as expected for an autosomal recessive disorder, but the proportion of females among probands diagnosed on the basis of clinical symptoms is 11% to 35%, rather than the expected 50% (3,4,45). In a large study, the prevalence of iron overload, as determined by liver biopsy or phlebotomy, was twice as frequent in males as females (46). This sex difference has been attributed to the lower degree of iron overload in women because of menstruation, pregnancy, and lactation.

Other possible modifiers include chronic blood loss (gastrointestinal bleeding, regular hematuria, helminthic or other parasitic infections) and regular blood donation, alcohol abuse, excessive iron intake, or vitamin C intake. Tannates, phytates, oxalates, calcium, and phosphates also modify HHC because they are known to bind iron and inhibit iron absorption (47). Chronic viral hepatitis B and C and metals such as zinc and cobalt may also influence expression of HHC (47–48). Iron modulates the course of hepatitis B (57), and iron reduction has been shown to decrease the severity of chronic hepatitis C while increasing the likelihood of response to antiviral therapy. Hepatitis C virus infection and *HFE* mutations have also been identified as risk factors for *porphyria cutanea tarda* (49).

Laboratory Tests

Serum Tests for Iron Status

The value of serum iron measures or *HFE* mutation analysis in screening for individuals at high risk for developing serious clinical manifestations of HHC is difficult to assess because of uncertainties about the natural history of the disease. Thus, the phenotype of interest must be specified before assessing the validity of each test for screening. For example, the phenotype might be defined by biochemical evidence of iron overload (e.g., hepatic iron index >1.9 or removal of more than 4 g of iron by quantitative phlebotomy), or by clinical symptoms compatible with iron overload in combination with biochemical evidence of iron overload, or by evidence of serious end-stage organ disease in combination with biochemical evidence of iron overload.

The marker for serum iron status most used is percent transferrin saturation (TS). This test can be used as a phenotypic screening test to identify persons with biochemical evidence of iron overload. The cutoff TS values recommended for screening have varied from 45% to 70% (1,14,15,50). With the use of data collected in family studies and screening trials, the performance of TS as a screening test (e.g., detection rate, false-positive rate, and positive and negative predictive values) has been estimated for different TS cutoff levels. For example, based on published parameters, screening at a TS cutoff level of 50% would identify about 94% and 82% of men and women with HHC, respectively, along with a number of false positives (about 6% of males and 3% of females screened). Assuming an HHC genotype prevalence of about 50 in 10,000, the odds of being affected given a positive result (OAPR) would be about 1 to 12 for males and 1 to 8 for females, corresponding to positive predictive values of 8% and 11%, respectively. Diagnostic testing (e.g., liver biopsy or quantitative phlebotomy) is recommended for persons with a positive screening result (either a single test result or persistently elevated TS) and no other identifiable explanation for increased body iron stores (e.g., chronic anemias, liver disease related to alcohol abuse or hepatitis). In persons diagnosed to have iron overload related to HHC by such a screening and diagnostic process, the probability of developing at least one clinical symptom can be estimated from family studies and screening trials to be about 50% to 70% for males and 40% to 50% for females. It is worth noting that most complications recorded in such studies were common and nonspecific clinical manifestations of the disease such as joint pain and diabetes. In the absence of control groups, the proportion of complications attributable to HHC is difficult to determine; as a result, the probability of developing clinical complications may be considerably lower.

Penetrance appears to be consistently lower in women than in men at all ages. However, as many as 40% of genetically susceptible younger individuals of both sexes do develop biochemical evidence of iron overload; many also have nonspecific symptoms compatible with early iron overload. A smaller proportion, not well defined, may develop serious complications such as diabetes, cirrhosis, or cardiomyopathy.

Ferritin is an intracellular iron storage protein and serum ferritin (SF) concentration significantly correlates with body iron stores (1 ng/mL = 10 mg of stored iron). SF values, but not TS values, are associated with HHC clinical signs, and SF concentrations are higher for those with clinical manifestations (13). SF has been used as a second screening test in many trials, and it can be very effective in reducing the number of false positives (46), if cutoffs appropriate for age and sex are used. Elevation of the SF concentration in HHC must be differentiated, however, from other liver disorders such as alcoholic liver disease, chronic viral hepatitis, and nonalcoholic steatohepatitis. Also, SF is an acute phase reactant and can be elevated as a result of inflammatory conditions.

HFE Gene Mutation Analysis

HFE mutation analysis identifies persons carrying one or two copies of either of the two known mutations, C282Y and H63D. Since the majority of clinically diagnosed probands are homozygous for C282Y, individuals with this genotype are considered to be at the highest risk for iron overload disease. However, approximately 20% of HHC cases occur in persons with other HFE genotypes, and as many as 7% have no identifiable mutation (12). The penetrance of the different HFE genotypes—that is, the likelihood that persons carrying a given HFE genotype will develop manifestations of iron overload—can only be roughly estimated from published data. The data suggest that a large proportion of individuals with the C282Y homozygous genotype will develop biochemical evidence of iron overload during their lifetime; we can only speculate how many will develop clinical symptoms related to iron overload (perhaps about half), or who will die from complications of iron overload (likely to be a small proportion).

A screening study at a health maintenance organization in southern California represents the only controlled study to evaluate penetrance of the *C282Y/C282Y* genotype (11). The study included 41,038 adults attending a health appraisal clinic (a clinic providing assessment of health status and prevention options, attended voluntarily) with a mean age of 57 years, of whom 152 subjects (0.4%) had the *C282Y/C282Y* genotype. Of these, 45 had previously been diagnosed with HHC (30%); for most, the diagnosis had been made on the basis of screening. The study evaluated 124 subjects with the *C282Y/C282Y* genotype, including all those not previously diagnosed with HHC and 17 for whom data were available prior to diagnosis. The TS was elevated in 75% of men and 40% of women, and SF was elevated in 76% of men and 54% of women with the *C282Y/C282Y* genotype. Compared with control subjects (22,394 white and Hispanic participants on whom questionnaire data were available and who did not have any *HFE* mutations), persons homozygous for C282Y were more likely to have a history of a "liver problem or hepatitis" (8% vs. 4%), elevated serum aspartate aminotransferase (8% vs. 4%), and elevated plasma collagen IV, a measure of mild liver fibrosis (26% vs. 11%), but were no more likely to have a history of fatigue, joint pain, impotence, skin pigmentation, or diabetes. Among the full cohort of 152 subjects with the *C282Y/C282Y* genotype, only one, an alcoholic, had a clinical history of end-stage HHC. Two others, out of 119 with complete data, had markedly abnormal laboratory values suggestive of severe liver fibrosis. On the basis of these data, the authors concluded that the likelihood of significant clinical disease in persons with the *C282Y/C282Y* genotype was 1%.

This study has a potential selection bias that could have resulted in an underestimate of penetrance: subjects were drawn from a preventive care setting, potentially selecting against patients with clinical disease. The limited clinical findings among the large group of subjects with the *C282Y/C282Y* genotype argues for low penetrance of the genotype, even if it ultimately proves to be above 1%. In keeping with this conclusion, the study found the prevalence of the *C282Y/C282Y* genotype was the same among older and younger subjects (11); high penetrance would be expected to result in premature mortality for some people with the genotype, resulting in a lower prevalence of the genotype at older ages. The penetrance of all other HFE genotypes is estimated to be much lower than that of *C282Y/C282Y* (10).

Implications of Genetic Testing

Screening for HHC using *HFE* mutation analysis could involve testing for both *HFE* mutations, or only for C282Y. If both mutations are tested, about 5% to 6% of persons of northern European descent will have a test result indicating the presence of two *HFE* mutations. However, about 0.5% of the general screened population with the *C282Y* homozygous genotype will be at high risk of iron overload. Another 2% of the general population will have the compound heterozygous (*C282Y/H63D*) genotype, but only about 1 in a 100 of these persons would be expected to develop significant iron loading. If testing is limited to C282Y, about 10% of the northern European population would be identified as heterozygote, but only 0.5% of the population homozygous for C282Y would be at high risk for iron overload. Either approach would identify the majority of persons at risk for hemochromatosis, though the risk for some persons would be low and difficult to quantify. In other populations—e.g., southern European—this screening approach may identify a smaller proportion of persons at risk. For either approach, costs and sequelae of screening are influenced by decisions concerning the provision of counseling and/or clinical follow-up. For example, the follow-up offered to all persons with the *C282Y* homozygous genotype should include counseling about the uncertainty of their prognosis, and the possibility that risk of clinical complications is low. For persons with other genotypes, risk of iron overload disease is known to be low, and appropriate counseling and follow-up have not been established. Decisions concerning the counseling needs of C282Y heterozygotes would have an important effect on the cost and outcome of a screening program, since these persons constitute a substantial proportion of the population (about 9% in populations of northern European descent). Clinical follow-up or counseling to address their potential risk for iron overload related to alcohol abuse or other risk factors, as well as the potential risk to family members, would be costly. In addition, there are currently no data to assess the value of such intervention.

Family-based detection represents an important alternative approach to identifying people with iron overload. When a diagnosis of HHC is made, it also identifies family members who represent a group with a markedly higher a priori risk of iron overload disease than the general population. Therefore, it is reasonable to consider assessment of iron status in relatives and to monitor them for symptoms suggestive of iron overload.

HFE genotyping provides a one-time test to determine which relatives of an identified proband have an increased risk of iron overload. These relatives can be offered ongoing surveillance, while others can be reassured. However, genotyping may also cause confusion about clinical status and adverse labeling, so the value of genotyping as a method for family-based detection of HHC is not entirely clear. Siblings of an affected person with the homozygous C282Y genotype have a 25% chance of sharing the same high-risk genotype; for siblings who do not share the genotype, this single test can greatly reduce the risk. However, HHC has occurred in some people with other *HFE* genotypes (e.g., *C282Y/H63D, C282Y/+*) (12), suggesting the need for caution in the interpretation of a "negative" test result. But even the implications of a "positive" result are not straightforward; current penetrance data make risk of disease hard to calculate even for relatives with a *C282Y/C282Y* genotype, and argue against making a diagnosis of HHC on the basis of genotype alone. In the uncommon instance of a proband with a different *HFE* genotype, genotypic studies of relatives are even more difficult to assess, given the very low penetrance of genotypes other than *C282Y/C282Y*.

Testing of offspring raises even more questions, because of the high carrier rate for *HFE* mutations (e.g., 9% for C282Y, 23% for H63D in populations of European descent) (12). If the parent with HHC is a C282Y homozygote, offspring have a 4.5% likelihood of inheriting the same genotype (calculated as: 100% chance of inheriting the C282Y allele from the affected parent × 9% chance that the other parent is a C282Y carrier × 50% chance of inheriting C282Y from the unaffected parent) and an 11.5% chance of inheriting a *C282Y/H63D* genotype. All other offspring will be C282Y carriers. Because disease occurs in middle age, there is no rationale for testing during childhood.

Genotype testing does not substitute for the serum iron studies needed to identify iron overload, and it could expose the family member to a premature diagnosis, unnecessary treatment, and the potential for stigma and discrimination. These considerations underscore the need for more information about the clinical penetrance of HFE genotypes in HHC, and also about effective ways to counsel patients after genetic testing to ensure an accurate understanding of the results.

Potential Benefits and Harms Associated with Genetic Testing for HFE *Mutations*

There are important potential benefits from early detection of affected HHC persons, including prevention of significant morbidity and mortality and long-term reduction in health-care costs for those who would otherwise suffer from serious medical complications of hemochromatosis. TS screening has been used successfully in pilot studies, suggesting that this is a feasible screening approach. At the same time, universal screening for HHC would expose a large number of persons to the possibility of adverse psychological, social, or economic consequences related to a diagnosis of HHC. As reviewed in previous sections, a majority of those identified through screening might remain healthy without treatment. The potential for loss of insurance or employment after a genetic diagnosis is a concern for consumers and policymakers (51–54). Legislative efforts to minimize such loss are being implemented (53), but the degree of protection they will provide is unknown. Although adverse outcomes after a diagnosis of HHC have been reported (55), no systematic study has been undertaken to further assess these outcomes. A genetic diagnosis may be stigmatizing, and has the additional effect of identifying a potential risk for family members. (56). The psychological burdens of a diagnosis of HHC may be reduced by counseling that the diagnosis does not imply a certainty of future disease, and that effective treatment is available to reduce future risk. Whether communication of this kind can change the stigmatizing potential of a diagnosis of HHC, or reduce the likelihood of discrimination, remains to be determined. These issues are not substantially different from those identified for other types of genetic testing (e.g., cystic fibrosis, cancer markers), and may influence decisions about the need for counseling procedures as part of a screening program. As with other aspects of HHC screening, judgments about the relevance of these issues to decisions about HHC screening must be made in the absence of definitive data.

Conclusions

Iron overload can be treated or prevented by phlebotomy, but treatment is often delayed, resulting in irreversible organ damage. Greater physician awareness of HHC may help reduce the morbidity and mortality of primary iron overload. HHC is usually diagnosed after a delay of several years, during this period care has been sought for the early nonspecific symptoms of the disorder (57). Some persons with HHC are not diagnosed until life-threatening complications are present, e.g., diagnoses have occurred after a liver transplant for end-stage cirrhosis (58). The delay in diagnosis of hemochromatosis suggests that physicians lack awareness of this disorder or have a low index of suspicion when symptoms compatible with the early stages of the disease are present and even sometimes when late complications are present. With early diagnosis, preventive therapy can be instituted in the form of regular phlebotomy. If treatment is begun before cirrhosis or diabetes has occurred, the prognosis is good. However, late and missed diagnoses lead to underutilization of this readily accessible preventive treatment.

A number of questions remain about the benefits and risks of identifying and treating asymptomatic persons at high risk for HHC. "Universal" or "population-based" screening refers to screening performed across an entire population of mainly asymptomatic individuals not referred for testing due to symptoms of the disease. This could be accomplished through public health screening programs, or as part of routine testing within primary health-care settings. This should be clearly distinguished from the alternative approach of "case finding" or "enhanced case detection," which could include iron status testing and/or *HFE* mutation analysis in targeted populations, such as persons at increased risk due to an affected family member or persons who present with clinical complaints consistent with a diagnosis of HHC. Generally accepted criteria for an effective population-based screening test include the following:

The disorder screened for must be well-defined and represent an important health problem. The natural history of the disorder should be understood. This is, in fact, a key question. Is the disorder for which we plan to screen HHC or iron overload? HHC, as defined by the *HFE* genotype, better fits the criterion of a "well-defined" disorder. HHC is a serious, treatable disorder with life-threatening complications. However, important questions about natural history, particularly age-related penetrance, remain, and recent data raise the possibility that very few people with the genotype will develop clinically significant disease. Iron overload, as defined by persistently elevated TS and diagnostic evidence of increased body iron stores, however, encompasses a broad range of disorders, and requires a complex diagnostic protocol to differentiate HHC from other acquired (e.g., chronic anemias, alcoholic liver disease) and inherited causes (e.g., juvenile hemochromatosis, non-*HFE* hemochromatosis). Because the natural history of hemochromatosis has not been systematically studied, questions remain concerning the persons most likely to benefit from early treatment.

The prevalence must be known and the disorder common enough that population-based screening will be cost-effective. Based on evidence from family studies, screening trials and mutation analysis, the prevalence of HHC, as defined by genotype, can be estimated to be about 50 to 60 per 10,000; lower estimates (3 to 19 per 10,000) are derived from autopsy studies and review of death records. Estimates of phenotype prevalence in screening trials, as defined by diagnostic testing (e.g., liver biopsy or quantitative phlebotomy) for iron overload, range from about 26 to more than 50 per 10,000. The prevalence of life-threatening complications due to HHC is not well established and may be much lower.

Suitable screening test(s) with known performance (detection rate, false positive rate, OAPR) must be available. As discussed before, two well-characterized screening tests are available, TS and *HFE* mutation analysis. Both tests are readily available (though there are licensing issues with regard to *HFE* testing in the US). While consensus may have to be reached on the distribution of TS measurements in homozygous, heterozygous and unaffected individuals, enough information is available to allow reasonably accurate prediction of detection rates and false positive rates using different cutoff levels of TS (alone or in conjunction with serum ferritin). A consistent set of data is available with regard to expected distributions of *HFE* genotype frequencies among patients and in the general population. However, screening performance cannot be fully assessed due to uncertainties about the age-related penetrance of HHC.

For individuals identified as screen positive, there must be an adequate and acceptable protocol for diagnosis and an effective treatment. Protocols for diagnosis and treatment of HHC are available. However, limited data are available on the outcome of treatment in persons with HHC who are asymptomatic at the time of diagnosis.

Adequate facilities must be available to support screening, diagnosis and treatment. It is likely that laboratories capable of supporting screening and diagnostic testing methodologies either exist now, or could be developed within a reasonable time frame. However, a number of logistical issues need to be considered further, including: the potential burden of time and cost to health care providers to educate patients about HHC, offer testing, and follow up on positive screens; the anticipated level of patient compliance with screening, diagnosis, and treatment; proposed mechanisms of long-term follow-up and maintenance of iron status or genotype information in patients' medical records.

Costs have been examined and are reasonable. TS-based screening is likely to be cost-effective even given unfavorable assumptions (e.g., low prevalence or rate of progression to serious disease, low compliance). Screening for the C282Y mutation may be cost-effective due to a lower positive rate than TS screening, with corresponding reduced costs and burden of intervention. However, published studies to date have not considered *all* costs of screening, both medical and societal; in particular, the physician and health system effort required for long-term follow-up and the personal psychological and economic consequences of screening have not been evaluated (47,50,59,60).

The approach must be considered ethical and must be acceptable to health care providers and consumers. Potential benefits from early detection of affected persons are significant, including prevention of significant morbidity and mortality and long-term reduction in health care costs. TS screening has been used successfully in pilot studies. Concern has been raised about potential harms of screening, including psychological morbidity, stigmatization and discrimination (e.g., insurance or employment) resulting from a diagnosis of HHC, particularly if the risk of clinical complications is low among those diagnosed by screening.

Health Policies Regarding Use of Genetic Information

At this time genetic testing for HFE mutations is not recommended for population-based screening for HHC (61,62) due to the uncertainty about disease prevalence and penetrance, the optimal care for asymptomatic persons found to carry HFE mutations, and the psychosocial impact of genetic testing. Mutational analysis of the HFE gene may be useful in confirming the diagnosis of HHC in persons with elevated iron measures or for identifying relatives at risk to develop HHC of patients with HHC due to the known mutations

Agenda for Future HuGE Research

The discovery of the HFE gene represents an important step in understanding the nature of HHC. However, much remains to be learned about this disorder. It is crucial to know the age- and sex-specific penetrance of the HFE mutations. Future efforts should also be directed towards the identification of environmental modifiers and assessment of their interactions with HFE genotypes. Moreover, as more and more genes involved in iron metabolism are being discovered, attempts should be made to understand the complex interplay of the HFE genotype and these genes. Future studies should be performed to assess the effectiveness of interventions to reduce disease burden in asymptotic individuals carrying susceptibility genotypes. In addition, more information is needed regarding the social, ethical, and psychological outcomes of genetic screening for HHC.

References

1. Witte DL, Crosby WH, Edwards CQ, et al. Practice guideline development task force of the College of American Pathologists. Hereditary hemochromatosis. Clin Chim Acta 1996;245:139–200.
2. Powell WL, George KD, McDonnell SM, et al. Diagnosis of hemochromatosis. Ann Intern Med 1998;129:925–993.
3. Niederau C, Fischer R, Purschel A, et al. Long-term survival in patients with hereditary hemochromatosis. Gastroenterology 1996;110:1107–1119.
4. Niederau C, Fischer R, Sonnenberg A, et al. Survival and causes of death in cirrhotic and in noncirrhotic patients with primary hemochromatosis. N Engl J Med 1985;313: 1256–1262.
5. Bomford A, Williams R. Long term results of venesection therapy in idiopathic haemochromatosis. Q J Med 1976;45:611–623.
6. Powell LW, Kerr JF. Reversal of "cirrhosis" in idiopathic haemochromatosis following long-term intensive venesection therapy. Australas Ann Med 1970;19:54–57.
7. Adams PC, Speechley M, Kertesz AE. Long-term survival analysis in hereditary hemochromatosis. Gastroenterology 1991;101:368–372.
8. Fargion S, Mandelli C, Piperno A, et al. Survival and prognostic factors in 212 Italian patients with genetic hemochromatosis. Hepatology 1992;15:655–659.
9. Steinberg KK, Cogswell ME, Chang JC, et al. Prevalence of C282Y and H63D mutations in the hemochromatosis (HFE) gene in the United States. JAMA 2001:285; 2216–2222.
10. Burke W, Imperatore G, McDonnell SM, et al. Contribution of different genotypes to iron overload disease: a pooled analysis. Genet Med 2000;2:271–277.
11. Beutler E, Felitti V, Koziol J, et al. Penetrance of the 845G → A (C282Y) HFE hereditary haemochromatosis mutation in the USA. Lancet 2002;359:211–218.
12. Hanson EH, Imperatore G, Burke W. HFE gene and hereditary hemochromatosis: a HuGE Review. Am J Epidemiol 2001;154:193–206.
13. Bradley LA, Haddow JE, Palomaki GE. Population screening for haemochromatosis: expectations based on a study of relatives of symptomatic probands. J Med Screen 1996;3: 171–177.
14. Phatak PD, Sham RL, Raubertas RF, et al. Prevalence of hereditary hemochromatosis in 16,031 primary care patients. Ann Intern Med 1998;129:954–961.
15. McDonnell SM, Hover A, Gloe D, et al. Population-based screening for hemochromatosis using phenotypic and DNA testing among employees of health maintenance organizations in Springfield, Missouri. Amer J Med 1999;107:30–37.
16. Yang Q, McDonnell SM, Khoury MJ, et al. Hemochromatosis-associated mortality in the United States from 1979 to 1992: an analysis of multiple-cause mortality data. Ann Intern Med 1998;129:946–953.
17. Brown AS, Gwinn M, Cogswell ME, et al. Hemochromatosis-associated morbidity in the United States: a analysis of the National Hospital Discharge Survey, 1979–1997. Genet Med 2001;3:109–111.
18. Simon M, Bourel M, Fauchet R, et al. HLA and "non-immunological" disease: idiopathic haemochromatosis [letter]. Lancet 1976;2:973–974.
19. Feder JN, Gnirke A, Thomas W, et al. A novel MHC class I-like gene is mutated in patients with hereditary haemochromatosis. Nat Genet 1996;13:399–408.
20. Merryweather-Clarke AT, Pointon JJ, Jouanolle AM, et al. Geography of HFE C282Y and H63D Mutations. Genetic Testing 2000;4:183–198.

21. Camaschella C, Roetto A, Cali A, et al. The gene TFR2 is mutated in a new type of haemochromatosis mapping to 7q22 Nat Genet 2000;25:14–15.
22. Montosi G, Donovan A, Totaro A, et al. Autosomal-dominant hemochromatosis is associated with a mutation in the ferroportion (SLC11A3) gene. J Clin Invest 2001; 108:619–623.
23. Feder JN, Penny DM, Irrinki A, et al. The hemochromatosis gene product complexes with the transferrin receptor and lowers its affinity for ligand binding. Proc Natl Acad Sci U S A 1998;95:1472–1477.
24. Waheed A, Parkkila S, Saarnio J, et al. Association of HFE protein with transferrin receptor in crypt enterocytes of human duodenum. Proc Natl Acad Sci U S A 1999; 96:1579–1584.
25. Parkilla S, Waheed A, Britton RS, et al. Immunohistochemistry of HLA-H, the protein defective in patients with hemochromatosis, reveals unique pattern of expression in hereditary hemochromatosis. Proc Natl Acad Sci U S A 1997;94:2534–2539.
26. Zhou XY, Tomatsu S, Fleming RE, et al. HFE gene knockout produces mouse model of hereditary hemochromatosis. Proc Natl Acad Sci U S A 1998;95:2492–2497.
27. Lebron JA, Bennet MJ, Vaughn DE, et al. Crystal structure of the hemochromatosis protein HFE and characterization of its interaction with transferrin receptor. Cell 1998;93:111–123.
28. Feder JN, Tsuchihashi Z, Irrinki A, et al. The hemochromatosis founder mutation in HLA-H disrupts beta2-microglobulin interaction and cell surface expression. J Biol Chem 1997;272:14025–14028.
29. Mura C, Raguenes O. HFE mutations analysis in 711 hemochromatosis probands: Evidence for S65C implication in mild form of hemochromatosis. Blood 1999;93:2502–2505.
30. Pointon JJ, Wallace D, Merryweather-Clarke AT, et al. Uncommon mutations and polymorphisms in the hemochromatosis gene. Genetic Testing 2000;4:151–161.
31. Jeffrey GP, Chakrabarti S, Hegele RA, et al. Polymorphism in intron 4 of HFE may cause overestimation of C282Y homozygote prevalence in haemochromatosis. Nat Genet 1999;22:325–326.
32. Sommerville MJ, Sprysak KA, Hicks M, et al. An HFE intronic variant promotes misdiagnosis of hereditary hemochromatosis. Letter. Am J Hum Genet 1999;65:924–926.
33. Noll WW, Belloni DR, Stenzel TT, et al. Polymorphism in intron 4 of HFE does not compromise haemochromatosis mutation results. Nat Genet 1999;23:271–272.
34. Elsea SH, Leykam V. HFE polymorphism and accurate diagnosis of C282Y hereditary hemochromatosis carriers. Blood 2000;95:2453–2455.
35. Beutler E, Gelbart T. A common intron 3 mutation (IVS3-48c → g) leads to misdiagnosis of the c.845G → A (C282Y) HFE gene mutation. Blood Cells Mol Dis 2000; 26:229–233.
36. Barton JC, Shih WW, Sawada-Hirai R, et al. Genetic and clinical description of hemochromatosis probands and heterozygotes: evidence that multiple genes linked to the major histocompatibility complex are responsible for hemochromatosis. Blood Cells Mol Dis 1997;23:135–145.
37. Moirand R, Jouanolle AM, Brissot P, et al. Phenotypic expression of HFE mutations: a French study of 1110 unrelated iron-overloaded patients and relatives. Gastroenterology 1999;116:372–377.
38. Brissot P, Moirand R, Jouanolle AM, et al. A genotypic study of 217 unrelated probands diagnosed as "genetic hemochromatosis" on "classical" phenotypic criteria. J Hepatol 1999;30:588–593.

39. Whitfield JB, Cullen LM, Jazwinska EC, et al. Effects of HFE C282Y and H63D polymorphisms and polygenic background on iron stores in a large community sample of twins. Am J Hum Genet 2000;66:1246–1258.

40. Levy JE, Montross LK, Andrews NC. Genes that modify the hemochromatosis phenotype in mice. J Clin Invest 2000;105:1209–1216.

41. Bulaj ZJ, Griffen LM, Jorde LB, et al. Clinical and biochemical abnormalities in people heterozygous for hemochromatosis. N Engl J Med 1996;335:1799–1805.

42. Edwards CQ, Skolnick MH, Kushner JP. Coincidental non-transfusional iron overload and thalassemia minor: association with HLA-linked hemochromatosis. Blood 1981; 58:844–848.

43. Edwards CQ, Griffen LM, Goldgar DE, et al. HLA-linked hemochromatosis alleles in sporadic porphyria cutanea tarda. Gastroenterology 1989;97:972–981.

44. Piperno A, Sampietro M, Pietrangelo A, et al. Heterogeneity of hemochromatosis in Italy. Gastroenterology 1998;114:996–1002.

45. Adams PC, Deugnier Y, Moirand R, et al. The relationship between iron overload, clinical symptoms, and age in 410 patients with genetic hemochromatosis. Hepatology 1997;25:162–166.

46. Niederau C, Niederau CM, Lange S, et al. Screening for hemochromatosis and iron deficiency in employees and primary care patients in Western Germany. Ann Intern Med 1998;128:337–345.

47. Barton JC, McDonnell SM, Adams PC, et al. Management of hemochromatosis. Hemochromatosis Management Working Group. Ann Intern Med 1998;129:932–939.

48. Burke W, Press N, McDonnell SM. Hemochromatosis: genetics helps to define a multifactorial disease. Clin Genet 1998;54:1–9.

49. Bonkovsky HL, Poh-Fitzpatrick M, Pimstone N, et al. Porphyria cutanea tarda, hepatitis C, and HFE gene mutations in North America. Hepatology 1998;27:1661–1669.

50. Adams PC, Gregor JC, Kertesz AE, et al. Screening blood donors for hereditary hemochromatosis: decision analysis model based on a 30-year database. Gastroenterology 1995;109:177–188.

51. Hudson KL, Rothenberg KH, Andrews LB, et al. Genetic discrimination and health insurance: an urgent need for reform. Science 1995;270:391–393.

52. Lapham EV, Kozma C, Weiss JO. Genetic discrimination: perspectives of consumers. Science 1996;274:621–624.

53. Rothenberg K, Fuller B, Rothstien M et al. Genetic information and the workplace: legislative approaches and policy challenges. Science 1997:275:1755–1757.

54. Holtzman NA, Watson MS (eds). Task Force on Genetic Testing. Promoting Safe and Effective Genetic Testing in the United States: Final Report. Baltimore, MD: Johns Hopkins University Press, 1998.

55. Alper JS, Geller LN, Barash CI et al. Genetic discrimination and screening for hemochromatosis. J Public Health Policy 1994;15:345–358.

56. Markel H. The stigma of disease: implications of genetic screening. Am J Med 1992; 93: 209–215.

57. McDonnell SM, Preston BL, Jewell SA, et al. A survey of 2,851 patients with hemochromatosis: symptoms and response to treatment. Am J Med 1999;106:619–624.

58. Kowdley KV, Trainer TD, Saltzman JR, et al. Utility of hepatic iron index in American patients with hereditary hemochromatosis: a multicenter study. Gastroenterology 1997;113:1270–1277.

59. Balan V, Baldus W, Fairbanks V, et al. Screening for hemochromatosis: a cost-effectiveness study based on 12,258 patients. Gastroenterology 1994;107:453–459.

60. Phatak PD, Guzman G, Woll JE, et al. Cost-effectiveness of screening for hereditary he-
    mochromatosis. Arch Intern Med 1994;154:769–776.
61. Burke W, Thomson E, Khoury M, et al. Hereditary hemochromatosis: gene discovery
    and its implications for population-based screening. JAMA 1998;280:172–178.
62. EALS International Consensus Conference on Haemochromatosis. Jury Document. J He-
    patol 2000;33:496–504.

29

Genetic testing of railroad track workers with carpal tunnel syndrome

Paul A. Schulte and Geoffrey Lomax

Carpal tunnel syndrome (CTS) is the most common peripheral compression neuropathy with an estimated prevalence of 2.1%. It has a multifactorial etiology involving systemic, anatomic, idiopathic, and ergonomic factors (1–5). Work-related activities have been strongly associated with CTS (4–6). Some of the highest rates of CTS occur in occupations with high work demand or extensive manual exertion such as automobile assembly and meat processing. CTS also occurs in individuals with various health conditions such as rheumatoid arthritis, thyroid disease, diabetes, late pregnancy, and rapid weight loss. CTS often occurs as a result of two hereditary conditions: hereditary neuropathy with liability to pressure palsies (HNPP) and familial amyloidotic polyneuropathy (FAP). Occupations and medical conditions associated with CTS are shown in Tables 29.1 and 29.2, respectively.

In 2001, the Burlington Northern Santa Fe Railroad Co. settled Equal Employment Opportunity Commission (EEOC) and union lawsuits over the company's genetic testing of approximately 20 railroad track workers with CTS on-the-job injury reports or compensation claims (7). The testing was performed, without the knowledge and consent of workers, to detect mutations and deletions associated with HNPP and FAP. However, the extent to which the railroad company's actions were within the scope and practice of the Federal Employers' Liability Act or other workers' compensation statutes or were subject to the Americans with Disabilities Act prohibitions against medical examinations of current employees was not adjudicated. It is not the function of this chapter to address the ethical and legal issues. Rather, beyond the ethical and legal issues two scientific questions arose: are genetic risks likely to be important in these cases of CTS; and is there a scientific rationale for testing these workers? In this chapter, those questions are addressed.

Table 29.1 Some Occupations Associated with Carpal Tunnel Syndrome

Meat packers	Platers
Sewing machine operators	Frozen food factory workers
Ski manufacturers	Packaging machine workers
Poultry processors	Electrical component assembly
Automobile assembly	Rock drillers
Grinders	Grocery checkers

Source: NIOSH, 1997.

Background

CTS is a compression neuropathy of the median nerve of the wrist (14). In the United States, CTS occurs with an estimated prevalence of 2.1% and an incidence of 3.46 cases per 1000 person–years in the general population (8,9). A large percentage of these cases are work related (5,10). The costs associated with CTS are estimated to be more than $2 billion per year (11,12). CTS constitutes 3% of all Workers' Compensation Insurance claims. The costs are often higher than the average claim filed under workers' compensation. CTS results in one of the largest numbers of lost workdays among occupational conditions (13).

In many cases, the cause of CTS is unknown, and this is referred to as idiopathic CTS. However, there is strong evidence of a positive association between exposure to a combination of risk factors (e.g., force and repetition and/or force and posture) and CTS (15). CTS has been associated with numerous medical conditions (Table 29.2) such as rheumatoid arthritis, thyroid disease, diabetes, and late pregnancy (16), although many of the studies on which such a list is based did not control for occupation. CTS is not known to be a predominantly genetic condition; however, two hereditary diseases, HNPP and FAP are known to exhibit CTS in some cases. There are practically no epidemiologic studies that have assessed both environmental and genetic risk factors for CTS in the same study.

Table 29.2 Medical Conditions Associated with Carpal Tunnel Syndrome

Anatomic	Inflammatory
Ganglion	Tenosynovitis
Neuroma	Hypertrophic synovium
Lipoma	Rheumatoid arthritis
Myeloma	Gout
Neuropathic	Dermatomyositis
Diabetes	Scleroderma
Alcoholism	Systemic lupus erythematosus
Amyloidosis	Alteration of fluid balance
	Pregnancy
	Myxedema
	Obesity
	Long-term hemodialysis

Source: Adapted from Sternbach (1999).

Genetic testing for HNPP was part of a protocol that the railroad company used to evaluate workers who reported an on-the-job injury (17,18). The test for FAP involved detection of variants of the protein, transthyretin (TTR), and was added to the test battery by the laboratory performing the assays. Ultimately, the protocol was applied to approximately 20 workers (males) who had gone on to file injury reports—compensation claims for work-related CTS under the Federal Employers Liability Act. The company indicated that the testing was performed to assist the company medical officer to determine whether CTS was related to work or some other nonwork factor including a genetic disorder (19). In 2001, the U.S. Equal Employment Opportunity Commission filed suit against the employer and ultimately achieved an agreement that required the employer to stop further genetic testing and refrain from using any of the testing information obtained (7,18).

Hereditary Neuropathy with Liability to Pressure Palsies

CTS is a common manifestation of HNPP, which is part of a heterogeneous group of demyelinating polyneuropathies. HNPP generally develops during adolescence (20). HNPP was first reported in 1947 in a family digging potatoes (21). It was also known as "bulb diggers" palsy. Although there have been a number of case reports since 1947, the first population-based study of HNPP was not published until 1997 (22). The prevalence of HNPP was evaluated in 69 patients from 23 unrelated families (diagnosed between 1978 and 1995) in a population of 435,000 in southwestern Finland through family and medical history, clinical, neurologic,·and neurophysiologic examinations and with documentation of a gene deletion (17p11.2). The prevalence of HNPP was estimated to be 16 in 100,000 (22).

In 1993, the genetic locus for HNPP was mapped to chromosome17p11.2-12, where it is often associated with a large 1.5 Mb (megabase pair) DNA deletion (20,23). In a study of 156 unrelated HNPP patients, 84% were found to have the 17p11.2 deletion and in 4.6% (6/131) of these, the deletion was de novo (24). However, higher frequencies of de novo deletions (25.6%) have been reported (25,26). The deletion of 17p11.2 appeared to be a reciprocal product of an unequal crossover involving Charcot-Marie-Tooth neuropathy type 1 (CMT1A) (23). The deletion at 17p11.2 includes the gene for peripheral myelin protein-22 (PMP22) (23,27). The gene for PMP22 spans approximately 40kb, and 27 distinct mutations have been identified in 35 unrelated patients (28). Of the 27 mutations, four were associated with HNPP phenotypes. Subsequently, investigators found deletions of 17p11.2 in patients in various countries (27,29–33).

The onset of HNPP is usually in childhood or adolescence. The clinical presentation of HNPP is broad and may range from clinically asymptomatic persons to those who present with recurrent palsies (20). HNPP usually occurs in the setting of a family history, indicating an autosomal dominant trait; however, sporadic cases have been described (24,34,35). Prior to 2001, the percentage of cases due to de novo deletion was not known. In 2001, a study of 14 consecutive unrelated index cases found that 3 (21%) were sporadic cases, due to a de novo deletion of 17p11.2

(25). There appears to be a direct relationship of gene dosage at the PMP22 locus with the phenotype palsies (20).

Animal models appear to exist for HNPP. Maycox et al. (36) found that transgenic mice expressing antisense PMP22 RNA showed modestly reduced levels of PMP22 together with a phenotype suggestive of HNPP. A striking movement disorder and a slowing of nerve conduction that worsened with age were observed in antisense homozygotes. Histologically, a subset of axons had thickened myelin sheaths and tomacula in young adult mice; significant myelin degeneration was observed in older animals (37). Nelis et al. (28) concluded that heterozygous mice correspond to HNPP patients with the 1.5 Mb deletion on 17p11.2. In both human and mouse, the myelin tomaculae are present, suggesting that the mice are useful animal models for HNPP (28).

Familial Amyloidotic Polyneuropathy

In addition to testing for PMP22, the laboratory marketing the genetic assay also used an assay for transthyretin (TTR), a plasma protein, on railroad track workers' blood specimens (38). Mutations in this plasma protein are observed in FAP. TTR is encoded by single gene on chromosome 18 (18q11.2-q12.1), of which more than 70 autosomal dominantly inherited point mutations, occurring at 51 different sites, have been described (39–42). The hereditary amyloidoses have been classified into four subtypes, two of which, familial amyloid polyneuropathy type 1 and 2 (FAP1 and FAP2) are associated with CTS. The incidence of TTR amyloidosis is unknown (43). Amyloidosis occurs in about 8 of 1,000,000 people. At diagnosis, FAP1 patients often have undergone carpal tunnel surgery. In FAP2 patients, there is more frequent carpal tunnel manifestation (39). In the United States, the gene frequency for FAP1 and FAP2 is estimated to be 1 in 1,000,000 to 1 in 100,000 (44). Hence, the prevalence of CTS in people with FAP1 and FAP2 would be less than the gene frequency, because not all the people with the gene develop CTS.

Epidemiologic Findings

The epidemiology of CTS can be seen to have two focal areas: studies of occupational risk factors and studies of nonoccupational risk factors (generally involving arthritis, obesity, pregnancy, diabetes, thyroid disease, hormone replacement, and corticosteroid therapy) (45–50). Genetic factors (other than race and gender) generally have not been included in epidemiologic studies of CTS (50). However, there have been a few studies of familial occurrence (50,51).

One issue in the evaluation of epidemiologic studies has been the definition of carpal tunnel syndrome. There are differing opinions about what constitutes appropriate diagnostic criteria (1,52,53). The debate focuses on whether electrodiagnostic study findings alone or with symptoms is the best criterion and, if in the absence of electrodiagnostic studies, specific combinations of symptom characteristics and physical examination findings are useful.

Occupational Studies

In a comprehensive review of more than 30 epidemiologic studies in the scientific literature in 1997, the National Institute for Occupational Safety and Health (15) concluded, and a committee of the National Academy of Sciences (54) subsequently confirmed, that there is strong evidence that a combination of workplace physical risk factors is associated with CTS. These included force and repetition and force and posture. Literature published since 1997 further supports this finding (5,45,55–57). Many of the studies in which a statistically significant association between individual or combinations of workplace physical factors was found controlled for potential confounders, such as age, sex, smoking, caffeine, alcohol, hobbies, body mass index (BMI), and medical conditions (8,15,55,58–61). Epidemiologic surveillance, nationally and internationally, has consistently indicated that the highest rates of CTS occur in occupations and job tasks with high work demands or extensive manual exertion (such as meat processors, poultry processors, and automobile assembly workers) (15); see Table 29.1 for some occupations associated with CTS. The prevalence of diagnosed carpal tunnel syndrome in U.S. workers has been estimated to be 53 per 10,000 (4).

Epidemiologic studies consistently link CTS to job tasks that involve a combination of risk factors (e.g., force and repetition, force and posture) (15). Railroad track maintenance is physically demanding and involves extreme manual exertion— the use of jackhammers and grinders, in some cases, for up to 14 hours a day. In one report, arising from tort litigation, that conflicts with much of the other literature, 900 railroad workers were randomly selected from a pool of 2500 Federal Employers' Liability Act (FELA) claimants, and their jobs were characterized into four categories previously described by Silverstein et al. (62). Jobs were classified into the following categories: I (low force/low repetition), 18.8%; category II (low force/high repetition), 20.7%; category III (high force/low repetition), 59.8%; category IV (high force/high repetition), 0.7% (14). The percentage of positive or borderline cases of CTS among all workers in each category (determined by an electrodiagnostic method of comparing median minus ulnar nerve digital latency differential) was as follows: category I, 47.9%; category II, 40.0%; category III, 43.4% and category IV, 50% (14). The report showed no association between occupational classification and CTS, but the investigators indicated that the small number of individuals in the most extreme job classification category IV (0.7%) may have obscured an occupational association. In other studies, it has been estimated that as much as 50% of all medically treated CTS is work-related (4,5,63). There is no published research that shows what portion of CTS is attributable to genetic factors. It is possible that gene–environment interactions are involved in the etiology of some CTS cases in the general population, but there are no data on this.

Nonoccupational Studies

Table 29.2 shows various medical conditions that have been associated with CTS. Additionally, older age and female gender have been identified as risk factors. Women in the general population have approximately three times the risk as men; men (8) and women in the same occupation appear to have similar risks (64). The strongest risk factor in one study of women was previous history of another musculoskeletal complaint for which consultation had been sought OR 1.98 (95% CI, 1.61–2.42). Autonomic disturbances have also been found in one study of CTS with increasing severity of electrophysiologic findings (65).

Familial CTS

Familial occurrence of CTS also has been documented (51,66–69). When carpal tunnel syndrome is inherited, it is often the manifestation of a systemic disease (70). Gossett and Chance (69) concluded that, in addition to linkage with familial amyloidosis and HNPP, patients may present with a familial CTS (McKusick number 115,430) (71). This familial CTS appears to be a rare but genetically distinct disorder. In a prospective study, a positive family history was predictive in 39.3% of cases with surgery for CTS who had median nerve slowing versus 13.3% of cases without these characteristics (51).

Risk Attributable to Genetic Factors

At the time of the testing, there were no published population data identified that confirmed that a genetic factor could explain the risk of CTS better than physical activities. In 2001, a study was published that examined 50 unrelated patients (age 18–76, mean age 50.5 years) diagnosed with CTS, all in need of surgical release, and none were found to have PMP22 deletions (72). Diagnosis of CTS was made by both clinical evaluation and electrodiagnostic methods. Exclusion criteria consisted of those with anatomic changes decreasing the available volume within the carpal tunnel; diagnoses that may result in the increased size of the carpal canal contents (including amyloidoses, rheumatoid arthritis, or edema); those diagnoses associated with soft tissue impingement (i.e., lipomas, hematomas, or urate crystal deposition); and other causes of peripheral mononeuropathies, such as diabetes mellitus. The authors calculated, based on their findings of no deletions, that the upper limit (95% confidence interval) of the prevalence of PMP22 deletion as a cause of CTS is approximately 6% (72). This analysis appears to assume a binomial distribution and uses a one-sided confidence interval that includes all 5% in the upper bound instead of 2.5%. The authors (72) concluded that the prevalence of HNPP in idiopathic CTS is unknown, but using estimated CTS incidences of 1% to 3.8% (8,10,73) and HNPP (0.04%) (74), they calculated that HNPP could be responsible for 1% to 4% of CTS. DNA-based testing for individuals with a negative family history for HNPP had been suggested (75,76) prior to the testing of railroad workers in 2000. The railroad company relied on the medical literature (e.g., 3,75) to support the contention that multiple causes of CTS, including HNPP, needed to be evaluated before determining work-relatedness.

Scientific Issues of Concern

The critical factors for evaluating predictive genetic tests are sensitivity, specificity, and predictive value (77) (see also Chapter 11). Other factors such as number needed to screen (NNS) and number needed to treat (NNT) are also useful. To assess these parameters, it is necessary to know the CTS risk for people with and without the genetic variant and particular exposure profiles. Prior to the testing of railroad track workers, no prospective studies have been published that assess the risk of CTS in people with 17p11.2 deletion, or TTR mutation. Until 2002, there were no population studies that assessed the relative risk or population attributable risk for genetic factors for CTS. In 2002, a study of twin women from the U.K. Adult Twin Registry (50) determined the relative genetic and environmental contributions to CTS. The genetic contribution was assessed using the variance component and regression methods and the heritability was adjusted for environmental confounders. The modeling resulted in a heritability estimate of 0.46 (95% CI, 0.34–0.58). The investigators reported that the study may have lacked the power to demonstrate that occupation in clerical and manual employment was a risk factor, since there were a small number of cases in these groups (50). There are no published data to establish the validity of susceptibility testing for CTS. Specifically, neither the positive nor negative predictive value has been established. The lack of data regarding the technical performance of the test and risk associated with a positive result would be a basis for rejecting the use of genetic testing except in the context of a research study. Further, in other examples where there is quantifiable risk corresponding with a genetic factor and occupational disease, authors have argued against screening on the basis of inadequate validation and poor predictive value (77–79).

For retrospective testing, this question then becomes analogous to the questions asked in historical cohort studies or in case-control studies. Like all cohort studies, historical cohort studies involve following groups with and without an exposure characteristic forward to determine if the risk for a health outcome is different in the two groups. The difference with historical cohort studies is that the start date of the study is in the past and determined retrospectively. In the case of the railroad workers, the question would be whether those with the 17p11.2 deletion or the TTR variant have a greater risk than those who do not. In a historical cohort study this would be assessed by the risk ratio. In a case-control study, cases of CTS and selected individuals without CTS would be cross-classified on the basis of the 17p11.2 deletion or TTR mutation. The association between the genetic variant and the disease would be assessed by using the odds ratio.

A search of the scientific literature prior to 2000 did not identify any studies of the risk ratio or odds ratio of 17p11.2 deletions or transthyretin mutations. The testing of approximately 20 railroad workers was not conducted as part of a case-control or prospective study, so there was no opportunity to calculate risk or odds ratios. The application of genetic tests to some cases, but not others, apparently was not defined in any identified research protocol or experimental design. The rationale provided by the company was that a case management protocol of CTS cases was developed and it included genetic testing of some workers. The prevalence of HNPP and FAP1 and FAP2 are believed to be relatively rare on a population basis, so the likelihood of finding a genetic variant in 20, or even 150 workers is very low. It is not known whether a person with either of these conditions would be likely to be long-term railroad track worker. The extensive physical demands of the job could lead workers who were subject to self-limited episodes of peripheral neuropathy to seek other employment. However, people with HNPP can have mild or no symptoms, and thus their condition could have little effect on their ability to work (25).

Potential Contribution of Genetic Information to Improved Health Outcomes

In the railroad case, the use of genetic testing was not for the purpose of improving health outcomes but for clarifying the contributing factors to CTS in workers' compensation claimants. Medical tests, including genetic tests, may provide pertinent information about nonwork factors that may contribute to causing CTS. Employers have a legal right to identify work and nonwork factors that contribute to CTS and thereby attempt to apportion causation. Under workers' compensation statutes, this can be done in some states without informed consent about the specific tests. In contrast, the conduct of genetic tests without informed consent is not condoned in any guidelines for genetic testing and, in fact, conflicts with the Guidelines of the Task Force on Genetic Testing (80) discussed later. The example, described in this paper, where workers were not informed that they were tested, reinforces concerns that genetic testing will be used to discriminate against or otherwise disadvantage workers.

In some situations, prospective genetic testing for PMP22 or TTR could benefit workers by providing them information with which to make employment decisions. However, at present the lack of information on predictive value and attributable risk does not support such use.

Conclusions

A review of this case indicates that neither the scientific basis nor validity of the PMP22 or TTR assay for CTS were adequately established before their use on railroad workers in 2000. The prevalences of HNPP and FAP are exceedingly low and unlikely to be a major contributory cause of work-related CTS. There are few data on the frequency of these variant genotypes in the population. The plan to use testing for these traits in the evaluation of railroad track workers with CTS is striking given the absence of evidence required to assess the use of the test in a workplace setting (e.g., absence of a database) to identify the role of genetic factors in CTS. There is no information indicating that equally exposed workers, with and without various genotypes, are at different risks of CTS. What data are available suggest that genetic factors play a very minor role, if any, in male railroad track workers. Ultimately, some genetic factors may be found that contribute along with occupational factors to CTS, but such information is not available at this time.

The role of genetic information in workers' compensation is an evolving question. Past practice has been that work-related disability could be generally compensated even when the source of the preexisting condition is not work related (81). There is no consistent record that demonstrates that the existence of genetic variants alone serve as pre-existing conditions; however, genetic information has been used in workers' compensation cases (82,83). State laws governing workers' compensation may provide an incentive for employers to use genetic screening tests. Iowa, New Hampshire, New York, and Wisconsin allow for consensual genetic testing for purposes of investigating workers' compensation claims. However, since the predictive value of PMP22 and TTR for CTS has not been demonstrated, these genotypes are not useful for retrospective assessment of causality in occupational populations. Before PMP22 and TTR variants could be viewed as pre-existing conditions for CTS, extensive information would be needed. This includes: (*1*) the frequency of the variants, (*2*) the absolute and relative risk of the association of variants of HNPP and FAP, respectively, (*3*) the frequency with which HNPP and FAP are related to CTS, (*4*) the predictive value of the tests, (*5*) the interaction between work related factors and genetic factors in the risk for CTS, and (*6*) the factors influencing the penetrance of the genetic factors in HNPP and FAP. Almost all of this information is lacking. In the interim, guidance is available from the Task Force on Genetic Testing (84). The Task Force concluded that in regard to genetic testing, four features are important:

- Assessment of validity of test is necessary before use.
- Formal validation is needed for each intended use of a genetic test.
- Data to establish clinical validity must be collected under investigative protocols.
- Investigative protocols for validation of genetic tests need IRB approval (80).

Without the validation information, the mere existence of genetic characteristic is not an indication of the nature of its role in multifactorial diseases such as CTS. Technologic advances in detection have outrun the ability to interpret and use the information obtained. Until appropriate interpretive research is conducted, use of genetic tests (for PMP 22 deletion and TTR mutations) to impute causality in railroad track workers with CTS claims is not warranted.

The pressure to use genetic tests without population validation appears to be increasing. In the case described here, premature testing was based on a presumption of informativeness of the test that was unsupported by data (85). In order to enhance the utility of genetic tests, it would be helpful if those conducting such tests would provide information on the prevalence of the genetic trait, the predictive value, and other information about the test's validity. When information on any of these test properties is lacking, the evidence gap should be disclosed. This information could be useful to decisionmakers considering genetic testing. Such testing efforts still require, as prerequisite, attention to the ethical, social, and legal protections that have been advanced by the Task Force on Genetic Testing and various professional organizations (86,87).

References

1. Hadler NM. Occupational Musculoskeletal Disorders, second edition. Philadelphia, PA: Lippincott, Williams, and Wilkins, 1999.
2. Sternbach G. The carpal tunnel syndrome. J Emerg Med 1999;17:519–523.
3. Derebery VJ. Determining the cause of upper extremity complaints in the workplace. Review. Occup Med 1998;13:569–582.
4. Tanaka S, Wild DK, Seligman PJ, et al. Prevalence of work-relatedness of self-reported carpal tunnel syndrome among U.S. workers: analysis of the occupational health supplement data of 1988 National Health Interview Survey 1988 National Health Interview Survey Data. Am J Ind Med 1995;27:451–470.
5. Davis L, Wellman H, Punnett L. Surveillance of work-related carpal tunnel syndrome in Massachusetts 1992–1997: A report from the Massachusetts Sentinel Event Notification System for Occupational Risks (SENSOR). Am J Ind Med 2001;39:58–71.
6. Hagberg M, Morgenstern H, Kelsh M. Impact of occupations and job tasks on the prevalence of carpal tunnel syndrome. Scan J Work Environ Health 1992;18:337–345.
7. EEOC settles ADA suit against BNSF for genetic bias. http://www.eeoc.gov/press/4-18-01.html. Accessed November 30, 2001.
8. Atroshi I, Gummesson C, Johnsson R, et al. Prevalence of carpal tunnel syndrome in a general population. JAMA 1999;282:153–158.
9. Nordstrom DL, DeStefano F, Vierkant RF, et al. Incidence of diagnosed carpal tunnel syndrome in a general population. Epidemiology 1998;9:342–345.
10. Tanaka S, Wild DK, Seligman PJ, et al. The US prevalence of self-reported carpal tunnel syndrome: 1988 National Health Interview Survey data. Am J Public Health 1994; 84(11):1846–1848.
11. Levine DW, Simmons BP, Koris MJ, et al. A self-administered questionnaire for the assessment of severity of the symptoms and functional status of Carpal Tunnel Syndrome. J Bone Joint Surg Am 1993;75:1585–1592.
12. Palmer DH, Hanrahan LP. Social and economic costs of carpal tunnel surgery. Instr Course Lect 1995;44:167–172.
13. Kish J, Dobrila V. Carpal tunnel syndrome in workers compensation: frequency, costs and claim characteristics. National Council on Compensation Insurance, Inc. Research Brief 1996;3(3):1–11.
14. Cosgrove JL, Chase PM, Mask NJ, et al. Carpal tunnel syndrome in railroad workers. Am J Phys Med Rehabit 2002;81:101–107.

15. NIOSH. Musculoskeletal disorders and workplace factors. A critical review of epidemiologic evidence for work-related musculoskeletal disorders of the neck, upper extremity, and low back. 1997. U.S. Department of Health and Human Services DHHS (NIOSH) Publication No. 97–141.

16. Dawson DM. Entrapment neuropathies of the upper extremities. N Engl J Med 1993; 329:2013–2018.

17. U.S. District Court: EEOC v Burlington N. Santa Fe Railway Company, Civ No 01-4013 MWB, (N.D. Iowa, February 8, 2001) (EEOC's Memorandum in support of petition for a preliminary injunction).

18. U.S. District Court: for the Eastern District of Wisconsin Equal Opportunity Commission v Burlington N. Santa Fe Railway Company, Civ No 02-C-0456, May 6, 2002.

19. Rose M. Letter to all employees from BNSF President and CEP Matt Rose. BNSF Today. http://www.busf.com/media/articles/2001/03/2001-03-0-a.html. Accessed August 22, 2001.

20. Chance PF. Overview of hereditary neuropathy with liability to pressure palsies. Ann NY Acad Sci 1999;883:14–21.

21. DeJong JSY. Over families met hereditarie disposite tot het optreten van neuritiden gecorreleard met migraine. Psychiat Neurol Bl 1947;50:60–76.

22. Meretoja P, Silander K, Kalimo H, et al. Epidemiology of hereditory neuropathy with liability to pressure palsies (HNPP) in south western Finland. Neuromuscular Disorders 1997;7:529–532.

23. Chance PF, Alderson MK, Leppig KA, et al. DNA deletion associated with hereditary neuropathy with liability to pressure palsies. Cell 1993;72:143–151.

24. Nelis E, Van Broeckhoven C, De Jonghe P, et al. Estimation of the mutation frequencies in Charcot-Marie-Tooth disease type: 1 and hereditary neuropathy with liability to pressure palsies: a European collaborative study. Eur J Hum Genet 1996;4:25–33.

25. Infante J, Garcia A, Combarros O, et al. Diagnostic strategy for familial and sporadic cases of neuropathy associated with 17p11.2 deletion. Muscle Nerve 2001;24:1149–1155.

26. Bort S, Martinez F, Palau F. Prevalence and parental origin of de novo 1.5-Mb duplication in Charcot-Marie-Tooth disease type 1A. Am J Hum Genet 1997;60:230–233.

27. Mariman ECM, Gabreels-Festen AAWM, Van Beersum SEC, et al. Prevalence of the 1.5-mb 17p deletion in families with hereditary neuropathy with liability to pressure palsies. Ann Neurol 1994;36:650–655.

28. Nelis E, Haites V, Van Broeckhoven C. Mutations in the peripheral myelin genes and associated genes in inherited peripheral neuropathies. Human Mutat 1999;13:11–28.

29. Silander K, Halonen P, Sara R, et al. DNA analysis in Finnish patients with hereditary neuropathy with liability to pressure palsies (HNPP). J Neurol Neurosurg Psychiat 1994; 57:1260–1262.

30. Umehara F, Kiwaki T, Yoshikawa IT, et al. Deletion in chromosome 17p11.2 including peripheral myelin protein-22 (PMP22) gene in hereditary neuropathy with liability to pressure palsies. J Neurol Sci 1995;133(1–2):173–176.

31. Gonnaud PM, Sturtz F, Fourbil Y, et al. DNA analysis as a tool to confirm the diagnosis of asymptomatic hereditary neuropathy with liability to pressure palsies (HNPP) with further evidence for the occurrence of de novo mutations. Acta Neurol Scand 1995;92:313–318.

32. Sessa M, Nemni R, Quattrini A, et al. Atypical hereditary neuropathy with liability to pressure palsies (HNPP): the value of direct DNA diagnosis. Med Genet 1997;34:889–892.

33. Lenssen PPA, Gabreels-Festen AAWN, Valentijn LJ, et al. Hereditary neuropathy with liability to pressure palsies. Phenotypic differences between patients with the common deletion and a PMP22 frame shift mutation. Brain 1998;121:1451–1458.

34. Lopes J, Ravise N, Vandenbergher A, et al. Fine mapping of de novo CMT1A and HNPP rearrangements within CMT1A-REPs evidences two distinct sex-dependent mechanisms and candidate sequences involved in recombination. Hum Mol Genet 1998;7:141–8.

35. Verhalle D, Löfgren A, Nelis E, et al. Deletion in the CMT1A locus on chromosome 17p11.2 in hereditary neuropathy with liability to pressure palsies. Ann Neurol 1994; 35:704–708.

36. Maycox PR, Ortuno D, Burrola P, et al. A transgenic mouse model for human hereditary neuropathy with liability to pressure palsies. Molec Cell Neuro Sci 1997;8:405–416.

37. OMIM 162500. Neuropathy, hereditary with liability to pressure palsies; HNPP. Accessed June 6, 2002, from http://www.ncbi.nlm.nih.gov/entrez/query.fcgi?db=omim.

38. Athena Diagnostics product tests–test descriptions. Accessed February 13, 2001, from http://www.athenadiagnostics.com/site/product_search/test_description_template.asp?id =152.

39. Hund E, Linke RP, Willig F, et al. Transthyretin-associated neuropathic amyloidosis. Pathogenesis and treatment. Neurology 2001;56(4):431–435.

40. Buxbaum JW, Tagoe CE. The genetics of the amyloidoses. Ann Rev Med 2000;51: 543–569.

41. Izumoto S, Younger D, Hays AP, et al. Familial amyloidotic polyneuropathy presenting with carpal tunnel syndrome and a new transthyretin mutation, asparagine 70. Neurology 1992;42(11):2094–2102.

42. Saraiva MJM. Transthyretin mutations in health and disease. Hum Mutat 1995;5:191–196.

43. Falk RH, Comenzo RL, Skinner M. The systemic amyloidoses. New Engl J Med 1997;337:898–909.

44. Benson MD. Amyloidosis. In Scriver CR, Beaudet AL, Sly WS, Valle D eds. The metabolic and molecular basis of inherited disease. New York, NY, McGraw-Hill. 1995;4159–4191.

45. Giersiepen K, Eberle A, Pohlabein H. Gender differences in carpal tunnel syndrome? Occupational and non-occupational risk factors in a population-based case-control study. Ann Epidemiol 2000;10:481.

46. Tanaka S, Wild DK, Cameron LL, et al. Association of occupational and nonoccupational risk factors with prevalence of self-reported carpal tunnel syndrome in a national survey of the working population. Am J Ind Med 1997;32:550–556.

47. Nordstrom DL, Vierkant RA, DeStefano F, et al. Risk factors for carpal tunnel syndrome in a general population. Occup Environ Med 1997;54:734–740.

48. Solomon DH, Katz JN, Bohn R, et al. Nonoccupational risk factors for carpal tunnel syndrome. J Gen Intern Med. 1999;14:310–314.

49. Ferry S, Hannaford P, Warskyj M, et al. Carpal tunnel syndrome: a nested case-control study of risk factors in women. Am J Epidemiol. 2000;151:566–574.

50. Hakim AJ, Cherkas L, El Zayat S, et al. The genetic contribution to carpal tunnel syndrome in women: a twin study, Arthritis Rheum 2002;47:275–279.

51. Radecki P. The familial occurrence of carpal tunnel syndrome. Muscle Nerve 1994;17:325–330.

52. Nathan PA, Meadows KD. Neuromusculoskeletal conditions of the upper extremity: are they due to repetitive trauma? Occup Med. 2000;15:677–693.

53. Rempel D, Evanoff B, Amadio PC, et al. Consensus criteria for classification of carpal tunnel syndrome in epidemiologic studies. Am J Public Health 1998;88:1447–1451.

54. National Academy of Sciences. Work-related musculoskeletal disorders. Report, workshop, summary, and workshop papers. Washington, DC: National Academy of Sciences, National Research Council. 1999. [OSHA Exhibit No. 26–37].

55. Leclerc A, Franchi P, Cristofari MF, et al. Carpal tunnel syndrome and work organisation in repetitive work: a cross-sectional study in France. Occup Environ Med 1998;55:180–187.

56. Lalumandier JA, McPhee SD. Prevalence and risk factors of hand problems and carpal tunnel syndrome among dental hygienists. J Dent Hyg 2001;75:130–134.

57. Yagev Y, Carel RS, Yagev K. Assessment of work-related risks for carpal tunnel syndrome. Isr Med Assoc J 2001;3:569–571.

58. Frost P, Anderson JH, Nielson VK. Occurrence of carpal tunnel syndrome among slaughterhouse workers. Scand J Work Environ Health. 1998;24(4):285–292.

59. Latko WA, Armstrong TJ, Franzblau A, et al. Cross-sectional study of the relationship between repetitive work and the prevalence of upper limb musculoskeletal disorders. Am J Ind Med. 1999;36:248–259.

60. Gorsche RG, Wiley JP, Renger RF, et al. Prevalence and incidence of carpal tunnel syndrome in a meat packing plant. Occup Environ Med 1999;56:417–422.

61. Rossignol M, Stock S, Patry L, et al. Carpal tunnel syndrome: what is attributable to work? The Montreal study. Occup Environ Med 1997;54:519–523.

62. Silverstein BA, Fine LJ, Armstrong TJ. Occupational factors in carpal tunnel syndrome. Am Ind Hyg Assoc J 1987;11:343–358.

63. Cummings K, Maizlish N, Rudolph L, et al. Occupational disease surveillance: carpal tunnel syndrome. MMWR Morb Mortal Wkly Rep 1989;38(28):485–489.

64. McDiarmid M, Oliver M, Ruser J, et al. Male and female rate differences in carpal tunnel syndrome injuries: personal attributes or job tasks? Environ Res 2000;8:23–32.

65. Verghese J, Galanopoulou AS, Herskovitz S. Autonomic dysfunction in idiopathic carpal tunnel syndrome. Muscle Nerve 2000;23:1209–1213.

66. Lambird PA, Hartman WH. Hereditary amyloidosis, the flex or retinaculum and the carpal tunnel syndrome. Am J Clin Pathol 1969;52:714–719.

67. Danta G. Familial carpal tunnel syndrome with onset in childhood. J Neurol Neurosurg Psychiatry 1975;38:350–355.

68. Sparkes RS, Spence MA, Gottlieb NL, et al. Genetic linkage analysis of the carpal tunnel syndrome. Hum Hered 1985;35(5):288–291.

69. Gossett JG, Chance PF. Is there a familial carpal tunnel syndrome? An evaluation and literary review. Muscle Nerve 1998;21:1533–1536.

70. Stoll C, Maitrot D,. Autosomal dominant carpal tunnel syndrome. Clin Genet 1998;54: 345–348.

71. McKusick VA. Mendelian inheritance in man, catalogs of human genes and genetic disorders. 12th edition. Baltimore, MD: Johns Hopkins University Press, 1998.

72. Stockton DW, Meade RA, Netscher DT, et al. Hereditary neuropathy with liability to pressure palsies is not a major cause of idiopathic carpal tunnel syndrome. Arch Neurol 2001;58:1635–1637.

73. Stevens JC, Sun S, Beard CM, et al. Carpal tunnel syndrome in Rochester, Minnesota, 1961–1980. Neurology 1988;38:134–138.

74. Skre H. Genetic and clinical aspects of Charcot-Marie-Tooth's disease. Clin Genet.1974;6:98–118.

75. Bird TD, Hereditary neuropathy with liability to pressure palsies. GeneReviews. Initial posting: 28 September 1998; Last revision: 27 June 2001. Accessed September 18, 2002 from http://www.geneclinics.org.

76. Tyson J, Malcolm S, Thomas PK, et al. Deletions of chromosome 17p11.2 in multifocal neuropathies. Ann Neurol 1996;39:180–186.

77. Holtzman NA. Medical and ethical issues in genetic screening—an academic view. Environ Health Perspect 1996;104S5:987–990.

78. Vineis P, Schulte P, McMichael AJ. Misconceptions about the use of genetic tests in populations. Lancet 2001;357:709–712.

79. Peto J, Houlston RS. Genetics and the common cancers. Eur J Cancer 2001;37S8: S88–S96.

African-American (*continued*)
   APOE allele frequencies, 370, 370*f*
   family aggregation, of sarcoidosis, 40
Age, *PON1* gene metabolic activity and, 308
Aging, Alzheimer disease prevalence and, 366
AHR, 166*t*
ALAD, 167*t*
Alcohol consumption
   breast cancer risk and, 452
   colorectal neoplasia and, 356
   esophageal cancer and, 150*t*
   folate metabolism and, 338
Aldosterone synthase gene, 344/T
   polymorphism, 250
ALL (acute lymphoblastic leukemia), 240–241,
   240*f,* 241*f*
Allele-disease associations, 134
Allele frequency
   ethnicity and, 212–213, 214*t*
   in family studies, 100
Allele specific oligonucleotide assay (ASO),
   207–208, 208*t*
Allele variants, 20, 22
Alzheimer disease
   age of onset, 368
   amyloid β deposition in, 369
   APOE ∈4 allele in, 367–374, 370f
   APOE gene and, 370–371
   APOE testing, 372, 373–374
   autopsy confirmation, 372
   brain changes in, 365–366
   candidate genes, 369
   definite, 365
   diagnostic criteria, 365
   diagnostic tests for, 371–372
   discrimination, 374
   environmental risk factors, 367
   frequency of, 366–377
   genetic epidemiology of, 367–369, 368*t*
   genetic factors in, 365
   genetic testing, 372
   incidence rate for, 366–367
   management, 373
   medical risk factors, 367
   possible, 365
   predictive testing, 373
   prevalence, 365, 366
   probable, 365
   progression, symptoms of, 365
   protective factors, 367
   survival, 366
   treatment, choinesterase inhibitors for, 366
American College of Medical Genetics
   (ACMG), 84–85
American Society of Human Genetics, 68
Americans with Disabilities Act, 511

Amyloid, 365
β-Amyloid protein, 365–36
Analytic framework, for evaluating
   interventions, 268, 269*f,* 270
Analytic sensitivity, 88
Analytic specificity, 88
Analytic validity
   in ACCE model system, 218, 218*f*
   analytic sensitivity and, 221
   analytic specificity and, 221–222
   definition of, 196, 197*t,* 218, 266
   determination of, 196, 197*t*
   for DNA-based tests, 88, 221
   in evaluating genetic tests, 221–223, 266,
   267*t*
   of qualitative *vs.* quantitative test, 221
   vs. clinical validity, 196
Angelman syndrome, 85
Angiotensin I-converting enzyme (ACE),
   insertion/deletion polymorphisms, 20,
   20*f*–21*f,* 22
Angiotensin I-converting enzyme gene *(ACE),*
   model-free linkage analysis, 50
Animal models
   development for chronic beryllium disease,
   394–395
   PON1 gene-OP toxicity association, 308–310
Anticoagulants, 80
Antiinflammatory agents, Alzheimer disease
   and, 367
Antiretroviral Adaptation trial (VIRADAPAT),
   236
Anti-retroviral therapy, for HIV-1 infection,
   477–478, 477*f*
APC. *See* Activated protein C
APE1, 167*t*
Apo AIV, 251
APOE gene
   allele frequencies, 370–371, 370*f*
   APOE ∈2 allele, 370, 370f
   APOE ∈3 allele, 370, 370f
   APOE ∈4 allele in, Alzheimer disease,
   367–374, 370f
   genotyping, 372–373
   polymorphisms, population-based research,
   61
   testing, 372, 373–374
Apolipoprotein B, 251
Apolipoprotein E gene. *See APOE* gene
APP gene, Alzheimer disease and, 127
Applied epidemiology, 8*t*
aPTT (activated partial thromboplastin time),
   324
ARIC (Atherosclerosis Risk in Communities),
   440–441
Aromatic amines, bladder cancer and, 150*t*

Arrays, 87
Arsenic, 150*t*
Arylesterase. *See* Paraoxonase
Ashkenazi founder mutations, 100, 101
Asians
    Alzheimer disease incidence and, 366–367
    APOE allele frequencies, 370, 370*f*
ASO (allele specific oligonucleotide assay),
        207–208, 208*t*
Association. *See also* Association studies
    with genetic disease, 38
    with genetic marker, 38
    true, 51
Association studies
    case-control design, 50–51
        analytic focus of, 51
        study group selection, 51
    chronic beryllium disease, 386–391, 387*t*
    confounding in, 96, 97*f*
    designs, 94–95, 94–101
        family-based, 98–101
        integration of, 101–103
        population-based, 95–97, 97*f*
    disadvantages of, 20
    family-based design, 50–51
    genome-wide, 22, 103
    hypothesis specification, reporting of, 169
    integrating genetic information in, 250–251
    interpretability, 95
    large-scale, 22
    limitations, 19–20
    Mendelian randomization, 95–96
    pharmacogenomic, 242*f*, 243–244, 243*f*
    PON1 gene and organophosphate pesticide
        toxicity, 308–310
    population stratification and, 51–52
    purpose of, 92
    single nucleotide polymorphisms, 19–20,
        20*f*–21*f*, 22
    study designs, 93
    transmission-disequilibrium test for, 52–53,
        52*t*
Atherogenesis, *APOE* gene and, 371
Atherosclerosis, 34, 42, 438
Atherosclerosis Risk in Communities (ARIC),
        440–441
ATM gene mutations, 454
Attributable fraction, for colorectal cancer, 199*t*
Autosomal recessive inheritance, 42

Bayesian clustering (latent class analysis), 97
Bayes Theorem, 136, 143–144
BCPT (Breast Cancer Prevention Trial), 209
Becker muscular dystrophy, 85
Belmont Report, 59, 64, 66

BeLPT (beryllium lymphocyte proliferation
        test), 385
Beneficence, 59
Benzene, 150*t*
Beryllium
    exposure, 150*t*, 383
        acute beryllium disease and, 383
        chronic beryllium disease and (*See*
            Chronic beryllium disease)
    properties of, 383
    uses of, 383, 384
Beryllium lymphocyte proliferation test
        (BeLPT), 385
Best practice policies, international, 85–86
Bias
    confounding (*See* Confounding)
    family controls and, 134–135
    information bias, 128*t*, 131–132
    selection, 128*t*, 129–131
    sibling controls and, 135
    sources of, 128–129, 128*t*
BIC (Breast Cancer Information Core), 459
Biological plausibilty
    in gene–disease and gene–gene associations,
        184–185
    in gene–environment interactions, 184–185
Biology
    paradigm shifts, 10, 10*t*
    samples (*See* Samples)
Biotransformation
    altered rates of, 146
    phase I, 146
    phase II, 146
    process of, 146
    xenobiotic, 145
Bladder cancer
    aromatic amines and, 150*t*
    gene effect modifiers of, 150*t*, 151*t*
    risk, *NAT2* genotype and, 136
Blood pressure
    heritability of, 438
    model-free linkage analysis, 50
    regulation, candidate genes for, 439*t*
Blood spots, as epidemiologic specimens, 79,
        80, 81*t*
Body mass index (BMI)
    complex segregation analysis, 42–43
    heritability, 41
Bonferroni method, 138, 139*f*
Brain, degenerative changes in Alzheimer
        disease, 365–366
BRCA1 gene
    assay, population impact, measurement of,
        209, 210*t*
    breast cancer risk and, 85, 452
    carriers, 453, 461–462

BRCA1 gene (*continued*)
  educational approaches, clinical trials of, 260
  epidemiologic findings, 455–459
  familial aggregation, 455–456
  family-based *vs.* population-based studies,
        59–60
  follow-up testing and/or intervention,
        228–229
  general cancer risk and, 453
  genotype-phenotype correlation, 457
  germline mutation, 453
  hereditary breast cancer and, 455–456
  interventions, evaluation of, 270–271
  linkage analysis, 453
  modifiers of, 457–458
  mutation, ovarian cancer risk and, 85
  N372H genotype of, 207–208, 208*t*
  ovarian cancer cluster region, 457
  ovarian cancer risk and, 453
  penetrance, 104, 456–457
    implications of, 460
    study designs for, 100, 101
  population-based studies, 456
  p53-responsive genes and, 454
  screening subjects, preferences of, 258–259
  structure, 453
  testing, 458–459
    clinical practice implications, 464–465
    economic issues, 464
    ethical issues, 464
    implications of, 461–462
    population-based, 465–466
    social implications, 464
BRCA2 gene
  breast cancer risk and, 85
  carriers, 461–462
  educational approaches, clinical trials of, 260
  epidemiologic findings, 455–459
  familial aggregation, 455–456
  follow-up testing and/or intervention, 228–229
  genotype-phenotype correlation, 457
  germline mutation, general cancer risk and,
        453
  hereditary breast cancer and, 455–456
  interventions, evaluation of, 270–271
  linkage analysis, 453
  modifiers of, 457–458
  ovarian cancer cluster region, 457
  ovarian cancer risk and, 85
  penetrance, 100–101, 456–457, 460
  population-based studies, 456
  p53-responsive genes and, 454
  screening subjects, preferences of, 258–259
  structure, 453
  testing, 458–459
    clinical practice implications, 464–465

  economic issues, 464
  ethical issues, 464
  implications of, 461–462
  population-based, 465–466
  social implications, 464
Breast cancer
  BRCA1/2 gene penetrance, 456–457
  candidate genes, 127, 454–455 (*See also*
        BRCA1 gene; *BRCA2* gene)
  clinical practice, 464–465
  deaths from, 451
  epidemiologic studies, 454
  family history, 452–455
  gene-environment interaction, 455
  genetic testing, 458–461
  heterocyclic amines and, 150*t*
  human Genome Epidemiology research
        agenda, 466
  lifetime risk for, 451
  management, 461–462
  prevention, primary, 462–463
  referral and mammography guidelines in
        U.K., 286–288, 287*t*
  risk assessment, 460–461
  risk factors, 451–452
  surveillance, 461–462, 464–465
  susceptibility, inherited, 452–455 (*See
        also BRCA1* gene; *BRCA2* gene)
Breast Cancer Information Core (BIC), 459
Breast Cancer Linkage Consortium, 101
Breast Cancer Prevention Trial (BCPT), 209
Breast-conserving therapy, 461–462
Bronchoalveolar lavage assessment, for chronic
        beryllium disease, 385
Bronchogenic carcinoma, gene effect modifiers
        of, 151*t*
Buccal cells, as epidemiologic specimens, 82*t*,
        83
Bulb diggers palsy. *See* Hereditary neuropathy
        with liability to pressure palsies
Butadiene, 150*t*

Calsequestrin, 29
Canadian Study of Health and Aging, 367
Candidate genes. *See also specific candidate
        genes*
  associations, 99
  breast cancer, 454–455 (*See also BRCA1
        gene; BRCA2* gene)
  for cardiovascular disease, 439–440, 439*t*
  case studies, 440–442
  gene–environment interaction and,
        442–445, 444*t*
  gene–gene interaction and, 445–446, 445*t*
  selection, 154–155

Candidate regions, 19
CAP (College of American Pathologists),
    84–85
Cardiochip, 33
Cardiovascular disease (CVD)
  Alzheimer disease and, 367
  candidate genes, 439, 439*t*
    gene–environment interaction and,
        442–445, 444*t*
    gene–gene interaction and, 445–446,
        445*t*
  CETP gene polymorphsim, 440–442
  gene-environment interaction, 437, 437*f*
  multifactorial etiology of, 437, 437*f*
  prediction, limitations of genetic information
        for, 446–447
  prevalence, 436
  race/ethnicity and, 436
  risk factors, 436–437
    genetic variation, 438
    heritability of, 438–439
Carpal tunnel syndrome (CTS)
  costs of, 512
  definition of, 514–515
  epidemiologic findings, 514–515
    nonoccupational studies, 516
    occupational studies, 515
  etiology, multifactorial, 511
  familial, 516
  health outcome improvement, genetic
        information for, 518
  hereditary conditions and (*See* Familial
        amyloidotic polyneuropathy;
        Hereditary neuropathy with liability
        to pressure palsies)
  idiopathic, 512
  incidence of, 512
  medical conditions associated with, 511,
        512*t*
  occupations associated with, 511, 512*t*
  predictive genetic tests, critical factors for,
        517
  prevalence of, 511, 512
  railroad track workers and, 511, 515,
        518–519
  retrospective testing, 517
  risk attributable to genetic factors, 516–517
  risk factors for, 512
  scientific literature review, 518
  workers compensation, genetic testing for,
        519–520
Case-case design. *See* Case-only design
Case-cohort studies
  confounding of, 177
  design, 96
Case-control-family study, 99, 101

Case-control studies
  for association analysis, 50–51
  compared with other designs, 173
  complete case ascertainment, 129, 130
  control selection in, 129
  critical appraisal, 180
  data analysis, 156
  drawbacks of, 95
  family-based, 175
    advantages of, 98
    disadvantages of, 98–99
  hospital-based
    case acertainment in, 130
    control selection, 130
    healthy subjects in, 130
    study base for, 130
    validity of, 130
  integrating genetic information in, 250–251
  measurement of gene–disease associations,
        197
  multistage sampling, 96
  nested, 127
  nested designs, 96
  population-based, 197
    advantages of, 130
    DNA collection in, 131
    population stratification and, 175–176
    study base for, 129
  population stratification, 96, 97*f*
  selection bias in (*See* Selection bias)
  sources of bias in, 128–129, 128*t*
  stand-alone, study base for, 129
  study base for, 129–130
Case-only design (case-case), 97
Case-parental control studies, 175–176
Case-parent-trio design, 98, 99
Case-sib design, 99
Causal inference, 183–186, 183*t*
CBD. *See* Chronic beryllium disease
CCR5Δ32 allele, in HIV-1 infection, 479–480,
        479*t*, 485–486
CCRF blockade, 486–487
CCR2-64I allele, in HIV-1 infection, 479*t*,
        480–481
CCR5 promoter, in HIV-1 infection, 479*t*, 480,
        485–486
CCR5-related haplotypes, in HIV-1 infection,
        481–483, 482*t*
CDC. *See* Centers for Disease Control and
        Prevention
cDNA (complementary DNA), 27
Celera Genomics Corporation, 17, 18
Centers for Disease Control and Prevention
        (CDC)
  guidelines for evaluating HuGE information,
        11–12

Centers for Disease Control and Prevention (CDC) (*continued*)
National Report on Human Exposure to Environmental Chemicals, 153
Office of Genomics and Disease Prevention Medical Literature Search, 179–180
quality control methods and, 84–85, 84*t*
Cerebral venous thrombosis, factor V Leiden and, 328
CETP (cholelsteryl ester transfer protein), 440
CETP gene
gene–environment interaction, 442–445, 444*t*
gene–gene interaction, 445–446, 445*t*
genotype, 235, 244
polymorphism
cardiovascular disease and, 440–442
non-insulin-dependent diabetes and, 441
CGAP (National Cancer Institute Cancer Genome Anatomy Project), 154
Charcot-Marie-Tooth neuropathy type I, 513
Chemokine genes, in HIV-1 infection, 483–484
Chemokine receptor expression, targeting for HIV-1 treatment/prevention, 486–487
Chemokine receptor gene polymorphisms
HIV-1 infection prognosis and, 485–486
HIV-1 infection therapeutic response and, 485–486
Chemokine receptor genes, role in HIV-1 infection, 479–481, 479*t*
Chemokine receptors, as HIV-1 infection co-receptors, 478–479
Chemoprevention, of breast cancer, 463
Chinese families, with orofacial clefts, 45–46
Chlorpyrifos
agricultural uses of, 307
environmental exposure, 306–307
toxicity, 305–306
Choinesterase inhibitors, for Alzheimer disease, 366
Cholelsteryl ester transfer protein (CETP), 440
Cholesterol. *See* High-density lipoprotein; Low-density lipoprotein
Cholineacetyltransferase, 366
Cholinesterase
inhibition, by organophosphate pesticides, 305
levels, in monitoring pesticide exposure, 306
Chronic beryllium disease (CBD)
animal model development for, 394–395
beryllium lymphocyte proliferation test for, 385
bronchoalveolar lavage assessment for, 385
case registry information, 385
epidemiology of, 385–386
gene associations, 150*t*, 386–391, 387*t*
gene testing, positive predictive value of, 395–398, 397*f*, 398*f*

genetic information, risks/benefits of, 393–396
historical aspects, 383–385
HLA-DPB1
interactions and, 391–393, 392*t*, 393*t*
population frequencies, 391, 392*t*
prevalence of, 386–391, 387*t*
treatment options, 395
Cigarette smoking, Alzheimer disease and, 367
Cleft lip, with/without cleft palate. *See* Orofacial clefts
CLIA (Clinical Laboratory Improvement Amendments), 33, 70, 84, 85
CLIAC (Clinical Laboratory Improvement Advisory Committee), 394
Clinical benefits, from genetic tests, 239
Clinical disorders. *See also specific clinical disorders*
definition, for ACCE model system, 217–218
natural history of, 226
Clinical genetics laboratories, practice standards, 85
Clinical Laboratory Improvement Advisory Committee (CLIAC), 394
Clinical Laboratory Improvement Amendments (CLIA), 33, 70, 84, 85
Clinical Molecular Genetics Society (CMGS), 86
Clinical practice guidelines
cost-effectiveness analysis and, 275–276
developing, 276–278, 276*t*, 278*t*
development of
bias in, 277–278
importance of, 278–279
quality, assessment of, 277–278, 278*t*
Clinical practice setting, genetic test use in, 277
Clinical trials. *See* Randomized controlled trials; *specific clinical trials*
Clinical utility, assessment of, consideration factors for, 271
Clinical validity, 224–226
in ACCE model system, 218–219, 218*f*
definition of, 218, 266
in evaluating genetic tests, 266, 267*t*
for new genetic tests, 266
uncertainty of, 268
vs. analytic validity, 196
Cloning, of *FRM1*, 411–412
Closed systems, 87
Clustering. *See* Familial aggregation
CMGS (Clinical Molecular Genetics Society), 86
Cohort studies
advantages, 96
comparison with other designs, 173

disadvantages, 96
    experimental, purpose of, 197
    population stratification and, 96–97, 97*f*
    purpose of, 197
    retrospective, purpose of, 197
College of American Pathologists (CAP),
    84–85
Colorectal cancer
    association studies
        A1298C polymorphism, 344
        C677T polymorphism, 340, 341*t*–343*t*,
            344
        methodology, 344, 349
    deaths from, 333
    DNA sequence information, 202
    environmental factors, 335
    epidemiologic studies, 198, 199*t*
    extent of disease, survival and, 334
    folate-*MTHFR* relationship, 356–357
    heterocyclic amines and, 150*t*
    high-risk groups, 334–335
    incidence/prevalence, 198
    incidence rates, 333–334
    lifestyle factors, 335
    prevalence, 333
    prevention
        folate and, 353–354
        increased folate intake for, 353–354
        MTHFR genotype and, 353
        primary, 356
    prognosis, 356–357
    risk factors for, 198, 199*t*, 335, 336*f*,
        337–338
    risk identification, value of, 4
    zinc and, 354–355
Colorectal polyps, causes of, 104
Common Rule
    consent disclosures, 68
    minimal risk definition, 64
    requirements, 59, 61–63
Community Programs for Clinical Research on
    AIDS (CPCRA), 236
Complementary DNA (cDNA), 27
Complex segregation analysis, 42–43
Computer Retrieval of Information on
    Scientific Projects (CRISP), 183
Confidentiality
    ACCE review and, 231
    breach of, 65–66
    definition of, 65
    safeguards, adequacy of, 66
Confounding
    age and, 177
    conditions for, 134–135
    description of, 132–133, 133*f*
    in gene-environment interaction, 176
    in genetic polymorphism studies, 128*t*

linkage disequilibrium and, 176
    population stratification, 96, 97*f*
    reporting, checklist for, 171*t*
Congenital malformations, 38
Congestive heart failure, 34
Connexin 26, 424
Consensus genotype, 207–208
Consent issues, 68–69, 231
Consistency, of gene–disease and gene–gene
    associations, 184
Control group, effect modification, 253
Control subjects, 134, 197
Coronary heart disease
    artery calcification, heritability of, 42
    genetic tests, 32
    multiple genes and, 38
    risk factors, 39
Cost-effectiveness analysis
    in clinical practice guidelines, 275–276
    of pharmacogenomic tests, 236–238, 237*t*
Costs, of genetic assay, 238
Cost-utility analysis, 237
Counseling, genetic
    clinical trials of, 259–260
    U.K. National Health Service and, 284
Counter-matching, 96
Cowden syndrome, 454
Cox partial likelihood approach, 104
CPCRA (Community Programs for Clinical
    Research on AIDS), 236
CRISP (Computer Retrieval of Information on
    Scientific Projects), 183
Critical appraisal, of single studies, 180–181
Cross-sectional studies, purpose of, 197
*αβ*Crystallin, 29
cSNPs, 19, 23
C677T polymorphism
    adenomatous polyps and, 344, 345*t*–348*t*
    colorectal cancer association studies, 340,
        341*t*–343*t*, 344
    functional consequences, 339
    gene–environment interactions, 349–350
    inflammatory bowel disease and, 344
CTS. *See* Carpal tunnel syndrome
CVD. *See* Cardiovascular disease
CXCR4 blockers, 486–487
CX$_3$CR1 gene, 483
CXCR4 gene, 483
CYP1A1, 166*t*, 180, 185
CYP1A2, 148–149, 151, 153, 166*t*, 185
CYP3A4, 151
CYP3A4, 166*t*
CYP2C9 polymorphism, warfarin dosing and,
        235, 242–243, 243*f*, 244*f*, 275
CYP2D6, 234
CYP2E1, 166*t*
Cystathionine *β*-synthase, 336*f*, 350

Cystic fibrosis, 38, 85
Cytochrome P450 enzymes, 146. *See also*
    specific CYP genes

Data analysis, pooled, 182
DCM (dilated cardiomyopathy), 29
Deception studies, 68
Decision analysis, 259
Deep vein thrombosis, as oral contraceptive
    side effect, 323
Dementia
    age at, 368
    Alzheimer disease and, 367
    APOE ε4 allele and, 371
Department of Health and Human Services
    (DHHS), 155
Depression, Alzheimer disease and, 367
DHHS (Department of Health and Human
    Services), 155
Diabetes mellitus
    type 2, 38
    type I, *HLA-DQ* polymorphisms and, 174
Diagnostic genetic tests, 9, 265. *See also*
    specific diagnostic genetic tests
Differential error, 131
Dilated cardiomyopathy (DCM), 29
Disclosure, of results, 231
Discrimination
    ACCE review and, 231
    Alzheimer disease, 374
    employment, *HLA-DPB1^{E59}*, 395
    genetic susceptibilities and, 66
    genetic testing and, 212–213, 214*t*
    HFE gene testing and, 503
    insurance, *HLA-DPB1^{E59}*, 395
Disease-association studies. *See* Gene–disease
    association studies
Diseases. *See also specific diseases*
    complex, models for, 103–105
    prevention strategies for, 4
DMD (Duchenne muscular dystrophy), 85
DNA
    double-stranded, 87
    molecular weight, 87–88
    single-stranded, 87
DNA amplification and genotyping, 87
DNA chips
    customized, 23, 25
    hybridization for genotyping, 23
    structure of, 23, 28*f*
DNA collection methods, 131
DNA extraction
    from blood spots, 88
    from buccal cells, 88
    and characterization, 86, 87
    methods, 87

purity, 87
    source, confirmation of, 88
    steps in, 87
    from transformed lymphocytes, 88–89
    from whole blood, 88
    yield, 87
DNA markers. *See* Genetic markers
DNA microarrays, 23, 25, 28*f*
DNA polymorphisms, 25, 147
DNA repair enzyme polymorphisms, 147
DNA tests
    analytic sensitivity of, 221–222
    assessment of (*See* ACCE model system)
    clinical sensitivity, 224
    clinical specificity, 224
    disorder of interest
        accessibility of remedy/action for, 227
        treatment availability for, 227
    economic benefits of, 229
    environmental modifiers and, 226
    evaluation of performance, 230–231
    external proficiency testing, 223
    facilities for, 229
    failure rates, 223
    false negatives, 221
    false positives, 221–222
        confirmatory testing for, 223
        resolution of, 224
    financial costs of, 229
    follow-up testing, health risks for, 228–229
    fragile X syndrome, 409–411
    genetic modifiers and, 226
    genotype/phenotype relationship and, 225
    internal QC program for, 222
    interventions, follow-up, health risks for,
        228–229
    justification for, 227
    for long-term monitoring, 230
    negative predictive value, 225
    personnel for, 229
    positive predictive value, 225
    prevalence of disorder and, 225
    qualitative *vs.* quantitative, 221
    quality assurance measures, 228
    racial/ethnic groups and, 225
    repeated measurements, 222
    results/data
        disclosure of, 231
        impact on patient care, 226
        ownership of, 231
    safeguards for, 231
    for socially vulnerable population, 227–228
    specimens
        range of, 223
        unsuitable, 223
    within- and between-laboratory precision,
        222

Dominance, 41

Dose–response relationship, in gene–disease and gene–gene associations, 184

Double-stranded DNA dye methods, 87

Down syndrome, 368, 371

Drug-gene interactions, surrogate markers, 243–245

Drug metabolism, genetic variation in, 275–276

Drug metabolizing enzymes, gene variants, 241–242, 242*t*

Duchenne muscular dystrophy (DMD), 85

Dyslipidemia, *APOE* and, 371

Dysmenorrhea, primary, 185

β-Dystrobrevin, 29

Early Hearing Detection and Intervention programs (EHDI), 423, 431–432

Early intervention services, for fragile X syndrome, 410–413

EBI (European Bioinformatics Institute), 18

Economic evaluations, health care, 236–238, 237*t*

Economic issues, in *BRCA1/2* testing, 464

EDTA (ethylenediamine tetraacetic acid), 80

Educational approaches
   clinical trials of, 259–260
   validated materials, for clinical disorders, 229–230

Effect dilution, 152–153

Effectiveness
   definition of, 285
   of genetic tests, 239
   vs. efficacy, 285

Effect modification, 252–253

EGP (Environmental Genome Project), 147

EHDI (Early Hearing Detection and Intervention programs), 423, 431–432

EMBASE searches, 179–180

Emphysema, gene effect modifiers of, 151*t*

Empirical-Bayes adjustments, for multiple comparisons, 138–139

Employment discrimination
   HFE gene testing and, 503
   HLA-DPB1^{E59}, 395

EMQN (European Molecular Genetics Quality Network), 86

Endometriosis, 185

Endotoxin, 150*t*, 153

Environmental exposure. *See also* Gene–environment interactions
   assessment
      reporting, checklist for, 170*t*
      reporting on, 174
   causal relationships, 127
   colorectal cancer and, 335

genetic polymorphisms and, 127

human genetic variation and, 145–147

impact of, 58

low-penetrance gene variants and, 68

NAT2 genotype and, 132

one exposure-many diseases principle, 213, 215*t*

Environmental Genome Project (EGP), 147

Environmental health, Web sites, 165–166

Environmental modifiers, DNA testing and, 226

Environmental risk factors, Alzheimer disease, 367

Environmental tobacco smoke (ETS), 151*t*

EPHX1 gene, 151, 166*t*

Epidemiological studies
   classical, 127 (*See also* Case-control studies; Cross-sectional studies; Prospective cohort)
   of colorectal cancer, 198, 199*t*
   designs for
      cohorts, 197
      cross-sectional, 197
      experimental cohorts, 197
      retrospective cohorts, 197
   genetic tests for, 196–197, 197*t*–199*t*

Epidemiology
   applied, 8*t*
   definition of, 6
   genetic (*See* Genetic epidemiology)
   human genome (*See* Human genome epidemiology)
   limitations of, 6
   methods of, 6–7
   molecular, 8*t*
   study applications, 8–9, 8*t*

Epoxide hydrolases, 146

EQA (external quality assessment), 83

EQA/PT component of quality assurance, 85–86

Equity of access, 285

Esophageal cancer, alcohol and, 150*t*

Estrogen, in oral contraceptives, deep vein thrombosis and, 323

Estrogen replacement therapy, Alzheimer disease and, 367

Ethical issues
   in *Belmont Report*, 64
   in *BRCA1/2* testing, 464
   commercialization, 72–73
   confidentiality, 65–66
   consent disclosures, 68
   consent for storage and future research, 68–69
   for epidemiologists, 58
   family-based *vs.* population-based studies, 59–60

Ethical issues (*continued*)
  in genetic testing, for hereditary
      hemochromatosis, 506
  group harms, 66–67
  human research participants
    deceased, samples from, 63
    protection of, 59
    regulations for, 61–63
  human tissue research, 60–61
  implications from DNA testing, 231
  informed consent, 67–68
  justice, 66
  privacy, 65–66
  protection of human participants in research,
      59–61
  reporting research results, 70–72
  in reporting research results, 70–72
  risk assessment, 64–65
  unlinking samples, 63–64
Ethnicity
  allele frequencies and, 212–213, 214*t*
  Alzheimer disease incidence and, 366–367
  APOE allele frequencies, 370, 370*f*
  825C and 825T alleles and, 442
  cardiovascular disease and, 436
  DNA testing and, 225
  factor V Leiden and, 272
  genetic testing discrimination and, 212–213,
      214*t*
  as selection criteria for hemoglobinopathy
      screening, 293–294
  self-reported information on, 135
  in study design
    bias potential and, 134
    population stratification and, 133–134,
        133*f*
  in study designs, 96–97
Ethylenediamine tetraacetic acid (EDTA), 80
ETS (environmental tobacco smoke), 151*t*
European Bioinformatics Institute (EBI), 18
European Molecular Genetics Quality Network
      (EMQN), 86
Europeans, *APOE* allele frequencies, 370, 370*f*
Evidence
  hierarchy of, 182
  quantitative synthesis of, 182
Evidence-based health services
  description of, 207
  foundation of, 200
  implementation of, 285–286, 286*t*
  in U.K. National Health Service, 284–288,
      286*t*, 287*t*
Exposure assessment, 154
Exposure misclassification, 154
Expression arrays, 25, 27–30, 29*f*
External quality assessment (EQA), 83

Factor V, 324–325
Factor Va, 324–325
Factor V Leiden
  acquired APC resistance and, 328–329
  carriers, 325
  cerebral venous thrombosis and, 328
  clinical decision making for, 272–273
  deficiency, 85
  geographic distribution, 326
  hormone replacement therapy and, 273,
      327–328, 328*t*
  interventions, risks of, 272
  oral contraceptive thromboembolism risk
      and, 273, 322, 329–330
    case-control study, 326–327, 326*t*, 327*t*
    discovery of, 322–323
  pregnancy and, 273–274
  prevalence, 272, 326
  race/ethnicity and, 272
  resistance to activated protein C and, 324,
      325
  screening, 272, 329–330
  structure, 325
  testing
    analytic validity of, 266, 267*t*
    guidelines for, 273
    limitations of, 277
  venous thromboembolism risk and, 272
  venous thrombosis and, 329
False positives
  associations, 96
  effectiveness of genetic tests and, 239
  psychological harm from, 276
  test findings
    Bayes Theorem and, 143–144
    prior probabilty of alternative hypothesis
        and, 136, 137*f*, 138
Familial adenomatous polyposis, colorectal
      cancer risk and, 198, 202, 334–335
Familial aggregation, 40
Familial amyloidotic polyneuropathy
  carpal tunnel syndrome and, 511, 512, 516
  genetic testing for, 513, 514
  subtypes of, 514
  transthyretin assay for, 513, 514
Familial cancer clinical services, 464
Familial hypercholesterolemia, 34, 438
Families, high-risk, 266
Family-based-association-tests (FBATs), 99
Family-based study design. *See also specific*
      family-based study designs
  advantages of, 98
  for association analysis, 50–51
  for complex disease, 104
  disadvantages of, 98–99
  nested-control studies and, 127–128

purpose of, 92
types of, 98–101
vs. population-based design, 59–60
Family controls, bias and, 134–135
Family history, of colorectal cancer, 199
FBATs (family-based-association-tests), 99
Federal Employers' Liability Act, 511, 513,
    515
Fenretinide, for breast cancer prevention, 463
Ferritin, serum, 495, 500
FEV$_1$ (forced expiratory volume in the first
    second), 153
Fluorescence in situ hybridization (FISH),
    quality control methods, 84–85, 84*t*
FMRP (fragile X mental retardation protein),
    402
Folate
    colorectal cancer and, prognosis for,
        356–357
    deficiency, colorectal cancer risk and, 335,
        336*f,* 337–338
    dietary supplements, 337
    increasing, additional health benefits, 356
    intake, increasing, 353–354
    metabolism, alcohol and, 338
    and *MTHFR* genotype, in colorectal cancer
        prevention, 353
    pathway, 336*f*
        colorectal cancer risk and, 335, 336*f,*
            337–338
        dietary factors in, 336*f,* 338
    supplementation
        adverse effects of, 354–355
        effects of, 353–354
Forced expiratory volume in the first second
    (FEV$_1$), 153
Fragile X mental retardation protein (FMRP),
    402
Fragile X syndrome
    carrier status, 412
    carrier testing, 414–416
    clinical overview, 402–403
    diagnosis by referral, 411, 414
    epidemiology of, 403
    in females, 403, 404, 407*t*
    FMR1 gene mutation, 402, 404, 413–414
    future directions
        for clinical practice, 413–414
        for population testing, 414–417
    gene–environment interactions, 409
    gene–gene interactions and, 409
    genetics of, 403–404
    health outcome, contribution of genetic
        information for, 410–413
    insurance coverage and, 413
    laboratory tests, 409–411

in males, 403, 404, 405*t–*406*t*
phenotype, clinical, 402–403
prevalence, 403
    full mutation, 404, 405*t–*407*t,* 408
    methylation status and, 410
    premutation, 408–409
    proficiency testing programs and, 85
    treatment, 413
Fredrich ataxia, 85

Gardner syndrome, 202
Gene characterization
    association studies for (*See* Association
        studies)
    complex multigene pathways, 104
    family-based designs and, 100
    studies, 9*t*
    study designs, integrating with gene
        discovery, 101–103
Genechip technology, disadvantages of, 29–30
Gene discovery
    continuum to disease prevention, 7–8, 8*f,* 8*t*
    familial aggregation and segregation analysis
        for, 92
    family-based designs and, 100
    in high-risk families, 7
    methods for, 92–94
    from selected population groups, 7
    studies, 9*t*
    study designs, integrating with gene
        characterization, 101–103
Gene–disease association studies
    bias sources in, 201
    case-control, recurrent problem in, 175
    causal inference, reporting, 183–186, 183*t*
    coherence, 186
    confounding of, 176
    epidemiologic studies of, 7
    estimating individual risk from population-
        based data, 199–200
    experimental support for, 185–186
    hypothesis specification, reporting of, 169
    linkage disequilibrium and, 176
    publication bias, 182–183
    reporting, check list for, 169, 170*t–*172*t*
    statistical issues, 177–179
Gene–drug interactions, 33–34
Gene–environment interactions
    assessment of
        confounding, 128*t,* 132–135, 133*f*
        genotype measurement error in, 128*t,*
            131–132
        selection bias in, 128*t,* 129–131
        sources of bias in, 127–129, 128*t*
    association studies

Gene–environment interactions (*continued*)
  bias sources in, 201
    causal inference reporting, 183–186, 183*t*
    confounding of, 176
    publication bias, 182–183
  bias and, 134
  biological plausibilty, 184–185
  breast cancer, 455
  in cardiovascular disease, 437, 437*f*
  case-parent-trio design for, 99
  dose-response relationship, 184
  exposure assessment, 154
  fragile X syndrome and, 409
  genetic effect modifiers and, 147–149,
      150*t*–151*t*, 151–152
  in hereditary hemochromatosis, 498–499
  HLA-DPB1, 391–393, 392*t*, 393*t*
  hypothesis specification, reporting of, 169,
      173
  methods of analysis, 178–179
  MTHFR gene, 349–350
  multiple genes and, 156–157
  NAT2-bladder cancer relationship and,
      148–149
  "one gene-one risk factor" approach and,
      156
  population stratification, minimizing,
      175–176
  reporting, 174
  reporting, check list for, 169, 170*t*–172*t*
  specificity, 185
  strength, 184
  study designs, 97, 111, 173
  subgroup analysis and, 135–136, 137*f*,
      138–140, 139*f*
  subject selection, 174–175
  temporality, 185
Gene–environment interaction studies,
    statistical issues, 177–179
Gene expression assays, 29–30
Gene families, Websites, 165
Gene–gene interactions
  association studies
    coherence, 186
    confounding of, 176
    experimental support for, 185–186
    publication bias, 182–183
  cardiovascular disease candidate genes,
      445–446, 445*t*
  fragile X syndrome and, 409
  in hereditary hemochromatosis, 498–499
  HLA-DPB1, 391–393, 392*t*, 393*t*
  hypothesis specification, reporting of, 169,
      173
  methods of analysis, 178–179
  MTHFR gene, 350, 351*t*

population stratification, minimizing,
    175–176
statistical issues, 177–179
study designs, 97, 173
subject selection, 174–175
GENEHUNTER, 49
Gene localization methods, from Human
    Genome Project, 18–20, 20*f*–21*f*
Gene mapping, 102
Gene mutations. *See also specific gene*
    mutations
  of xenobiotic metabolism enzymes, 146
Gene penetrance, 239
Gene polymorphisms. *See* Polymorphisms
Gene prevalence. *See* Prevalence
Generalizability, assessment of, 253–254, 254*t*
Generalized Estimating Equations methods,
    103
Genes, environmental health-relevant,
    166*t*–167*t*
GeneTests database, 195
Gene therapy
  limitations of, 34
  prospects, 34
  success of, 34
Genetic architecture, 201
Genetic diseases, risks for, 4
Genetic effect modifiers. *See also specific*
    genetic effect modifiers
  of *BRCA1/2* gene, 457–458
  DNA testing and, 226
  examples of, 147–149, 150*t*–151*t*, 151–152
  in hereditary hemochromatosis, 499
Genetic epidemiology
  of Alzheimer disease, 367–369, 368*t*
  applications, 8*t*, 9
  challenges for, 200–201
  studies, 8*t*, 9
  study designs, 39–40
    when genes are measured, 43–53, 44*f*, 47*f*,
        48*t*, 52*t*
    when genes are not measured, 40–43
Genetic information
  anonymizing, 62, 63–64
  application of, 9*t*
  clinical use/impact of, evaluating, 256–261
  complementary uses of, 247
  evaluation of, 9
  health policies for, 506
  impact on health outcome, *MTHFR* gene,
      352–353
  integrating into clinical trials, 254–256
  pharmacogenomic implications, 33–34
Genetic information Websites, 165–166
Genetic labeling, 4

Genetic markers. *See also specific genetic markers*
definition of, 43–44
model-based linkage analysis of, 44–46, 44*f*
Genetic predictors of study outcomes, integrating genetic information in, 251–252
Genetic risk factor profiles, 33
Genetics, definition of, 7
Genetic studies, commercial gain from, 72
Genetic susceptibility, ionizing radiation and, 212, 213*f*
Genetic tests
for Alzheimer disease, 372
analytic validity of, 196, 197*t*, 266, 267*t*
available, 256
basic principles of, 32–33
clinical benefits from, 239
for clinical usage, before outcome study availability, 271–275
clinical utility of, 196–197, 197*t*
clinical validity of, 196, 197*t*
database for, 195
disclosure of results to patient, 70
educational approaches, clinical trials of, 259–260
effectiveness of, 239, 285
for epidemiologic studies, 195–197, 197*t*–199*t*
evaluation, 12
of clinical validity, 266, 267*t*
of test properties, 264–266, 267*t*, 268
evidence-based approach
description of, 207
discrimination, 212–213, 214*t*
genotype combinations, 211–212, 212*t*, 213*f*
one disease-many genotypes principle, 213–214, 213*f*, 215*t*
one exposure-many diseases principle, 213, 215*t*
penetrance and, 208
population impact, measurement of, 209–210, 210*t*
prevalence and, 207–208, 208*t*
risk-benefit relationship, 210–211, 211*t*
fragile X syndrome, 409–411
guidelines/regulations for, 83–86
impact of, 32
of minors, in absence of direct intervention benefits, 430
population impact, measurement of, 209–210, 210*t*
predictive, prevention strategies for, 195–196
properties, accurate reporting of, 277
for risk assessment, 195–196

validity of, 79
for viral/tumor cell genotype, 235–236
Genetic variants
adverse drug reactions and, 234
clinical outcome and, 268
of drug metabolizing enzymes, 241–242, 242*t*
effect on drug safety/efficacy, 234
environmental exposures and, 145–147
frequency of penetrance and, 208
prevalence of, 240–242, 240*f*, 241*f*, 242*t*
role in human disease, 168
sensitivity/specificity for, 266
Gene transfer, 34
Genome Database, 39
Genome mapping, 18
Genome scans, for identifying candidate regions for complex traits, 49–50
"Genomes to Life," 18
Genome-wide association, 19, 177
Genomic control, 97
Genomic markers, types of, 79. *See also specific genomic markers*
Genomic medicine, vision of, 3–6, 5*t*, 6*t*
Genomics
definition of, 7–8
for preventive medicine, 203
Genomic technologies. *See also specific genomic technologies*
association studies, 19–20, 20*f*–21*f*, 22
for complex diseases, 22–23, 24*f*–31*f*, 25, 27–30, 32
DNA microarrays, 23, 25, 28*f*
microsatellite markers, 18–19
single nucleotide polymorphisms, 19–20, 20*f*–21*f*, 22
for SNP genotyping, 22–23, 24*f*–27*f*
Genotyped proband design, 100
Genotype/phenotype relationship, 225
Genotype prevalence studies
bias sources in, 201
population stratification, minimizing, 175–176
reporting, check list for, 169, 170*t*–172*t*
statistical issues, 177–179
study design, 173
Genotypes
analytic validity of, 173–174
assessment of
DNA extraction and characterization for, 87–89
guidelines/regulations, 83–86, 84*t*
quality control for, 83
reporting, 173–174
specific recommendations, 86–87
specimen selection for, 79–80, 81*t*–82*t*, 83

Genotypes (*continued*)
  combination of, 211–212, 212*t*, 213*f*
  consensus, 207–208
  high-risk, 60
  measurement error, 131–132
  one disease-many genotypes principle,
      213–214, 213*f*, 215*t*
  risk-associated, 79
  susceptibility, joint probability distribution
      of, 206*t*
Genotyping techniques
  allele specific oligonucleotide, N372H
      genotype of *BRCA1* gene, 207–208,
      208*t*
  cost, in large cohorts, 96
  HLA-DPB1^{E69}, 393–394
  Invader Cleavase, N372H genotype of
      BRCA1 gene, 208, 208*t*
  methods, selection of, 155
  RsaI digest, N372H genotype of *BRCA1*
      gene, 208, 208*t*
  samples, collection of, 155
  Taquman, N372H genotype of *BRCA1* gene,
      207–208, 208*t*
  validity, reporting, checklist for, 170*t*
  validity of, genetic information quality and,
      254–255
GENSCAN, 18
GJB2 gene
  connexin 26 and, 424
  testing, for nonsyndromic hearing loss, 423
GJB2 variants
  hearing loss and, 424–425
    age of onset and, 425, 427–428
    environmental factors and, 428–429
    phenotypic outcome and, 428
    population-specific epidemiologic data,
        424–425, 426*f*, 427*t*
    syndromic, 429–430
    testing for, 423
  testing
    clinical utility of, 430, 431
    decision-making for, 430–431
    laboratory techniques for, 429–430
Glucocorticosteroid remediable aldosteronism
    (GRA), 438
Glucose-6-phosphate dehydrogenase, 234
Glucuronosyltransferases, 146
Glutathione S-transferase polymorphisms,
    179–182
Glutathione S-transferases (GSTs), 146, 149
GNB3 gene
  gene–environment interaction, 443–445, 444*t*
  hypertension, obesity and, 441–442
GSTM1 gene
  assay, 210, 210*t*

genotype, 151
relevance to environmental health, 166*t*
GSTP1 gene, 167*t*
GSTT1 gene, 167*t*
GSTT1 null genotype, 175

HAART (highly active-anti-retroviral therapy),
    236
Halomethanes, 150*t*
Haplotypes
  analysis of, 22, 178
  CCR5-related, in HIV-1 infection, 481–483,
      482*t*
  definition of, 46, 94
  mapping methods, 94
Hardy-Weinberg equilibrium, 131
Harms
  group, 66–67
  vs. wrongs, 65–66
Hay dust, 150*t*
Hazard function, 103
HCM (hypertrophic cardiomyopathy), 29
HDL cholesterol. *See* High-density
    lipoprotein
Health disparities, 231
Health effects studies, incorporating in
    polymorphisms, 152–153
Health Insurance Portability and Accountability
    Act (HIPAA), 413
Health services research, 8*t*
Hearing loss
  age of onset, 423, 425, 427–428
  congenital, *GJB2* variants and, 427
  environmental factors, *GJB2* variants and,
      428–429
  genetic evaluation of, 423 (*See also GJB2*
      gene, testing)
  GJB2 variants and (*See GJB2* variants,
      hearing loss and)
  health outcomes, effect of genetic
      information on, 430–431
  identification, EHDI programs for, 423,
      431–432
  locus heterogeneity of, 423
  noncongenital prelingual, *GJB2* variants and,
      427
  nonsyndromic, *GJB2* testing for, 423
  phenotypic outcome, factors in, 428
  postlingual, *GJB2* variants and, 427–428
  prevalence of, 423
  syndromic, *GJB2* variants and, 429–430
Heart disease. *See also* Cardiovascular disease
  Alzheimer disease and, 367
  congenital, 38
Helsinki Heart Study, 250

Hemoglobinopathies
  clinical features, 288–289
  population screening, 288–289, 290*t*–291*t*,
      291–294
  screening
    acceptability of, 292
    ethnicity as selection criteria for, 293–294
    recommendations in United Kingdom,
        289, 290*t*–291*t*
    selective strategies for, 291–292
    universal strategies for, 291–292
Hemoglobin S/C, 85
Hemostasis
  candidate genes, 439*t*
  teleological description of, 323–324
Hepatocellular carcinoma, aflatoxin B_1 and,
    149, 150*t*, 151, 153
Hereditary hemochromatosis
  carriers, 503
  clinical overview, 495–496
  diagnosis, 495, 503–505
  epidemiology, 496
  future HuGE research agenda, 506
  genetic testing
    ethical issues in, 506
    information, health policies for, 506
    risks/benefits of, 503–506
  genotype prevalence, 498
  HFE gene, 495, 497
    allelic variants, 497–498
    C282Y mutation, 496, 497–498
    H63D mutation, 496, 497, 498
    interactions, 498–499
  iron overload in, 504
  laboratory tests
    HFE gene mutation analysis, 500–501
    for iron status, 499–500
  natural history of, 504–505
  non-*HFE* gene related, 497
  penetrance, 500, 501
  prevalence, 505
  proficiency testing programs and, 85
  psychological burden of diagnosis, 503–504
  screening, 504
    cost-effectiveness of, 505, 506
    performance issues, 505
  screening methods, 495
  treatment, 505
  treatment of, 495–496
Hereditary neuropathy with liability to pressure
    palsies (HNPP)
  age of onset, 513
  animal models, 513–514
  carpal tunnel syndrome and, 511, 512,
      516–517
  clinical presentation of, 513–514

genetic loci for, 513
genetic testing for, 513
prevalence of, 513
Hereditary nonpolyposis colorectal cancer
    (HNPCC), 199, 202, 334–335
Hereditary syndromes, hearing loss in, 423
Heritability (h^2)
  definition of, 41
  estimation, approaches for, 41–42
  gene therapy success and, 34
Heterocyclic amines (HCAs), 104, 150*t*
Heterogeneity in linkage, genetic information
    interpretation and, 255
HFE gene
  C28Y mutation, 500–501, 502
  future HuGE research agenda and, 506
  H63D mutation, 500–501
  mutation analysis, 500–501
    risks/benefits of, 503–506
  testing, implications of, 502–504
High-density lipoprotein (HDL)
  cardiovascular disease and, 437, 438
  CETP gene polymorphism and, 440–443
  estrogen and, 323
  transport, cholesteryl ester transfer protein
      and, 440
Highly active-anti-retroviral therapy (HAART),
    236
High-penetrance genes, 60
HIPAA (Health Insurance Portability and
    Accountability Act), 413
Hirschsprung disease, family aggregation of, 40
Hispanics, Alzheimer disease incidence and,
    366–367
HIV (human immunodeficiency virus), 236
HIV-1 infection. *See also* Acquired
    immunodeficiency syndrome
  anti-retroviral therapy
    for prevention, 477–478, 477*f*
    for treatment, 478
  asymptomatic, 475
  CCR5-related haplotypes in, 481–483,
      482*t*
  chemokine co-receptors, 478–479
  chemokine genes in, 483–484
  chemokine receptor genes and, 479–481,
      479*t*
  epidemiology, 476
  HIV-1 RNA levels, 475
    for treatment monitoring, 478
  non-chemokine genes in, 484–485
  pathogenesis, 475
  prevention of, 477
  prognosis, chemokine receptor gene
      polymorphisms and, 485–486
  sexual transmission of, 476, 477

HIV-1 infection (*continued*)
  symptoms of, 475
  treatment, chemokine receptor gene
      polymorphisms and, 485–486
  treatment/prevention, targeting chemokine
      receptor expression for, 486–487
  vaccine strategies, 487
HLA class I loci, in HIV-1 infection, 484–485
HLA-DP $_1$, 167*t*
HLA-DPB1^{E59}, testing, positive predictive
      value of, 395–398, 397*f*, 398*f*
HLA-DPB1 gene
  interactions and, 391–393, 392*t*, 393*t*
  population frequencies, 391, 392*t*
  prevalence of, 386–391, 387*t*
  testing
      availability of, 393–394
      information from, risks/benefits of,
          393–396
HLA-DP variants, chronic beryllium disease
      and, 386–391, 387*t*
HLA-DQ polymorphisms, type I diabetes
      mellitus and, 174
HNPCC (hereditary nonpolyposis colorectal
      cancer), 199, 202, 334–335
HNPP. *See* Hereditary neuropathy with liability
      to pressure palsies
Hormone replacement therapy, factor V Leiden
      and, 327–328, 328*t*
Hormones, female, prothrombotic mutation
      and, 326–327, 326*t*, 327*t*
HuGE. *See* Human genome epidemiology
Human genome discoveries
  applications, 3–6, 5*t*, 6*t* (*See also specific*
      applications)
  concerns about, 4
  Web-based newstories on, 3, 4*t*
Human genome epidemiology (HuGE)
  definition of, 3, 7–10, 8*f*, 8*t*
  future research agenda, 506
  literature reviews, 11
  peer-reviewed literature, 9–10
  standards, need for, 10–12
  studies, 9, 9*t*
      candidate gene selection, 154–155
      check list for reporting, 169, 170*t*–172*t*
      heterogeneity of, 168
      incorporating genetic effect modifiers in,
          154–156
      meta-analysis of, 11–12
      published, number of, 168
      reporting, 186–187
      standards, need for, 10–12
Human Genome Epidemiology Network
  goals of, 53–54
  research agenda, for breast cancer, 466

Human Genome Project
  gene localization methods from, 18–20,
      20*f*–21*f*
  Genome Database, 39
  goals of, 39
  working draft sequence, 17–18
Human immunodeficiency virus (HIV), 236
Human leukocyte antigen, 43
Human research participants
  definition of, 61
  protection of, 59
  regulations for, 61–63
Human tissue research, 60–61
Huntington disease, 38, 85, 208
Hypercholesterolemia, Alzheimer disease and,
      367
Hypertension
  Alzheimer disease and, 367
  GNB3 gene and, 441–442
  GNB3 gene–environment interaction,
      443–445, 444*t*
  model-free linkage analysis, 50
Hypertrophic cardiomyopathy (HCM), 29

IBD (identical by descent), 46–47, 47*f*, 48–49,
      94
IBS (identical by state), 46–47, 47*f*
ICER (incremental cost-effectiveness ratio),
      237, 238
IDEA (Individuals with Disabilities Education
      Act), 411
Identical by descent (IBD), 46–47, 47*f*, 48–49,
      94
Identical by state (IBS), 46–47, 47*f*
IL-4 gene, in HIV-1 infection, 484
IL10 gene, in HIV-1 infection, 484
Immunoreactive trypsinogen (IRT), 221
Implementation, of evidence-based practice,
      285–286, 286*t*
Incremental cost-effectiveness ratio (ICER),
      237, 238
Individuals with Disabilities Education Act
      (IDEA), 411
Inflammation, candidate genes, 439*t*
Inflammatory bowel disease, C677T
      polymorphism and, 344
Information bias, 128*t*, 131–132
Information sources
  database searches, 179–180
  unpublished, inclusion in reports, 180
Informed choices, 230
Informed consent
  for genetic testing, 67–68, 155
  requirements, 230
  Websites, 166

INR (International Normalized Ratio), 242–243, 244
Insertion-deletion polymorphisms, 79
Institutional Review Board (IRB)
 group harms and, 66–67
 informed consent and, 67–68
 minimal risk concept and, 64–65
 protection of human research participants, 59
 research result reporting and, 70
 unlinked samples and, 63–64
Insulin signaling, candidate genes, 439*t*
Insurance
 coverage, fragile X syndrome testing and, 413
 discrimination
  HFE gene testing and, 503
  HLA-DPB1^{E59}, 395
Integration of genetic information
 into data analysis, 250–253
 into design of randomized clinical trial, 248–249
 eligibility criteria and, 248
 a priori adjusted analysis of data, 249
Interleukin-4 polymorphisms, population-based research, 61–62
Internal quality control, for DNA tests, 222
International Human Genome Sequencing Consortium, 17
International Normalized Ratio (INR), 242–243, 244
Interventions
 evaluation of (*See also under specific diseases*)
  analytic framework for, 268, 269*f,* 270
  for *BRCA1/2* mutations, 270–271
  for factor V Leiden, 270, 272–275
  evidence-based, implementation of, 285–286, 286*t*
  risk/benefit analysis of, 276–277, 276*t*
Invader Cleavase technology, N372H genotype of *BRCA1* gene, 208, 208*t*
Ionizing radiation
 as genetic modifier, 151*t*
 genetic susceptibility and, 212, 213*f*
Iron
 in hereditary hemochromatosis, 495, 499
 overload
  in hereditary hemochromatosis, 495
   biochemical evidence of, 496
   treatment of, 495–496
  non-*HFE* gene-related, 497
 overload, in hereditary hemochromatosis, 504
IRT (immunoreactive trypsinogen), 221
Ischemic heart disease, Alzheimer disease and, 367

Justice, 59, 66

Kaiser Permanente Women Twins Study, 41
Kin-cohort design, 99–101

Laboratories, quality control requirements, 83–86
Latent class analysis (Bayesian clustering), 97
LD. *See* Linkage disequilibrium
LDL cholesterol (low-density lipoprotein), 251, 437, 438
Lead, 150*t*
Left ventricular mass index reduction, 252
Legal issues
 breach of confidentiality/privacy, 65–66
 commercialization, 72–73
 Common Rule requirements, 59, 61–63
 consent disclosures, 68
 consent for storage and future research, 68–69
 implications from DNA testing, 231
 informed consent, 67–68
 in reporting research results, 70–72
 research regulations, 61–63
 unlinking samples, 63–64
Lewy bodies, 368
Liddle's syndrome, 438
Lifestyle factors
 for breast cancer, 462
 for colorectal cancer, 335
Li-Fraumeni syndrome, 454
Linkage analysis
 contributions of, 127
 between families, 93
 within families, 93
 further sequencing techniques, 94
 for major susceptibility genes, 92–93
 model-based, 44–46, 44*f,* 93–94
 model-free, 19, 46–50, 47*f,* 48*t,* 93, 94
 multipoint, 93
 purpose of, 92
 resolution limit of, 94
 study designs, 93
 two-point, 93
Linkage disequilibrium (LD)
 definition of, 51, 94
 gene–disease associations and, 176
 genetic information interpretation and, 255
 single nucleotide polymorphisms and, 19
Lipid metabolism
 APOE and, 371
 candidate genes, 439*t*
Lipocortin, 29
Lipopolysaccharide (LPS), 150*t,* 153

Lipoprotein and Coronary Atherosclerosis
        Study, 252
Lisinopril, 252
Logarithm of the odds score (Lod score)
    definition of, 45
    in gene discovery, 93–94
    linkage analysis, 45–46
        disadvantages of, 46
        efficiency of, 46
        vs. model-free linkage methods, 46
Low-density lipoprotein (LDL), 251, 437, 438
Low-density lipoprotein receptor (LDLR), 438
Low-penetrance genes
    definition of, 92
    family-based vs. population-based studies, 60
    research results, disclosure of, 70
    risk assessment for, 64–65
    variants, environmental exposures and, 68
LPS (lipopolysaccharide), 150t, 153
Lumican, 29
Lung cancer, gene effect modifiers of, 151t
Lymphocytes, EBV-transformed, 80, 81t, 83

Macular degeneration, 185
Magnetic resonance imaging (MRI), breast
        cancer, 461
Major susceptibility genes, definition of, 92
Malaria prevalence, 289
Mannose-binding lectin (MBL), 485
MAPMAKER/SIBS, 49
Mastectomy, prophylactic, 462–463
Maximum likelihood analysis, 100
MBL (mannose-binding lectin), 485
MBL gene, in HIV-1 infection, 485
McKusick, Victor, 39
Medical outcomes, definition of, 276
Medical risk factors, Alzheimer disease, 367
Medicine
    paradigm shifts, 10, 10t
    predictive, 33
Medline searches, 179–180
Memory decline, APOE ε4 allele and, 371
MEN-2 (multiple endocrine neoplasia-2), 85
Mendelian inheritance, 42
Mendelian randomization, 95–96
Mental activity, Alzheimer disease and, 367
Mental retardation, in fragile X syndrome, 402
6-Mercaptopurine (6-MP), 235, 253
Mercury, 150t
Meta-analysis, of observational studies, 182
Methionine, 336f, 338
Methionine synthase, 336f, 350
Methionine synthase reductase, 336f, 350
Methylation status, fragile X syndrome and, 410
Methylene tetrahydrofolate reductase, 85

5,10-Methylenetetrahydrofolate reductase
        (MTHFR)
    in folate metabolism, 336f, 338–339
    gene (See MTHFR gene)
5-Methyltetrahydrofolate, 335
S-Methyltransferase (TPMT), 235
Methyltransferases, 146
Microsatellites. See Short tandem repeats
Minimal risk, definition of, 64
Misclassification, 131–132
Model-based linkage analysis, 44–46, 44f
Model-free linkage analysis
    identical by descent, 46–47, 47f, 48–49
    identical by state, 46–47, 47f
    midpoint analysis, 49
    of sibling pairs, 47–50, 48t
Mod score, 100–101
Molecular epidemiology, 8t
Molecular methods
    ACMG guidelines, 85
    guidelines/regulations for, 83–86, 84t
    proliferation of, 86
    quality control for, 83
Molecular Probes, 87
Moore v. Regents of the University of
        California, 72
MORE (Multiple Outcomes of Raloxifene
        Evaluation), 463
6-MP (6-mercaptopurine), 235, 253
6-MP therapy, for ALL, TPMT genotyping
        and, 240–241, 240f, 241f
MRI (magnetic resonance imaging), breast
        cancer, 461
M34T allele, 432
MTHFR (5,10-methylenetetrahydrofolate
        reductase), in folate metabolism,
        336f, 338–339
MTHFR gene
    colorectal cancer and, prognosis for,
        356–357
    gene–disease association studies, 340,
        341t–348t, 344, 349
    gene–environment interactions, 349–350
    gene–gene interactions, 350, 351t
    genotype, need for randomized controlled
        trials, 355–356
    information, impact on health outcome,
        352–353
    location of, 339
    polymorphisms, 350, 352
        A1298C, 339, 344
        C677T (See C677T polymorphism)
        prevalence of, 339–340
        test methods for, 352
Multiethnic populations, 97
Multilocus genotyping assay, 33

Multiple comparisons, adjustment methods for, 135–136, 137*f*, 138–140, 139*f*
Multiple endocrine neoplasia-2 (MEN-2), 85
Multiple gene interactions, chronic disease and, 199–200
Multiple Outcomes of Raloxifene Evaluation (MORE), 463
Multiple Risk Intervention Trial, 250
N-Myc proto-oncogene amplification, 236
Myocardial infarction, customized DNA chips, 23, 25
Myotonic dystrophy type 1, 85

N-Acetyltransferases (NATs), 146
NAT1 gene
  bladder cancer and, 152
  polymorphism, 166*t*
NAT2 gene
  bladder cancer risk and, 136, 148–149, 152
  environmental exposure and, 132
  polymorphism, 166*t*
National Academy of Science/National Research Council World War II Veteran Twins Registry, 41–42
National Bioethics Advisory Commission (NBAC), 60–63
  consent for storage and future research, 68–69
  disclosure of research results, 71–72
  group harms and, 67
  informed consent and, 67–68
  minimal risk concept and, 64, 65
  unlinking samples and, 63–64
National Cancer Institute (NCI), Cooperative Family Registries for Breast and Colorectal Cancer Research, 102
National Cancer Institute Cancer Genome Anatomy Project (CGAP), 154
National Center for Biotechnology Information (NCBI), 18
National Health and Nutrition Examination Survey (NHANES), 62, 153
National Institute for Neurological and Communication Disorders and Alzheimer disease and Related Disorders Association (NINCDS-ADRDA), 371–372
National Institute of Environmental Health Sciences (NIEHS), 147
National Institutes of Health (NIH)
  Department of Energy Task Force on Genetic Testing, 196–197, 197*t*
  evaluation of genetic tests, 12
  guidelines for evaluating HuGE information, 11–12

Native Americans, *APOE* allele frequencies, 370, 370*f*
NBAC. *See* National Bioethics Advisory Commission
NCBI (National Center for Biotechnology Information), 18
Nested case-control studies, 173
Neuritic plaques, 365–366
Neutropil pathology, in Alzheimer disease, 366
Newborn screening, for fragile X syndrome, 416–417
Newstories, Web-based, 3, 4*t*
NHANES (National Health and Nutrition Examination Survey), 62, 153
N372H genotype, of *BRCA1*, 207–208, 208*t*
NIH. *See* National Institutes of Health
NINCDS-ADRDA (National Institute for Neurological and Communication Disorders and Alzheimer disease and Related Disorders Association), 371–372
Nitro-polycyclic aromatic hydrocarbons, 151*t*
Nondifferential error, 131
Non-insulin-dependent diabetes, *CETP* gene polymorphsim and, 441
Nonparametric linkage analysis. *See* Linkage analysis, model-free
NQO1 gene, 166*t*
Nucleic acid amplification, quality control methods, 84*t*
Null hypothesis, 45, 136, 137*f*, 138
Nurses Health Study, 337–338

Obesity
  breast cancer risk and, 452
  candidate genes, 439*t*
  complex segregation analysis, 42–43
  GNB3 gene and, 441–445, 444*t*
  heritability, 41
Obstructive pulmonary disease, gene effect modifiers of, 151*t*
Oceanians, *APOE* allele frequencies, 370, 370*f*
Oltipraz (OPZ), 153
OMIM, 154
Online Mendelian Inheritance in Man, 39
OOCR (ovarian cancer cluster region), 457
Oophorectomy, bilateral prophylactic, 463
Oral contraceptives
  deep vein thrombosis and, 323
  factor V Leiden screening and, 329–330
  factor V Leiden-thromboembolism risk, 273, 322, 330
    case-control study, 326–327, 326*t*, 327*t*
    discovery of, 322–323

Organization for Economic Co-operation and
    Development (OECD), 85
Organophosphate pesticides
  cholinesterase inhibition, 305
  environmental exposure, 306–307
  as genetic modifier, 150t
  toxicity
    clinical manifestations of, 305–306
    deaths from, 306
    epidemiologic data on, 306
    paraoxonase polymorphisms and, 305
    PON1 gene screenings for (See PON1
      gene, screenings)
  types of, 307 (See also Chlorpyrifos)
Orofacial clefts
  association analysis, 52
  CYP1A1 gene variants and, 185
  model-based linkage analysis, 45–46
  multiple genes and, 38
Ovarian cancer cluster region (OOCR), 457
Ovarian cancer risk, 127
  BRCA1 gene and, 453
  BRCA1/2 gene penetrance, 456–457
Ownership, of DNA test data, 231
Ozone, 150t

PAHs (polycyclic aromatic hydrocarbons), 104,
    151t
Parametric linkage analysis. See Linkage
    analysis, model-based
Paraoxonase (PON1), 307
Paraoxonase gene. See PON1 gene
Parental controls, 98
Parkinson's disease, 41–42, 185
Patient quality of life, 239–240
PCBs, 150t
PCR. See Polymerase chain reaction
Pedigrees
  clinic-based collections, 101
  sequential sampling of, 101–102
Penetrance
  definition of, 45, 208
  determining factors for, 208
  family-based designs and, 100
  frequency of variant and, 208–209
  incomplete, 45
  parameters, for LOD score linkage analysis,
    46
  for quantitative traits, 45
  statistical model of, 103
Peptide-mass fingerprint analysis (PMF), 30,
    30f–31f
Peripheral myelin protein-22 (PMP22), 513, 516
Pharmacogenomics (pharmacogenetics)
  association studies, 242f, 243–244, 243f
  clinical benefits from, 239
  definition of, 234
  future studies, 245
  gene variant prevalence, 240–242, 240f,
    241f, 242t
  history of, 234
  implications of genetic information, 33–34
  pharmacodynamic strategies, 235–236
  pharmocokinetic strategies, 235
  surrogate markers, 243–245
  test evaluations
    comparator treatment strategies, 238
    cost-effective analysis, 236–238, 237t
    cost offsets, 238
    direct test costs, 238
    effectiveness of genetic tests and, 239
    gene prevalence and, 240
    patient quality of life and, 239–240
  testing, achieving benefits from, 275–276
Pharmacokinetic models, 104
Phase I enzymes, 146
Phase II enzymes, 146
Phenocopy, 45
Phenotype-genotype discrepancy, 101
Phenotypes
  definition of, 39
  with genetic components, 42
Phenylketonuria, penetrance, 208
Physical activity, Alzheimer disease and, 367
Picogreen, 87
Pilot trials, 228
PinPoint assay, 22–23, 24f
PMP22 (peripheral myelin protein-22), 513, 516
POF (premature ovarian failure), 412
Polycyclic aromatic hydrocarbons (PAHs), 104,
    151t
Polymerase chain reaction (PCR)
  fragile X syndrome, 410
  microsatellite markers and, 18
  quality control methods, 84t
  selection of, 155
Polymorphisms. See also specific
    polymorphisms
  analytic methods, 86
  for association analysis, 51
  definition of, 146
  disease-causing, 369
  DNA, 25
  dose–response modifications from, 147–148
  environmental exposures and, 127
  environmental health-relevant, 166t–167t
  epidemiologic studies of, sources of bias in,
    128–129, 128t
  functional, 146
  incorporating in health effects studies,
    152–153

markers (*See* Short tandem repeats; Single
  nucleotide polymorphisms)
microsatellite markers and, 18
nonfunctional, 146
in non-Hispanic whites, 96–97
in protein coding region, 25
risk associated with, 156
silent, 146
single nucleotide, 19–20, 20*f*–21*f,* 22
Polypropylene, 80
PON1 enzyme (paraoxonase), 307
PON1 gene
  epidemiologic data, further need for,
    315–316
  gene–gene interaction, 445–446, 445*t*
  as genetic modifier, 167*t*
  location of, 307
  metabolic activity, variations in, 308
  metabolism of oxidized lipids, 307
  polymorphisms, organophosphate pesticide
    toxicity and, 305
  screenings
    analytic validity of, 310–311
    clinical validity of, 310–311, 312*t*
    predictive value of, 314
    relative risk assessment, 311–314, 313*t*
  status, association with OP toxicity, 308–310
  workplace genetic testing programs, 314,
    315
Population-based case-control-family design,
  101
Population-based data, estimating individual
  risk from, 199–200
Population-based studies
  BRCA1/2 gene, 465–466
  case control design, 95–97, 97*f*
  cohort design, 95–97, 97*f*
  data, for preventive medicine, 201–203
  on existing biologic samples, 61–62
  for fragile X syndrome, 414–417
  group harms, 66–67
  integration of, 102, 104
  on new biologic samples, 61
  purpose of, 79
  samples for, 61–63
  specimens, blood spots as, 80, 81*t*
  vs. family-based studies, 59–60
Population risk characterization, 8–9
Population stratification
  association analysis and, 51–52, 96, 97*f*
  conditions for, 134
  control of, 134–135
  description of, 133–134, 133*f*
  genetic effect modifier studies and, 152
  integration of genetic information and, 249
  minimizing, 175–176

protecting against, 98
reporting, checklist for, 171*t*
results of population-based case-control
  studies and, 175–176
Positional cloning, contributions of, 127
Positive predictive value, *HLA-DPB1*^{E59},
  395–398, 397*f,* 398*f*
Power, statistical, 177
Prader-Willi syndrome, 85
Precision, 88
Predictive genetic tests, 4
Predictive medicine, 33
Predictive modeling of disease risk, integrating
  genetic information in, 251–252
Pregnancy
  factor V Leiden and, 273–274
  first trimester screening for
    hemoglobinopathies, 293
Prejudices, 66
Premature ovarian failure (POF), 412
Prenatal diagnosis/screening
  fragile X syndrome, 409, 414–417
  hemoglobinopathies, 292–293
Presenilin-1 gene *(PS1),* Alzheimer disease
  and, 367–369, 368*t*
Prevalence
  for colorectal cancer, 199*t*
  definition of, 207
  in setting, DNA testing and, 225
Prevention
  genetic information for, 9
  genomic approach, commercial marketing of,
    4–5
Prevention strategies
  accumulating knowledge on, 270
  availability, randomized controlled trials and,
    258
Privacy
  ACCE review and, 231
  definition of, 65
  invasion of, 65–66
  safeguards, adequacy of, 66
Privacy Act, 70, 72
Probands, 99–100, 102
Proficiency testing (PT), 83, 85
Prognostic significance, of genetic test, 239
Proportional hazards model, 104
Prospective cohort studies
  advantages of, 96
  complete case ascertainment, 129
  disadvantages of, 96
  population stratification and, 96–97, 97*f*
  selection bias in (*See* Selection bias)
  sources of bias in, 128–129, 128*t*
  study base in, 129
Protein C system, 324

Proteomics
  methods of, 30, 30*f*–31*f*
  potential for, 32
Prothrombin, 85
Pseudocholinesterase, 234
Pseudosiblings, 98
PTEN gene mutations, 454
Publication bias, 11, 182–183
Public health
  gene discovery applications and, 6, 6*t*
  new opportunities/challenges for, 201–203
Public health contribution, of *PON1* gene
      status screenings, 310–314, 312*t*

Q360H polymorphism, 251
Quality-adjusted life-year (QALY), 237–238
Quality assurance, international efforts, 85–86
Quality assurance measures, for DNA testing,
      228
Quality control
  guidelines/regulations for, 83–86
  for molecular methods, 83
  proficiency testing and, 85

Race
  cardiovascular disease and, 436
  factor V Leiden and, 272
  GNB3 gene 824C and 825T alleles, 442
  study designs and, 96–97
Racial groups, DNA testing and, 225
Railroad track workers, carpal tunnel syndrome
      and, 511, 515, 518–519
Raloxifene, for breast cancer prevention, 463
Randomized controlled trials (RCTs)
  data analysis, integrating genetic information
      into, 250–254, 254*t*
  design, 247, 254, 254*t*
    change, genetic information and,
        247–248
    integrating genetic information into,
        248–249
  design, considerations for, 258–259
  eligibility criteria, genetic information and,
      248
  endpoints, genetic information interpretation
      and, 255
  evaluating clinical use/impact of genetic
      information, 256–261
  genetic test
    acceptability for target population, 257
    as strong disease determinant, 257–258
    methods, accurate/reproducible, 257
  MTHFR genotype, 355–356
  multiple comparisons, 256

outcomes, genetic information interpretation
    and, 255
population stratification, genetic information
    and, 249
prerequisites for, 257–258
prevention interventions, availability of, 258
a priori adjusted analysis of data, genetic
    information and, 249
results/findings
  generalizability, assessment of, 253–254,
      254*t*
  replication of, 256
study populations
  identifiable, 257
  randomized, nonrandomized uses of,
      255–256
  selection, genetic information and, 248
    subgroup analysis, 247
    surrogate outcomes, genetic information
        interpretation and, 255
    treatment interventions, availability of,
        258
    use of genetic information in, 254–256
RANTES gene, in HIV-1 infection, 483–484,
      486
RCTs. *See* Randomized controlled trials
Reagent blanks, 87
Recall bias, 95
Recombination
  definition of, 44
  example of, 44–45, 44*f*
  frequency, 45
Relative risk
  for colorectal cancer, 199*t*
  estimates, biasing, 96
Reliability, 88
Reporting, 186–187
  of human genome epidemiologic studies
    check list for, 169, 170*t*–172*t*
    hypothesis specification, 169, 173
  of research results, 70–72
  systematic review, 179–186
    abstraction of data, 181
    critical appraisal, 180–181
    interpretation, 182–186, 183*t*
    synthesis of evidence, 181–182
Reproducibility, 88
Research. *See also specific research*
  family-based, 59–60
  future, consent for, 68–69
  human participants
    protection of, 73
    rights of, 67
    selection of, 66
  human tissue, 60–61
  population-based, 59–60

reporting results of, 70–72
results, clinical validity of, 70
subject to human participants regulations, 61–63
Respect for persons, 59
Restriction enzymes, 88
Restriction fragment length polymorphisms (RFLPs), 43–44, 79, 155
Retrospective likelihood approach (mod score), 100–101
RFLPs (restriction fragment length polymorphisms), 43–44, 79, 155
RhD, 85
Risk
assessment, 64–65
from breach of confidentiality, 65–66
identification, genetic tests for, 32
from invasion of privacy, 65–66
minimal, definition of, 64
Risk-benefit relationship, genetic test, 210–211, 211*t*
Risk factors, genetic, identification of, 79
Risk stratification, 33
RsaI digest, N372H genotype of *BRCA1* gene, 208, 208*t*
RTECS, 154

SAM (S-adenosylmethionine), 335, 336*f*, 337–338
Samples
from deceased participants, 63
existing, research on, 61–62, 70
future research, consent for, 68–69
new, research on, 61
publicly available, 62
storage, consent for, 68–69
unidentified, 62
unlinked or anonymizing, 62, 63–64
Sarcoidosis, family aggregation of, 40
SCD. *See* Sickle cell disease
Scottish Intercollegiate Guidelines network, 180
Screening
selective strategies, for hemoglobinopathies, 291–292
tests, evaluating, 265
universal strategies, for hemoglobinopathies, 291–292
SEER Program data, 334
Selection bias, 128*t*, 129–131
Selective estrogen receptor modulators (SERMs), 463
Selective strategies, for hemoglobinopathy screening, 291–292
Semi-Bayes adjustments, for multiple comparisons, 138–139

Sensitivity
analytic, 221
clinical, 224
of genetic tests, 239
Sensitivity of test, for genetic variants, 266
SERMs (selective estrogen receptor modulators), 463
Sex differences, in hereditary hemochromatosis, 499
Short tandem repeats (STRs), 18–19, 79
Sibling controls, 98, 135, 175
Sib-pair design, model-free linkage analysis, 47–50, 48*t*
Sickle cell disease (SCD)
clinical features, 288
early detection, 289
lifetime health service costs in United Kingdom, 289
penicillin prophylaxis, 289
prevalence in United Kingdom, 288
proficiency testing programs and, 85
screening
acceptability of, 292
ethnicity as selection criteria for, 293–294
in first trimester, 293
implementation of, 292–293
Single-gene disorders, voluntary Laboratory Accreditation Program, 85
Single nucleotide polymorphisms (SNPs)
cataloging/mapping, 168
Consortium Web site, 147
DNA chip analysis, 23, 25, 28*f*
DNA sequence variation, 19*f*–20*f*, 20
as genomic marker, 44, 79
genotyping, 22–23, 24*f*–27*f*
high-throughput genotyping, 22–23, 24*f*, 26*f*–27*f*
identification/cataloging of, 147
incidence of, 147
PinPoint assay, 22–23, 24*f*
in protein coding regions, 19, 23
Smoking
Alzheimer disease risk and, 371
NAT2-bladder cancer relationship and, 148–149
one exposure-many diseases principle, 213, 215*t*
SNPs. *See* Single nucleotide polymorphisms
Social issues
commercialization, 72
group harms, 66–67
implications
of *BRCA1/2* testing, 464
in clinical practice guidelines, 275–276
from DNA testing, 231

Social issues (*continued*)
  protection of human participants in research, 59–61
  reporting research results, 70–72
  risk assessment, 64–65
Socioeconomic status
  adjustment for, 133
  Alzheimer disease and, 367
Solvents, halogenated, 150*t*
Southern blot analysis, fragile X syndrome and, 410
Specificity
  analytic, 221–222
  clinical, 224
  of genetic tests, 239
  for genetic variants, 266
Specimens, for epidemiologic studies
  analysis, purposes of, 86
  blood spots, dried, 79, 80, 81*t*
  buccal cells, 82*t*, 83
  EBV-transformed lymphocytes, 80, 83 81*t*
  selection of, 79–80, 81*t*–82*t*, 83
  whole blood, 79–80, 81*t*
Spinal muscular atrophy, 85
Spinocerebellar atrophy, 85
St. Jude's Children's Research Hospital Protocol Total XII, 253
Statistical issues
  adjustment methods for multiple comparisons, 135–136, 137*f*, 138–140, 139*f*
  association study design and, 94–101, 97*f*
  complex disease models and, 103–105
  gene discovery methods and, 93–94
  integration of discovery and characterization, 101–103
  methods of analysis, 178–179
  multiple testing, 177–178
  null hypothesis, 136, 137*f*, 138
  power, 177
  reporting, checklist for, 172*t*
Stigmatization, 66, 231
Strength of association, in gene-disease and gene–gene associations, 184
Stroke, Alzheimer disease and, 367
Stromal cell derived factor-1 (SDF-1), in HIV-1 infection, 484
Study base, selection bias and, 129
Study designs. *See specific study designs*
Study population definition, in evaluating genetic test, 265
Study subject selection reporting, checklist for, 171*t*
Subgroup analysis
  adjustment methods for multiple comparisons, 135–136, 137*f*, 138–140, 139*f*

gene–environment interactions and, 135–136, 137*f*, 138–140, 139*f*
  problems with, 135
Subject selection, 174–175
SULT1A1, 166*t*
Surrogate markers
  in drug-gene interactions, 243–245
  for effect modification, 252–253
Surrogate outcomes, genetic information interpretation and, 255
Surveillance, Epidemiology, and End Results (SEER) registries, 95
Survival analysis, 103
Survival function, 103
Susceptibility genes, identifying in families, 200–201
Susceptibility to disease, single nucleotide polymorphisms and, 22
Systematic review
  critical appraisal, 180–181
  identification of studies, 179–180
  interpretation, 182–186, 183*t*
  reporting
    abstraction of data, 181
    synthesis of evidence, 181–182
    specification of issue, 179

Tamoxifen, for breast cancer prevention, 463
Taquman assay, N372H genotype of *BRCA1* gene, 207–208, 208*t*
Tay-Sachs disease, 220
TCDD, 150*t*
T1317C polymorphism, 352
TDT (transmission-disequilibrium test), 52–53, 52*t*, 98
Thalassaemias
  clinical features, 288–289
  lifetime health service costs in United Kingdom, 289
  prevalence in United Kingdom, 288
  screening
    acceptability of, 292
    in first trimester, 293
    implementation of, 292–293
Thiopurine S-methyltransferase, 253
Thrombosis, candidate genes, 439*t*
TLR4, 153, 167*t*
TNF", 167*t*
Tobacco smoke, 151*t*
TP53 gene germline mutations, 454
TPMT (S-methyltransferase), 235
TPMT genotyping, for pediatric ALL, 240–241, 240*f*, 241*f*
Traits
  binary disease, 103
  complex, 38

criteria for, 39
genetic factors for, 39
heterogeneous etiology of, 39
qualitative, 41
quantitative, 41
genetic influences on, 38–39
quantitative
penetrance for, 45
transmission-disequilibrium test for, 53
simple, 38
Transferrin saturation, 495, 499–500, 506
Transgenic mouse model, development for
chronic beryllium disease, 394–395
Transmission-disequilibrium test (TDT), 52–53,
52*t*, 98
Transthyretin (TTR), 513
Traumatic head injury, Alzheimer disease and,
367
Treatment interventions
accumulating knowledge on, 270
availability, randomized controlled trials and,
258
Turcot syndrome, 202
Twins
concordance rates for, 41–42
dizygotic, 38
heritability estimations and, 41
monozygotic, 38
Two-dimensional gel electrophoresis, of protein
spots, 30, 30*f*–31*f*
2 × 2 table analysis, 156
2 × 4 table analysis, 156

UDP (uridine diphosphate), 146
U.K. Cancer Family Study Group, 286–288,
287*t*
Ultraviolet light, 151*t*
Ultraviolet radiation, 147
United Kingdom
Clinical Pathology Accreditation, 86
external quality control, 86
National Health Service
administration of, 283–284
evidence-based health services in,
284–288, 286*t*, 287*t*
implementing new genetic services,
294–295
population hemoglobinopathy screening,
288–289, 290*t*–291*t*, 291–294
purpose of, 283
specialist genetic services in, 284
reporting of diagnostic results, 83
United States
Department of Energy, 18
Preventive Services Task Force, 278
quality control requirements, 83–84

Universal strategies, for hemoglobinopathy
screening, 291–292
University of California at Santa Cruz, 18
Uridine diphosphate (UDP), 146

Validity, of hospital-based case-control studies,
130
Variable number of tandem repeat
polymorphisms (VNTRs), 44
Variance components analysis, 42
VDR, 167*t*
Vegetable consumption, colorectal cancer risk
and, 335
Venous thromboembolism risk, 272–273. *See
also* Factor V Leiden
Venous thrombosis, as multicausal disease,
329
VIRADPT (Antiretroviral Adaptation trial),
236
Viral genotype, genetic testing for, 235–236
Vitamin B$_6$, 336*f*, 338
Vitamin B$_{12}$
deficiency, 354–355
in folic acid pathway, 336*f*, 338
VNTRs (variable number of tandem repeat
polymorphisms), 44

Wald *p* value, 178
Warfarin dosing, *CYP2C9* polymorphism and,
235, 242–243, 243*f*, 244*f*, 275
Websites
environmental health-related, 165–166
on working draft sequence, 18
Whole blood specimens, for epidemiologic
studies, 79–80, 81*t*
"Whole genome shotgun" approach, 18
Women's Health Initiative, 462
Workers' Compensation Insurance, 512
Workplace genetic testing programs, for *PON1*
gene, 314, 315
Wrongs, *vs.* harms, 65–66

Xenobiotic biotransformation, 145
Xenobiotic metabolism enzymes (XMEs)
polymorphisms
functional, 146–147
role of, 147
silent, 146
XPD(ERCC2), 167*t*
XPF, 167*t*
XRCC1, 167*t*

Zinc, colorectal cancer and, 354–355